Studies in Surface Science and Catalysis 3

PREPARATION OF CATALYSTS II
Scientific Bases for the Preparation of Heterogeneous Catalysts

Studies in Surface Science and Catalysis

Volume 1 **Preparation of Catalysts I.** Scientific Bases for the Preparation of Heterogeneous Catalysts. Proceedings of the First International Symposium held at the Solvay Research Centre, Brussels, October 14—17, 1975. Second Impression
edited by **B. Delmon, P. Jacobs and G. Poncelet**

Volume 2 **The Control of the Reactivity of Solids.** A Critical Survey of the Factors that Influence the Reactivity of Solids, with Special Emphasis on the Control of the Chemical Processes in Relation to Practical Applications
by **V.V. Boldyrev, M. Bulens and B. Delmon**

Volume 3 **Preparation of Catalysts II.** Scientific Bases for the Preparation of Heterogeneous Catalysts. Proceedings of the Second International Symposium, Louvain-la-Neuve, September 4—7, 1978
edited by **B. Delmon, P. Grange, P. Jacobs and G. Poncelet**

Explanation of the cover design

The figure gives a pictorial representation of surface analysis techniques. Eight basic input probes are considered, which give rise to one or more of four types of particles that leave the surface carrying information about it to a suitable detector. The input probes can be particle beams of electrons, ions, photons, or neutrals or non-particle probes such as thermal, electric fields, magnetic fields or sonic surface waves. All of the input probes (with the exception of magnetic fields) give rise to emitted particle beams, i.e. electrons, ions, photons, or neutrals. The various surface analysis techniques can therefore be classified according to the type of input probe and the type of emitted particle (e.g. electrons in, ions out; thermal in, neutrals out, etc.). In analyzing the emitted particles, one can consider four possible types of information; identification of the particle, spatial distribution, energy distribution and number. Any or all of these forms of information are then used to develop a better understanding of the surface under study.

Studies in Surface Science and Catalysis 3

PREPARATION OF CATALYSTS II

Scientific Bases for the Preparation of Heterogeneous Catalysts

Proceedings of the Second International Symposium, Louvain-la-Neuve, September 4—7, 1978

Editors

B. Delmon, P. Grange, P. Jacobs and G. Poncelet

ELSEVIER SCIENTIFIC PUBLISHING COMPANY
Amsterdam — Oxford — New York   1979

ELSEVIER SCIENTIFIC PUBLISHING COMPANY
335 Jan van Galenstraat
P.O. Box 211, 1000 AE Amsterdam, The Netherlands

Distributors for the United States and Canada:

ELSEVIER/NORTH-HOLLAND INC.
52, Vanderbilt Avenue
New York, N.Y. 10017

Library of Congress Cataloging in Publication Data
Main entry under title:

Preparation of catalysts II.

   (Studies in surface science and catalysis ; 3)
   Includes bibliographical references and index.
   1. Catalysts--Congresses. I. Delmon, Bernard.
II. Series.
QD505.P722   541'.395   79-13927
ISBN 0-444-41733-8

ISBN 0-444-41733-8 (Vol. 3)
ISBN 0-444-41801-6 (Series)

© Elsevier Scientific Publishing Company, 1979
All rights reserved. No part of this publication may be reproduced, stored in a retrieval system or transmitted in any form or by any means, electronic, mechanical, photocopying, recording or otherwise, without the prior written permission of the publisher, Elsevier Scientific Publishing Company, P.O. Box 330, 1000 AH Amsterdam, The Netherlands.

Printed in The Netherlands

CONTENTS

Organizing Committee	VIII
Foreword	IX
Acknowledgements	XI
Financial Support	XIII

The design of catalysts (D.L. Trimm) — 1

Effect of impregnation and activation conditions of Al_2O_3/CuO supported monolith catalysts in the reduction of NO (M. Engler, K. Unger, O. Inacker, C. Plog and M. Seidl) — 29

Effect of pretreatment on activity, selectivity and adsorption properties of a Fischer-Tropsch-catalyst (G. Lohrengel, M.R. Dass and M. Baerns) — 41

The influence of impregnation, drying and activation on the activity and distribution of CuO on α-alumina (M. Kotter and L. Riekert) — 51

A study of the chemisorption of chromium(VI), molybdenum(VI) and tungsten(VI) onto γ-alumina (A. Iannibello, S. Marengo, F. Trifiro' and P.L. Villa) — 65

Dispersion and compound formation in some metathesis catalysts and fluorine containing alumina's, studied by XPS and Laser-Raman spectroscopy (F.P.J.M. Kerkhof, J.A. Moulijn, R. Thomas and J.C. Oudejans) — 77

Reduction of silica supported nickel catalysts (J.W.E. Coenen) — 89

Interaction of nickel ions with silica supports during deposition-precipitation (L.A.M. Hermans and J.W. Geus) — 11.

Crystallite size distributions and stabilities of homogeneously deposited Ni/SiO_2 catalysts (J.T. Richardson, R.J. Dubus, J.G. Crump, P. Desai, U. Osterwalder and T.S. Cale) — 131

The preparation and pretreatment of coprecipitated nickel-alumina catalysts for methanation at high temperatures (E.C. Kruissink, L.E. Alzamora, S. Orr, E.B.M. Doesburg, L.L. van Reijen, J.R.H. Ross and G. van Veen) — 143

Controlled catalyst distribution on supports by co-impregnation (E.R. Becker and T.A. Nuttall) — 159

Multicomponent chromatographic processes during the impregnation of alumina pellets with noble metals (L.L. Hegedus, T.S. Chou, J.C. Summers and N.M. Potter) — 171

Factors controlling the retention of chlorine in platinum reforming catalysts (S. Sivasanker, A.V. Ramaswamy and P. Ratnasamy) — 185

Preparation of alumina or silica supported platinum-ruthenium bimetallic catalysts (G. Blanchard, H. Charcosset, M.T. Chenebaux and M. Primet) — 197

VI

Preparation of catalysts by adsorption of metal complexes on mineral oxides (J.P. Brunelle) — 211

Analysis of steps of impregnation and drying in preparation of supported catalysts (V.B. Fenelonov, A.V. Neimark, L.I. Kheifets, and A.A. Samakhov) — 233

Some mechanistic correlations between impregnation and activation operations for the preparation of high-selectivity supported metal catalysts (H. Shingu and T. Inui) — 245

The impregnation and drying step in catalyst manufacturing (G.H. van den Berg and H. Th. Rijnten) — 265

Study of the interaction of Fe_2O_3-MoO_3 with several supports (L. Cairati, M. Carbucicchio, O. Ruggeri and F. Trifiro) — 279

Relationship between average pore diameter and selectivity in iron-chromium-potassium dehydrogenation catalysts (Ph. Courty and J.F. Le Page) — 293

Preparation of highly-dispersed ruthenium on magnesium oxide supports : comments on the advantages of non-aqueous catalyst preparations (L.L. Murrell and D.J.C. Yates) — 307

Catalyst activation by reduction (N. Pernicone and F. Traina) — 321

The activation of iron catalyst for ammonia synthesis (A. Barański, M. Łagan, A. Pattek, A. Reizer, L.J. Christiansen and H. Topsøe) — 353

Preparation and characterization of small iron particles supported on MgO (H. Topsøe, J.A. Dumesic, E.G. Derouane, B.S. Clausen, S. Mørup, J. Villadsen and N. Topsøe) — 365

Ruthenium catalysts for ammonia synthesis prepared by different methods (A. Ozaki, K. Urabe, K. Shimazaki and S. Sumiya) — 381

Limiting factors on the structural characteristics of Ru/SiO_2 and $RuFe/SiO_2$ catalysts (L. Guczi, K. Matusek, I. Manninger, J. Király and M. Eszterle) — 391

Preparation aspects of Ru-supported catalysts and their influence on the final products (A. Bossi, F. Garbassi, A. Orlandi, G. Petrini and L. Zanderighi) — 405

High surface area oxide solid solution catalysts (A.P. Hagan, M.G. Lofthouse, F.S. Stone and M.A. Trevethan) — 417

Tentative classification of the factors influencing the reduction step in the activation of supported catalysts (B. Delmon and M. Houalla) — 439

Preparative chemistry of cobalt-molybdenum/alumina catalysts (A. Iannibello and P.C.H. Mitchell) — 469

The influence of the support on Co-Mo hydrodesulfurization catalysts (H. Topsøe, B.S. Clausen, N. Burriesci, R. Candia and S. Mørup) — 479

Study of some variables involved in the preparation of impregnated catalysts for the hydrotreatment of heavy oils (O. Ochoa, R. Galiasso and P. Andreu) — 493

Preparation and properties of thiomolybdate graphite catalysts
(G.C. Stevens and T. Edmonds) — 507

Catalyst stabilization and deactivation compared with catalyst
preparation (R.E. Montarnal) — 519

Preparation and properties of monodispersed colloidal metal hydrous
oxides (E. Matijević) — 555

Process for the production of spherical catalyst supports
(R.M. Cahen, J.M. André and H.R. Debus) — 585

Methods of saturation with alkali ions. Influence of the properties
of oxides (R. Hombek, J. Kijenski and S. Malinowski) — 595

Mechanism of formation of a catalytically active phase in the
reaction of CrO_2Cl_2 with silica gel (D. Mehandjiev, S. Angelov and
D. Damyanov) — 605

The use of transport reactions for dispersing supported species
(R. Haase, H.-G. Jerschkewitz, G. Öhlmann, J. Richter-Mendau and
J. Scheve) — 615

Synthesis of new catalytic materials : metal carbides of the
group VI B elements (L. Leclercq, K. Imura, S. Yoshida, T. Barbee
and M. Boudart) — 627

A novel method for the preparation and production of skeleton
catalysts (J. Petró) — 641

MINISYMPOSIUM ON CATALYST NORMALIZATION

Normalization of catalyst test methods (L. Moscou) — 659

Mesopore determination from nitrogen sorption isotherms:
fundamentals, scope, limitations (J.C.F. Broekhoff) — 663

Metal surface area and metal dispersion in catalysts
(J.J.F. Scholten) — 685

Progress report on the work of the SCI/IUPAC/NPL working party on
catalyst reference materials (C.C. Bond, R.L. Moss,
R.C. Pitkethley, K.S.W. Sing and R. Wilson) — 715

Organization and functions of ASTM Committee D-32 on catalysts
(A.H. Neal) — 719

Measurement of the activity of solid state catalysts
(G.K. Boreskov) — 723

The Council of Europe Research Group on Catalysis (E.G. Derouane) — 727

Concluding remarks — 729
List of participants — 735
Author index — 761

ORGANIZING COMMITTEE

Members
: Dr. S.P.S. ANDREW, Imperial Chemical Industries Ltd., Great Britain.
: Dr. H. BOHLBRO, Haldor Topsøe, Denmark.
: Dr. R. CAHEN, Labofina S.A., Belgium.
: Prof. J.W.E. COENEN, Unilever Research, The Netherlands.
: Prof. B. DELMON, Université Catholique de Louvain, Belgium.
: Prof. B. DEROUANE, Université de Namur, Belgium.
: Prof. J.W. HIGHTOWER, Rice University, Houston, Texas, U.S.A.
: Dr. P. LAMBERT, Union Chimique Belge, Belgium.
: Dr. A. LECLOUX, Solvay et Cie., Belgium.
: Dr. J.L. LE PAGE, Institut Français du Pétrole, France.
: Dr. R. MONTARNAL, Institut Français du Pétrole, France.
: Dr. L. MOSCOU, AKZO-Chemie Nederland, The Netherlands.
: Prof. E. PLUMAT, Université Libre de Bruxelles, Belgium.

Şecretaries
: Dr. P. GRANGE, Université Catholique de Louvain, Belgium.
: Dr. P.A. JACOBS, Katholieke Universiteit Leuven, Belgium.
: Dr. G. PONCELET, Université Catholique de Louvain, Belgium.

FOREWORD

As a result of the First International Symposium on "The Scientific Bases for the Preparation of Heterogeneous Catalysts" it clearly appeared to the local organizers that a large international audience was interested by the topics treated. However, the scope of the symposium was probably too large, and possibly did not allow sufficient discussion and interaction between participants.

It was clear from the beginning that for the Second Symposium a smaller number of scientific domains involved in the preparation of industrial heterogeneous catalysts should be discussed.

The scientific program was set up by an Organizing Committee formed by representatives of industries, universities and research institutes. This committee was composed by experts in the field of catalyst preparation from different countries.

At the first meeting of the Organizing Committee, it was felt that the symposium should focus on two unit processes, IMPREGNATION and ACTIVATION of supported catalysts, and more particularly on : the chromatographic effect, transport in pores, calcination, activation by reduction and sulfidation, carrier effects and compound formation. It was also decided that new trends in the preparation of real catalysts should be included in the scientific program as they were in the First Symposium.

Over 90 papers treating these topics were submitted. The Organizing Committee at a second meeting selected 36 communications which fitted most closely the imposed topics. It was decided not to discuss the selected unit processes separately but rather to structurate the program sessions around distinct groups of catalysts treating impregnation and activation. The symposium topics were introduced by four plenary lectures and three extended communications, each of them delivered by scientists who, with a remarkable cooperation spirit, accepted to adapt their contribution to the audience and to the subject and style of the Symposium.

The Organizing Committee also felt that the standardisation of methods of catalysts characterization remained a topic of much interest, and that progress in international agreement and cooperation was important. Encouraged by the success of an almost informal discussion included in the program of the First Symposium, it was found necessary to devote half-a-day to catalyst normalisation. Three scientific papers and reports from national and international committees were scheduled.

A round-table discussion was planned to evaluate the importance of a scientific approach for the impregnation and activation of catalysts. The discussion also focused on the role of this symposium in fostering scientific research in the field and for enabling a better interaction between industrial and fundamental investigation. A summary of these discussions is given by the Symposium Chairman in his concluding remarks.

Approximately 350 participants representing 30 nations attended the Symposium. Almost 60% of the participants came from industry.

IUPAC accepted to sponsor the symposium. In his introductory remarks IUPAC's President, Prof. G. Smets, mentioned that this sponsorship is only awarded provided three prerequisites are fulfilled:
. fair geographical representation of the contributors
. quality of the geographical site
. quality of the organisation.

There is no doubt that the first condition was met. Whether the two other ones were fulfilled, the appreciation is left to those who attended the symposium.

If the participants agree that the organisation of the symposium went on smoothly, this is the result of a high number of personal efforts, which the local organizers would like to acknowledge.

P. GRANGE
P.A. JACOBS
G. PONCELET

ACKNOWLEDGEMENTS

   In the name of the Organizing Committee, we want to thank the Rector of the Université Catholique de Louvain, Mgr. E. Massaux, who agreed that the Second International Symposium could be held in Louvain-la-Neuve. For all the facilities generously provided by the university, we also sincerely thank the Rector.

   The Organizers acknowledge the sponsorship of IUPAC and the benevolent appreciation of its President for the project of this 2nd International Symposium. The introductory remarks of IUPAC's President, Prof. G. Smets, were also gratefully appreciated by the local Organizers, the Scientific Committee and all those who participated to the set-up of the Symposium.

   We also thank the Société Chimique de Belgique, Division de Catalyse, for sponsoring the Symposium, and in particular its President, Dr. A. Lecloux, for his welcome address.

   The local Organizers are pleased to mention the exceptional and enthousiastic participation of the members of the Organizing Committee. Their work and contribution go well beyond what usually is expected from members of scientific committees of symposia: they determined the topics of the Second Symposium and selected the papers, they structurated the scientific program, acted as session chairmen, and animated the discussions. This symposium was really the collective achievement of all the members of the Organizing Committee. To all of the members of the Organizing Committee, we convey our deepest gratitude.

   While preparing these Proceedings, we learned the death of one member of the Organizing Committee. All those who knew Dr. R. Montarnal appreciated his human qualities, enthousiasm and scientific competence.

   Special thanks are due to Dr. R. Cahen for his efficient help in contacting several scientists of industry and collecting financial contributions.

   The efforts of Dr. L. Moscou and Prof. E. Derouane, who successfully took in charge the complete organization of the Minisymposium on Catalyst Normalisation, are also greatly appreciated.

   We also would like to thank Prof. D.L. Trimm, Dr. J.P. Brunelle, Dr. N. Pernicone and Prof. E. Matijevic for their highly stimulating lectures.

The Organizers also acknowledge all those who submitted a paper, including those whose contribution could be not accepted. Special thanks are due to the authors of the papers included in the Proceedings. They were the artisans of the success of the symposium.

Two departments of the university deserve to be congratulated: the 'Relations Extérieures de l'Université de Louvain' (REUL), and in particular Mrs. F. Bex and her team, and the 'Service de Logement' headed by Mr. E. Vander Perre.

Finally, we want to acknowledge the help of all those of the 'Groupe de Physico-Chimie Minérale et de Catalyse' who, in charge of different organisational aspects before and during the symposium, contributed to its success: Dr. M.A. Apecetche, Dr. N. De Keyzer, Dr. F. Delannay, Dr. P. Gajardo, M. Genet, Dr. P. Gérard, Dr. J. Lemaître, A. Mathieu, L. Petit, C. Pierard, D. Pirotte, M. Rodriguez, P. Scokart and Dr. M.L. Somme-Dubru.

FINANCIAL SUPPORT

The Organizing Committee gratefully acknowledges the "Ministère de l'Education Nationale et de la Culture Française" and the "Fonds National de la Recherche Scientifique" for their financial guarantee.

The pecuniary security provided by the Société Chimique de Belgique is also acknowledged.

Financial support was also contributed by the following catalysts manufacturers and companies:

AKZO Nederland, b.v., The Netherlands
BASF, W. Germany
Catalysts and Chemicals Europe, Belgium
Condea, W. Germany
Cyanamid of Great Britain, England
Essochem Europe, Belgium
Haldor Topsøe, Denmark
Harschaw Chemie, The Netherlands
Imperial Chemical Industries Ltd., England
Johnson Matthey and Co. Ltd., England
Mallinckrodt Inc., Calsicat Division, U.S.A., and Catalysts International, France
Métallurgie Hoboken-Overpelt, Belgium
Montedison, S.p.A., Italy
Procatalyse, France
Solvay et Cie., Belgium
Südchemie AG, Sparte Girdler Katalysatoren, W. Germany
Unilever, The Netherlands
Universal Matthey Products Ltd., England
U.O.P. Processes, U.S.A.

Labofina S.A. has contributed to the costs of the printing of the announcing booklets.

The organizers are indebited to these companies.

THE DESIGN OF CATALYSTS*

D.L. TRIMM
Laboratory of Industrial Chemistry, The University of Trondheim, Norway

Realisation of the importance of catalyst preparation to the activity, selectivity and life of a catalyst has led to increasing interest in the scientific basis of different preparations. This, in turn, has led to improved performance, and to the necessity of defining carefully what catalyst should be prepared - rather than optimising a catalyst that can be prepared. This paper describes a general approach to the problem of recognition of which catalyst should be prepared for a given reaction.

The design of a catalyst is discussed in terms of a flow diagram in which means of identification of major and minor constituents of a catalyst are interrelated. Factors influencing the choice of the chemical constituents are considered together with the choice of a support, which often depends more on the desired physical characteristics of the catalyst. Both theoretical concepts and empirical observations can be combined in the overall design to give a limited number of possibilities for experimental testing. Finally, correlation of suggestions from the design with catalyst preparation is considered briefly, with particular reference to impregnation and activation.

INTRODUCTION

It is well known that the activity and selectivity of a heterogeneous catalyst depends on the inherent activity of the components, on the physical structure of the catalyst and on the operating conditions for the reaction. This has led to attention being focused on catalyst preparation, since this is a major point for control of chemical composition and physical structure of the catalyst. For a long time catalyst preparation was regarded as one of the last strongholds of alchemy, but the advent of modern methods of surface analysis (ref. 1), coupled to increasing scientific knowledge, have elevated the subject from an art to a science (refs. 2, 3). As a result, it begins to be possible to manufacture a catalyst to a wide variety of given specifications.

Although our capability is far from perfect, this possibility raises new

---
* Published in Pure Appl. Chem., 50, 9-10 (1978), p. 1147.

questions. In particular, given that we may know how to prepare a catalyst, it is necessary to ask which catalyst should be prepared. Since this question covers factors such as chemical composition, physical structure and mechanical strength, the answer may well involve optimisation of several non-related parameters. While the bulk of the proceedings of this meeting will be concerned with how to prepare catalysts, this paper is focused primarily on how to decide which catalyst to prepare.

As can be seen from figure 1, the interacting requirements of a catalyst complicate the decision. As a result, it is not surprising that most successfull catalysts have been developed on an empirical basis. In recent years, however, procedures have been developed to place the selection of catalysts on a more scientific footing. Generally, these involve a theoretical study in which general knowledge of catalysis, arguments by analogy and various theoretical concepts are applied to a particular problem (refs. 4, 5, 6, 7).

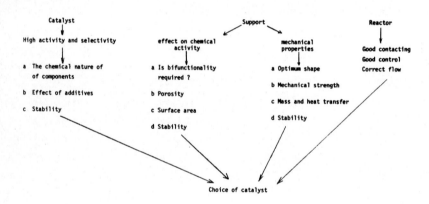

Figure 1. Factors influencing catalyst selection.

It is regrettable that our understanding of catalysis is still not good enough to guarantee success, but the approach has been found to be very useful in explaining why a given catalyst is active and, in some cases, in successfully predicting new catalysts (ref. 5). Insufficient knowledge is available to be independent of experimental testing, but the approach acts as a valuable guide to experimentation. As a result, the time needed for catalyst identification and testing can be significantly decreased, for the investment of only a short time on catalyst design.

The procedure has been discussed in a few papers in recent years (refs. 4, 5, 6, 7), usually with emphasis on a given reaction. A more general approach is attempted in the present paper, with particular emphasis on those aspects of design that are of interest in the context of catalyst preparation. The broader framework of the complete design is presented in reference 5.

Intitial stages of the design.

The design process discussed below is summarised in scheme 1, and various entry points to the design are indicated. The extent and depth of a study depends on the effort warranted by a particular application and, in this article, the full design will be discussed.

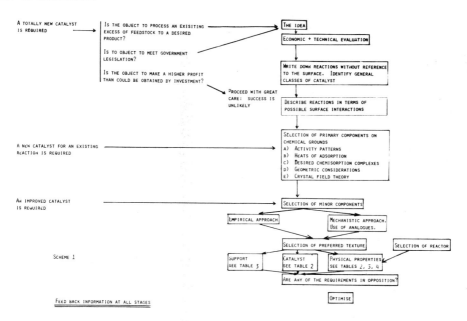

Scheme 1

Feed back information at all stages

The starting point for the design requires the definition of the objective and a statement, in the form of a chemical reaction network, of desirable or undesirable reactions which may occur (ref. 16). Thermodynamic and economic calculations can then be used to establish the most attractive route, and general types of catalyst that should favour individual reactions can be identified. This reaction network provides the basic outline of the design, to which it is possible to apply the catalyst selection criteria described below.

Selection of the chemical basis of the catalyst: major components

At this point we are concerned with the identification of the major chemical components of the catalyst, and factors such as mechanical strength, surface area and porosity will be considered at a later stage.

Experience has shown that one difficulty can arise at this point. Because of our scientific training, we tend to think linearly, rather than to consider all of the possibilities that exist. It is better to consider (and possibly to reject) a wide variety of reactions rather than to ignore one totally new approach.

The tools that are used to identify possible catalysts range from empirical observations to theoretically based calculations. The first step in the process is to translate the reaction mechanism in terms of adsorbed intermediates and surface reaction: this allow close definition of what is required from the catalyst.

To illustrate this process it is useful to consider a specific example, such as the production of limonene (ref. 8). This is a member of the terpene family that have considerable importance as solvents. A combination of thermodynamic and economic calculations showed that the reaction sequence was most attractive.

The scheme involves hydrogenation and dehydrogenation (catalysed by metals (ref. 9) or metal oxides (ref. 10)) as well as alkylation. This is normally carried out over acidic catalysts (11), but an oxidative coupling is also possible (ref. 12).

For the acid catalysed reaction the surface reaction may be summarised as (refs. 11, 13).

The surface oxidation reaction, on the other hand, may be written as (refs. 5, 12):

$$C_3H_6 + O^2 - M^{n+} - O^{2-} \longrightarrow O^{2-} - \underset{M^{(n-1)+}}{\overset{CH_2-CH-CH_2}{\text{---}}} - \underset{(O)}{\overset{H -}{\text{---}}}$$

$$\underset{H_2C-CH-CH_2}{\bigcirc} + O^{2-} - M^{n+} - O^{2-} \longrightarrow O^{2-} - \underset{M^{(n-1)+}}{\overset{H_3C-\bigcirc}{\text{---}}} - \underset{(O)}{\overset{H -}{\text{---}}}$$

$$\underset{\downarrow}{\bigcirc} - CH_3$$

$$2^-O-M^{(n-1)+} -(OH)^- + O^{2-} - M^{(n-1)+} - (OH)^- \longrightarrow \bigcirc\!\!=$$

$$+$$
$$2(O^{2-} - M^{(n-1)+} -(OH)^-)$$

Of course, there will be many other possibilities, but these simplified schemes serve to illustrate the transcription of the reactions to the surface.

It is now possible to discuss different means of identifying active catalysts.

a) **Activity patterns.** By far the most useful guide, where they are available, are patterns of activity for reactions of the same type. These may vary from the simple pattern well known for the decomposition of nitrous oxide (figure 2) (ref. 10), to the more complicated patterns for hydrogenation (figure 3) (ref. 13) and oxidation (figure 4) (refs. 15, 16). Generally it is possible to find an activity pattern for a class of solids which catalyse a type of reaction. Activity patterns for metals catalysing reactions involving hydrogen are shown in figure 3 and for metal oxides involved in oxidation are shown in figure 4.

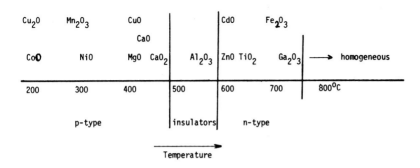

Figure 2. Activity pattern for nitrous oxide decomposition (Temperature at which different oxides catalyse the decomposition of N_2O).

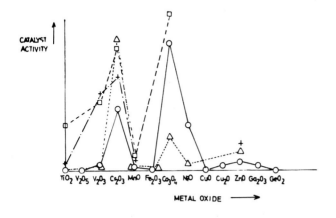

Figure 3. Activity pattern for reactions involving hydrogen
○ H_2/D_2 exchange at $80°C$
+ propane dehydrogenation ($550°C$)
△ ethylene hydrogenation (-120 to $400°C$)
□ cyclohexene disproportionation (200 to $450°C$)

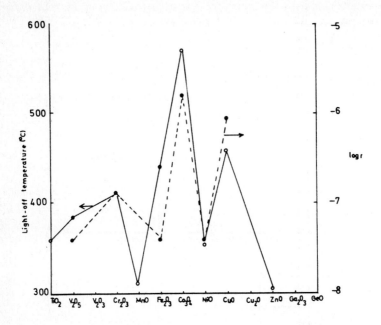

Figure 4. Activity patterns for oxidation
— light off temperature for Pt/Rh gauzes doped with metal oxide and used to catalyse the oxidation of ammonia
--- log rate of oxidation of propylene at 300°C.

A less quantitative pattern has been developed for metal oxides involved in acid catalysed reactions (figure 5) (ref. 17), while the activity of a bifunctional catalyst can be established with the aid of diagrams such as figure 6 (ref. 18). The position of the desired compound on the diagram determines the strength of the metal and acid functions desired.

It should be remembered that activity patterns give only an idea of the relative activity of catalysts, unless they refer to the specific reaction in question.

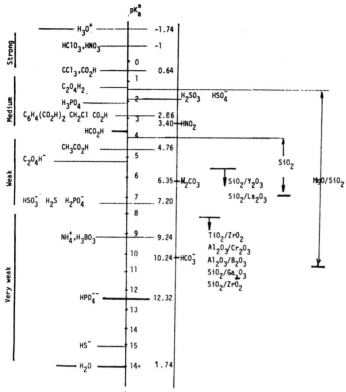

Figure 5. Orders of acidity of different materials.

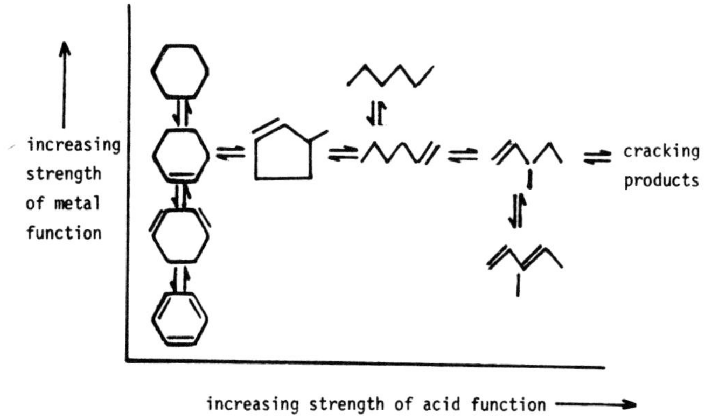

Figure 6. Bifunctional catalysis

b) _Correlations of activity with bulk properties of the catalyst._ In the past there have been several attempts to relate catalytic activity with the bulk properties of solids. Although these have fallen into disrepute with the realisation that the surface may have very different properties to the bulk, it is possible to obtain valuable pointers for catalyst design from the approach. Thus, for example, the concept of percentage d bonding (ref. 19) is open to severe criticism, and yet these are some systems in which it is possible to relate catalytic activity with this factor (refs. 20, 21). Similarly, attempts to relate adsorption and catalysis with bulk semiconductor properties of oxides have been reasonably successful in explaining some patterns of catalytic activity (ref. 10), and should certainly not be ignored at this stage of the design. However, more accurate predictions should be obtained by considering the gas-solid interface.

c) _Predictions from heats of adsorption._ In a few cases it is possible to predict the most active catalyst on the basis of heats of adsorption. This approach is more theoretically justified in that, if adsorption is too strong, then a gas will not be displaced from the surface or will not react. If adsorption is too weak, the residence time of the adsorbed gas on the surface will be too short to favour reaction. Using arguments of this type, it has been possible to identify the most active catalyst for the hydrogenation of nitrogen (ref. 22) and for ethylene (ref. 18), and similar arguments can be used for analogous reactions. Prediction of optimal catalysts does depend, however, on the availability of heats of adsorption data.

A somewhat similar approach has been used for oxidation catalysis, where various properties that indicate the strength of adsorption of oxygen have been correlated with catalytic activity. Boreskov (ref. 23) has measured the heat of adsorption of oxygen on some oxides and have also measured isotopic oxygen exchange between gas and oxides (ref. 24): their results are summarised in figure 7. Moro-oka (ref. 25), on the other hand, relates catalytic activity to the heat of formation of the bulk oxide (figure 8). It will be seen that all of these measurements may have a satisfactory theoretical explanation, even when the authors were not aware of this. Certainly the plots offer a guide to the selection of the most active catalyst, although it must be remembered that activity and selectivity in oxidation are often inversely related.

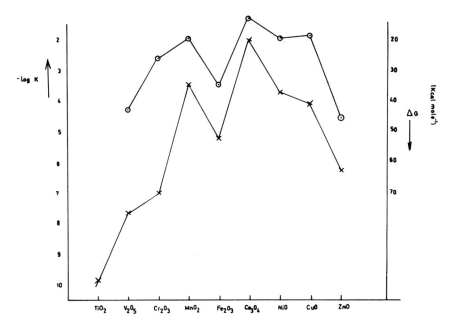

Figure 7. Patterns of behaviour of oxides
⊙ Metal-oxygen bond energy
✗ log rate constant for oxygen exchange on the surface

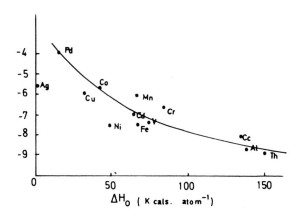

Figure 8. Oxidation of propylene as a function of heats of formation/oxidation atom.
Rates measured at 300°C.

d) **Predictions on the basis of geometric considerations.** One of the more useful methods of predicting catalytic activity arises from consideration of geometric factors. The concept that the geometry of the catalyst can affect activity has been recognised for many years, and formed the basis of the multiplet theory of catalysis (ref. 26). It is particularly useful in prediction, partially because data is readily available and can be easily applied, and partly because the predictions turn out to be reasonably accurate (refs. 5, 6).

Predictions are made on the basis of matching bond lengths of adsorbed species with crystal parameters of catalyst. Thus, for example, alkylation reactions to produce limonene could favour the products

2.09 Å
2.21 Å

4.33 Å

where x is an adsorption site.

Assuming that the products come from an adsorbed cyclo-olefin and adsorbed propylene, it is possible to generate a table of distances between adsorption centres which are desired or undesired, depending on the modes of adsorption and the reaction path:

Mode of adsorption		Distance between adsorption centres (Å)	
Propylene	Cyclodiene	Desired*	Undesired*
pi-allyl	pi	1.98	2.51
pi	pi	2.15	2.51
pi-allyl	pi-allyl	2.92	1.45, 3.84
pi	pi-allyl	3.26	1.91, 2.13, 3.84, 4.33.

*In terms of a reaction leading to limonene.

Comparison with the lattice parameters of metals and of metaloxides (figures 9 and 10) gives a strong indication of which metals or metal oxides can be considered and which should be avoided.

These type of arguments can be easily extended to consider terraces or edges on the catalyst surface. One apparent difficulty lies in the fact that the surface structure does not always minor the bulk (refs. 1, 27), but the fact remains that predictions made on this basis are often accurate.

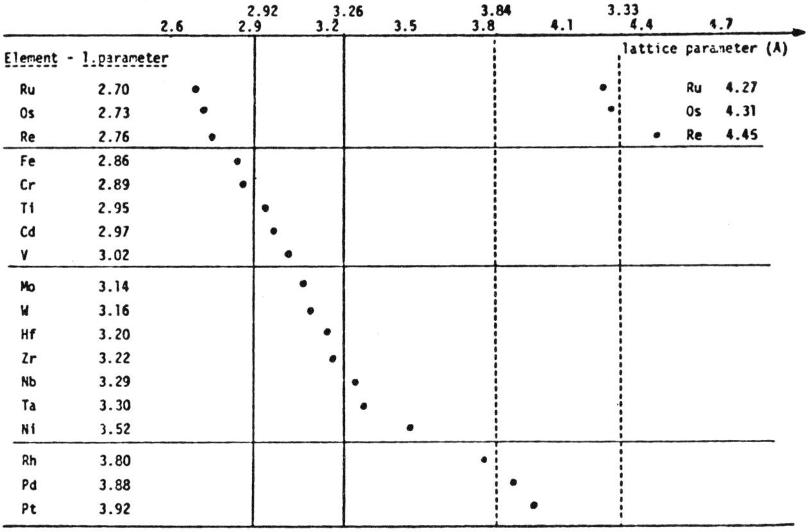

Element	l.parameter								lattice parameter (Å)
		2.6	2.92 / 2.9	3.26 / 3.2	3.5	3.84 / 3.8	4.1	3.33 / 4.4	4.7
Ru	2.70	•						•	Ru 4.27
Os	2.73	•						•	Os 4.31
Re	2.76	•						•	Re 4.45
Fe	2.86	•							
Cr	2.89	•							
Ti	2.95		•						
Cd	2.97		•						
V	3.02		•						
Mo	3.14			•					
W	3.16			•					
Hf	3.20			•					
Zr	3.22			•					
Nb	3.29				•				
Ta	3.30				•				
Ni	3.52					•			
Rh	3.80						•		
Pd	3.88						•		
Pt	3.92						•		

Lattice parameters of transition metals correlated to desired (——) and undesired values (----) for limonene production.

Figure 9. Lattice parameters of transition metals correlated to desired (——) and undesired (---) values for production of limonene.

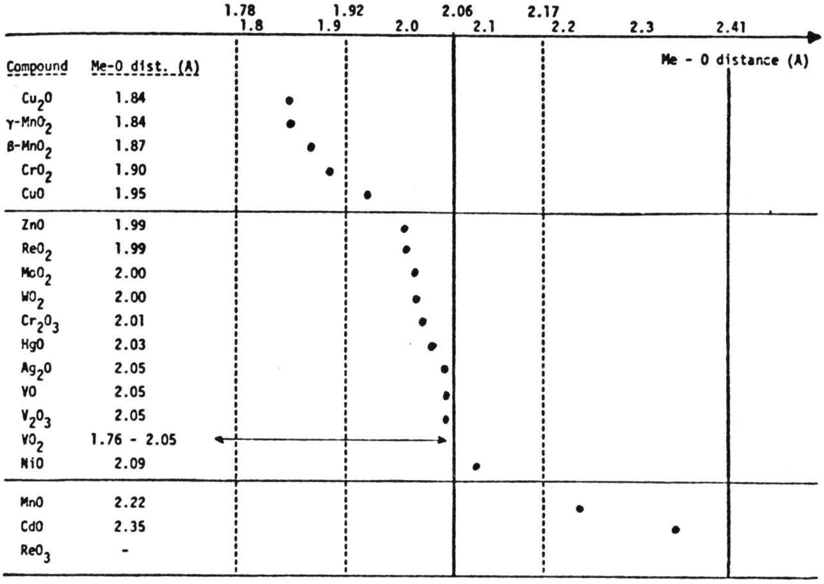

Compound	Me-O dist. (Å)							Me – O distance (Å)
		1.78 / 1.8	1.92 / 1.9	2.0	2.06 / 2.1	2.17 / 2.2	2.3	2.41
Cu$_2$O	1.84	•						
γ-MnO$_2$	1.84	•						
β-MnO$_2$	1.87	•						
CrO$_2$	1.90		•					
CuO	1.95		•					
ZnO	1.99			•				
ReO$_2$	1.99			•				
MoO$_2$	2.00			•				
WO$_2$	2.00			•				
Cr$_2$O$_3$	2.01			•				
HgO	2.03			•				
Ag$_2$O	2.05				•			
VO	2.05				•			
V$_2$O$_3$	2.05				•			
VO$_2$	1.76 – 2.05	←———————→						
NiO	2.09				•			
MnO	2.22					•		
CdO	2.35						•	
ReO$_3$	–							

Cation – anion distances, in Å, of oxides correlated to desired (——) and undesired values (-------) for limonene production.

Figure 10. Lattice parameters of metal oxides correlated to desired (——) and undesired (---) values for limonene production.

e) **Nature of chemisorbed complexes.** Even from the simplified reactions on the surface listed above, it is obvious that the direction of reaction must be very dependent upon the nature of the adsorbed complex. Since there is usually more than one form of adsorption that is possible, catalysts should be selected which can be expected to favour the desired form.

Thus, for example, the formation of a pi bond requires the overlap of a filled, bonding pi orbital from an olefin with an empty sigma (dz^2) orbital of a metal if it is to be strong. Back donation from occupied d_{xy}, d_{yz} orbitals of the metal to the empty pi antibonding orbitals of the olefin is also desirable. Considering, for example, square pyramidal coordination on the surface, such adsorbed species can only be formed on d^1, d^2, d^3 metals if only the D orbitals are involved in bonding (V (2), (3), (4), Ti (1), (2), (3), Cr (3), (4), (5), etc) and on d^8, d^9, d^{10} metals when both D and P orbitals are involved (Fe(0), Co(0), (1), Ni (0), (1), (2), Cu (1), (2) and Zn (2)).

Similar arguments can be applied to other adsorbed species to produce a table of solids that can adsorb different reactants or products in the desired form. Thus, for example, in considering the oxidation of olefins, possibilities of adsorption of reactants can be summarised as in table 1.

Table 1.

No. of d electrons	0	1	2	3	4	5	6	7	8	9	10	s^1	s^2
		Ti(3)							Pd(2)		Sn(4)	Zn(1)	Sn(2)
		V(4)	V(3)	V(2)		Fe(3)	Fe(2)		Pt(3)		Sb(5)		Sb(3)
		Cr(5)	Cr(4)	Cr(3)	Cr(2)		Co(3)	Co(2)			Cu(2) Cu(1)		
		Mo(5)	Mo(4)			Mn(3)	Mn(2)		Ni(3)	Ni(2)	Te(6)		
		W(5)	W(4)										
Adsorption of olefins													
(pi bonded)		+	+	+	+				+	+	+		
(sigma bonded) d_z^2						+	+	+	+	+		+	+
Oxygen (radical)					+	+	+	+	+	+			
(pi bonded)		+	+	+					+	+	+	+	+
(donor lone pair) (atoms)	+	+	+	+	+	+	+	+	+	+	+	+	+

This type of approach is very useful in limiting the number of catalysts that should be considered in the design. It does not take too much time since, once available, such a table is widely applicable.

Similar arguments can be applied to one other, more complex, method of assessing the importance of different chemisorbed complexes. This involves the application of molecular orbital calculations to chemisorption and catalysis. The calculations are complex, and are not to be undertaken lightly. However, several papers have appeared in recent years which assess the probability of finding a given adsorption form on a catalyst (refs. 28, 29). One particular useful review of molecular orbital calculations of chemisorbed molecules and intermediates in heterogeneous catalysis has been published by Beran and Zagradnik (ref. 28), covering mainly Russian work up to the second half of 1975. For the normal catalyst design, individual calculations of this type are probably unrewarding: where such information is available, it can be used to good effect.

The dependence of the formation of chemisorbed species on the directional properties of bonds emerging from a surface in an approach which combines the present concept with geometric effects. In an elegant paper covering the possibility of adsorption on different crystal faces, Bond (ref. 30) has considered the fact that e_g and t_{2g} orbitals are spatially directed. As a result, location of particular chemisorbed species in positions favourable to reaction can be envisaged, and predictions based on this were found to be accurate. Regrettably, for the catalyst design, this approach is of limited value, in that we know little about how we should prepare a catalyst with a desired structure, and even less about how we should stabilise that structure during reaction.

There is no doubt that consideration of desired and undesired chemisorbed complexes can be of great importance to a catalyst design. Success depends on how accurately the reaction has been transcribed to the surface (both desired and undesired reactions) and how feasible the proposed surface reactions are. This can only be determined by advanced surface analysis (ref. 1) or, by analogy, from experimental testing of proposed catalysts.

f) Crystal field theory. Relatively modern theories advanced to explain the behaviour of inorganic complexes have been found to offer a good description of chemisorption and catalysis: these include the crystal field and the ligand field theories (ref. 31). The basis of the theories lies in the fact that d orbitals are known to have directional properties and, if a transition metal ion is associated with ligands, the energies associated with these orbitals can vary. The nature of the ligand and the nature of the complex (high spin or low spin) can obviously affect the energies, but the geometry of the complex, as dictated by the coordination, is very important. Now the chemisorption of a reactant on a metal ion centre can also be described as the formation of a complex, whether the ion is isolated (say in solution) or is located on the surface of a crystal matrix. Obviously the geometry will be more constrained by the general form of the matrix, but the principle is the same.

The energy changes that occur on formation of such "complexes" depend on many factors, of which the crystal field stabilisation energy is one. The five degenerate d orbitals of the free ion are split by crystal fields of different symmetry, and the amount of the splitting is measured in terms of an energy parameter, D_q, which is usually obtained from optical data. Chemisorption, the addition of a ligand to a complex, results in a change in geometry of the complex, for example from square pyramid to octahedron or from tetrahedron to square pyramid to octahedron. This, in turn, alters the crystal field stabilisation energy and calculations show that a characteristic twin peak pattern is obtained (figure 11). What is interesting is that this pattern is similar to the patterns of chemisorption and catalytic activity for many metals and oxides (figures 3 and 4). Since catalysis involves reaction of the chemisorbed complex (either to a new chemisorbed molecule or to a desorbed product), it is not surprising that the same effect can be seen in both cases. Indeed, if catalysis involves transfer of an electron, rearrangement of the coordination to a different complex (which has the electrons distributed to give overall lower energy) could well be an important driving force for the reaction.

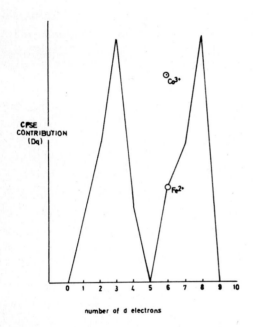

Figure 11. Calculation of crystal field stabilisation energy for strong field complexes.

Arguments on this basis provide a useful diagnostic tool for prediction and provide a sound theoretical basis for many of the observed activity patterns. A full description of the approach is given in references 15, 31 and 32 but, for the purpose of design, the information summarised in figure 11 is very useful indeed.

Obviously all the above approaches are very interactive, and the importance of a given approach depends on data that is available. Some idea of the inter-

relationships between approaches can be obtained, however, by consideration of
the line diagram shown in scheme 2.

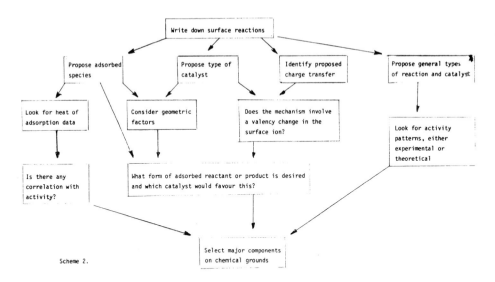

Scheme 2.

Application of the different approaches summarised above can be expected to
suggest several possibilities for the basis of the catalyst, and these can only
be distinguished by experimental testing. Assuming that the primary prediction
has suggested about twenty catalysts, of which two or three are found to be
interesting, it is then necessary to consider how the catalytic activity/selectivity
can be improved by the addition of minor components.

Selection of the chemical basis of the catalyst b) minor components.

The objective of this part of the design is easily stated, in that the catalyst
is performing less well in some respect(s) than is desired. The question then is
how to improve the performance.

Two approaches are possible. The simplistic approach is easy to apply and often
produces results. Thus, for example, if the reaction produces chemicals via a
reaction path involving an excess of one reagent (say oxygen), addition of a minor
component designed to decrease the amount of oxygen adsorbed is easily carried out
(ref. 12). Although such an approach is pragmatic, it usually works.

The second approach is more intellectually stimulating and almost certainly
will work: the disadvantage is that is usually takes considerable time and effort.
The basis of the method is to delve deeply into the mechanism of the reaction, on
the grounds that understanding the mechanism allows optimal fine tuning of the
catalyst. This is generally correct, but there is a closed circle in the sense
that a catalyst must be of considerable interest to warrant the necessary attention

and yet the catalyst may only be of sufficient interest once it has been improved. As a result, detailed studies of this kind are usually carried out only for catalysts which are in current use, but which could be improved.

There are, in fact, two ways of studying the mechanism in order to fine tune the catalyst. The most widely used way is to study reactions on the surface, using recently developed analytical techniques. Of these, electron spin resonance and electron paramagnetic resonance spectroscopy (ref. 33), together with infra red spectroscopy (ref. 34) have proved particularly useful, while electron spectroscopy (ref. 1) holds out much hope for the future. It should also be noted that isotopic labelling experiments, while not fashionable, can be very revealing (ref. 35).

The second method is less direct, but appears to be very interesting. It involves studies of analogues of the catalyst, in which it is possible to control, for example, the location or valency of one of the components of the original catalyst. Several such systems have now been identified, varying from solid solutions to compounds such as scheelites (ref. 36), perovskites (ref. 37), palmierites (ref. 39) and tungstates (ref. 40). The approach and it's benefits can be illustrated from a brief consideration of two of these.

The scheelite structure is represented in figure 12 (a). The general formula is AMO_4, with the M cation being tetrahedrally coordinated to oxygen. About 100 compounds of the general structure have been identified, and some phase diagrams have been presented (ref. 41).

As a catalyst analogue, they are very interesting in that defects, consisting of A cation vacancies, can be introduced in relatively high concentrations (up to one-third of the total of A). Alternatively, it is possible to replace A cations by a second cation, B, to give the general structure

$A_x B_y \emptyset_z MO_4$

where $\emptyset$ = defect and $(x+y+z) = 1$. As a result, it is possible to generate an analogue which contains two cations at known position and in known coordination, with or without controlled amounts of defects.

This analogue was used to advantage in an elegant study of the oxidation of olefins by Sleight and Linn (ref. 36). Investigating the mode of action of bismuth molybdate, the analogue showed that defects promoted the formation of allyl radicals, while the role of bismuth appears to be mainly to replenish the active site with oxygen. Coupled with other direct observations of the catalyst (refs. 42, 43), the role of additives in the system was established in terms of the various processes that occured during the overall reaction.

Similar arguments were used to develop perovskite catalysts, but here there is additional interest in that the materials are interesting catalysts in their own right (refs. 37, 38). The crystal structure is very close to cubic (figure 12 (b)), containing large anions (A = La, Nd, Pr, Ca, Sr or Ba) and B ions held in an octahedral configuration. In addition to being useful analogues, this series of compounds is of great interest in the automobile exhaust gas clean up field, where Ru and Rh are stabilised in the perovskite structure.

There is no doubt that detailed studies are needed to understand the mode of action of additives, and that such studies can also be used to predict which additives could be useful. However, the time needed to be invested is high, and one must be sure that the catalyst will be useful before undertaking the effort. Where this is not certain, it is probably best to use a blend of empiricism (based, where possible, on similar catalysts) and as deep a level of thought as is possible, based on the proposed mechanism. This latter process has been illustrated for different catalysts designs (refs. 4, 5, 6, 12).

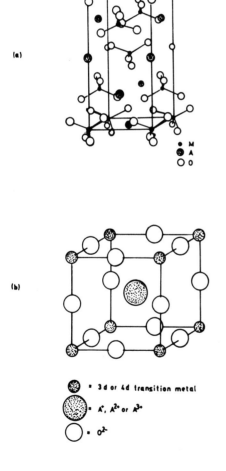

Figure 12. Crystal structures of scheelites (a) and perovskites (b).

Selection of the preferred form of the catalyst.

The morphology of a catalyst is of interest to catalyst preparation for two reasons, which may be called micro and macro effects. Micro effects is used as a general term covering the desired crystallinity, surface area, porosity etc of the catalyst, while macro effects covers such factors as pellet size and strength. With a few notable exceptions, macro effects have not received the attention in the literature that they deserve, since mechanical breakdown is a very common cause of catalyst replacement. However, micro and macro effects, although showing individual characteristics, are closely tied together. This is clearly shown, for example, in the case of alumina, where variation in crystallite size, localised variation in phase change or heating, and homogenity (in crystallite dimensions) were found to have a major effect on attrition and crush strength (ref. 44).

Andrew has presented a good, but too short, review on macro effects in catalysts (ref. 45), in which the desired properties are related by the diagram shown in figure 1. While correctly identifying physical strength characteristics as being very important, the relation with other factors important in catalysis does not emerge from the review. Thus, for example, the strength of catalysts is related to surface area and porosity, but these factors also have a large influence on activity and selectivity, in that they influence mass and heat transfer in the system.

Considering the problem in terms of catalyst preparation, we would expect that a high surface area should give highest activity. However high surface areas are difficult to prepare, are difficult to maintain (because of the possibility of sintering) and are associated with high porosity. This may introduce mass transfer limitations and will certainly give rise to a weaker catalyst.

The problem can be neatly illustrated by the oxidation of methane to formaldehyde and to carbon dioxide. Formaldehyde is unstable and, at the temperatures needed to oxidise methane, will further oxidise to carbon dioxide very easily. Since both reactions are exothermic, the catalyst temperature will tend to rise and this will favour over-oxidation and catalyst sintering. As a result, it is necessary to remove formaldehyde from the active catalyst rapidly, and to equalise temperature in the catalyst bed at as low a value as possible. This requires a low porosity catalyst with good thermal conductivity. For similar reasons, platinum-rhodium gauzes are widely used for the oxidation of ammonia and novel geometries have been developed for methanation catalysts, including the idea of plating them on the surface of a heat exchanger (ref. 46). A general guide to selection of the properties of the catalyst is given in table 2. It should be emphasised that this table is general, since particular systems may show individual effects. Thus, for example, mass transfer limitations may be desirable, in that they improve selectivity (ref. 47) or even, in some cases, increase rate (ref. 48).

Table 2.

The reaction produces	Temperature control* is		Diffusional effects are		Surface area**	Porosity**	Thermal conductivity
	Desirable	Unnecessary	Desirable	Undesirable			
Terminal products such as CO_2, CH_4 etc.	✓		✓		Medium	Medium : maxima in PSD is 50-100 Å	High
		✓		✓	High	High, provided temperature is not too high. Low if temperature rise is very large	Any value
Two products concurrently, one of which is desired	✓		✓		Medium	Medium : maxima in PSD at 50-100 Å	High
	✓			✓	Medium	Low porosity or very wide pores	High
Two products consecutively, the first of which is desired	✓		✓		Medium	Medium : maxima in PSD at 50-100 Å	High
	✓			✓	Medium	Low porosity or wide pores	High
A product, but there is a potential poison in the feed or produced	✓		✓		Medium	Medium. Pores must be such either not to allow the poison to enter the pores or to avoid pore blocking by poison accumulation	
	✓			✓	Medium		High
A very high temperature rise	✓				Low	Non porous	

*For any reason; it is assumed that the catalyst is stable at the highest temperature liable to be observed
**Consistent with desired strength

The variation in porosity and surface area that can be achieved using a pure catalyst is limited by the preparation methods available and by the fact that such materials tend to sinter rapidly. As a result, it is usual to introduce the desired characteristics primarily though the use of a suitable support. Here, too, there are several other factors which can influence the choice. These may be discussed with the aid of table 3.

Table 3. Choice of support

Chemical Factors.

Is the support required to show catalytic activity ?
Are chemical interactions with the catalyst possible ? If so, are these desired or undesired ?
Can the support interact with reactants or products ? Is this desired or undesired ?
How resistant is the support to poisoning ?
Can the catalyst be deposited on the support in the desired form ?
Does the support induce a particular coordination geometry on the catalyst (52, 53, 54).
Is the support stable under operating conditions ?

Physical Factors.

What is the desired surface area and porosity ?
What is the desired thermal conductivity ?
Is the support mechanically strong (45) ? Would this be affected by deposition of poisons, such as carbon ?
Is the support stable under the operating conditions ?
What is the desired form of the pellet ?

Recent studies have revealed that chemical interactions between the catalyst and the support may be more important than was previously thought. Of course, it has been long established that bifunctional catalysis is of major industrial importance (ref. 49), and that chemical interaction between the support and the catalyst may be desirable (ref. 50) or undesirable (ref. 51). What has emerged comparatively recently is that the support may be able to induce a given geometry on a catalyst without formal chemical interaction, and that this will influence adsorption and catalysis. Such effects have been reported for supported silver (ref. 52) and for supported metal oxides (refs. 53, 54), and may be widely spread. No general guide lines for catalyst design can, as yet, be listed.

However, it is probably the physical properties of the support which are primarily responsible for it's selection, provided that these are consistent with desired chemical properties. Both mechanical strength and porosity/surface area are important, as well as stability with respect to temperature etc. In addition, we have to consider the environment in which the catalyst will be used, both with respect to stability and with respect to the reactor. A general guide to the factors that influence the choice of the latter is given in table 4.

TABLE 4.

REACTOR	ADVANTAGES	DISADVANTAGES	FORM OF CATALYST PELLETS
GAS-SOLID REACTOR (TUBULAR)	WIDELY USED AND WELL UNDERSTOOD	TEMPERATURE CONTROL MAY BE DIFFICULT	PELLETS. EFFECTIVENESS FACTOR ~ 0.6; GOOD THERMAL STABILITY
FLUID BED	GOOD TEMPERATURE CONTROL. USEFUL WHERE CATALYST NEEDS FREQUENT REGENERATION	BAD MIXING. CATALYST ATTRITION. DIFFICULT TO OPERATE.	FLUIDISABLE PARTICLES (E.G. 40-70 MICRONS): GOOD ATTRITION STABILITY
TRICKLE BED	ALLOWS GOOD GAS-LIQUID-SOLID CONTACTING GOOD TEMPERATURE CONTROL.	DIFFICULT TO OPERATE. FOAMING. SPLASHING.	SMALL PARTICLES: OPEN POROSITY: HIGH SURFACE AREA.
HOMOGENEOUS CATALYST REACTOR	MAY GIVE GOOD SELECTIVITY AT LOW TEMPERATURES	MAY BE DIFFICULT TO SEPARATE PRODUCTS AND CATALYST. ALTHOUGH "ANCHORED" CATALYSTS ARE NOW FASHIONABLE (55, 56).	
SLURRY REACTOR	GOOD TEMPERATURE CONTROL	MAY INVOLVE DIFFICULTIES IN GAS-LIQUID-SOLID CONTACTING	

Considerably more is known of micro effects in catalysis, although these may be of more academic interest than applied. This results from the fact that although we begin to know how to prepare a catalyst with a given structure, we know little about how to retain this structure under reaction conditions. This may be illustrated by considering supported metal catalyst.

It is generally desirable to optimise the use of the metal by preparing it in as high a dispersion as possible and, in practice, this usually means depositing the metal in small crystallites on a support. Now, theoretically, this can have other advantages, in that small crystallites are most likely to show geometrical effects (ref. 57) in their catalytic action. The concept of structure sensitive and structure insensitive reactions is well established (refs. 58, 59), and some assessment can be made of the nature of a given reaction, using the criteria outlined in table 5. Although it should be emphasised that this table has no theoretical justification, experimental observations indicate that the greater the number of criteria satisfied by a given reaction, the greater the probability that a reaction is demanding or is facile. Similarly, on a more theoretically sound base (ref. 57), it can be shown that the number of edge and corner sites on a catalyst will be much higher with smaller crystallites.

Table 5.

Structure insensitive (facile) reactions may be

a) addition reactions or elimination reactions
b) reactions involving a large decrease in free energy
c) reactions involving reactants with lone pair electrons or pi bonds or strain energy
d) reactions that do not require a multifunctional catalyst
e) reactions occuring on an active catalyst whose lattice parameters do not change with dispersion

Structure sensitive (demanding) reactions

a) may occur on certain sites, e.g. N_2 chemisorbs only on W(111) and NH_2 is produced only on W(111)
b) involve single C-C bond breakage
c) may involve reactants with no lone pair electrons, pi bonds or strain energy
d) need a multifunctional catalyst
e) need a less active catalyst whose lattice parameters change with dispersion
f) may involve reactants with unpaired electrons (e.g. NO).

In practice, there are two difficulties with this approach. With the exception of catalytic reforming, in which the overall reaction involves both facile and demanding reactions (ref. 49), most reactions of industrial interest are, in fact, facile. Secondly, given that a reaction is demanding, it can be very difficult to achieve or to maintain a given geometry in a catalyst prepared on anything but the smallest scale, since reorganisation of the catalyst - either within the particle (ref. 60), or between particles (refs. 61, 62) - is often rapid under reaction conditions - or even during preparation and activation.

Gross rearrangements (sintering) can be prevented to some extent by the use of spacers. Conventionally these involve a high melting point non-deleterious metal salt which is co-precipitated on a support together with the catalyst, thereby acting as a physical barrier to agglomeration. However, there are other forms of spacer, although they are not considered in these terms. Thus, for example, a solid solution not only dictates the geometry of the solute ion (ref. 63) but also acts to separate solute ions. Similarly, an alloy distributes one metal in another, and the resulting effect can be due either to dilution or to chemical or electronic interaction (refs. 64, 65). For the correct properties, one component could also be regarded as a spacer for the other.

The overall design.

A general programme for the design of a catalyst has been discussed above, but this is only part of the overall design process which has been described in detail elsewhere (refs. 5, 6). Although the discussion has been presented only briefly, it does attempt to review the areas important in the context of catalyst preparation. Before the combination of these ideas with catalyst preparation is considered, it is worthwhile emphasising some points.

Regrettably, we have insufficient knowledge to ensure that the catalyst design is absolutely correct, and experimental testing must be carried out. However, it should be stated that, on the thirty-odd occasions that the author has carried out a catalyst design, the procedure has shown up a catalyst that has been found - either previously or subsequently - to be active for the reaction under consideration It should be hastily added, however, that several inactive catalysts have also been suggested by the design. As a reasonable assessment, the design procedure offers a guide to experimentation which can often be successful and which requires the investment of only a little time. As the knowledge and experience of the designer improves, so the accuracy of the design can also be expected to improve.

Secondly it is necessary to emphasise the feed back cycle in the design. Experimental testing is necessary at various points, and the results of these experiments can be used to modify the conceptional basis of the design. Thus, for example, if experiments show that reaction path A leads to a more desirable product spectrum than reaction path B, the design can be adjusted to put more weight on reactions of type A.

Thirdly it must be remembered that there are always factors that have not been considered. Catalysis is a complex subject, involving inter-related phenomena from a wide variety of fields. As a result, a set of experiments, carried out for a given reason, may give the "wrong" results because of a second factor which has not been considered. Perhaps the most obvious case of this was work carried out on the design of a catalyst to convert propylene to benzene (ref. 5): at the end of the design it was discovered that changing economics would make the reverse reaction more attractive ! This, again, emphasises the importance of feedback at all stages.

With these comments in mind, it is now possible to consider the interaction of catalyst design and catalyst preparation.

Catalyst design and catalyst preparation.

Finally, it is necessary to consider correlation of the suggestions from the design procedure with what can be practically achieved by catalyst preparation. In deference to the main theme of the meeting, this will be limited to impregnation and activation.

Impregnation is widely used as the first step in the preparation of a catalyst, and involves the deposition of a metal salt on a support: this metal salt can be converted into the active catalyst during subsequent processing, activation being one of the major steps.

As with many often apparently simple processes, impregnation may be complex. Three general types may be identified. In the first, the support is soaked in the metal salt solution, and the excess liquor evaporated: this generally leads to an even distribution of metal salt throughout the pore system which can be penetrated by the original solution. This need not include micropores in the support. In the second, it is possible to pre-treat the metal salt in solution before impregnation. Thus, for example, a colloidal suspension of metal could be prepared, subsequent impregnation resulting in the concentration of the colloid at the exterior surface of the support. Thirdly, and perhaps of most interest in the context of control, impregnation could result from a chemical reaction between the salt and surface groups on the support. This can be illustrated for the preparation of a platinum on silica catalyst.

Silica is known to have a complex surface structure, in which there are a number of different hydroxyl sites of varying reactivity (ref. 66). Impregnation by chlorplatinic acid involves hydrolysis at some of these sites to produce platinum-anion-silica groups at the surface (ref. 67). In these terms, the possibilities of control are large. Thus, for example, chlorplatinic acid can be easily converted to $(PtCl_x(OH)_{6-x})^{2-}$ (ref. 68), and the composition of the salt could be adjusted to favour reactivity with only one type of surface hydroxyl group. Secondly, the hydroxyl groups on silica could be poisoned, either with respect to their reactivity (by pre-treating with compounds of different basicity) or with respect to their location (by using bases in which steric effects may limit the penetration of the pore system). As a result, it is possible to deposit the metal salt in varying density and at different locations on the surface.

From the above, it is seen that the amount and location of the catalyst can be controlled, at least in principle, during impregnation, although subsequent processing may redistribute the catalyst (see below). Interaction with the catalyst design is possible in the selection of the active catalyst (major and minor constituents) and in the use of mass and heat transfer considerations to decide where the catalyst should be located. In addition, possibilities of sintering could dictate the desired strength of catalyst-support bonding, although this may be harder to achieve in practice.

In the present context, activation may be considered as the conversion of the as-deposited salt to the desired catalyst by thermal treatment in the presence of a gas: this may, or may not, involve reactions such as oxidation, reduction or sulphidation. The conditions and efficiency of the process are easy to assess, in terms of the nature of the desired conversion, although complete activation may be more demanding than these calculations would indicate (ref. 69).

Two major problems may arise during activation. The first of these involves the rearrangement of the catalyst or the support: either may be desired or undesired. Rearrangement of the support generally arises as a result of phase transformations in the material, causing collapse of pore structure and decrease in surface area. This is often accelerated by impurities (including the catalyst) (refs. 70, 71), and by the ambient gas (ref. 72): during activation, this is usually undesirable. Estimates have been published of the stability of common supports (refs. 44, 45, 70) but, regrettably, these are widely dispersed in the literature and a comparative review is badly needed. However, it is possible to identify conditions under which rearrangement may occur.

Rearrangement of the catalyst during activation may be desired if the catalyst migrates to support surface sites where it can be located more firmly. The resulting catalyst can be expected to be more stable during it's working life. If, on the other hand, rearrangement leads to catalyst agglomeration and to decreased surface area or increased particle size, then the process should be avoided. This means either activating under conditions where rearrangements is minimal or, if this is impossible, introducing a second component (such as a spacer) that minimises rearrangement.

The choice of such a component depends, to some extent, on the mechanism of rearrangement. As a general rule, rearrangement can involve surface diffusion (ref. 73), volume diffusion (ref. 74) or evaporation-condensation (ref. 75), the important process depending on the temperature. Use of a spacer minimises volume diffusion, but surface diffusion or evaporation-condensation can only be prevented by the addition of a second component to the catalyst (as, for example, is the case with Pt/10 % Rh gauzes used for the oxidation of ammonia). It is important to remember that, although a second component may be added to stabilise the catalyst, it may affect the overall catalytic behaviour of the solid.

As a result of recent work on sintering (refs. 76, 78), it is possible to predict with some accuracy the conditions to be avoided. From the viewpoint of catalyst design, however, it is usually sufficient to use the old rule that surface rearrangements are liable to be important at ca. 0.3 x melting point and volume rearrangement at ca. 0.5 x melting point. If conditions are critical, it may be necessary to calculate the temperature rise that could occur in a catalyst bed, and here a simple one dimensional model will usually suffice (ref. 79): the temperature in a pellet is unlikely to rise markedly (less than 20°C) above the ambient in the particular portion of the bed (ref. 80).

The second factor that may be important during activation is the possibility of catalyst-support interactions. These, too, may be desirable (e.g. chromia-alumina (ref. 50)) or undesirable (e.g. nickel alumina (ref. 51)), and this can be established from the design. Fortunately, the conditions under which such solid-solid interactions are possible are usually available (refs. 74, 81, 82), although there has been some consternation in recent years to discover that interaction between platinum and alumina is possible (ref. 83).

Although the purpose of the catalyst design is primarily to decide which catalyst to prepare, it is seen that interaction between the design procedure and catalyst preparation is possible and can be productive. It is certainly possible to recognise which factors are open to control and to predict the probable effects of such control during preparation. Catalyst design is still far from perfect, but it can offer a logical guide to experimentation that can considerably shorten the time needed to develop a new catalyst.

ACKNOWLEDGEMENTS

The author acknowledges, with gratitude, valuable discussions with Prof. J.W. Coenen.

REFERENCES

1. L. Lee, Characterisation of metal and polymer surfaces: vol 1 and 2, Academic (1977).
2. Proc. Int. Symp. on Preparation of Catalysts. Ed.: B. Delmon, Elsevier (1976).
3. M.S. Borisova, V.A. Dzis'ko and L.G. Simonova, Kin. and Cat. 15(1974)425.
4. D.A. Dowden, C.R. Schnell and G.T. Walker, Proc. IV Internat. Congr. on Catalysis (Moscow)(1968)201.
5. D.L. Trimm, Chem. and Ind. (London)(1973)1012.
6. D.A. Dowden, Chem. Eng. Prog. Symp. Ser. 63 no 73 (1967)90.
7. D.A. Dowden, La Chimica e l'Industria, 55(1973)639.
8. T.W. Østlyngen, Diplom report, Univ. of Trondheim (1977).
9. J.R. Anderson, "Structure of metallic catalysts", Academic (1975).
10. O.V. Krylov, "Catalysis by non metals", Academic (1969).
11. F.C. Whitmore, Ind. Eng. News, 26(1948)668.
12. M. Goldwasser and D.L. Trimm, Acta Chem. Scand. Ser. B, in press.
13. B.S. Greenfelder, H.H. Voge and G.M. Good, Ind. Eng. Chem. 41(1949)2573.
14. D.A. Dowden, N. Mackenzie and B.M.W. Trapnell, Proc. Roy. Soc. A 237, (1956)245.
15. D.A. Dowden, Cat. Revs. Sci. Eng. 5(1972)1.
16. J.A. Busby and D.L. Trimm, Chem. Eng. J. 13(1977)149.
17. J.E. Germain, "Catalytic Conversion of Hydrocarbons", Academic (1969).
18. G.C. Bond, "Heterogeneous Catalysis: Principles and Applications". Oxford Chem. Ser. (1974).
19. Adv. in Catalysis 7(1955).
20. A.G. Daglish and D.D. Eley, Actes 2me Congr. Int. Catal. Paris (1961)1615.
21. T.J. Gray, N.G. Masse and H.G. Oswin, Ibid 1697.
22. G.C. Bond, "Catalysis by Metals", Academic (1962).
23. G.K. Boreskov, Kin. and Cat. 8(1967)878.
24. V.V. Popovskii and G.K. Boreskov, Kin. i Kat. 1(1960)566.
25. Y. Moro-Oka, Y. Morkawa and A. Ozaki, J. Catal. 7(1967)23.
26. A.A. Balandin, Russ. Chem. Revs., (1962)589.
27. G.A. Somorjai, J. Catal. 27(1972)453.
28. S. Beran and R. Zagradnik, Kin. and Cat. 18(1977)299.
29. R.A. Van Santen and W.M.H. Sachtler, Surface Sci. 63(1977)358.
30. G.C. Bond, Disc. Farad. Soc., 14(1966)200.
31. J.A. Duffy, "General Inorganic Chemistry", 2nd. ed.: Longmans (1974).
32. D.A. Dowden and D. Wells, Actes 2me Congr. Int. Catal. (Paris) (1961)1499.
33. J.H. Lunsford, Adv. in Cat. 22(1972)265.
34. R.P. Eischens, and W.A. Pliskin, Adv. in Ca. 10(1958)2.
35. M. Ozaki, "Isotopic Studies of Heterogeneous Catalysis", Academic (1977)
36. A.W. Sleight, and W.J. Linn, Annals New York Acad. of Sci., 272(1976)22
37. R.J.H. Voorhoeve, J.P. Remeika and L.E. Trimble, Ibid, p. 3.
38. R.J.H. Voorhoeve, J.P. Remeika and L. Trimble, Mat. Res. Bull 9(1974)1393.
39. J.M. Longo and L.R. Clavenna, Annals New York Acad. Sci. 272(1976)45.
40. S. De Rossi, E. Iguchi, M. Schiavello and R.J.D. Tilley, Z. Phys Chem. 103(1976)193.
41. A.W. Sleight, K. Aykan and D.B. Rogers, J. Solid State Chem. 13(1975)231.
42. J.M. Peacock, M.J. Sharp, A.J. Parker, P.G. Ashmore and J.A. Hockey, J. Catal. 15(1969)373, 379, 387, 398.
43. P.A. Batist, B.C. Lippens and G.C.A. Schmit, J. Catal. 5(1966)55, 64.
44. R. Gauguin, M. Graulier and D. Papee, "Catalysts for the Control of Automotive Pollutants", Adv. in Chem. 143(1975)147.
45. S.P.S. Andrew, "Catalyst Handbook", Wolfe Scientific Texts (1970).
46. G.A. Mills and F.W. Steffgen, Cat. Rev. Sci. Eng. 8(1973)159.
47. J.M. Thomas and W.J. Thomas, "Introduction to the Principles of Heterogeneous Catalysis", Academic (1967)
48. J. Wei, Adv. in Chem. 148(1975)1.
49. M.J. Sterba and V. Haensel, Ind. Eng. Chem. Prod. Res. Dev. 15(1976)3.
50. P.B. Weisz, C.D. Prater and K.D. Ritthenhouse, J. Chem. Phys. 21(1953)2236.
51. J.R.H. Ross, Surface and Defect Properties of Solids, 4(1975)34.

52 K.K. Kakati and H. Wilman, J. Phys. D., Appl. Phys. 6(1973)1307.
53 V.A. Schvets and V.B. Kazansky, Kin. and Cat. 25(1972)123.
54 M. Goldwasser and D.L. Trimm, in preparation.
55 R. Ugo, Cat. Revs. Sci. Eng., 11(1975)225.
56 Yu. I. Yermakov, Cat. & Rev. Sci. Eng. 13(1976)77.
57 R. Van Hardeveld and F. Hartog, Surface Sci. 15(1969)189.
58 C. Bernardo and D.L. Trimm, Carbon 14(1976)225.
59 M. Boudart, Proc. VI Internat. Congr. on Catal. (London)(1976)1.
60 B.J. Cooper, B. Harrison and E. Shutt, S.A.E. paper 770367 (1977).
61 A.E.B. Presland, G.L. Price and D.L. Trimm, J. Catal. 26(1972)313.
62 R.M.J. Fiederow and S.E. Wanke, J. Catal. 43(1976)34.
63 F.S. Stone, Adv. in Catalysis 13(1962)1.
64 V. Ponec, Cat. Rev. Sci. Eng. 11(1975)41.
65 R.L. Moss and L. Whalley, Adv. in Catalysis 22(1972)115.
66 J.B. Peri, J. Catal. 41(1976)227.
67 J.P. Brunelle, A. Sugier and J.F. Le Page, J. Catal. 43(1976)273.
68 E.H. Archibald and W.A. Gale, J. Chem. Soc. 121(1922)2849.
69 S.P. Noskova, M.S. Borisova, V.A. Dzisko, S.G. Khisanieva and Yu.A. Alabuzhev, Kin. and Cat. 15(1974)527.
70 D.A. Dowden, I. Chem. Eng. Symp. Ser. 27(1968)18.
71 N.M. Zaidman, V.A. Dzisko, A.P. Karnaukhov, L.M. Kefeli, N.P. Krasilenko, N.G. Koroleva and I.D. Ratner, Kin. and Cat. 10(1969)313.
72 A.V. Kiselev and Yu. S. Nikitin, Kin. and Cat. 4(1963)562.
73 G.E. Rhead, Surface Sci., 15(1969)353.
74 P. Wynblatt and N.A. Gjostein, Prog. Solid State Chem. 9(1975)21.
75 G.C. Fryburg, Trans. A.I.M.E., 233(1973)1986.
76 P.C. Flynn and S.E. Wanke, J. Catal. 34(1974)390.
77 C.G. Granqvist and R.A. Buhrman, J. Catal. 42(1976)477.
78 P.C. Flynn and S.E. Wanke, Cat. Rev. Sci. Eng. 12(1975)93.
79 O. Levenspiel, "Chemical Reaction Engineering", John Wiley (1962).
80 D.L. Trimm, J. Corrie and R.D. Holton, Chem. Eng. Sci. 29(1974)2009.
81 R.W.G. Wyckoff, "Crystal Structures", Interscience (1966).
82 J. Mellor, "A comprehensive Treatise on Inorganic and Theoretical Chemistry", Longmans (1922-37).
83 F. Dautzenberg, Lecture: Rideal Conference: April 1977.

DISCUSSION

D. CHADWICK : You raised the question of the importance of defects in relation to catalytic activity. Recent experimental results in surface science suggest that defects do indeed play an important role in surface reactions and that selectivity, for example, may be related to step/kink density, in which case it seems surprising that one can obtain activity patterns with simple parameters such as lattice spacing.

D.L. TRIMM : It is true that defects play an important role in determining catalytic activity. It should be emphasized that each one of the factors discussed in the paper may, or may not, be important. As a result one should use lattice spacing arguments (or any other approach) as an indicator only. Experimental testing is the only way to check that an indicator is (or is not) accurate. One is, however, on much stronger ground when it is advantageous to study the mechanism of the reaction in greater detail, in order to investigate possible secondary components. At this stage the role of defects and the importance of different sites can be clarified. It should be emphasized, however, that rearrangements under operating conditions may make it difficult to optimise the surface geometry, even when the desired surface structures have been established

M. MARTAN : Did you really study the direct synthesis of limonene from toluene and propylene ?

D.L. TRIMM : A design study for the catalytic production of limonene was carried out. No experimental testing has been completed. I suspect that the partial hydrogenation of toluene will not be possible using a heterogeneous catalyst.

EFFECT OF IMPREGNATION AND ACTIVATION CONDITIONS OF Al_2O_3/CuO SUPPORTED MONOLITH CATALYSTS IN THE REDUCTION OF NO

M. ENGLER and K. UNGER
Institut für Anorganische Chemie und Analytische Chemie
Johannes-Gutenberg-Universität Mainz, 65oo Mainz
O. INACKER, C. PLOG and M. SEIDL
Dornier System GmbH, 799o Friedrichshafen

SUMMARY

Reduction of NO in the presence of CO and Ar was examined on two series of monolithic sheets carrying an active CuO/Al_2O_3 layer. In the impregnation procedure the ratio of Al_2O_3 to CuO was varied in a wide range. The texture of the layer and the dispersion of CuO therein was controlled by mercury porosimetry, scanning electron microscopy and secondary ion mass spectrometry (SIMS). It could be established that the surface concentration of CuO determined by SIMS measurements is the most decisive quantity of the catalyst correlating linearly with the conversion of NO.

INTRODUCTION

The need of catalytic converters to control the emission of automobile exhausts gave rise to a wide-spread activity in the preparation of optimum designed catalysts (ref. 1,2). Due to the specific requirements such as mechanical, thermal and thermal shock stability supported ceramic monoliths made of cordierite were found to be superior as carriers compared to pellets arranged in a fixed bed. As the automobile exhaust constitutes of a large variety of pollutants to be removed different catalytic systems were proposed utilizing oxidation, reduction and so-called three-way catalysts (ref. 2). In this context considerable interest has focussed on the reduction of NO as one hazardous component of exhaust.

This study deals with the reduction of NO in the presence of CO and by means of an oxidation catalyst composed of CuO/Al_2O_3 which is supported on a cordierite monolith. The purpose of the work was to study the effect of the geometric properties of the layer and the dispersion of CuO therein on the conversion of NO at constant reaction conditions.

2. Experimental

2.1 Starting Materials for Catalyst Preparation

The catalyst carrier was a monolith of type AlSiMag 795 (American Lava Corp., Chattanooga, Tenn., USA) with about 233 holes/inch^2. The specific surface area according to BET was determined to 0.1 m^2/g. The specific pore volume measured by the intruded volume of mercury up to 4.000 bar amounts to 0.064 ml/g. The pore volume distribution is of bimodal type ($D_{max(1)}$ = 10 - 40 nm; $D_{max(2)}$ = 3 000 nm) but 94 % of the total specific pore volume is distributed in pores larger than 2 500 nm. Since the coating of the whole monolith or pieces of it leads to a largely inhomogeneous distribution of the catalyst along the channels the monolith was sliced into sheets of 40 x 10 mm in dimensions. These sheets coated at both sides were used in all further investigations such as conversion measurements, mercury porosimetry, scanning electron microscopy (SEM), secondary ion mass spectrometry (SIMS) etc.

The starting material supporting the active coating was a finely divided non-porous gamma-alumina of type Alu C (Degussa, Hanau, West Germany) of mean particle size dp = 20 nm and of specific surface area S_{BET} = 100 m^2/g. The catalytic component CuO was made from copper(II)nitrate (cryst. pure, E. Merck, Darmstadt, West Germany).

2.2 Catalyst Preparation

The monolithic sheets were coated in one step by controlled dipping into a homogeneous and stable suspension made of Alu C, copper(II)nitrate and ethanol as dispersing agent. The impregnated sheets were dried at 410 K for 8 hours and finally calcined at 873 K for one hour.

2.3 Examinations

The specific surface area and the pore structure data of the untreated and coated sheets were determined by means of nitrogen sorption techniques and mercury porosimetry (ref. 3). The mass increase of the sheets was monitored by weighing before and after coating. The bulk concentration of copper in the sheet was measured titrimetrically after dissolution of CuO in concentrated sulfuric acid. Conversion measurements were made in a reactor constructed by Dornier System GmbH, Friedrichshafen, West Germany (ref. 4). Conditions were: gas composition 10 sccm NO, 1430 sccm Ar, pressure 400 Torr, temperature 623 K. A special probe transfer system developed by Dornier System GmbH (ref. 5) connected to the reactor permitted to transfer the catalyst from the reactor into ultra high vacuum chamber of the SIMS apparatus (type 102, Balzers, Liechtenstein) without exposing it to the atmosphere. SEM measurements were made with a model MSM 5 apparatus of International Scientific Instruments Inc., Japan.

3. Results and Discussion

3.1 Variables in Catalyst Preparation

Two series of catalysts were prepared. In series I the mass of deposited alumina was varied while holding the content of CuO constant. This was performed by dipping the sheets in suspensions of increasing content of Alu C (0.5 to 5.0 % (w/w)) which were 0.8 molar on copper.

In series II the mass of deposited alumina was kept constant while the content of CuO was varied in the range between 0,2 to 6,6 % (w/w) by dipping the sheets in suspensions of increasing molarity of copper (II) nitrate from 0,04 to 1,6 mol/l while the content of Alu C was maintained at 1 % (w/w).

In this way it was possible to vary
(i) the degree of dispersion of a constant amount of cupric oxide within the active alumina layer (s. series I)
(ii) the amount of cupric oxide dispersed in a constant mass of deposited alumina (s. series II)

3.2 Pore structure and Thickness of the Coating

The deposited layer made of Alu C alone is regarded as a close-packed assembly of three-dimensionally linked non-porous alumina particles forming a regular pore structure. Independent of the mass of the mounted alumina the most frequent pore diameter derived from mercury porosimetry measurements amounts to 20 - 35 nm which is one order smaller than that of the uncoated monolith. In the presence of copper(II)nitrate as a second component of the suspension CuO particles are formed through the calcination procedure. The question arises in which way the CuO particles are arranged within the porous alumina layer. It can be assumed that at a high ratio of alumina to CuO the CuO particles are uniformly dispersed within the alumina whereas at a low ratio CuO forms clusters resulting in a lower degree of dispersion. By means of mercury porosimetry measurements it was evidenced in both series that the most frequent pore diameter remained constant at 20 - 35 nm and hence was independent of the amount of CuO embedded in the porous layer.

A linear relationship was established between the content of Alu C in the starting suspension and the mass deposited at constant content of CuO in series I and between the molar concentration of $Cu(NO_3)_2$ in the starting suspension and the mass deposited at constant content of alumina in series II (s. Table I).

Table I : Properties of CuO/Al_2O_3 supported layers in series I and II, respectively

series I

content of Alu C in the suspension % (w/w)	total mass deposited % (w/w)	molar ratio Al_2O_3/CuO
0	3.0	-
0.5	3.8	0.01
1.0	4.1	0.02
1.5	4.2	0.07
2.0	4.5	0.09
2.5	5.1	0.23
3.0	5.4	0.21
4.0	5.6	0.35
5.0	6.0	0.66

series II

concentration of CuO in the suspension mol/l	total mass deposited % (w/w)	molar ratio Al_2O_3/CuO
0	0.46	-
0.04	0.51	1.00
0.08	0.53	0.89
0.20	1.26	0.55
0.40	2.37	0.19
0.80	3.93	0.07
1.20	5.90	0.10
1.60	6.93	0.08

The thickness of the layer, d_s, can be formally estimated by the equation

$$d_s = \frac{\left[V_{p(Hg)} - V_{p(Hg, macro)}\right] + \frac{m_{Al_2O_3}}{\varrho_{Al_2O_3}} + \frac{m_{CuO}}{\varrho_{CuO}}}{S_{BET}} \quad (1)$$

where $V_{p(Hg)}$ is the total specific pore volume of the layer, $V_{p(Hg, macro)}$ the amount of $V_{p(Hg)}$ which penetrates macropores larger than 2 500 nm, $m_{Al_2O_3}$ and m_{CuO} the mass of the deposited alumina and CuO, respectively, $\varrho_{Al_2O_3}$ and ϱ_{CuO} the respective true densitites and S_{BET} the specific surface area of the parent monolithic sheet determined by means of the BET method. Comparison of the d_s values of series I for instance reveals that d_s varies from 140 to 340 nm and hence d_s amounts to about one tenth of the most frequent pore diameter of the native monolith.

3.3 Dispersion of CuO Within the Porous Layer

The degree of dispersion of CuO on the sheets of series I was studied at various amounts of alumina and constant content of CuO ($3,9 \pm 0,2$ % (w/w)). In Fig. 1 the variation of the concentration of Cu in atom percent determined with SIMS following a method developed by Plog (ref. 6) is shown as a function of the depth of the layer expressed in numbers of monolayer units of Cu.

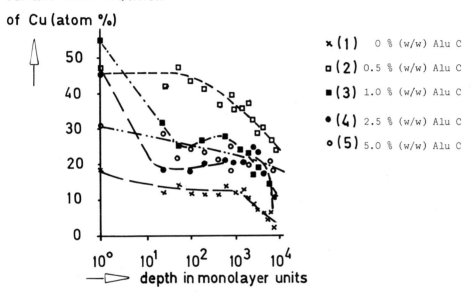

Fig. 1 Depth profile of Cu at different amounts of alumina

The impregnation of the sheet with CuO alone without a supporting alumina layer (Curve (1) in Fig. 1) gives the lowest surface concentration. The SEM photograph taken of this particular sheet shows relatively large clusters of CuO of 20 nm in diameter distributed at the surface (s. Fig. 2a).

The addition of alumina drastically increases the degree of dispersion of CuO, a fact which is evidenced by the higher surface concentration of Cu shown by curves 2 - 4. The dispersion effect due to alumina starts to become noticeably at 1 %(w/w) of Alu C in the suspension whereas at higher contents of Alu C the surface concentration of copper decreases again. By going from the outside to the interior of the layer, the shape of the depth profile of Cu is seen to be specifically dependent on the amount of alumina in the range considered.

Fig. 2 : SEM photographs of sheets coated in different ways
(magnification 1000 x left hand side, magnification 7000 x right hand side)
a) native monolithic sheet
b) sheet coated with 3.0 % (w/w) CuO without alumina
c) sheet coated with 0.5 % (w/w) alumina without CuO
d) sheet coated with 3,0 % (w/w) CuO and 3,0 % (w/w) alumina according to a suspension containing 5 % (w/w) Alu C and 0.8 mol/l $Cu(NO_3)_2$

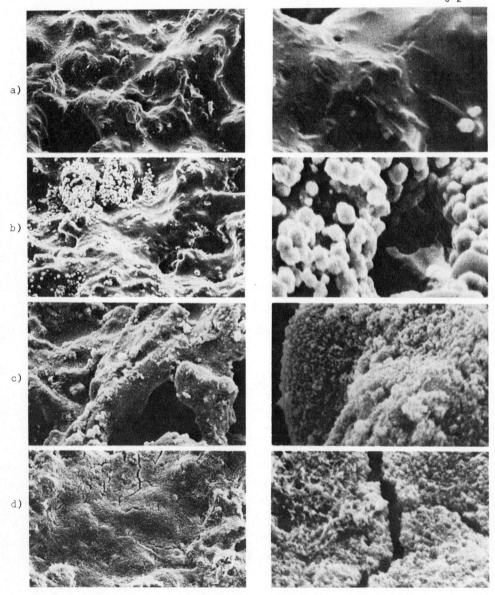

3.4 Effect of the Impregnation Conditions on the Conversion of NO

3.4.1 Amount of Mounted Alumina at Constant Content of CuO vs. Conversion of NO

In Fig. 3 the conversion of NO is plotted as a function of the amount of alumina in the starting suspension. The sheet without alumina yields the lowest conversion. At about 1 % (w/w) of Alu C the conversion goes through a maximum which is consistent with the results of Fig. 1 which shows the highest surface concentration of CuO in this region. At higher amounts of alumina the conversion decreases and finally approximates a constant level. As shown in Fig. 3 the shape of the curve conversion vs. amount of alumina is in concert with that of surface concentration of Cu vs. amount of alumina. With regard to conversion the optimum amount of Alu C in the suspension is 1 % (w/w).

3.4.2 Concentration of CuO in the Layer at a Constant Amount of Alumina vs. Conversion of NO

In this case the amount of alumina was maintained at 0.46 % (w/w) whereas the bulk concentration of CuO varied between 0.2 - 6.6 % (w/w). At low CuO concentrations in the range between 0.05 and 1.91 % (w/w) the conversion increases remarkably with the CuO content. At about 2.0 % (w/w) CuO the curve shows a downward inflection and conversion increases to a lower extent with increasing content of CuO. The course of the function can again be explained in context with the variation of the surface concentration which was measured by means of SIMS on the respective sheets. Again the two curves conversion vs. copper content and surface concentration vs. copper content are in concert with one another (s. Fig. 4). Finally Fig. 5 shows an approximately linear relation between the NO conversion and the surface concentration of copper.

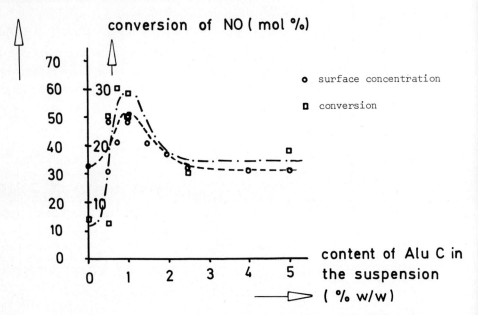

Fig. 3 : Dependence of the conversion of NO and surface concentration of Cu, respectively, on the amount of alumina in the suspension

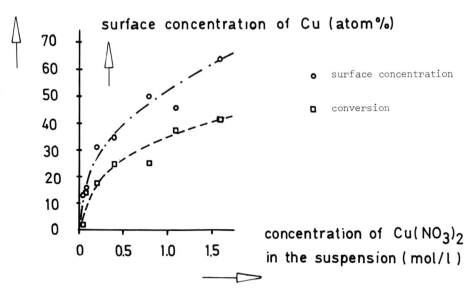

Fig. 4 : Dependence of conversion of NO and surface concentration of copper on the concentration of $Cu(NO_3)_2$ in the starting suspension

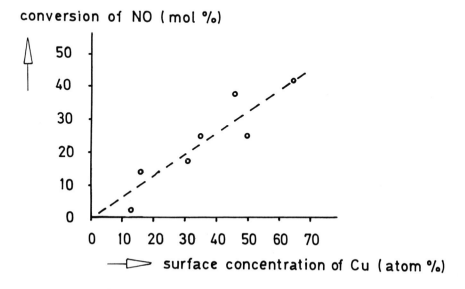

Fig. 5 : Conversion vs. surface concentration of Cu (data obtained at sheets of series II)

4. Concluding Remarks

The results reveal that

(i) the reaction of NO preferably takes place at the outer surface of the catalytic layer and

(ii) the highest surface concentration of CuO which also gives the highest conversion is obtained at a molar ratio CuO/Al_2O_3 of 100 to 200 in the layer.

REFERENCES

1 E. Koberstein and W.M. Weigert, Angew. Chem. 88 (1976) 657
2 J.W. Hightower, in B. Delmon, P.A. Jacobs and G. Poncelet (Editors), Preparation of Catalysts, Elsevier, Amsterdam, 1976, p. 615.
3 S.J. Gregg and K.S.W. Sing, Adsorption, Surface Area and Porosity, Academic Press, London, 1967.
4 M. Seidl, O. Inacker, C. Plog and E. Steinheil
  in Proceedings of 3. Symposium über Gas-Oberflächenwechselwirkung, Meersburg, 1975, Part II, BMVG-SBWT 76-20, p. 29.
5 M. Seidl and O. Inacker
  to be published
6 C. Plog
  to be published

DISCUSSION

S.P.S. ANDREW : The author's measurement suggests that the activity is determined by the composition of the outermost layer of the outside coating. This would imply that the reaction is very fast. Could it be that the surface composition of inner layers also take part in the reaction ? If so, does it not imply that surface composition of these layers is also important ? Is the reaction really so fast ?

J.W. HIGHTOWER : Is it possible that the maximum observed in Fig. 3 may be due to a dilution of the active ingredient as the Cu is dispersed within the larger amount of alumina and is therefore not available to the reactants in this strongly diffusion limited reaction?

O. INACKER : Both questions are closely related. They concern the interpretation of our results. The paper shows only that a correlation between the conversion rate and the surface concentration of the active component can be established. For this type of catalyst we don't find 100% NO conversion under the above reaction conditions and the applied catalyst arrangement, but for other catalysts we can achieve 100% conversion under the same conditions. From that we exclude that the reaction is very fast. We consider two possibilities for the interpretation of our results. The reaction rate may be limited by pore diffusion, and the reaction rate is determined mainly by the highest concentration of the active component close to the gas-solid interface, which can be concluded from the depth profiles. The copper on the gas solid interface may also have a much higher catalytic activity than the copper within the cupric oxide alumina layer. But applying the approach of Dornier System (5), this question can principally be answered.

L. GUCZI : Did you investigate the stability of the layer on the cordierite, i.e. how fast does the catalytic layer leave the monolith surface ?

O. INACKER : We have not investigated the longtime mechanical stability of the active layer but fast deterioration processes occurring within several hours. Copper is removed slowly from the active layer via volatile compounds. Additionally a fast deterioration of the catalyst takes place due to a redox-process or to structural changes within active layer.

V. FATTORE : All your date were obtained using copper nitrate solutions. Did you try other copper salts ? Do you think that changing the impregnating copper salt, you can find a different dependence of the conversion of NO on the surface concentration of Cu and on the amount of alumina in the suspension ? (Fig. 3).

O. INACKER : We have used other impregnation procedures too : e.g. water as solvent and $|Cu(NH_3)_4|^{2+}$ as copper ions. We found different activities and surface concentrations of copper compared to the same copper and alumina concentration in ethanol. Within the limits of this research program, there was no time to establish a correlation for a series of catalysts with variable copper and alumina concentration.

P.G. MENON : During the use of the catalyst in the reaction, does the surface concentration of copper change ? Do you find any surface enrichment or surface depletion of copper on the used catalyst relative to the surface concentration of copper on the fresh catalyst?

O. INACKER : We investigated the aging of the catalyst and found that a fresh catalyst has a higher copper concentration on the surface than a used catalyst. The copper content decreases during the reaction. Copper leaves the catalyst via the gas phase.

M.V. TWIGG : The impregnation procedure described involves the use of ethanolic solutions of cupric nitrate. Such solution can be dangerously explosive and should be diluted with a large excess of water before disposal.

O. INACKER : We used only small quantities of that solution and we were careful, so we had no explosion.

C.J. WRIGHT : May I point out that the two disadvantages of SIMS, namely the requirement for vacuum conditions and the frequent possibility of non-uniform sputtering are both overcome with Rutterford Back Scattering. This technique would have given a quantitative, and non-destructive profile of the copper present in the catalyst you have described.

M. SEIDL : Certainly Rutherford Back Scattering in some cases would give valuable informations additional to SIMS, but we are not sure how far our rough and structured surfaces limit the possibilities of this method. We think also that Rutherford Back Scattering hardly can work in a catalytic reactor at technical gas atmospheres and pressures. So a sample transfer in an idealized gas atmosphere should always be necessary. Because the change of the gas composition leads to a change of the surface composition by gas-surface interactions, we think a transfer in vacuum is better. If you have enough experience of non-uniform sputtering processes by SIMS, it is not such a big problem.

EFFECT OF PRETREATMENT ON ACTIVITY, SELECTIVITY AND ADSORPTION
PROPERTIES OF A FISCHER-TROPSCH-CATALYST

G. LOHRENGEL, M.R. DASS, M. BAERNS
Lehrstuhl für Technische Chemie
Ruhr-Universität Bochum, Bochum, G.F.R.

ABSTRACT
The oxide precursor of a Fischer-Tropsch catalyst was thermally treated at 300°C and subsequently reduced with hydrogen at 300°C (type A) and 500°C (type C) before being used for the synthesis. The surface area and pore size distribution was only slightly effected by this pretreatment but activity and selectivity as well as adsorption properties of the catalyst were changed. Decrease of activity is related to diminishing activity of a single active site at higher reduction temperature. Higher selectivity of catalyst type C towards olefins and short-chain hydrocarbons is suggested to be due to difference in adsorption properties. To elucidate the change of activity, selectivity and adsorption properties, surface composition of the two types of catalyst was studied by ESCA.

INTRODUCTION
Catalysts for Fischer-Tropsch-synthesis are obtained by either coprecipitating suitable catalytic compounds, or impregnating a support with them, or mixing the appropriate oxides and treating this catalyst precursor thermally and subsquently reducing it with hydrogen or its mixture with carbon monoxide. Different conditions of these treatments can greatly change activity and selectivity. Certain aspects of the effect of pretreatment on such catalysts have been studied by several investigators. R.B. Anderson [1] noted a decrease in surface area and an increase in average pore diameter as a result of increasing temperature of thermal treatment performed in the range from 100 to 550°C. H. Kölbel and G. Leuteritz [2] further observed that the presence of potassium carbonate diminishes the crystallization of Fe_2O_3 up to a temperature of 400°C when the catalyst precursor is thermally treated. It

was further seen by Kölbel and Schneidt [3] that the thermal pretreatment influences the primary structure of a catalyst precursor containing potassium carbonate: the addition of 0,2 % potassium carbonate resulted into an increase of 60 % in surface area and of 50 % in the heat of adsorption for the reversible CO adsorption. This addition also stepped up the olefin production by a factor of three.

An influence of the reduction temperature on activity and selectivity of a Fischer-Tropsch catalyst has been recently observed by one of the present authors [4]: The same catalyst precursor also used in the present study was composed of unsupported oxides of mainly iron, manganese besides those of zinc, copper and traces of vanadium, aluminium and silicium; it was further modified by potassium carbonate. The pelletized mixture of these compounds (pellet dimensions: d = 3,7 mm, l = 6,2 mm) were tempered for 20 hours in an argon atmosphere at 300°C and reduced with hydrogen for 50 hours at 300°C (type A) and 500°C (type C). The two catalysts A and C showed only slightly different structural data, but their activity and selectivity were significantly different.

The data on surface area, pore size distribution and pore volume which were obtained by nitrogen adsorption at 77 K and mercury porosimetry are given in Table 1. The results indicate that only the surface due to micropores is slightly reduced.

TABLE 1
Structural data of the catalyst

	Catalyst A	Catalyst C
BET-surface m^2/g	14,8 $\pm$ 2	12,0 $\pm$ 2
Surface distribution between pore radii nm < 1,5 1,5 to 50 50 to 7500	5,5 $\pm$ 2,2 6,7 $\pm$ 0,4 2,6	2,3 $\pm$ 2,1 7,1 $\pm$ 0,2 2,6
Pore volume cm^3/g between pore radii nm 1,5 to 50 50 to 7500	0,017 0,424	0,014 0,447
Average pore radius (nm) between pore radii 0 to 7500 1,5 to 50 50 to 7500	60 4,6 326	77 3,8 344

Activity and selectivity measurements have been performed with the two types of catalyst by means of an internal recycle reactor. Reaction temperature was varied between 233 and 335°C and pressure between 5 and 15 bars. Detailed information on experimental techniques and results are outlined in [5]. Selected data on activity and selectivity which are characteristic for the two catalysts are specified in Tables 2 and 3.

TABLE 2

Average selectivity $\bar{S}$ of catalyst A and C

$$\bar{S} = \frac{\text{moles of hydrocarbon of carbon number n} \times \text{n}}{\text{moles of CO converted to hydrocarbons}}$$

Component	Catalyst A		Catalyst C	
	$\bar{S}$ %	$C_n^=/C_n$	$\bar{S}$ %	$C_n^=/C_n$
C_1	6,7	-	8,0	-
$C_2^=$	6,2	4,8	7,5	6,8
C_2	1,3		1,1	
$C_3^=$	7,8	6,0	9,1	8,3
C_3	1,3		1,1	
$C_4^=$	6,4	4,9	6,7	9,6
C_4	1,3		0,7	
C_{5+}	69,1	-	65,5	-
T_R °C	233-271		309-335	
$P(H_2)$ bar	3,6-3,7		3,7-3,8	
$P(CO)$ bar	3,5-3,7		3,4-3,6	
$X(CO)$ %	ca. 15		ca. 25	

Although the average selectivities as given in Table 2 refer only to a limited range of conversion X(CO) it could be shown that they are independent of X(CO) up to values of about 65 %; furthermore it should be mentioned that the selectivities for both the catalyst types A and C were only very slightly effected by reaction temperature, thus, the selectivity change of the two catalyst types is mainly due to the different pretreatment [4].

The activities shown in Table 3 have been calculated on the basis of the experimental data given in Table 4 for comparable temperatures.

TABLE 3

Activities of catalyst A and C; (P_{H_2} = 3,6 bar, P_{CO} = 3,7 bar)

	r_{CO}	10^{-3} moles/g-cat·h		
Temperature °C	250	270	310	330
Catalyst A	3,7	8,9	30[+)	88[+)
Catalyst C	0,4[+)	1,1[+)	6,7	15,3

[+) extrapolated, since no tests were performed at this temperature

It is obvious that catalyst A exhibits a greater activity than C. From table 4 it can be derived that the activation energies corresponding to catalyst A are lower than those of C. The different activities and selectivities as outlined above certainly cannot be explained by the slight differences in the structural data of catalysts A and C.

TABLE 4

Apparent activation energies of CO-consumption and hydrocarbon formation

$$r = r_o \cdot \exp(-E_a/RT), \quad r_o = r_o^o \cdot f(P_{H2} \cdot P_{CO})$$

Component	Catalyst A		Catalyst C	
	r_o [1]	E_a [2]	r_o [1]	E_a [2]
CO	$8{,}9 \cdot 10^7$	105 ± 5	$3{,}9 \cdot 10^8$	121 ± 6
CH_4	$4{,}0 \cdot 10^8$	127 ± 6	$1{,}6 \cdot 10^9$	144 ± 12
C_2H_4	$1{,}2 \cdot 10^7$	113 ± 7	$4{,}8 \cdot 10^6$	118 ± 6
C_2H_6	$7{,}0 \cdot 10^6$	119 ± 10	$8{,}8 \cdot 10^5$	119 ± 7
C_3H_6	$2{,}8 \cdot 10^7$	118 ± 13	$8{,}1 \cdot 10^6$	122 ± 10
C_3H_8	$1{,}2 \cdot 10^5$	101 ± 12	$1{,}8 \cdot 10^5$	114 ± 10
C_4H_8	$8{,}0 \cdot 10^6$	115 ± 13	$2{,}9 \cdot 10^6$	119 ± 10
C_4H_{10}	$6{,}8 \cdot 10^4$	100 ± 11	$3{,}0 \cdot 10^5$	118 ± 10
P_{CO} bar	3,5 - 3,7		3,4 - 3,6	
P_{H2} bar	3,6 - 3,7		3,7 - 3,8	
X_{CO} %	12 - 19		22 - 28	
T °C	233 - 271		309 - 335	

[1] mol/h g-cat   [2] kJ/mol

In order to obtain an insight into the different behaviour of catalyst types A and C their adsorption properties were investigated. For adsorption hydrogen and carbon monoxide were used as reactants and ethylene as an important intermediate which can be further hydrogenated to ethane or incorporated into chain growth. The primary objective of the adsorption was to determine the strength of adsorption in terms of heat of adsorption and the number of active sites on the surface and hence, the turn-over-number for CO consumption as a measure of catalyst activity.

EXPERIMENTAL

The adsorption measurements were performed with a vacuum microbalance (Sartorius No. 4433) with a maximum sensitivity up to 0,001 mg. The balance was connected to a vacuum system where an ultimate vacuum of 10^{-6} torr was attainable with an oil diffusion pump which was backed up by a mechanical rotary pump . The residual gas analysis in the system was made with the help of a quadrupole mass spectrometer (Balzers QMG 311) being directly attached to the microbalance. This gas analysis was designed to check on the concentration of gases in the system. No compounds other than hydrogen and traces of nitrogen and water were detected. Before starting the adsorption experiments the catalyst in its reduced state was treated with hydrogen at 523 K and 1 bar for approximately 15 hours. Hereafter the system was evacuated to a pressure of 10^{-6} torr and the catalyst was heated up to 568 K for few hours till its weight was constant. The gases used showed the following purity: 99,999 vol % H_2, 99,995 vol % CO and 99,950 vol % C_2H_4. Adsorption isotherms were measured up to final pressure of 800 torr. For hydrogen the isotherms were taken in the temperature range from 77 K to 523 K, for carbon monoxide from 77 K to 473 K, and for ethylene from 195 K to 523 K. Adsorption measurements were not possible between $5 \cdot 10^{-1} - 10^{-4}$ torr due to Knudsen forces, which interfered with the balance in this pressure range. At constant pressure and temperature the increasing weight of the catalyst was noted with time. Two adsorption processes were observed which could clearly be differentiated into a fast and a slow process. The fast process came to an end within seconds while the slow process went on for hours without reaching an equilibrium. In the present work, only the fast process which came to equilibrium is dealt with in detail.

RESULTS

The primary results of the adsorption measurements are shown in Figures 1a to 1c in form of the isobars for carbon monoxide, ethylene and hydrogen for the two types of catalysts. At low temperatures the amount of gas adsorbed decreases in all cases with increasing temperature. Partiallay this decrease is followed by an increase at higher temperatures which again may be followed by another decrease at a constant pressure. For catalyst A the isobars for carbon monoxide have a corresponding minimum at about 330 K while for C it occurs at about 380 K. For ethylene only catalyst A shows a minimum at about 473 K while catalyst C exhibits no minimum up to 523 K. The hydrogen isobars reach a minimum at 330 K for both the catalysts. For catalyst A a maximum is obtained at 473 K while for catalyst C there is no such maximum up to 523 K.

The heats of adsorption which are given in Table 5 for the three gases have been determined by using Clausius-Clapeyron equation for the descending parts of the isobars. The numerical values have an accuracy of about ± 10 %. It should be mentioned that within experimental accu-

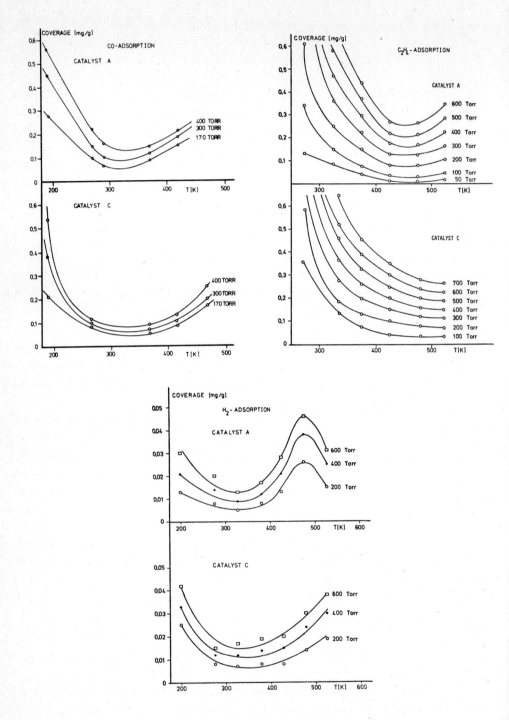

Fig. 1a to 1c: Adsorption isobars for CO, C_2H_4 and H_2

racy the heat of adsorption is independent of coverages between 0,9 and $5 \cdot 10^{13}$ molecules/cm^2. With respect to the heat of adsorption of carbon monoxide and hydrogen no significant difference between types A and C

TABLE 5
Heats of Adsorption

Adsorbent	Catalyst-Type	Temperature Range (K)	Heat of Adsorption (kJ/mole)
CO	A	195 - 273	8,9
	C	195 - 273	9,3
C_2H_4	A	273 - 473	14,7
	C	273 - 523	7,5
H_2	A	77 - 325	3
		473 - 523	35
	C	77 - 325	4,5

of the catalyst was noted; for ethylene, however, catalyst A yields a greater heat of adsorption than C. It should be mentioned that the heats of carbon monoxide adsorption are comparable to the values of about 10 kJ/mole obtained earlier by Kölbel and Roberg [6] on iron catalysts; however, their values of about 8 kJ/mole for hydrogen are high as compared to the present data. The difference may be due to the presence of various oxide compounds besides iron.

The active sites defined as the number of molecules adsorbed per surface area have been determined for the various gases by extrapolation of the observed isotherms to higher pressures corresponding to maximum coverage of the surface. This could be done for hydrogen by applying Langmuire equation for non-dissociative adsorption for all temperatures and pressures. For carbon monoxide this method was only applicable at 423 K. Freundlich isotherms have been used for ethylene adsorption. The results are given in Table 6; within experimental accuracy no significant difference in the numbers of active sites is observed for the three adsorbed species.

TABLE 6
Number of active sites per surface area ($10^{14} \cdot cm^{-2}$)

Catalyst Type	CO	C_2H_4	H_2
A	0,7 ± 0,2	2 ± 1	1,4 ± 0,3
C	0,8 ± 0,2	1 ± 1	1,5 ± 0,7

Turn-over-numbers for CO-consumption corresponding to the reaction rates as listed in Table 3 have been calculated using an average number of active sites of $1,5 \cdot 10^{14}$ cm^{-2}. The results shown in Table 7 indicate that the different reduction temperatures of catalysts A and C resulted into different activities of a single site; i.e. the activity decrease of the catalyst by higher reduction temperature is not due to a change of the number of sites per surface area but to their different activities.

TABLE 7
Turn-over-numbers for consumption of CO; (P_{H_2} = 3,6 bar, P_{CO} = 3,7 bar)

Temperature °C	250	270	310	330
Catalyst A	0,03	0,07	0,23+)	0,66+)
Catalyst C	0,004+)	0,01+)	0,06	0,14

+) compare Table 3

DISCUSSION

Adsorption behaviour of the two catalyst types A and C is different in so far as at higher temperatures additional adsorption processes may occur as shown by the various isobars. If such a second adsorption is observed, its onset temperature is always lower for catalyst A as compared to C. The second process of carbon monoxide adsorption starts at about 330 K for catalyst A and only at 380 K for C. Similarly for ethylene it begins at 470 K for catalyst A and no such process takes place in catalyst C up to 520 K, while for hydrogen it occurs at 330 K for both the catalysts. These results indicate that at comparable temperatures adsorption activity of catalyst A is greater than that of C. In accordance herewith the turn-over-numbers show that the activity of an individual site is less for catalyst C than for A. This is in agreement with the facts that the activation energies for the various reactions are less lowered by catalyst C than by A.

The selectivity difference of the two catalysts with respect to olefin formation may be related to the different heats of ethylene adsorption obtained. The stronger adsorption of ethylene on catalyst A most probably leads to greater activation of the adsorbed molecule and hence, chain formation and further hydrogenation to ethane is facilitated. This is experimentally verfied for catalyst A in comparison to C by the formation of less olefins and the greater probability of chain growth which has been shown elsewhere (4).

CONCLUDING REMARKS

Applying two different reduction temperatures to the oxide precursor

of a Fischer-Tropsch-catalyst resulted in different selectivity and
activity behaviour as well as adsorption properties. The catalytic and
adsorption properties are in certain respect complementary to one
another. As a supplementary indication for the different catalytic performance preliminary ESCA-studies were performed, which did not reveal
any significant differences in surface composition of catalyst types
A and C. Tentatively it is therefore suggested that the two reduction
temperatures may result in the formation of different catalytic surface
compounds in between the individual components.

ACKNOWLEDGEMENTS

Thanks are due to Dr. Hartig for providing the ESCA spectra. Financial support for this investigations has been given by Ruhrchemie AG
within a project sponsored by the Ministry of Research and Technology
of the Federal Republic of Germany.

REFERENCES
1 R.B. Anderson, Catalysis Vol. IV, Ed. P.H. Emmett, New York, 1956,
  p. 119
2 H. Kölbel, G. Leuteritz, Z. Elektrochem. 64 (1960) 437
3 H. Kölbel, D. Schneidt, Erdöl Kohle 3 (1977) 139
4 A. Zein El Deen, J. Jacobs, M. Baerns, ACS Symp. Series 65, Ed.
  W. Weekmann jr., Dan Luss (1978) p. 26
  (Minor differences between some of the results given in [4] and those
  of Tables 1 to 4 of this work are due to additional data which have
  become available)
5 A. Zein El Deen, J. Jacobs, M. Baerns, Chem. Ing. Techn., submitted for publication
6 H. Kölbel, H. Roberg, Ber. Bunsenges. Phys. Chem. 81 (1977) 634

DISCUSSION

S.P.S. ANDREW : Fischer-Tropsch catalyst is a complex formulation
with a complex reaction product including the production of high
boiling point waxes. Could a different activity of the K_2CO_3 as a
result of different preparation procedures cause a different formation of waxes and hence a different catalyst activity ? Have you
made similar measurements but in the absence of K_2CO_3 ?

M. BAERNS : The results reported in our paper have been performed
with a catalyst of constant composition: the amount of K_2O was not

changed. We have however, experimental evidence that the absence of K_2O in a catalyst of otherwise similar composition results in shorter chain length but also in a higher proportion of saturated hydrocarbons.

J.F. LEPAGE : With respect to the distribution of products, what is the relative importance of additives to your iron based catalyst, especially Mn and K_2O.

M. BAERNS : To our experience the addition of manganese, which may be considered as a promotor, to an iron-based F.T. catalyst results into a change of the product distribution towards shorter chain length of the hydrocarbons. The addition of K_2O increases the olefins but simultaneously the chain length is also increased. This has been outlined before, for instance by Dry et al. (J. Catal. $\underline{15}$, 190, 1969).

R. KIEFFER : What is the importance of CO_2 in your F.T. reactions ?

M. BAERNS : CO_2 is most probably not formed by a primary reaction but by the water shift conversion as a secondary reaction between H_2O, produced by the F.T. synthesis, and CO. To a lesser degree CO_2 may also originate from the Boudouard-reaction. However, the deposition of carbon on the catalyst is not appreciable as compared to the amount of CO_2 formed. It should be mentioned that the selectivity of the CO-conversion towards CO_2 amounted to about 45 to 50% under all conditions applied to the catalyst in this study.

THE INFLUENCE OF IMPREGNATION, DRYING AND ACTIVATION ON THE ACTIVITY
AND DISTRIBUTION OF CuO ON α-ALUMINA

M. Kotter and L. Riekert
Institut für Chemische Verfahrenstechnik der Universität Karlsruhe,
Karlsruhe (G.F.R.)

ABSTRACT
The influence of the distribution of the active component on the activity of impregnation-type porous catalysts is demonstrated for CuO on alumina, activity being measured for oxidation of CO. The viscosity of the impregnating solution affects the distribution of the catalyst-precursor ($Cu(NO_3)_2$) after impregnation and drying. The uniformity of the distribution of the active component and thereby the activity of the finished catalyst depend considerably on the procedure of impregnation and drying.

INTRODUCTION
Ideally an impregnation-type catalyst should be a porous structure consisting of an inert carrier-material, which is covered everywhere with the active component. Such porous catalysts are generally treated as homogeneous pseudo-continua in theoretical work (e.g. /1/), aiming at an understanding of simultaneous reaction and mass transfer in the porous solid. It appears doubtful, however, if the assumption of uniformity is generally justified with respect to real impregnation-type catalysts, which sometimes fall somewhat short of the ideal, since the active component does not uniformly cover the porous structure, the active surface being smaller than the total surface. Two types of inhomogeneous distribution of the active component can be distinguished :

(a) macroscopic inhomogeneity, the average density of the active
    component being different for different layers of the pellet,
    for example when the active component is concentrated at the
    core or near the outer surface;

(b) microscopic or surface inhomogeneity, the pore walls of the carrier material being not homogeneously covered everywhere by the active component.

Both types of inhomogeneity will simultaneously influence the dynamics of a chemical reaction taking place in the catalyst.

Macroscopic inhomogeneity is encountered frequently in catalysts that have been prepared by impregnation, drying and activation and has been documented in the literature for a few cases /2,3,4/; microscopic inhomogeneity seems to prevail in general in such catalysts; real impregnation-type catalysts are thus in general far from the ideal. In the present paper some experimental methods that characterize the distribution of the active component will be reviewed briefly and the results of experiments aimed at improving the method of preparation will be reported.

CHARACTERIZATION OF THE DISTRIBUTION OF THE ACTIVE COMPONENT

Visual characterization of the distribution of the active component is possible if it's reflectance or morphology is different from that of the carrier, either by inspection, light-microscopy or electron-microscopy. It is difficult, however, to quantify the results of visual inspection in detail.

The observation of equilibria between a porous catalyst and a fluid surrounding it (e.g. adsorption, mercury-penetration) will always average over the entire space of the pellet and can therefore not give information about the spatial distribution of solid components. For example the determination of total surface area (from nitrogen-adsorption) and of the surface area of the active component (from chemisorption) will give a measure of the fraction of the surface which is covered by the active component, but not about it's distribution in space.

Certain quantitative information about the distribution of the active component (or rather its accessibility) can be obtained from observations of the kinetics of a catalytic reaction in the porous solid, as we have shown previously /5/. An apparent diffusivity of the reactants with respect to the active component can be obtained from the temperature-dependence of the rate of reaction /6/ and compared to the effective diffusivity of the reactant in the porous structure, the ratio of the two quantities being a measure of the accessibility of the active component relative to that of the porous structure in general. However, kinetic data which characterize catalyst performance can not be predicted exactly from other observations describing the distribution of the active component in a more or less qualitative way.

CuO ON ALUMINA : PREPARATION AND PROPERTIES
CuO an γ-alumina

CuO was deposited on γ-alumina in the conventional way : the carrier (spheres of γ-alumina, diameter 6 mm, specific surface area 240 m^2/g, porosity 45 %) was soaked in 3 m $Cu(NO_3)_2$-solution for 14 h, dried at 350 K for 14 h and subsequently heated to 820 K for 24 h to convert $Cu(NO_3)_2$ to CuO. The resulting catalyst has a BET-surface area of 142 m^2/g, however, the surface area of the active component amounted to only about 7 % of this value, as shown by CO-chemisorption. The kinetics of the oxidation of CO (less than 0.1 mol % in air) was studied on this catalyst in a gradientless recirculation reactor. The reaction is first order in CO; its rate is described by a rate coefficient k (in sec^{-1})

$$- \frac{1}{V_k} \frac{dn_{CO}}{dt} = k\, c_{CO} \tag{1}$$

where V_k is the volume of the catalyst pellet and c_{CO} the volume-concentration (in mol m^{-3}) of CO in the gas phase at the surface of this pellet.

The temperature-dependence of this rate-coefficient is shown in figure 1. The point of intersection (a) of the high-temperature and the low-temperature Arrhenius-line indicates the onset of a diffusion-limitation at a higher rate (or temperature) as would be expected from an independent determination of the effective diffusivity of CO in the structure ($D_{eff} = 1.9\, 10^{-2}\, cm^2/sec$), assuming a macroscopically uniform distribution of CuO (dotted line) /5/. The active component which covers only about 7 % of the surface in the pores must therefore be more accessible to the reactants than the porous structure on the average.

The rather high surface area of this particular carrier is thus not utilized catalytically and a material of much smaller surface area should be equally suitable as a carrier for a catalyst of comparable activity, provided a uniform distribution of the active component in the porous structure can be obtained. Such a carrier would be α-alumina which also has the advantage of higher thermal and mechanical stability.

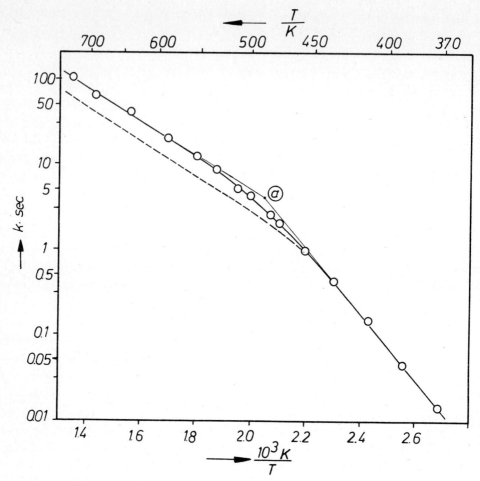

Fig. 1. Rate coefficient k of CO-oxidation of CuO/γ-Al$_2$O$_3$ as function of temperature. Dotted line : computed values of k for D_{eff} = 1.9 10^{-2} cm^2/sec, assuming uniform distribution of CuO

CuO on α-alumina

The carrier material used for the preparation of CuO-catalysts on α-Al$_2$O$_3$ was characterized by the following data : cylinders, 6 mm diameter, specific surface area 6 m^2/g, porosity 60 %, pore size distribution see figure 2. According to its x-ray-diffraction-pattern the ideal hexagonal structure of α-Al$_2$O$_3$ was not yet fully developed, this type of high temperature alumina is sometimes designated Θ-Al$_2$O$_3$.

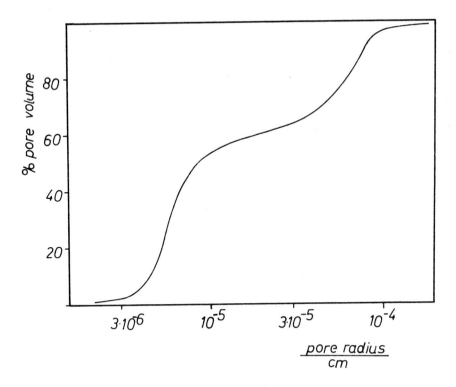

Fig. 2. Pore volume distribution of $\alpha\text{-}Al_2O_3$ from mercury-porosimetry

Preparing a catalyst on this carrier by the same sequence of impregnation, drying and calcination as described above leads to a rather disappointing result. The active component (3.5 wt % CuO) is concentrated in a shell on the outside surface of the structure (figure 3a). A much better dispersion of the active component was obtained by impregnating with an aqueous solution containing 3 mol/l of $Cu(HCOO)_2$ saturated with NH_3, drying at 350 K and calcination at 800 K (figure 3b).

The two methods of preparation (a) and (b) are different in the following respect :
(a) impregnation with solution of $Cu(NO_3)_2$:
    viscosity of impregnation solution 1.5 cP; decomposition of catalyst precursor $Cu(NO_3)_2$ to CuO occurs around 500 K and is endothermic; intermediate $Cu(NO_2)_2$ is volatile /7/.
(b) impregnation with ammoniacal solution of $Cu(HCOO)_2$:
    viscosity of impregnation solution 3.5cP; decomposition of catalyst precursor $Cu(HCOO)_2$ to CuO occurs around 470 K and is exothermic; $Cu(HCOO)_2$ is less volatile than $Cu(NO_2)_2$.

Fig. 3. CuO on α-Al$_2$O$_3$
(a) impregnation with 3 m Cu(NO$_3$)$_2$, (b) impregnation with 3 m Cu(HCOO)$_2$ saturated with NH$_3$

The catalyst precursor must be uniformly distributed in the pores immediately after impregnation when the material is still wet in either case (a) or (b), since α-Al$_2$O$_3$ is inert and has no capacity for ion-exchange, contrary to γ-Al$_2$O$_3$.

The nonuniform distribution must therefore arise during drying and/or during decomposition of the different precursors to CuO. Examination of the distribution of the crushed cylinders after drying showed that in case (a) the precursor Cu(NO$_3$)$_2$ was already concentrated near the surface of the porous solid before decomposition to CuO and that this effect was more pronounced at lower drying temperatures (lower evaporation rates). A similar distribution of the precursor-salt after impregnation and drying has already been observed by Boreskov /8/.

It appears that the nonuniform distribution of the active component in case (a) is mostly due to nonuniform distribution of the precursor Cu(NO$_3$)$_2$, which is already established after drying; although an additional migration of Cu in the form of Cu(NO$_2$)$_2$ during calcination is also conceivable. A nonuniform distribution of the initially dissolved and uniformly distributed salt can only arise due to flow of the solution in the porous structure or due to diffusion in the solution. The driving forces of the migrations must be a consequence of the evaporation of solvent and they will be capillary forces in the first case, concentration gradients in the second case. Viscous flow due to capillary forces is probably responsible for the observed migration of dissolved $Cu(NO_3)_2$ towards the surface of the pellet in our case. Rates of viscous flow in the impregnating solution as well as of diffusion in the liquid are roughly inversely proportional to the viscosity of the liquid. The difference in the viscosity of the impregnating solutions therefore appeared important with respect to the distribution of the catalyst precursor after drying, since a higher viscosity should lead to a more uniform distribution, as was observed. In order to verify this hypothesis a series of catalysts was prepared by impregnation with solutions of 2 m $Cu(NO_3)_2$, where different small amounts of Hydroxyethylcellulose (HEC) had been added in order to increase the viscosity of the impregnating solution. An α-alumina carrier with exceptionally low surface area (< 0.1 m^2/g) and hence a high average pore-radius (10^{-5}m) was used for these experiments. The impregnated pellets were dried and calcined for 1 h at 500 K and heated to 800 K to remove the HEC. The results are shown in figures 4 and 5. All catalysts contained 0.5 wt% CuO.

The viscosity of the solution used for impregnation influences the macroscopic homogeneity and also the catalytic activity (Fig. 5)

It is in this way possible to achieve roughly the same catalytic activity as was observed for 10 % CuO on high-surface γ-Al_2O_3 with the more rugged α-Al_2O_3 catalyst carrier (Fig. 6).

It should be mentioned, however, that the macroscopic distribution of the active component is not the only property which changes with the carrier and the method of preparation. The activation energy of the CO-oxidation is different by 35 % for CuO on α-Al_2O_3 and on γ-Al_2O_3, respectively, as can be seen from figure 6. Addition of HEC to the impregnating solution also changes the habit of the deposits of CuO in the porous structure (Fig. 7).

The method of preparation influences simultaneously the macroscopic and the microscopic inhomogeneity of the resulting catalysts and it is not possible - at least at present - to differentiate between both effects with respect to catalyst performance.

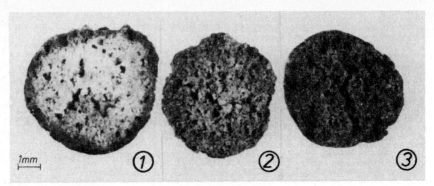

Fig. 4. Photographs of cross section of CuO on α-Al_2O_3 prepared by impregnation with 2 m $Cu(NO_3)_2$ solution of different viscosities : (1) no HEC added; $\nu \cong 1.5$ cP; (2) 1 wt% HEC; $\nu = 15$ cP; (3) 2 wt% HEC; $\nu = 100$ cP

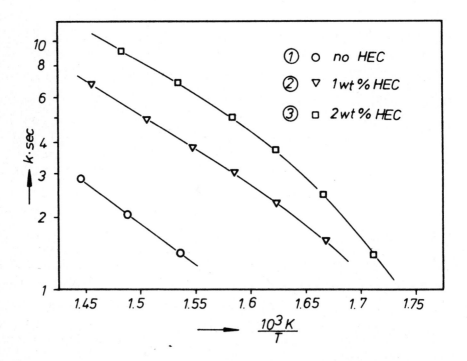

Fig. 5. Volume-specific rate coefficient k of CO-oxidation on CuO/α-Al_2O_3 catalysts as function of temperature; designation as in figure 4

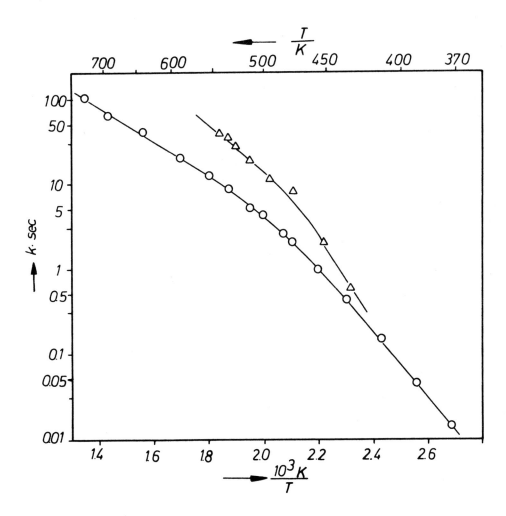

Fig. 6. Volume-specific rate coefficient k of CO oxidation on CuO :
△:3.5 wt % CuO on α-Al_2O_3 (6 m^2/g) ex formiate, dried and activated at 520 K
O:10 wt % CuO on γ-Al_2O_3 (240 m^2/g) ex nitrate

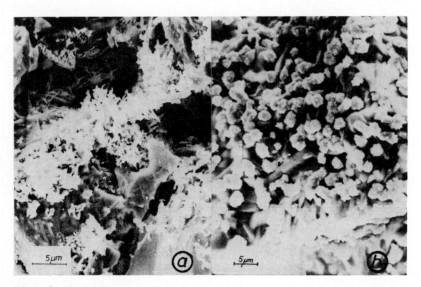

Fig. 7. Deposits of CuO in porous $\alpha\text{-Al}_2O_3$ as observed by scanning electron microscopy. (a) impregnation with 2 m $Cu(NO_3)_2$ with 2 wt% HEC added; (b) impregnation with 2 m $Cu(NO_3)_2$, no HEC added

REFERENCES

1. R. Aris, "The Mathematical Theory of Diffusion and Reaction in Permeable Catalysts", Oxford University Press, Oxford 1975
2. R.C. Vincent and R.P. Merrill, J.Catalysis 35, (1974), 206
3. H.C. Chen and R. Anderson. J. Catalysis 43 (1976), 200
4. J. Cervelló, J.F. Garcia de la Banda, E. Hermana and F. Jiménez, Chem.Ing.Tech. 48 (1976), 520
5. M. Kotter, P. Lovera and L. Riekert, Ber.Bunsenges.phys.Chem. 80 (1976), 61
6. B. Kadlec, J. Michálek and A. Simecek, Chem.Eng.Science 25 (1970) 319
7. K. Stern, J.Chem .Ref.Data, 1 (1972), 757
8. G. Boreskov, in "Preparation of Catalysts" (D. Delmon, P. Jacobs and G. Poncelet, eds.), p. 222, Elsevier Scientific Publishing Company, Amsterdam-Oxford-New York, 1976

DISCUSSION

B. DELMON : Concerning Fig. 6, have you an explanation for the apparently different activation energies (especially in the low temperature region) ? Is this still related to diffusional processes, or do you attribute the difference to chemical modifications of the active species by some interaction with the support ?

L. RIEKERT : The different activation energies of CO-oxidation observed with different carriers (α vs. γ-Al_2O_3) indicate that besides the macroscopic distribution other properties of the active component (e.g. texture, incipient compound formation) depend on the type of carrier and/or the method of preparation. See also Fig. 7 of the paper.

E. MATIJEVIC : The addition of nonionic polymers affects the nucleation stage of the precipitation of metal oxide. As a result the primary particle size of the precipitated solid will decrease as the polymer concentration increases. A minimum concentration of the additive is needed to initiate this effect, which indeed may be due to a change in viscosity. Entirely different observations are made if ionic polyelectrolytes are used. This was clearly demonstrated in our studies of the effects of nonionic and ionic Dextrans on the formation of monodispersed chromium hydroxides.

L. RIEKERT : Any addition to the impregnating solution can possibly have an effect on nucleation, crystallization and also on solid-state-reactions taking place during calcination. This possibility can never be excluded and should be kept in mind - compare also Fig. 7 in the paper. However, we believe that our experiments show conclusively that the macroscopic distribution of the active component (ex-nitrate) on α-Al_2O_3 is critically affected by the addition of HEC. The most obvious and simple explanation seems to be that the increase in viscosity hinders capillary flow during drying.

H.W.J. NEUKERMANS : When you change the viscosity of the dipping solution, do'nt you change the volume of liquid absorbed, or did you take special precautions to always have the same copper content?

L. RIEKERT : We took care to exclude this effect in our experiments (Figs. 4 and 5) by using a carrier with rather wide pores. The same amount of liquid was taken in all cases, independent of the viscosity of the impregnating solution.

W.D. ROBINSON : Have you considered or used fast drying - using reduced pressure of microwave - to minimize the transfer of Cu from the inside of the pellet ? Was it, if used, successful ?

L. RIEKERT : We did observe a more homogeneous macroscopic distribution of the active component as the rate of drying was increased. This effect was to be expected because the distribution must depend on the interaction of two rate-processes : removal of the solvent by evaporation and capillary-flow towards narrow pores at the pellet-surface. It is the ratio between the rates of these two processes which seems to be important with respect to the resulting distribution of the active component on inert carriers.

H. SCHARF : Why is $CuO-\alpha-Al_2O_3$ more active than $CuO-\gamma-Al_2O_3$ despite the lower surface area of $\alpha-Al_2O_3$?

L. RIEKERT : Because the surface area of the active component - not the total surface area - determines catalyst activity.

J. SCHEVE : Is it possible that wetting of the surface has a main effect on your process, because formamide can be formed during annealing of copper formate with ammonia ? This substance is well known by catalyst manufacturers to improve wetting of surfaces. In the same way the other organics you used for impregnation may act.

L. RIEKERT : Any change of the composition of the impregnating solution will of course change all its physical properties. However, since the alumina carriers are completely wetted by all solutions which have been used and since the contact-angle can not fall below zero, I would not expect wetting of the support to be of critical influence.

J.J.F. SCHOLTEN : In many cases where impregnation methods are applied, adsorption of ions or molecules from the solution to the surface may be used as well. Is there any adsorbability of Cu^{++} or especially $Cu(NH_3)_n^{++}$ ions with respect to the surface of α- and γ-alumina ?

L. RIEKERT : There was no additional uptake of $Cu(NO_3)_2$ on our $\alpha-Al_2O_3$ supports beyond pore-volume filling. Thus the $\alpha-Al_2O_3$ carriers can be considered as chemically inert towards the catalyst precursor $Cu(NO_3)_2$. Also in the case of $|Cu(NH_3)_n|(HCOO)_2$ the amount of Cu on the finished catalyst corresponded to the amount expected for pore-volume filling of the α-Al_2O_3 support.

G.H. van den BERG : What is the influence of increasing the viscosity of the impregnation solution on the penetration of the micropores? Or does your carrier not contain any micropores ?

L. RIEKERT : There are no micropores in the α-Al_2O_3-carriers which we used, as shown from physisorption-isotherms and electron-microscopy.

A STUDY OF THE CHEMISORPTION OF CHROMIUM(VI), MOLYBDENUM(VI) AND TUNGSTEN(VI) ONTO γ-ALUMINA

A. IANNIBELLO[1], S. MARENGO[1], F. TRIFIRO'[2] and P. L. VILLA[3]

[1] Stazione sperimentale per i Combustibili, San Donato Milanese, Italy
[2] Istituto di Tecnologie chimiche speciali dell'Università, Bologna, Italy
[3] Istituto di Chimica industriale del Politecnico, Milano, Italy

ABSTRACT

In the pulse and step addition of ammonium dichromate, ammonium paramolybdate and ammonium paratungstate solutions to γ-alumina a large increase of the pH was observed in the liquid phase which suggests a common mechanism of interaction with the basic sites of γ-alumina. This mechanism involves the decondensation of the isopolyanion in the presence of alumina followed by the chemisorption of monomeric species. The most relevant difference between the three metal ions is the limiting amount of the chemisorbed oxospecies on γ-alumina: CrO_3 3.28; MoO_3 7.72; WO_3 19.2 wt. %.

The electronic and Raman spectra of the chemisorbed molybdenum(VI) and tungsten(VI) at low coverage can be interpreted as due to oxospecies in tetrahedral co-ordination. The relevance of the adsorption phenomena in the preparation of the commercial catalyst is finally pointed out.

1 INTRODUCTION

The oxides of the sixth group of elements promoted with other components and supported onto γ-alumina represent an important class of commercial catalysts. The most frequent method of incorporation of these elements is impregnation; in this technique of preparation adsorption phenomena are involved, but their relative importance compared to the solid state reactions occurring during thermal treatments of the catalysts has not been extensively investigated [1-6, 14]. On the other hand the study of these adsorption phenomena may give useful informa tion for the rationalizing of the catalyst preparation. Moreover the presence of chemisorption phenomena, if relevant, offers the possibility of selectively preparing catalysts, i.e. of preparing samples of catalysts containing only surface phases at various degrees of coverage; this may be useful also for the understanding of the catalytic process [5c]. In this paper we report the results of an investigation on the chemisorption processes which occur when γ-alumina is impregnated with ammonium dichromate, ammonium paramolybdate and ammonium paratungstate aqueous solutions. In the experimental approach the liquid chromatographic technique [5a, 5b] was used to point out the chemistry of the interaction also from a quantitative point of view, together with a characterization of the chemisorbed species by Raman and electronic spectroscopies.

2 EXPERIMENTAL SECTION

2.1 Materials

γ-alumina AKZO Chemie CK 300 (particle diameter 0.1-0.3 mm); other chemicals are Carlo Erba, pure reagent grade.

2.2 Experimental procedure

Pulse addition experiments were carried out as described elsewhere [5a]. The column was filled with 4.2 g of γ-alumina; bidistilled water was used as a carrier (20-40 mL/h); 30-100 μL pulses of ammonium paratungstate, ammonium paramolybdate, ammonium dichromate, chromium nitrate and chromic acid aqueous solutions containing $6-50 \cdot 10^{-6}$ mol Me(VI) were injected at 20-80°C.

The composition of the effluent was recorded with a refractive index monitor, LDC Model 1107, followed by an UV detector, LDC Model 1520, operating at 254 nm, followed by a flow pH-meter, Metrohom Model E 532 with a 5-mm outside diameter glass electrode plugged inside a teflon cell with free volume of 300 μL.

Step addition experiments [5b, 5c] were carried out at 20, 60 and 80°C on a column 200 mm long, 4 mm diameter, filled with 1.5 g of γ-alumina. The inlet flow rate was 25 mL/h; bidistilled water was initially fed, followed by the isopolyanion ammonium salt solution containing 0.038 mol Me(VI)/L. After each run the column was washed with 40 mL of bidistilled water at the same temperature of the adsorption experiments, to eliminate non adsorbed Me(VI).

Adsorption experiments were carried out at 20 and 60°C pouring 4.5 g of γ-alumina into a 500 mL flask containing 450 mL of vigorously stirred aqueous solution of the isopolyanion ammonium salt (10^{-2} mol Me(VI)/kg).

2.3 Analysis

Mo(VI) and W(VI) in liquid and solid phases were determined by X-ray fluorescence using a Philips Model PW 1540 spectrophotometer. Cr(VI) was determined by atomic adsorption using a Perkin-Elmer Model 306 spectrophotometer.

The electronic spectra were determined by a Cary 15 spectrophotometer with a Model 1511000 attachment for diffuse reflectance spectra. The Raman spectra were recorded by a Cary 83 spectrophotometer.

3 RESULTS AND DISCUSSION

3.1 Pulse addition experiments

Fig. 1 shows the apparent retention volume, the pH and the area of the peak (as revealed by the UV monitor) of the outlet pulses vs. the number of pulses of ammonium dichromate and ammonium paramolybdate.

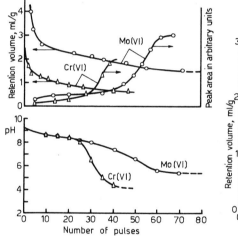
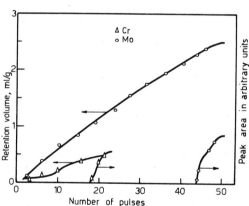

Fig. 1. Retention volume, peak area (UV monitor) and pH of the outlet pulse vs. the number of equimolecular pulses (57 μmol Me(VI)) of ammonium dichromate and ammonium paramolybdate solutions onto γ-alumina in neutral environment

Fig. 2. Retention volume and peak area (UV monitor) vs. the number of equimolecular pulses (57 μmol Me(VI)) of ammonium dichromate and ammonium paramolybdate solutions in acidic environment (pH 2.7)

No signal in the outlet was observed for the first 2-3 pulses suggesting the complete chemisorption of the sample injected. Apart from differences in the retention volumes, common features of the subsequent pulses were:

i) the high value of the pH of the initial outlet pulses compared to the inlet value (ammonium dichromate solution: pH 3.7; ammonium paramolybdate solution: pH 5.4); the initial value decreases in the subsequent pulses; ii) the presence of oxy anions in the outlet also at low value of coverage of γ-alumina and in increasing quantity; iii) the decrease of the retention volume.

The presence of Cr(VI) and Mo(VI) in the outlet basic solution, or, in other words, the partial irreversible chemisorption of these species also at low value of coverage, indicates an equilibrium adsorption controlled by the H_3O^+ depletion in the solution and an acid-base mechanism of interaction. Indeed by using a 10^{-3} M HNO_3 or HCl solution as a carrier a complete adsorption of Cr(VI) and Mo(VI) was recorded together with the presence in the outlet pulse of the ammonium ion (Fig. 2). The variation of the response of the adsorption system to the successive pulses reflects the surface heterogeneity of alumina.

In table 1 the limiting amount of Cr(VI) and Mo(VI) irreversibly adsorbed onto γ-alumina is reported. In this table data on the chemisorption of chromium nitrate and chromic acid (frequently used in the preparation of chromia-alumina catalysts) are also included. In the pulse addition of these species in the presence of bidistilled water complete adsorption was observed. Recalling that chromium nitrate is an acid, also the adsorption of this species can be described in terms of acid-base interaction, though the irreversibly adsorbed quantity is smaller (1.13 wt. % CrO_3).

TABLE 1

Me(VI) content of γ-Al_2O_3 at maximum coverage in pulse addition

Pulse	$(NH_4)_2Cr_2O_7$		H_2CrO_4	$Cr(NO_3)_3$	$(NH_4)_6Mo_7O_{24}$	
Carrier	H_2O	HNO_3 10^{-3}M	H_2O	H_2O	H_2O	HCl 10^{-3}M
wt. % MeO_3	2.32	1.7	3.28	1.13	7.19	7.72
μmol Me(VI) / $m^2 Al_2O_3$	1.29	0.94	1.82	0.63	2.77	2.98

In considering the different amount of Cr(VI) and Mo(VI) irreversibly adsorbed onto γ-alumina, it must be kept in mind that pulse addition was not carried out up to the point of maximum coverage (i.e. complete recovery in the outlet pulse of the sample injected). In fact in proximity of the surface saturation a slow desorption from weaker sites, particularly evident in the chromium incorporation, is observed. The difference in the amount of transition metal ions chemisorbed can be partially attributed to these facts. However the low amount of Cr(VI) adsorbed in the presence of HNO_3 probably reflects also a competing adsorption of NO_3^- anion.

The results of the pulse addition of W(VI) were not reported in Fig. 1. In the first pulses of W(VI) at 60°C an increase of the pH of the liquid effluent from the column was observed, indicating a common interaction mechanism for W(VI), Cr(VI) and Mo(VI). However the relative high response of the UV monitor at 254 nm indicates the presence in the outlet pulse of W(VI) in a condensed form [7], which must be considered an unstable species at the pH measured (8.9). We may deduce that the equilibrium conditions were not reached in the liquid phase, mainly due to the slow decondensation of the isopolytungstate species [8].

As the response of the adsorption system is under kinetic control, we considered it of little significance to carry out pulse addition up to the maximum coverage.

3.2 Step addition experiments

Fig. 3 shows the transient value of the pH in the effluent of the column vs. the hexavalent species added/m^2 of γ-alumina. In these experiments increases of pH higher than 9.5 were detected. The profile of the pH curve may be related to the spectrum of basic sites of γ-alumina. In table 2 the amount of Cr(VI), Mo(VI) and W(VI) irreversibly adsorbed is reported. We have no data on the thermodinamics of these systems; from the results obtained it appears that in changing from Cr(VI) to W(VI) the equilibrium is shifted in the sense of the adsorption.

We also observed [7] that in the step addition of ammonium paratungstate the outlet signal was strongly dependent on temperature.

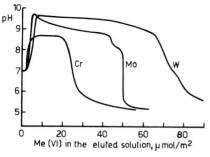

Fig. 3. pH meter response in the step addition of ammonium dichromate, ammonium paramolybdate and ammonium paratungstate to γ-alumina. Temperature: 60°C

3.3 Batch adsorption experiments

Fig. 4 shows the change of the pH and of the content of the transition metal ions in the liquid phase when 450 mL of a solution of ammonium dichromate, ammonium paramolybdate and ammonium paratungstate (10^{-2} mol Me(VI)/kg) are brought into contact with 4.5 g of γ-alumina at 20°C. Adsorption of W(VI) at 60°C is also reported. It appears that at 20°C equilibrium is obtained in a few minutes for Cr(VI) and Mo(VI) while the rate of adsorption of W(VI) is much lower in agreement with the pulse experiments. In fact the adsorption of W(VI) is far from complete at 20°C after 6 h: indeed the observed quantity at equilibrium should be greater at 20°C than at 60°C.

The smaller but significant increase of the pH observed in the batch adsorption experiments further substantiates an adsorption mechanism which involves the decondensation of polyanions followed by the adsorption of monomeric species:

$$Al-OH + H_3O^+ + MeO_4^{--} \rightleftharpoons Al-O-MeO_3^- + 2H_2O$$

The metal ion content of γ-alumina is reported in table 2.

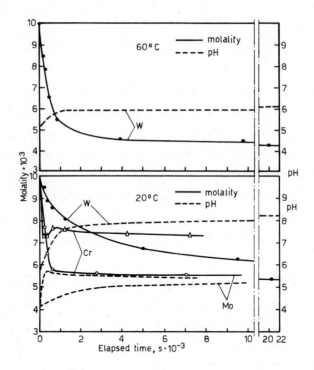

Fig. 4. Me(VI) concentration in the liquid phase and pH of the solution vs. time in batch adsorption experiments from ammonium dichromate, ammonium paramolybdate and ammonium paratungstate solutions

TABLE 2

Me(VI) content of γ-Al_2O_3 at the end of step addition and batch adsorption

	Step addition (1)			Batch adsorption (1)			
	Cr(VI)	Mo(VI)	W(VI)	Cr(VI)	Mo(VI)	W(VI) 20°C	W(VI) 60°C
wt. % MeO_3	3	7.7	19.2	1.97	5.75	9.31	11.95
μmol Me(VI) / $m^2 Al_2O_3$	1.70	2.96	4.60	1.10	2.20	2.20	2.86

(1) See text

3.4 Electronic and Raman spectra

Fig. 5 shows the electronic spectra of Cr(III) and Cr(VI) on γ-Al_2O_3. In all the calcined samples the absence of the 600 nm band showed that no Cr(III) was present: thus alumina stabilizes the highest valence state of chromium as already known from the literature [9].

Moreover samples obtained by addition of Cr(VI) solution onto γ-alumina and calcined at 550°C show only minor modifications in respect to the same samples only dried at 110°C. Such a trend was also observed in the electronic and Raman spectra of all Mo(VI)-Al_2O_3 and W(VI)-Al_2O_3 samples, thus evidentiating the stability of the chemisorbed species.

Mo(VI)-Al_2O_3 and W(VI)-Al_2O_3 samples with low content of transition metal ion showed spectra with a maximum of absorption at about 250 nm. At higher contents a shoulder at 320 nm is observed which could be attributed to another type of centres (Fig. 6).

The high fluorescence and the low signal did not allow the recording of the Raman spectra of Cr(VI) chemisorbed on γ-Al_2O_3, while for Mo(VI) and W(VI) significant spectra were obtained only for contents not smaller than 5% MoO_3 and 2% WO_3 (Fig. 7). The spectra of all the samples display the absence of crystallographically well defined phases [7] as we also checked by X-ray patterns; they are characterized by a strong band (at 950 cm^{-1} for Mo(VI) and 960 cm^{-1} for W(VI)) shifted to higher wavelengths with the increase of Mo(VI) and W(VI) content.

These spectra remind of the spectra of aqueous solutions. Unlike the solutions, Mo(VI) and W(VI) adsorbed on γ-Al_2O_3 display a broader band which reflects the surface heterogeneity, and a shift of the maximum to somewhat higher wavenumber. The position of the maximum of the most intense band, particularly at a low cover-

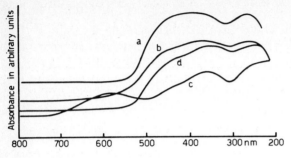

Fig. 5. Electronic spectra of Cr(VI)-Al$_2$O$_3$ (3% CrO$_3$) dried at 110°C (a) and calcined at 600°C (b); of Cr(III)-Al$_2$O$_3$ (1.13% CrO$_3$) dried at 110°C (c) and calcined at 700°C (d)

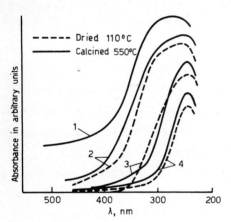

Fig. 6. Electronic spectra. (1) 7.7% MoO$_3$ on Al$_2$O$_3$; (2) 19.2% WO$_3$ on Al$_2$O$_3$; (3) 9.31% WO$_3$ on Al$_2$O$_3$; (4) 3.05% WO$_3$ on Al$_2$O$_3$

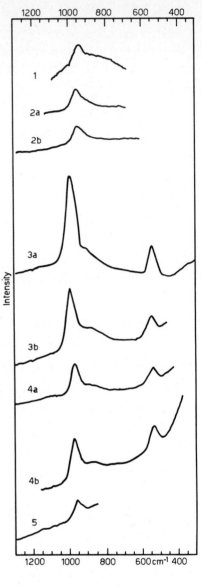

Fig. 7. Raman spectra. (1) 5.75% MoO$_3$ on Al$_2$O$_3$ dried at 110°C; (2a) 7.7% MoO$_3$ on Al$_2$O$_3$ dried at 110°C; (2b) 7.7% MoO$_3$ on Al$_2$O$_3$ calcined at 550°C; (3a) 19.2% WO$_3$ on Al$_2$O$_3$ dried at 110°C; (3b) 19.2% WO$_3$ on Al$_2$O$_3$ calcined at 550°C; (4a) 9.31 WO$_3$ on Al$_2$O$_3$ dried at 110°C; (4b) 9.31% WO$_3$ on Al$_2$O$_3$ calcined at 550°C; (5) 3.05% WO$_3$ on Al$_2$O$_3$ calcined at 550°C

age, could suggest an octahedral co-ordination of the transition metal ion (aqueous solutions of polymolybdate and tetrahedral molybdate exhibit a strong band at 940 cm^{-1} [10] and 897 cm^{-1} [11] respectively; polytungstate and tetrahedral tungstate at 952 cm^{-1} [12] and 931 [11] respectively). However this conclusion contrasts with the results of the adsorption experiments, of the electronic spectra and finally with the literature data which suggest the formation of aluminium molybdate and tungstate-like species in tetrahedral co-ordination [13, 14].

Therefore we attribute the intense band to the symmetric stretching of Me(VI)O$_4^=$ in tetrahedral co-ordination, while the shift to higher wavelength could reflect the distortion from T_D symmetry and an increase of the Me(VI)=0 double bond character [7].

4 CONCLUSION

The pulse and step addition experiments clearly prove the interaction of the sixth group of transition element oxides with γ-Al$_2$O$_3$ and evidence the role of basicity of γ-Al$_2$O$_3$ surface.

These experiments suggested a common interaction mechanism involving the decondensation of isopolyanion in the presence of alumina followed by the chemisorption of monomeric species. The most significant differences are observed in the quantities of transition metal ion chemisorbed.

The relevance of the chemisorption phenomena on the preparation of commercial chromia-alumina catalysts is moderate because of the higher chromium content of these commercial catalysts.

Instead in the MoO$_3$-WO$_3$-γ-Al$_2$O$_3$ systems the chemisorption processes which occur during the impregnation may be determining factors in the preparation of industrial catalysts.

In the case of Mo(VI) the amount of irreversibly chemisorbed Mo(VI) (about 8 wt. % MoO$_3$) must be compared with the content of a typical commercial HDS catalyst (10-12% MoO$_3$). This observation becomes more pressing in the WO$_3$-γ-Al$_2$O$_3$ system when it is considered that the amount of chemisorbed W(VI) (19-20 wt. % WO$_3$) is in the range of WO$_3$ content of a commercial catalyst (20-21%).

The presence of surface phases, i.e. of phases highly dispersed and not revealed by X-rays, was directly detected for chromia-alumina catalysts and is practically established for molybdena-alumina. As regards the WO$_3$-Al$_2$O$_3$ system, the general consistence of adsorption experiments, of spectroscopic data and of

X-ray pattern suggests that in a typical commercial γ-alumina the limiting amount of W(VI) which can be dispersed as a surface phase is in the range of 19-20% WO_3.

ACKNOWLEDGEMENTS

Dr. Paolo Tittarelli carried out X-ray determinations. Prof. Alberto Girelli is thanked for helpful suggestions and for the discussion of our work. One of the authors (P. L. Villa) also thanks the Consiglio Nazionale delle Ricerche for financial support.

REFERENCES

1 G. P. Poole and D. S. McIver, Advances in Catalysis, 17 (1967) 223.
2 (a) J. H. Ashley and P. C. H. Mitchell, J. Chem. Soc. (A), (1968) 2821; (1969) 2730; (b) N. P. Martinez, P. C. H. Mitchell and P. Chi plunker, in P. C. H. Mitchell (Ed), Proc. Climax 2nd Int. Conference on the Chemistry and Uses of Molybdenum, Climax Molybdenum Co. Ltd., London, 1977, p. 164.
3 J. Sonnemans and P. Mars, J. Catal., 31 (1973) 209.
4 G. C. A. Schuit and B. C. Gates, A. I. Ch. E. J., 19 (1973) 417.
5 (a) A. Iannibello and F. Trifirò, Z. Anorg. Allg. Chem., 413 (1975) 293; (b) A. Iannibello, S. Marengo and F. Trifirò, Chimica e Industria, 57 (1975) 676; (c) A. Iannibello, S. Marengo, V. Berti, P. L. Villa and F. Trifirò, Can. J. Chem. Eng., 55 (1977) 747.
6 F. Trifirò, Annali Chimica, 64 (1974) 377.
7 A. Iannibello, P. L. Villa and S. Marengo, J. Phys. Chem., submitted for publication.
8 P. Souchay, Ions Minéraux Condensés, Masson & CIE, Paris, 1969, p. 54.
9 J. Deren, J. Haber and J. Siechowski, in W. M. H. Sachtler, G. C. A. Schuit and P. Zwietering (Eds.), Proc. 3rd Int. Congr. Catal., Amsterdam, July 20-25, 1964, North-Holland Publishing Co., Amsterdam, 1965, p. 993.
10 J. Aveston, E. W. Anacker and J. S. Johnson, Inorg. Chem., 3 (1964) 735.
11 W. P. Griffith, J. Chem. Soc. (A), (1970) 286.
12 J. Aveston, Inorg. Chem., 3 (1964) 981.
13 (a) P. Biloen and G. I. Pott, J. Catal., 30 (1973) 169; (b) G. I. Pott and W. M. J. Stork, in B. Delmon, P. A. Jacobs and G. Poncelet (Eds.), Preparation of Catalysts, Elsevier, Amsterdam, 1976, p. 537.
14 A. Iannibello and P. C. H. Mitchell, 2nd Int. Symp. Scientific Bases for the Preparation of Heterogenous Catalysts, Louvain-La-Neuve, September 4-7, 1978, in press.

DISCUSSION

R. GALIASSO : 1) Do you observe the formation of a white precipitate during your pulse experiment with ammonium hepta-molybdate solution ? 2) How did you measure the % of MoO_3 irreversibly adsorbed if this precipitate is formed ?

A. IANNIBELLO : 1) We did not observe any precipitate in pulse addition experiments. In step addition experiments, we observed the presence of Al^{3+} cations in the last fraction of the effluent solution; these cations gave rise, by standing, to the precipitation of aluminium molybdate species. 2) Surface area and pore volume of the original γ-alumina were little changed after pulse and step addition of Mo (VI) (γ-alumina : 180 m^2/g and 0.55 ml/g; 7.7% MoO_3 on $\gamma\text{-}Al_2O_3$: 172 m^2/g and 0.51 ml/g); moreover, X-ray diffraction pattern indicated that in the $MoO_3\text{-}Al_2O_3$ sample no other phase was present but $\gamma\text{-}Al_2O_3$. Therefore we concluded that all Mo (VI) is present as chemisorbed species.

K. HABERSBERGER : I suppose that you have extended these studies also to the simultaneous adsorption of two or more components of this type, e.g. molybdate and vanadate, in order to obtain a better insight into the preparation of supported *mixed* oxide catalyst.

A. IANNIBELLO : We are not yet dealing with this problem in a systematic fashion. Our preliminary data suggest that in adsorption systems in which competing oxoanions are present, kinetic factors can seriously affect the time at which a true adsorption equilibrium is reached. Compare, for instance, the equilibrium distribution of Cr(VI), Mo(VI) and W(VI) between liquid and solid phases with the respective adsorption rates.

K.R. KRISHNAMURTHY : The limiting amount of chemisorbed oxo species in γ-alumina is in the order $CrO_3 > MoO_3 > WO_3$. Can you provide any explanation for this observed pattern ?

A. IANNIBELLO : The thermodynamics of these systems, except some preliminary equilibrium adsorption studies, have not been deeply investigated. At present we can only speculate that the observed adsorption equilibria mainly reflect the large difference in the activity of Cr(VI), Mo(VI) and W(VI) in aqueous environment.

J.L. LEMAITRE : At low pH values did you observe dissolution of alumina and/or formation of tungstoaluminate ions ?

A. IANNIBELLO : There is no doubt that at low pH alumina dissolves. Aluminium cations in the solution have been observed when a large excess of dilute aqueous solution of ammonium dichromate or ammonium paramolybdate are contacted with γ-alumina. Concerning W(VI), we recall that the pH of aqueous solutions of ammonium dichromate or ammonium paramolybdate are contacted with γ-alumina. Concerning W(VI), we recall that the pH of aqueous solutions of ammonium paratungstate is about 6-6.4; in this range of pH we do not expect significant dissolution of alumina.

In our opinion, in the impregnation of γ-alumina with paratungstate aqueous solutions, the formation of aluminum tungstate species (mainly as surface phase), rather than of tungsten aluminate species, must be expected.

DISPERSION AND COMPOUND FORMATION IN SOME METATHESIS CATALYSTS AND FLUORINE CONTAINING ALUMINA'S, STUDIED BY XPS AND LASER-RAMAN SPECTROSCOPY

F.P.J.M. Kerkhof, J.A. Moulijn, R. Thomas and J.C. Oudejans
Institute for Chemical Technology, University of Amsterdam, Amsterdam.

ABSTRACT

We have studied the nature of the interaction between carrier and promoter for a number of catalysts. It is concluded that surface compounds are formed as well on silica as on γ-alumina. The interaction of alumina with the promoter is stronger, resulting in better dispersion, while on silica the formation of crystallites is observed at low promoter content.

INTRODUCTION

Many catalytic systems consist of a high-surface area support and a well dispersed promoter. However, the nature of the interaction between promoter and carrier is not always uderstood and can differ in various systems.

A number of interactions between carrier and promoter can take place in the formation of the precursor of the active site (ref. 1, 2).
- Diffusion of the promoter into the carrier (solid solution).
- Formation of a new bulk compound between support and promoter.
- Creation of a molecular dispersion of the promoter on the carrier (monolayer).
- Growth of promoter crystallites on the support.

We have studied the formation of the precursor for a number of catalysts. The transformation of precursor into the final active site is not dealt with here.

Metathesis catalysts

A typical example of the metathesis reaction is the conversion of propene into an equimolar mixture of ethene and 2-butene. This reaction is catalysed by e.g. tungstenoxide on silica or alumina and rheniumoxide on alumina. We have studied these catalysts by several techniques and obtained valuable information from laser-Raman spectroscopy (ref. 3, 4).

In the case of tungstenoxide on silica two tungsten species could be identified viz. crystalline tungstentrioxide with the main Raman lines at 809 and 715 cm^{-1} and a compound with a broad band at 970 cm^{-1}. This band was tentatively ascribed to a surface compound consisting of distorted tungstenoxide octahedrons.

We tried to correlate these species with the results of reduction experiments.

Further the presence of the band at 970 cm^{-1} as a function of catalyst synthesis and calcination temperature was studied.

In the case of rheniumoxide on alumina we concluded that in these catalysts rhenium is present as a monolayer of rheniumoxide tetrahedrons. Even at high rheniumoxide content (18 wt %) no other species like octahedrally coordinated rhenium or Al/Re/O compounds could be found. In the case of tungstenoxide on alumina we found that up to 15 wt % tungstenoxide only a band at 970 cm^{-1} was present and therefore the formation of tungstenoxide crystallites could be excluded. It will be shown that XPS confirms the formation of tungstenoxide crystallites in the case of tungsten/silica catalysts and the presence of a monolayer on the rheniumoxide/alumina catalysts.

Fluorinated alumina's

It is well know that incorporation of fluorine enhances the activity of γ-alumina in polymerization, cracking, isomerization and the disproportionation of toluene (ref. 5, 6). The fluorine in these catalysts can be present as:
- Surface groups.
- AlF$_3$ crystallites (α and β modification).
- Aluminum hydroxyfluorides.

By using XPS, we could show, that for catalysts prepared by impregnation of boehmite with a solution of ammoniumfluoride in aqueous ammonia, up to ± 6 wt % fluorine no AlF$_3$ crystallites were present (ref. 7).

The binding energies of F(1s) electrons in AlF$_3$ and in surface fluorine were 687.5 and 685.7 eV respectively. The binding energy of the Al(2p) electrons in AlF$_3$ was 77.3 eV, whereas a value of 75.0 eV was found in γ-alumina. By measuring the XPS spectra we therefore could examine the presence of crystallites and surface fluorine from a shift in the binding energy of the Al(2p) and F(1s) electrons.

Besides XPS, X-ray diffraction (XRD) was used to study the compound formation in fluorinated alumina's prepared by co-impregnation of Al(NO$_3$)$_3$ and NH$_4$F on γ-alumina and silica.

The use of XPS-intensities

The binding energies of the W(4f) and and Si(2p) electrons of some tungstenoxide on silica catalysts are reported elsewhere (ref. 8). We showed that differential charging caused a line broadening of the XPS peaks. With a new sample preparation technique the line broadening could be avoided and e.g. the broad peak of the W(4f) electrons resolved in a sharp doublet.

In spite of the line broadening the intensities of the signals can be used to measure the dispersion of the promoter.

Angevine et al. (ref. 9) have proposed a model which predicts the relative XPS intensity of the promoter in the case of monolayer and crystallite formation. One of the assumptions of this model is the presence of the promoter on a semi-infinite support. We think taht this assumption is not always justified and therefore the prediction of the support intensity by the model of Angevine can be too high.

The thickness of the support can be estimated from a simple model. If the support is envisaged as a thin square plate, the thickness d is given by $d = 2 \rho^{-1} S^{-1}$ (ρ = true density of the support and S is the surface area). For a typical silica support with ρ = 2200 kgm^{-3} and S = 350 m^{2}g^{-1} this results in d = 2.6 nm. Because this is in the order of magnitude of the mean free path of the electrons in the solid most of the carrier is "seen" by XPS. In the case of a monolayer catalyst also no loss of promoter signal occurs and therefore the XPS intensity can be predicted from the bulk atomic ratio and the photo-electron cross sections. When a thin support is assumed it can easily be proven that the model of Angevine reduces to this prediction (ref. 10).

EXPERIMENTAL

Preparation of the catalysts

<u>Tungstenoxide on silica and alumina by aqueous impregnation.</u> These catalysts were prepared by adding the carrier e.g. Grace silica (190 μm) to an aqueous solution of ammoniummetatungstate ($(NH_4)_6H_2W_{12}O_{40}$). After removal of the water in a rotating film evaporator the sample was dried overnight at 390 K. Than the samples were fluidized in dry air at 820 K during two hours. In the case of the tungstenoxide promoted silica's the white powders turned yellow during this procedure. At higher promoter content this effect became stronger. In case of the tungstenoxide on alumina catalysts this effect was not observed and the samples were white before and after calcining. A tungstenoxide/silica catalyst was prepared by aqueous impregnation with silico-tungsten acid. ($SiO_2 \cdot 12WO_3 \cdot 26H_2O$).

<u>Tungstenoxide on silica-aerogel.</u> The preparation of silicagel with excellent optical properties is given by Peri (ref. 11). We have synthesized the silica-aerogel and tried to impregnate these supports by passing tungstenhexachloride vapour in dry nitrogen over the silica at 470 K. It proved that most of the tungsten was present on the outside of the silicaplate and no uniform dispersion could be obtained.

<u>Tungstenoxide on silica with solutions of tungstenhexachloride in methanol.</u> Before impregnation the support is contacted with dry methanol during 40-50 days. During this period the methanol is refreshed 4 to 5 times. After decanting the methanol, tungstenhexachloride in methanol is added. When tungstenhexachloride is added to the methanol, hydrogenchloride is evolved, probably resulting from the formation of hexamethoxytungsten. The methanol is removed in an autoclave under supercritical conditions (P = 9-11 MPa, T = 533 K). After this, the blue samples were calcined under the same conditions as described before.

<u>Tungstenoxide and rheniumoxide on alumina.</u> These impregnations are similar to the first described procedures and are reported in detail by Kapteijn et al. (ref. 12) and Thomas et al. (ref. 3).

<u>Fluorine containing alumina's.</u> These samples were prepared by adding γ-alumina

(Ketjen alumina grade B, calcined in dry air during 16 hours by the method given before) to an aqueous solution of aluminumnitrate and ammoniumfluoride in the molar ratio 1 : 3. After drying in the same way as mentioned above catalysts were calcined during 16 hours.

Laser-Raman spectroscopy

The spectra were recorded on a Jeol-JRS-1 or Ramanor HG2S spectrometer. The details of the experimental procedures are given by Thomas et al. (ref. 3). Duplicate runs of some samples were run on both spectrometers.

X-ray photoelectron spectroscopy

The spectra were recorded on a AES-200 spectrometer using AlKα and MgKα radiation. The X-ray source power was 180 watt. The samples were powdered and mounted on adhesive tape. Further details are given elsewhere (ref. 8).

RESULTS AND DISCUSSION

Laser-Raman spectroscopy and reducibility of some metathesis catalysts

The Laser-Raman spectra of the tungsten on alumina and tungsten on silica catalysts prepared by the aqueous impregnation with ammoniummetatungstate are given elsewhere (ref. 3). The spectra of the silica catalysts are characterized by the lines of tungstenoxide at 809 and 715 cm^{-1} and a broad band at 970 cm^{-1}. In case of a 15 wt % tungstenoxide on alumina catalyst only a broad band at 970 cm^{-1} is observed. The fact that at least two tungsten species are present on the silica catalysts is confirmed by reduction experiments (ref. 1). At low tungsten contents a rather large fraction of the tungsten is present as a hardly reducible compound whereas at higher contents the easier reducible tungstenoxide is formed. The amount of easier reducible material is plotted as a function of the Raman intensity of the tungstenoxide line at 809 cm^{-1} in figure 1. The amount of hardly reducible material versus the intensity of the band at 970 cm^{-1} is given in figure 2.

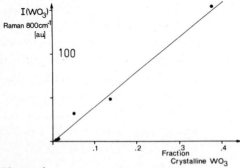
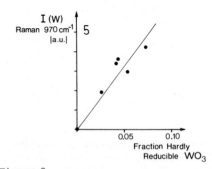

Figure 1.
Intensity of the Raman line of tungstenoxide as a function of the amount of easily reducible material.

Figure 2.
Intensity of the Raman band at 970 cm^{-1} as a function of the amount of hardly reducible material.

In figure 1 is shown that there is a good correlation between the amount of crystalline material measured by laser-Raman spectroscopy and by reduction experiments. The correlation in figure 2 suggests that the band at 970 cm^{-1} can be ascribed to a hardly reducible compound. This is in agreement with the fact that for tungstenoxide on alumina, which is harder to reduce than tungstenoxide on silica, only a band at 970 cm^{-1} is present (ref. 13).

By comparing the slopes of the lines in figure 1 and 2 it can be concluded that the Raman activity of tungstenoxide is about six times higher that the Raman activity of the hardly reducible material.

The presence of tungstenoxide crystallites was also detected with electronmicroscopy. The electronmicrographs of a 12 and 41 wt % catalyst showed that the size of the crystallites is in the same order of magnitude in both cases (±20 nm). This is in agreement with estimates from the line broadening of the XRD lines. It was also revealed that on the 41 % catalyst more crystallites were present than would be expected from an interpolation of the number of crystallites present on the 12 % catalyst. This corresponds with the fact that on a 12 wt % catalyst about half of the tungsten is present as crystalline material whereas on the 41 % catalyst 90 % of the tungsten is crystalline.

In the spectra of the catalysts prepared by impregnation with tungstenhexachloride in methanol no intense lines could be observed. Except for bands at about 800 cm^{-1}, 700 cm^{-1} and in some cases a weak one at 970 cm^{-1} no other bands were present. Because these catalysts are quite active and have a good dispersion, we conclude that on these catalysts part of the tungsten is present as a Raman inactive species. Brown et al. (ref. 14) showed that for molybdenum on alumina catalysts a treatment with water vapour resulted in the appearance of a Raman band which was absent in a regenerated catalyst. Whether the formation of a Raman active species upon water or hydroxyl coordination can be observed in our catalysts, will be investigated.

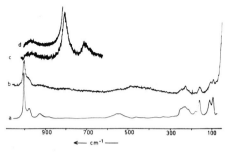

Figure 3.
Laser-Raman spectra of silicotungsten acid; pure (a), on silica before calcining (b), on silica after calcining (c+d). A, b and c recorded at 100 cm^{-1}/min; d at 20 cm^{-1}/min.

The laser-Raman spectrum of a catalyst prepared by aqueous impregnation of silica with silicotungsten acid is shown in figure 3. This figure shows that the lines of the spectrum of the uncalcined catalyst are the same as for the unsupported silicotungsten acid. After calcining the same spectrum is found as in the case of the catalysts prepared with ammoniummetatungstate viz. tungstenoxide lines at 809 and 715 cm^{-1} and a broad band at 970 cm^{-1}.

Before calcining, the laser-Raman spectrum does not reveal a band at 970 cm^{-1}. This suggests that the interaction compound is formed during the calcination. Also when a catalyst was prepared by impregnation of a different silica (Mallinckrodt) with ammoniummetatungstate these lines were found. We therefore conclude that when a silica catalyst is prepared by aqueous impregnation with a tungsten compound, as well tungstenoxide as an interaction compound is present after calcining.

Sintering effects

We tried to vary the size of the tungstenoxide crystallites by heating the catalysts in air at several temperatures. Figure 4 gives the laser-Raman spectra of a 12 wt % tungstenoxide catalyst after several hours at 920 K. The figure shows that heating of the catalyst results in the disappearance of the peak ascribed to the surface compound. After prolonged heating the lines of tungstenoxide become more intense. The diffraction patterns of these catalysts are shown in figure 5.

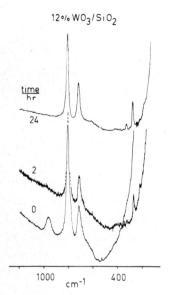

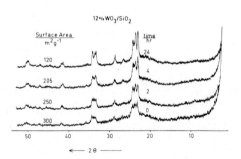

Figure 4.
Laser-Raman spectra of a 12 wt % tungstenoxide on silica catalyst before and after heating in air at 920 K.

Figure 5.
X-ray diffraction patterns of a 12 wt % tungstenoxide on silica catalyst, before and after heating in air at 920 K.

The better crystallinity upon heating is shown here by the decreasing peak width. This effect is more pronounced after heating at 1070 K. In this case crystallization of as well tungstenoxide as the carrier is observed. This is shown by the XRD-pattern of the catalysts (figure 6). After two hours the lines of tungstenoxide are observed and after 24 hours also the XRD-pattern of crystobalite (α-quartz) is found. It is striking that the crystallization of the carrier is far less when no tungstenoxide is present. This is shown in figure 7 which gives the XRD-patterns of the support after heating at 1070 K. The main effect of heating in this case is the decrease of the surface area from 300 to 160 m^2g^{-1}. The increase of the particle size of the silica can be estimated by a decreasing width of the silica band at $2\theta=22°$ (from 1.3 to 1.5 nm)

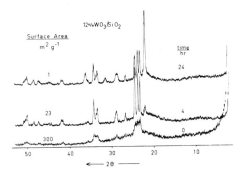
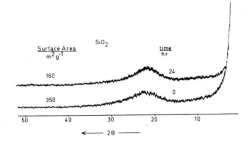

Figure 6.
X-ray diffraction pattern of a 12 wt % tungstenoxide on silica catalyst before and after heating in air at 1070 K.

Figure 7.
X-ray diffraction patterns of silica before and after heating in air at 1070 K.

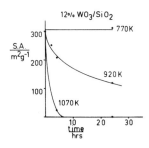

Figure 8.
Surface area of a 12 wt % catalyst as a function of calcination time and temperature.

Because of the decrease in small angle scattering (ref. 15) this increase in particle size is apparently due to the disappearance of the smallest silica particles.

The loss of surface-area is represented in figure 8. From this figure it can be concluded that heating at the normal calcination temperature has no influence on the surface area, while a sharp decrease is found upon heating at 1070 K. After 24 hours at 1070 K the surface area of the catalyst is practically zero, whereas the support without tungstenoxide still has half of its orginal surface area. When an alumina catalyst is heated at 1020 K, nor a decrease in surface area nor the disappearance of the surface compound with a Raman band at 970 cm^{-1} is found.

Fluorinated alumina's

The XRD-patterns of a number of calcined catalysts, prepared by co-impregnation of NH_4F and $Al(NO_3)_3$ on γ-alumina are shown in figure 9. By measuring the XPS spectra of these catalysts the presence of surface bound fluorine atoms can be measured (ref. 7). By combining XPS and XRD we could compose table 1 which indicates the presence of several compounds in uncalcined and calcined samples.

It can be concluded that up to 10 wt % F, fluorine is present only in surface groups. From 10 to 40 wt % F aluminumhydroxyfluorides are formed whereas at higher fluorine content β-AlF_3 is observed. The co-impregnation was based on the possible formation of β-AlF_3 on the catalyst. Without support indeed the formation of β-AlF_3 was observed after calcining. Also on a 15 wt % fluorine on silica catalyst the formation of β-AlF_3 was favoured. The difference in formation of aluminumhydroxy-fluorides on silica and alumina, will be the subject of further investigation.

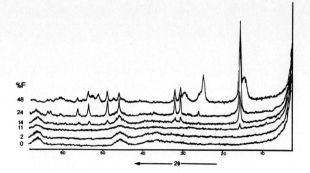

Figure 9. XRD-patterns of some fluorine containing alumina's.

TABLE 1

Compounds in fluorine containing alumina's

% F	Compounds present (XRD)		Surface fluorine present (XPS signal at 685.7 ev)
	before calcination	after calcination	after calcination ☆
1	–	–	+
2	–	–	+
5	$AlF_{1.65}OH_{1.35}$	–	+
6	"	–	+
7	"	–	+
11	"	$AlF_{1.96}OH_{1.04}$	+
12	"	"	
14	"	"	? F(1s) of the hydroxy-fluorides (687.5 ev) is dominant.
24	$AlF_{1.96}OH_{1.04}$	"	
48	"	$AlF_{1.96}OH_{1.04}$ + β-AlF_3	

☆ Before calcination not measured.

XPS-intensities

The relative intensities of promoter and support of the metathesis catalysts is plotted as a function of the bulk atomic ratio in the figures 10, 11 and 12.

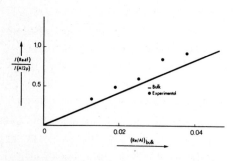

Figure 10.
XPS intensity ratio of the rhenium-oxide on alumina catalysts.

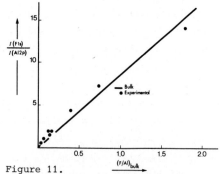

Figure 11.
XPS intensity ratio of the fluorine containing alumina catalysts.

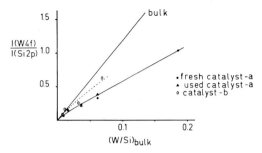

Figure 12.
XPS intensity ratio of the tungstenoxide on silica catalysts.
a = catalysts prepared by aqueous impregnation
b = catalysts prepared by imprgnation with a solution of tungstenhexachloride in methanol.

In these figures also the prediction based on the atomic bulk ratio and the photoelectron cross section of Scofield (ref. 16) are given.

In the case of rheniumoxide on alumina and fluorinated alumina's these predictions agree well with the experimental values. This is rather surprising, considering the fact that XPS is a surface technique.

In the introduction we showed that the thickness of the support is in the same order of magnitude as the electron escape depth and therefore only a small loss of signal of the carrier is to be expected. In the case of a monolayer catalyst no loss of intensity of the promoter electrons will occur. These two facts result in a simple model predicting the intensities of monolayer catalysts.

Figure 10 confirms the idea of Olsthoorn and Boelhouwer (ref. 17) and of Kapteijn et al. (ref. 12) that rheniumoxide on alumina is a monolayer catalyst up to a rhenium content of 18 wt %. The same conclusion can be drawn for the fluorinated alumina's up to 10 wt % fluorine. At higher fluorine content crystallites are formed and a decrease in XPS intensity is observed.

The fact that the experimental values are lower than is predicted from the bulk ratio for tungstenoxide on silica is explained by the formation of crystallites from which only a part is "seen" by XPS. Figure 12 shows that the catalysts prepared from a solution of tungstenhexachloride in methanol have a higher XPS intensity. We therefore conclude that on these catalysts the dispersion of the tungstenoxide is better. This is confirmed by the fact that the XRD-patterns of the tungstenhexachloride/methanol catalysts show the presence of crystalline material at a higher tungsten content than the catalysts prepared by aqueous impregnation.

It might be possible that the reason for the better dispersion of the tungstenhexachloride / methanol catalysts is caused by an interaction of the hexamethoxytungsten with the methoxy groups present on the silica after equilibration with methanol. Obviously, the interaction of the hydroyl groups of the silica with the tungsten compound in water is smaller and therefore the formation of crystallites is favoured.

CONCLUDING REMARKS

In this paper we showed that alumina supported catalysts have a stronger interaction with the promoter than silica.

The strong interaction also seems to result in a better stability of the compound formed between carrier and promoter e.g. the tungsten/silica systems segregates at high temperature in α-quartz and tungstenoxide, whereas the interaction compound between alumina and tungstenoxide is not decomposed.

We showed how XPS can be used as a tool to observe monolayer and crystallite formation. Whereas rheniumoxide on alumina is a monolayer catalyst consisting of ReO_4^- tetrahedrons on the surface, crystallites are formed in the case of tungstenoxide on silica.

ACKNOWLEDGEMENTS

We are indebted to Dr G. Sawatzky and A. Heeres (Laboratory for Physical Chemistry, University of Groningen) for usefull discussion on the XPS spectra. Thanks are also due to Dr B. Koch and W. Molleman (Department of X-ray Spectroscopy and Diffraction, University of Amsterdam). We are also indebted to Dr J. Medema, Dr D.J. Stufkes and Dr V.H.J. de Beer for help in recording and interpretation of Ramanspectra. This study was supported by the Netherlands Foundation for Chemical Research (S.O.N.) with financial aid from the Netherlands Organization for the Advancement of Pure Research (Z.W.O.).

REFERENCES

1. F.P.J.M. Kerkhof, R. Thomas and J.A. Moulijn, Recl. Trav. Chim.,Pays-Bas 96(11), 1977, M121.
2. T. Jansen, P.C. van Berge and P. Mars in "Preparation of Catalysts" edited by B. Delmon, P.A. Jacobs and G. Poncelet, Elsevier, Amsterdam, 1976.
3. R. Thomas, J.A. Moulijn and F.P.J.M. Kerkhof, Recl. Trav. Chim., Pays-Bas, 96(11), 1977, M134.
4. F.P.J.M. Kerkhof, J.A. Moulijn and R. Thomas, submitted for publication.
5. V.C.F. Holm and A. Clark, Ind. Eng. Chem., Prod. Res. Develop. 2(1963)38.
6. R. Covini, V. Fattore and N. Giordano, J. Catal. 7(1967)126.
7. F.P.J.M. Kerkhof, H.J. Reitsma and J.A. Moulijn, React. Kinet. Catal. Lett., 7-1(1977)15.
8. F.P.J.M. Kerkhof, J.A. Moulijn and A. Heeres, submitted for publication.
9. P.J. Angevine, J.C. Vartuli and W.N. Delgass in the Proceedings of the Six International Congress on Catalysis, volume 2, p 611 edited by G.C. Bond, P.B. Wells and F.C. Tompkins, The Chemical Society, Burlington House, London 1976
10. F.P.J.M. Kerkhof and J.A. Moulijn, to be published.
11. J.B. Peri, J.Phys.Chem., 70(1966)2937.
12. F. Kapteijn, L.H.G. Bredt and J.C. Mol, Recl.Trav.Chim Pays-Bas 96(11),1977 M139.
13. P. Biloen and G.T. Pott, J.Catal. 30(1974)169.
14. F.R. Brown, L.E. Makovsky and K.H. Rhee, J. Catal. 50(1977)162.
15. M.H. Jellinek and I. Frankuchen, Adv. Catal. 1(1948)257.
16. J.H. Scofield, J. Electron Spectrosc. Relat. Phenom. 8(1976)129.
17. A.A. Olsthoorn and C. Boelhouwer, J. Catal. 44(1976)197.

DISCUSSION

D. CHADWICK : Firstly, have you considered using a flood gun to eliminate differential charging ? Secondly, in the case of the XPS results on fluorine containing catalysts it would be interesting to plot the F(2s)/F(1s) ratio against bulk composition. Since these photolines have different kinetic energies and therefore different sampling depths, such a plot should provide an independent check on your method.

F.P.J.M. KERKHOF : 1) The apparatus we used was not equipped with a flood gun. Moreover, using a flood gun the sample may get a negative charge and this charging can be not-uniform too.
2) Using AlK_α X-rays, the kinetic energies of the F(2s) and F(1s) electrons are 1.4 and 0.8 keV respectively. Therefore the difference in escape depth will be about 30% (2.0 and 1.5 nm). Because of this fact it can be expected that at high fluorine contents (crystallite formation) the value F(2s)/F(1s) will slightly increase.

M. HOUALLA : You have elegantly shown that the fraction hardly reducible WO_3/SiO_2 is related to the intensity of the Raman band at 970 cm^{-1}. On the other hand, during your study of the sintering effect, you report the disappearance of this band by heating at 920 K. Would this disappearance correlate with an absence of the fraction unreducible ?

F.P.J.M. KERKHOF : We have one indication of the fact that indeed catalysts without a band in the Raman spectra at 970 cm^{-1} are easier to reduce, viz. catalysts which were reduced at 920 K and reoxidized at 650 K reduced much faster than the fresh catalysts and show no band at 970 cm^{-1}. We hope to obtain more evidence on the relation between reducibility and surface composition by application of T.P.R.

P.G. ROUXHET : We have recently investigated (P.O. Scokart, S.A. Selim, J.P. Damon and P.G. Rouxhet, J. Colloid Interface Sci. (1978), in press) fluorinated aluminas prepared by impregnation of γ-Al_2O_3 by NH_4F, using XPS and other techniques. At low F content, the surface of alumina is modified. When the F content increases above about 5%, fluorine contributes to make AlF_3 particles which are separated from alumina and develop a low surface area. We have also investigated the acidity properties of the modified alumina surfaces.

For products outgassed at 600°C, hydroxyls remaining on the surface have a stronger acidity than hydroxyls of alumina; stronger non-protonic sites are also observed. For products calcined at high temperature, stored in the laboratory atmosphere and reactivated at 300°C, strong protonic sites are observed; they are responsible for the isomerizing activity of the catalysts and are tentatively assigned to water molecules held by the strong non-protonic sites observed after outgassing at higher temperature.

F.P.J.M. KERKHOF : Your result for the fluorine distribution as surface groups and as crystalline material in catalysts treated with NH_4F is similar to our results for these catalysts (Kerkhof,F.P.J.M., Rettsma, N.J. and Moulijn, J.A., Reakt. Kinet. Catal. Lett., 7-1 (1977), 15). Recently we have studied fluorinated aluminas by infrared spectroscopy, poisoning experiments and activity measurements, and also found a correlation between the number of protonic sites and catalytic activity. These results will be reported in the near future.

F. TRIFIRO : You have attributed the octahedral coordination to the species of tungsten present at the surface of γ-Al_2O_3 by Raman spectroscopy. In my paper at this symposium I have attributed the tetrahedral coordination by the analysis of electronic spectra. Are the Raman spectra diagnostic for the attribution of coordination of tungsten in oxide systems ?

F.P.J.M. KERKHOF : Generally, the Raman technique gives more direct information on the coordination of an atom. Moreover the bands in the reflectance spectra are rather broad. Therefore, we prefer the use of Raman spectroscopy in identifying compounds present in a catalyst. Structural information is obtained by comparing spectra of the catalyst with spectra of reference samples, e.g. the spectra of the W/SiO_2 and W/Al_2O_3 catalysts are both characterized by the presence of a band at 970 cm^{-1}. This band is also the strongest one in the spectrum of ammonium meta-tungstate which is composed of distorted WO_6 octahedra. Therefore we concluded the octahedral coordination of the W-atom, at least at higher metal loading. A tetrahedral coordination at low metal content (<3wt %) cannot be excluded.

REDUCTION OF SILICA SUPPORTED NICKEL CATALYSTS

J.W.E. COENEN
Unilever Research Vlaardingen, P.O. Box 114, 3130 AC Vlaardingen
The Netherlands

SUMMARY

The difficult reducibility of silica supported nickel catalysts has often been ascribed to chemical interaction with formation of basic silicates. From a review of literature data and from our own work many arguments for the importance of silicate formation will be derived. A structural model for silica supported nickel catalysts made by precipitation has been evolved. Nickel silicate is seen as a "glue" layer between the silica and nickelhydroxide, which forms an epitaxial overlay on the silicate. In the first stage of the reduction or in a precalcining stage the hydroxide breaks up in an array of small oxide crystallites, still epitaxially linked to the silicate. In subsequent reduction each oxide particle is reduced to one nickel crystallite, still epitaxially linked to the silicate glue layer.

As often reported in the literature we also find a strong inhibition of the reduction by water vapour. The kinetics of reduction can be described by an overall activation energy of about 90 $kJ.mol^{-1}$, slightly increasing at higher degrees of reduction. The rate is close to third order in unconverted oxide. The surface area of the nickel is an almost linear function of the degree of reduction.

Arguments are put forward for the assumption that in the reduction the nucleation is rate controlling. The most important reason for the low reduction rate is the low yield of reduced material per nucleus, due to the fine oxide dispersion. To explain the fast rate decline – third order – surface heterogeneity possibly partly due to silicate has to be invoked as well as a particle size distribution for the supported oxide. It proved possible to explain the rate behaviour in a reasonably quantitative way.

1. INTRODUCTION

Dispersions of various nickel compounds on silica supports, prepared by a great variety of impregnation and precipitation methods, after their activation by high temperature reduction with a reducing gas, find widespread application as catalysts. Although the very extensive literature is hardly coherent and often contradictory, there is general agreement at least on two important issues:
- A much finer nickel dispersion is obtained by reduction of silica supported

nickel compounds than with the same material without support. Average nickel crystallite sizes as small as 0.5 nm have been reported (1) for successful preparations, individual crystallites containing 10 atoms or less. Such fine dispersions in which virtually all nickel atoms are exposed and available for catalytic action, are completely unattainable without the aid of a support. Even if such fine dispersions could be prepared without support they would be highly unstable and their recovery from a reaction mixture would be next to impossible.
- Reduction to the metallic form, which is the active form for most applications, is much more difficult for the supported material. For comparable rates the silica supported material requires reduction temperatures at least 200°C higher than the unsupported material. Complete reduction even at much higher temperature and long reaction time is often difficult to achieve.

A structural model and a compatible mechanism are clearly required to explain these observations.

2. THE IMPORTANCE OF NICKEL SILICATE FORMATION

2.1. Published data

Already in 1946 de Lange and Visser (2) postulated the occurrence of chemical interaction between silica supports and nickel compounds, with formation of basic silicates, as the reason for difficult reducibility of silica supported nickel catalysts. Many later investigations (3 - 8) substantiated silicate formation. By quantitative x-ray analysis it was shown that more nickel is bound as silicate with lower nickel to silica ratios (9,10).

Longuet (3) showed in 1947 that nickel hydroxide is highly corrosive to glass, so that nickel antigorite is formed from nickel hydroxide and pyrex glass under hydrothermal conditions. Under the conditions of catalyst preparation a similar reaction between silica support and nickel compounds should then be possible. The highly characteristic structure of kieselguhr, a commonly used siliceous support material, is often not found back in electron micrographs of the catalyst. Fig.1 gives striking evidence for this interaction for a catalyst where complete destruction of the guhr did not take place. On the left a structure from the native guhr is shown, on the right the same structure from a catalyst after reduction and extraction

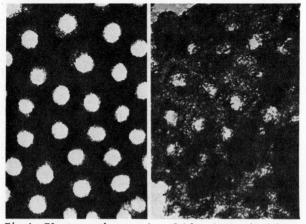

Fig.1. Electron micrographs of kieselguhr (left) and same recovered from reduced catalyst (right)

of nickel metal with carbon monoxide. In the latter picture it can be seen that the silica surface has been extended considerably by the outgrowth of thin silicaplates remaining from the antigorite formed intermediately. The BET surface area was increased by a factor of more than 20.

As shown before (6,9,10) there is a close structural relation between nickel hydroxide and antigorite. The latter can be formed from the former by replacing on one side of the hexagonal nickel layer the hexagonal array of hydroxyl ions by a network of interlinked SiO_4-tetrahedra. Thus a layer of antigorite froms an ideal "glue" between the silica support and an overlay of nickelhydroxide, which fits perfectly on the hydroxide side of the antigorite layer. In the dehydration of the hydroxide prior to reduction a shrink occurs, so that the hydroxide layer breaks up into small oxide blocks. It is logical to assume that these retain epitaxial relation with the supporting antigorite ([111] perpendicular). In the ensuing reduction the epitaxial bonding of the nickel metal crystallites is retained. Dalmai et al (11) proved that in Ni/SiO_2 obtained by reduction of nickel antigorite the supposed [111]-orientation of the nickel indeed preponderates. Due to the smaller Ni-Ni distance in fcc nickel metal this induces strain. Coenen (9,10) found about 0.4% enlarged lattice parameter for 3.5 nm nickel crystallites in Ni/SiO_2-catalysts. Recently Sharma (12) interpreted satellite peaks in ferromagnetic resonance spectra of such catalysts (13,14) as evidence of the same amount of strain in the nickel metal. From a quantitative x-ray analysis of passivated reduced Ni/SiO_2-catalysts it was found that the amount of nickel oxide was proportional to the nickel surface area before passivation, indicating that the oxide found was present on the surface of the nickel crystallites and was formed in passivation (9,10). Thus in the non-passivated reduced catalyst no nickel oxide was present and incomplete reduction was located in the silicate foundation structure. We are thus led to believe that an oxide particle, derived from hydroxide by dehydration, once its reduction is initiated, is reduced completely.

At the high reduction temperatures, generally applied, the resulting system - small nickel crystallites strongly bonded by residual antigorite to the support - will strive towards minimum free energy, in this case a combination of maximum bonding surface and minimum exposed metal surface. Thus the idea of hemi-spherical nickel crystallites was born. Obviously the demands of the fcc-lattice will induce faceting of the spherical surface.

For this model confirming evidence was found. For a series of 22 silica supported nickel catalysts Coenen and Linsen (10) found that the crystallite sizes derived from x-ray line broadening were virtually equal to those calculated from the nickel surface areas with the hemisphere model. Moreover the total surface area expressed per gram SiO_2 of a reduced catalyst decreased, on extraction of nickel metal with CO (cf.fig.1), by an amount equal to half the nickel surface area, as expected for the model.

2.2. Reactivity of nickel in the unreduced catalyst

In view of the manifest importance of silicate formation and its probable relevance also for reduction behaviour we devised a simple test for nickel reactivity in unreduced catalysts. This test comprises extraction of the catalyst with a 0.2 molar solution of EDTA at 55°C. The reaction proved virtually zero order in complexing agent in the concentration range 0.1-0.4 m. For one catalyst of which the reduction kinetics will be discussed later the progress of nickel extraction and of reduction at 410°C are shown in fig. 2. The two curves are surprisingly similar. The close correlation between reactivity towards EDTA and reducibility is further demonstrated in fig. 3 where standard reducibility - degree of reduction attained under standardized conditions - is plotted against a similar extractibility for a series of catalysts comprising different methods of precipitation and supporting silica types.

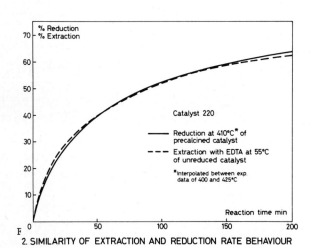

2. SIMILARITY OF EXTRACTION AND REDUCTION RATE BEHAVIOUR

From the data presented in 2.1. it is clear that interaction between nickel compounds and silica with intermediate silicate formation is most certainly involved. The extent of silicate formation depends on

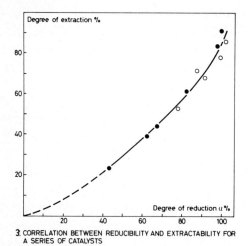

3. CORRELATION BETWEEN REDUCIBILITY AND EXTRACTABILITY FOR A SERIES OF CATALYSTS

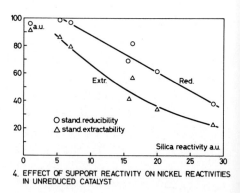

4. EFFECT OF SUPPORT REACTIVITY ON NICKEL REACTIVITIES IN UNREDUCED CATALYST

precipitation conditions and on the reactivity of the silica used as support. The latter can be roughly assessed by a simple standardised solubility test of the silica in alkaline solution. The data in fig.4 for five catalysts precipitated by the same method on different silica supports further demonstrate the importance of chemical interaction with the support. The relevance of these data will be discussed after consideration of the reduction behaviour. So far we can only note that certain features in the unreduced catalyst, which may be chemical -type of nickel compound- or physical -pore structure with attendant mass transport limitation- have closely similar effects in reduction and extraction.

3. STRUCTURE CHANGES DURING DEHYDRATION OF THE UNREDUCED CATALYST

Although this is not immediately evident the crystal lattices of nickel hydroxide, nickel oxide and nickel metal have a rather pronounced interrelation. Nickel hydroxide has a layer structure of hexagonal symmetry consisting of sandwiches of two close packed hydroxyl layers with a similar hexagonal nickel layer in between, perpendicular to the hexagonal c-axis. Viewed along the [111] - axis of the fcc NaCl-type structure of NiO we find regular alternation of hexagonal layers of oxygen and nickel ions. Viewed along the same axis of the fcc nickel lattice we find a succession of hexagonal nickel layers. The dehydration step does not require a second reactant, hydrogen, which the reduction does. Hydroxyl ions and protons are located in close proximity in the hydroxide lattice. It is thus not surprising that the dehydration step is already quite fast at $200^{\circ}C$. From kinetic studies of separate dehydration and of reduction of the calcined catalyst we found the dehydration step to be about five times faster than the reduction. As will be discussed in chapter 4 water acts as an inhibitor for the reduction. We may thus conclude that also when catalyst reduction is performed in a single process the dehydration step will precede the reduction. It will then be useful if we can visualise the structure of a catalyst after decomposition before reduction.

Of unreduced catalyst 58, predried to constant weight at $120^{\circ}C$, we heated samples for 6h at 300, 375, 450 and $550^{\circ}C$ and we measured weight losses and nitrogen adsorption isotherms. From observed weight losses we calculated the shrink in volume due to decomposition of nickel hydroxide to oxide and loss of adsorbed water. These expected volume changes (table 1, line 1) may be compared with changes in pore volume (line 2) derived from the saturation value of the isotherms (p/p_o = 0.96)(line 3). Apart from a discrepancy for the $450^{\circ}C$ sample the agreement is quite good, indicating that in this particular catalyst no significant structural changes occur in the studied temperature range, apart from the direct effect of dehydration.

The nitrogen adsorption isotherms were converted to V_a/t-plots according to De Boer and Lippens (15) and these are shown in fig.5. The isotherm of the dried starting material in this presentation is a close approximation to a straight line up to $p/p_o \sim 0.7$, showing thus only slight deviation from the universal isotherm for oxidic materials (16) up to porewidth of about 2.5nm. The slope indicates a surface area of 239 m^2/g Ni.

The isotherms of the calcined samples all show an initial straight line in the V_a/t-plot, which near relative pressure 0.7 changes to a new slope close to that of the plot for the original sample.

To interpret these observations we will assume that the entire support surface in the unreduced catalyst is coated with a reasonably uniform layer of nickel hydroxide presumably bonded to the silica by a thin silicate layer. By heating this hydroxide layer is decomposed into NiO, which induces a shrink with development of cracks. The sharp break in the V_a/t-plots of fig. 5 for all igni-

Table 1. Catalyst 58, calcined, resp. reduced at four temperatures

Treatment temp. °C			120	300	375	450	550	
Pore volume increase from weight loss				0.19	0.24	0.31	0.33	1
Pore volume increase from isotherm				0.20	0.24	0.23	0.29	2
Pore volume from isotherm			0.53	0.73	0.77	0.76	0.82	3
S_t			239	473	436	405	297	4
S_t' (1.9 nm)	(a)		220	230	260	238	222	5
Nickel oxide	L			3.5	4.3	6.3	10.0	6
	H			1.6	1.6	1.3	1.1	7
	D	$=\sqrt[3]{L^2 H}$		2.7	3.1	3.7	4.8	8
Nickel metal	D	(b)		2.5	2.8	3.4	4.4	9
	D_H	(c)		2.9	3.6	4.2		10
	D_X	(c)		2.7	3.5	4.0		11

Areas in m^2, volumes in ml, weights in g, all per gram total nickel.

a. Isotherm for 300°C also shows 0.8 nm cracks which were also taken into account in calculation
b. From D_{NiO} assuming one oxide crystal yields one metal crystal
c. Detd. on same catalyst directly reduced in H_2 at temp. shown.

ted samples at about t = 0.95 nm indicates cracks of uniform 1.9 nm width: as soon as the thickness of the absorbed nitrogen layer reaches 0.95 nm with com-

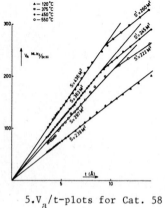

5. V_a/t-plots for Cat. 58 heated to four temperatures

plete filling of the cracks the remaining surface area drops to a value of about 240 m^2/g Ni, about equal to the area of the original dried catalyst. Table 1 also shows the most important data from the nitrogen adsorption isotherms: pore volumes V_p, surface areas S_t from the initial slopes of the V_a/t-plots and S_t' from the slopes above t = 0.95 nm are shown. Assuming that the array of oxide blocks re- - sulting from the dehydration is monodisperse we can calculate from these data and the known total volume of the oxide the dimensions L and H of the oxide blocks. These are shown in lines 6 and 7 of table 1. The oxide crytallite size $D_{NiO} = \sqrt[3]{L^2 H}$ is given in line 8.

With the assumption of one nickel crystallite to result from reduction of one oxide crystallite we then calculated $D_{Ni} = D_{NiO} \cdot \sqrt[3]{\rho_{NiO}/\rho_{Ni}}$ (line 9). These are then compared to the nickel crystallite sizes for the same catalyst directly reduced at the same series of temperatures, as determined from hydrogen adsorption (D_H) and from x-ray line broadening (D_X). (Lines 10 and 11).

In view of the rather drastic simplifying assumptions involved in the computations one should not attach too absolute value to the data obtained. Still the overall picture looks quite reasonable. First we note the close agreement between the nickel crystallite sizes derived from hydrogen adsorption D_H and x-ray line broadening D_X, which further strengthens the model discussed earlier.

Comparing now these values with the value D (line 9) derived from the oxide crystallite size which followed from the isotherm interpretation, we are struck by their close parallelism and even their reasonable absolute agreement. But even the differences can find an acceptable explanation. We feel strongly inclined to conclude that the assumption that one oxide crystallite produces one nickel crystallite, holds good for all nickel oxide derived from nickelhydroxide which is epitaxially attached to the support. However, such hydroxide which because of local excess of precipitating agent remained unsupported or simply occluded in the precipitate, may not sinter too badly in the dehydration step but will sinter drastically upon reduction to the metal (melting point oxide 2163°K, metal 1726°K), especially at the higher reduction temperatures well above the Tamman temperature of the metal). This would explain that the nickel crystallite sizes calculated from the oxide dispersion deviate the more from the actual nickel crystallite size as the reduction temperature is higher, because of an as yet undefined fraction of badly supported material.

4. EXPERIMENTAL OBSERVATIONS OF REDUCTION BEHAVIOUR

4.1. The importance of water vapour

We will discuss the very extensive literature on reduction of both unsupported and supported nickel oxide later. One conclusion from all published data is that water vapour has a strong retarding effect and our data confirm these findings.

We reduced a catalyst at temperatures 350 to 500°C with dry hydrogen and with 20% water vapour, other conditions were standardized. The result is shown in Fig. 6. Especially below 450°C the

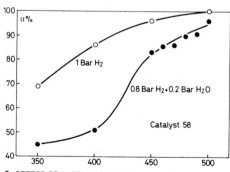

6. EFFECT OF WATER VAPOUR IN REDUCING GAS

inhibiting effect of water vapour is dramatic.

We now recall what we have said before about the importance of silicate formation and the effect of hydrothermal treatment. The effect shown in fig. 6 may at least partly be due to increased silicate formation. To separate the effects we pretreated a catalyst at three temperatures in nitrogen containing 3% water vapour during 1 hour. The effects are shown in fig. 7, as obtained after subsequent reduction in dry hydrogen. Line 1 gives the degree of reduction after a "standard" reduction, line 2 after a prolonged reduction. In both cases we find a lower α than without pretreatment. Also is shown (line 3) the activity for benzene hydrogenation, expressed per unit weight nickel metal. As will appear later, the latter quantity is almost proportional to the specific nickel metal surface area. We must thus conclude that with the pretreatment the reducibility has decreased but the average crystallite size of the reduced metal is also smaller. Both may be ascribed to increased silicate formation.

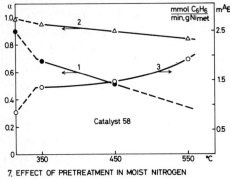

7. EFFECT OF PRETREATMENT IN MOIST NITROGEN

We now turn our attention to normal reductions. Fig. 8 illustrates a curious observation, which again focusses on the importance of water vapour. At three temperatures we did reductions such that the product of hydrogen flow rate and reduction time was constant - 50 m^3/kg Ni. We then find that - except for very short reduction times - the degree of reduction is independent of either time or flow rate. In the decomposition and subsequent reduction a considerable amount of water is set free so that the average water vapour pressure in the reducing gas is always significant. The effect is further illustrated in fig. 9 as a function of total flow and of temperature. We find again the two conflicting effects: costly measures like the use of much hydrogen and high temperatures admittedly produce a higher degree of reduction but at the same time larger nickel crystallites.

8. INDEPENDENCE OF DEGREE OF REDUCTION α OF FLOW RATE AND TIME FOR GIVEN TOTAL VOLUME OF GAS

4.2. The kinetics of reduction

Two experimental set-ups were used for kinetic studies. In a simple apparatus reduction was studied with hydrogen flowing over a bed of catalyst of thickness 0.5 or 1 cm. Parallel experiments were broken off after different times and the catalyst was dissolved in 4 n sulphuric acid after degassing in vacuum at reduction temperature to eliminate adsorbed hydrogen. The degree of reduction was calculated from the volume of hydrogen evolved and the amount of nickel in solution. The procedure was very tedious if a reduction was to be followed over a complete span of reduction degrees. Later a more sophisticated apparatus was built with continuous mass flow and dew point monotoring so that the degree of reduction could be continuously monitored. To eliminate the effect of adsorbed water we used in these experiments catalyst which was pre-ignited at reduction temperature. For a limited number of broken off runs the degree of reduction was measured

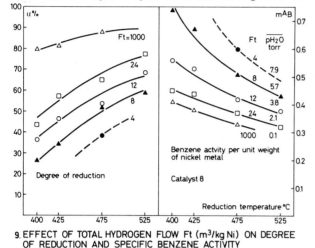

9. EFFECT OF TOTAL HYDROGEN FLOW Ft (m³/kg Ni) ON DEGREE OF REDUCTION AND SPECIFIC BENZENE ACTIVITY

as described to calibrate the apparatus. In either apparatus a large number of runs were done and representative data are shown in fig. 10.

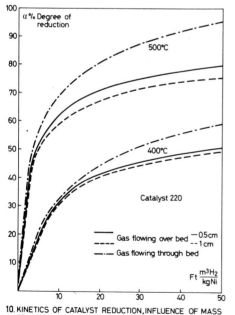

10. KINETICS OF CATALYST REDUCTION, INFLUENCE OF MASS TRANSPORT AND TEMPERATURE

As expected and reported in the literature there is clearly some transport limitation. A thinner bed gives a better result but the difference is not spectacular. The combined effect of pre-ignition, which eliminates about two thirds of the water, and of hydrogen flowing through the bed - impracticable on a technical scale for a powder of a few μm particle size - is more noticeable. We note for later reference that the effect of lower effective water vapour pressure becomes more pronounced at higher degrees of reduction which has also been reported in the literature.

Since conditions are clearly better defined for the pre-ignited catalyst and flow of hydrogen through the bed we will use the upper curves and similar ones for intermediate temperatures for a detailed kinetic analysis.

We have seen that in most silica supported catalysts prior to reduction part of the nickel is bonded as hydroxide/oxide, part as silicate. It is then of interest to see whether the activation energy shifts to higher values as the reduction progresses. From the kinetic plots the times required for the degree of reduction to progress from 0 to 30, from 30 to 40% etc. were read and their logarithms are plotted against reciprocal temperature as inverted Arrhenius plots in fig. 11. The slopes represent E_a/R. The resulting activation energies for the average rates of the successive α-spans do show a small but significant tendency to increase above 40% reduction, as shown in fig. 12. Obviously the apparent activation energy must be a very complex quantity and we will postpone discussion until later. With little success we tried fitting the rate data to numerous equations proposed in the literature for reduction of nickel oxide. Surprisingly a good fit is obtained with a simple third order rate equa-

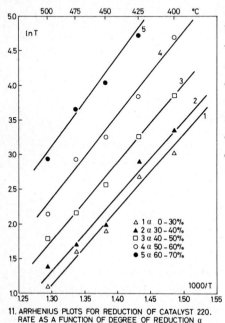

11. ARRHENIUS PLOTS FOR REDUCTION OF CATALYST 220. RATE AS A FUNCTION OF DEGREE OF REDUCTION α

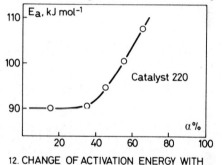

12. CHANGE OF ACTIVATION ENERGY WITH DEGREE OF REDUCTION α

tion: $-(d\beta/dt) = k\beta^3$ in which β is the fraction of unconverted oxide, $\beta = 1-\alpha$. The experimental data are plotted in the integrated form $\beta^{-2} = 1 + 2kt$ in fig. 13. We will come back to this empirical correlation in a later discussion. The reaction is obviously too complex to attach basic significance to this apparently simple rate behaviour. From the slopes of the third order plots an overall activation energy of 101 kJ/mol can be derived in agreement with the data of fig. 12.

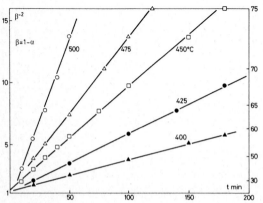

13. THIRD ORDER RATE PLOT FOR REDUCTION OF PRE-IGNITED CATALYST 220

4.3. The result of reduction. The reduced catalyst

The sole purpose of catalyst reduction is the generation of a large nickel surface area and thereby production of a high catalytic activity. The data on nickel surface area presented here were obtained by measuring the hydrogen adsorption at $20°C$ and 1 Bar hydrogen after evacuation of the freshly reduced catalyst at reduction temperature during 2 hours. For reasons explained elsewhere(10) an equilibration time of 16 hrs was allowed for the adsorption, which was then assumed to represent a monolayer of atomically adsorbed hydrogen with 1 H per exposed nickel atom. For the latter an average area of $0.0633 nm^2$ was assumed. Thus $S_{Ni} = 3.41 V_m$ m^2/g Ni, V_m in ml H_2 at STP per gram total nickel.

For catalytic activity assessment we used hydrogenation of benzene in a continuous flow system. Details were described elsewhere (1). Unless otherwise stated hydrogenation was done at $70°C$, benzene partial pressure was 0.1 Bar, total pressure 1 Bar. Results are expressed as A_B, mmol benzene converted per minute and per gram total nickel or as $_mA_B$, expressed per gram metal. Clearly $_mA_B = A_B/\alpha$. We show a few representative results.

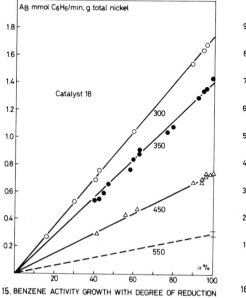

14. DEVELOPMENT OF NICKEL SURFACE AREA AND BENZENE ACTIVITY

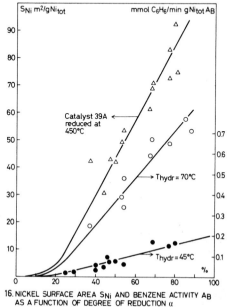

15. BENZENE ACTIVITY GROWTH WITH DEGREE OF REDUCTION

16. NICKEL SURFACE AREA S_{Ni} AND BENZENE ACTIVITY A_B AS A FUNCTION OF DEGREE OF REDUCTION α

In fig. 14 combined results on catalyst 58 are presented. Especially the results on benzene hydrogenation are striking: we find a linear increase of A_B with degree of reduction. This implies that successive fractions of nickel formed have the same benzene activity $_mA_B$ per unit weight of metal. Although the surface areas scatter more the conclusion is similar: a pseudo-linear increase of metal surface with α. Together the data suggest that the nickel formed in reduction is close to mono-disperse and that the crystallite size is larger at higher reduction temperature. The latter conclusion was already indicated by the results of the decomposition study, discussed in chapter 3.

Fig. 15 gives benzene activity data for another catalyst, which confirm the earlier observations. This catalyst appears to have a finer nickel dispersion since it has a value of $_mA_B$ which is 1.54 times larger than that for catalyst 58 at all temperatures. Fig. 16 gives combined data for yet another catalyst. Benzene activities were determined both at 45 and at 70°C. The data confirm the earlier picture with the exception that now the best straight lines both for the surface areas and for the benzene activities have an intercept with the axis. We will postpone detailed discussion of the results to the next chapter.

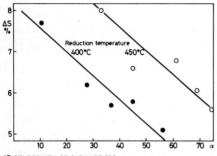

17. SELECTIVITY OF CATALYST 220
Increase in stearic acid content ΔS in standard hydrogenation of soya bean oil.

Next the question may be asked how the metallic nickel produced in partial reduction is distributed in the catalyst particles. An average particle of about 3 nm contains about 10^8 nickel crystallites. Are we to assume that a spherical reaction front progresses into the particle or is reduced material randomly distributed? Hydrogenation of fatty oils is often mass transport limited, which may impair selectivity. This is caused by an accumulation of desired intermediate product in the pore system of the catalyst particle core, which is then hydrogenated to undesired fully saturated product, since more highly unsaturated raw material is depleted in the pore system. This manifests itself in the hydrogenation of e.g. soya bean oil by formation of fully saturated stearic acid too early in the hydrogenation. Test hydrogenations under standardised conditions to a fixed end iodine value (degree of unsaturation) can thus be used to assess selectivity. If now in the progress of catalyst reduction a pseudo-spherical reduction front progresses in each particle –as has been found for much larger nickel oxyde compacts (17-20) – catalysts with low degree of reduction should have a better selectivity than at higher degree of reduction. Fig. 17 shows the result for two temperatures of reduction. High ΔS indicates low selectivity. We find that selectivity improves with degree of reduction. By taking into account that at the higher reduction temperature the crystallite size is larger and thus the nickel surface area per unit mass of nickel metal is smaller we can let the two lines merge into a single one, indicating that selectivity improves at higher nickel surface area. But this is a familiar finding in fat hydrogenation (21): selectivity always improves with use of more catalyst or more active catalyst. We may then conclude that the data do not indicate progressive penetration of the reduction front into the catalyst particles so that the distribution of reduced material in partially reduced catalyst is most probably random.

5. CONCLUSIONS. A TENTATIVE EXPLANATION OF OUR OBSERVATIONS

5.1. Summing up of experimental data

Intermediate silicate formation, with sometimes drastic modification of supporting silica, is well established and appears very important. The silicate is a bonding layer between the support and hydroxide/oxide/metal through an epitaxial relation and by inhibiting sintering it aids formation of small crystals of the oxide and of the metal. On the other hand it makes reduction more difficult.

EDTA extraction shows remarkably parallel behaviour with the progress of reduction, which may indicate that the chemical character of the nickel to be reduced governs both extraction and reduction.

Decomposition to the oxide to a large extent precedes reduction. In a suitable experimental set up we could observe that in reduction with hydrogen the light green unreduced catalyst first turned black (oxide with excess oxygen), then light green (stoichiometric NiO) and finally black (nickel metal).For one catalyst we showed that the nickel containing layer (hydroxide + basic carbonate) breaks up into small oxide crystallites. The latter are larger at higher decomposition temperature.Basic carbonate decomposes to oxide almost as easily as hydroxide (56).

In the reduction water vapour plays an important role. Clearly reduction without water is virtually impossible since it is a product of the reaction. Its partial pressure will go down with increasing α. Some hydrothermal treatment occurs at higher water pressures, resulting in slower reduction, presumably increased silicate formation and smaller ultimate nickel crystallites. Fenomenologically the reduction is about third order in unconverted oxide, the overall activation energy is about 90kJ. mol^{-1} increasing somewhat at higher degrees of reduction.

Both the nickel area and A_B increase pseudo-linearly with α. Generally the S_{Ni}/α-line has an intercept with the α-axis, sometimes also the A_B/α-line. The increase of S_{Ni} with α is steeper at lower reduction temperature, indicating smaller crystallites. There is no indication of a spherical reaction front penetrating into the catalyst particles (~ 10 μm).

5.2. Relevant literature data

The literature on reduction of nickel oxide, as such and supported, is extensive. We will quote some of the more important findings:

Nickel oxide, NiO,is an interesting substance. Unless prepared at very high temperature it contains excess oxygen, which only disappears at about $1000°C$. The excess oxygen charge is compensated by presence of Ni^{3+} ions. The crystal structure is close to cubic, NaCl-type but only above $275°C$ it is really cubic.

Between 250 and 275°C there is a λ-transformation, involving an antiferromagnetic/paramagnetic transition and a discontinuity in specific heat, thermal expansion, lattice type, electrical conductivity and significantly the reduction rate shows a maximum (23-29,50). NiO with excess oxygen is black, the stoichiometric form is yellow-green.

As for many solid state reactions, reduction of nickel oxide requires nucleation, followed by propagation of the metal/oxyde reaction front. As a result oxyde reduction curves are often sigmoid. The induction time t_i depends on pretreatment (high T pre-ignition gives longer t_i (30)) and on reduction conditions (shorter t_i at higher reduction temperature (18,29,30) and higher hydrogen pressure (31)).

Many investigators note the inhibiting effect of water vapour on reduction and most of them indicate that it is especially the nucleation which is inhibited. (22, 37-41). Lattice faults are deemed important for nucleation (34-36) and the relative inactivity of nickel oxide ignited at high temperature is ascribed mainly to slow nucleation.

By various treatments nucleation could be artifically aided: incorporation of metallic nickel, platinum and palladium accelerate reduction by creating nuclei. Gold and silver are completely inactive. Curiously, incorporation of copper is also very effective although metallic copper hardly adsorbs hydrogen (30,31,19,35, 38, 41-46). Thus the rates of nucleation and propagation could be separated (19, 35, 44-46). We will come back later to the quantitative data thus obtained.

The kinetics of reduction were extensively investigated. The systems studied however, were generally very remote from our interest (large oxide particles, macroscopic oxide compacts) and it is thus not surprising that the equations derived and fitted to published data are practically useless for our observations. Mass transport was often found to be partly or entirely rate determining (17-19, 47-49). Observed activation energies vary widely: 42 - 134 kJ.mol^{-1}. Lower values can generally be ascribed to mass transport limitation. (18,19,22,26,30,35,39,53, 55). For reaction front propagation agreement is good: 117-120 kJ.mol^{-1} (30,35,39). For nucleation the spread is much wider, reflecting the importance of the detailed structural perfection of the lattice and its surface: 117-188 kJmol^{-1} (26,30,19, 35, 44-46).

Although the literature on reduction of supported catalysts is also quite voluminous only a limited number of papers gives fundamental information, relevant to our present considerations. The importance of silicates and their difficult reducibility was already referred to (3-10, 54)

Of particular interest are the observations of Delmon et al (58,57) who like Nowak (59) observe acceleration of reduction by incorporation of other metals. Delmon indicates the crucial importance of the high degree of dispersion of supported nickel oxide, which may make effective nucleation more difficult and may inhibit propagation of reaction fronts.

Telipko et al (60) found evidence from their observations in reduction of nickel on kieselguhr for two states of supported nickel varying widely in reducibility. Zapletal et al (61) found strong retardation of reduction by water vapour. They also found a pseudo-linear increase of benzene activity with α and higher A_B for a given α if moisture bearing hydrogen is used for reduction.(see fig. 7 and 9)

5.4. A tentative contribution to understanding

Based on the separate determinations of nucleation and propagation rates by Delmon et al (44), Frety et al (35,45,46) and Yamaguchi et al (19) we concluded that propagation of the reaction front in the small separated oxide crystallites at the usual reduction temperatures is very fast on the time scale of catalyst reduction, so that an oxide crystallite, once nucleated "sparks" into the metallic state. This then should mean that the reduction rate we observe is really the nucleation rate. We verified this idea by plotting the nucleation rates, observed for pure nickel oxide by Delmon et al (44) in an Arrhenius plot together with the nucleation rate calculated for catalyst 220 from the reduction rate at $400°C$ and the estimated nickel oxide surface area, in an Arrhenius plot. The result is shown in fig. 18. We admit that the span of extrapolation is irresponsible, nevertheless the result is encouraging. We recall that nucleation rates per unit surface area may vary by at least an order of magnitude, depending on the perfection of the oxide lattice and the water vapour pressure. A fit within a factor of three may then be lucky chance but it is still inspiring.

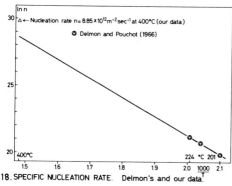

18. SPECIFIC NUCLEATION RATE. Delmon's and our data.

We recall again the picture we developed of silica supported nickel catalysts after dehydration but before reduction: an array of very small oxide crystallites "glued" to the silica support. If -for the moment- we assume the oxide to be mono-disperse, and one oxide crystallite to produce one nickel crystallite then with proceeding reduction degree the nickel surface area will grow linearly, which is close to what we observe.

Thus encouraged we now consider reduction rate behaviour. For a monodisperse system of nickel oxide particles of linear size D, or for a narrow particle size cut from a size distribution, the rate equation is worked out-retaining the assumption that nucleation is fully rate determining - in table 2. At first sight it would appear that the large exposed oxide area would result in fast nucleation, so that our basic assumption that nucleation is rate determining would be invalidated. In fact the nucleation rate is fast but the follow up is slow, because the crystals are small per nucleation event the yield is very small because the reaction front cannot jump to another oxide crystal. The result remains that what we essentially observe is the nucleation rate. From the equations in table 2 we can see that for the simple case of a single particle size D and a nucleaction rate which is constant per unit surface area (n_D) the rate is first order in unconverted oxide. The first order rate constant equals $f.n_D.D^2$ which goes far towards explaining why supported catalysts reduce slowly. The situation is remarkably similar to crystallisation of fats in emulsion, recently discussed by van den Tempel (62) who found that very much more undercooling was required to obtain observable crystallization rates. Returning to rate behaviour we note first order behaviour for our simplified system, whereas experimentally we observe third order in unconverted oxide.

We then must scrutinise our assumptions. It is highly unlikely that the oxide in our catalyst is really mono-disperse. Entropy certainly requires a size distribution. Secondly not all oxide crystallites will be of equal perfection, so that also n_D is unlikely to be constant. For the moment we will ignore such practical considerations as the fact that the water vapour pressure declines in the course of reduction – and water is a nucleation poison – and the possible effect of silicate, which being more difficult to reduce will be "saved" to the last. In table 3 we have set out what happens if we assume a particle size distribution and at the same time assume n_D, the specific nucleation rate, to be some obscure function of D.

For our thought experiment, which empatically is not meant to represent a reality for the catalyst involved, we chose a log-normal crystallite size distribution as shown in fig. 19, which was chosen in such a way that the volume-average crystallite size of the oxide was 2.5 nm, this value admittedly inspired by the nickel surface area development of the catalyst 220, reduced at 400°C. Still leaving n_D constant the effect of the crystallite size distribution, which is cer-

Table 2
Reduction Kinetics of highly dispersed oxide

Initial number of oxide particles $_oN_D$ of size D

Surface area $S_D = N_D f D^2$ f and g
Mass $m_D = N_D \rho g D^3$ shape factors

Nucleation rate n_D nuclei/sec. m^2 of particle size D

Partial nucleation rate = $N_D f D^2 n_D$ nuclei/sec

Assume nucleation rate determining, yield per nucleus $\rho g D^3$

Partial reduction rate = $N_D f D^2 n_D \times \rho g D^3$ kg/sec

Degree of reduction (α) rate for size D

$$\frac{d\alpha_D}{dt} = \frac{N_D f D^2 n_D \rho g D^3}{_oN_D \rho g D^3} = (1-\alpha_D) f D^2 n_D$$

Define $\beta_D = 1-\alpha_D = N_D / _oN_D$

$$-\frac{d\beta_D}{dt} = \beta f D^2 n_D \qquad \beta_D = \exp(-f n_D D^2 t)$$

Thus for monodisperse system and n_D constant decline of β is first order in β

Experimentally we find order in β to be 3 (fig. 13)

Table 3
Effect of Oxide Crystallite Size Distribution on Reduction Kinetics

F_D = number fraction of size D

Instantaneous degree of reduction α_D of size D at time t

$$\alpha_D = 1 - \exp(-f n_D D^2 t)$$

Total α at time t equals

$$\alpha = \frac{\int_o^\infty (1-\exp(-f n_D D^2 t)) F_D^3 g \rho dD}{\int_o^\infty F_D^3 g \rho dD}$$

$$\beta = 1 - \alpha = \frac{\int_o^\infty \exp(-f n_D D^2 t) F_D^3 dD}{\int_o^\infty F_D^3 dD}$$

Result plotted for F_D a log normal distribution, $n_D = kD^{3/2}$ in fig. 20 .B as third order rate plot.

tainly not very wide and may well be realistic, already works in the desired direction. Due to the D^2-dependence of the "rate constant" the largest crystallites will be reduced preferentially. This in turn entails a faster decline of the rate with degree of reduction than for the first order reaction expected for the mono-disperse system. Computer simulation showed that for the chosen distribution the apparent reaction order is close to two in unconverted oxide. An unrealistically wide distribution would have to be chosen to simulate an apparent order of three, as required by the experimental data.

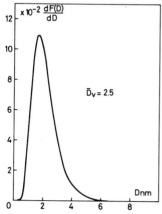

19. LOG NORMAL PARTICLE SIZE DISTRIBUTION (numbers)

We recall now that we quoted many observations which indicate that the nucleation is promoted by lattice faults, impurities, etc. It is logical then to assume that the oxide surface, composed of the surfaces of the oxide crystallites, is heterogeneous with respect to nucleation rate. Obviously we must expect those oxide crystallites which happen to have a high nucleation tendency to disappear faster than those with a greater perfection and purity. The size preference and the nucleation preference may be expected to be interlinked for the following reasons: A larger crystallite has obviously a greater chance of containing an impurity atom to serve as a nucleating center than a small crystallite. Similarly a dislocation or vacancy has a better chance of annealing out in a small crystal. Thus the largest crystallites, which we expected already to be the first to be reduced, have an added reason for doing so because of a higher nucleation chance. We can incorporate this idea in the equations of tables 2 and 3 by assuming n_D to contain the particle size D to a power yet to be defined.

It appears useful at this stage to emphasize again that the computation we will do is less an analysis of rate data but rather an illustrative test of a theory, which aims to explain at the same time the low absolute rate of reduction and the fast decline of the rate with degree of conversion.

The computation comprises the following steps: we define a lognormal particle size distribution F(D) thus that the distribution on complete reduction produces the correct nickel surface area. The volume-average size D_v = 2.5 nm. The distribution is cut in size spans of equal width - quasi-monodisperse - for which the progress of reduction -decline of α - is calculated with first order kinetics. Successive complete computations are done for powers of D in the rate constant ranging from 2 (n_D constant) to 4 ($n_D \sim D^2$). $\alpha = f(t)$ is then calculated by integrating over the distribution according to the last equation in table 3. For each computation (power of D in rate constant) β^{-2} is plotted against t. A linear plot is obtained for $n_D = kD^{3/2}$. Conversion of computer time to real time makes the computed plot coincide with the experimental plot of fig. 13. Fig.20 shows the result. The crystallite volume distribution for the ignited catalyst and 3 stages in reduction is shown in A. There is clear shift to smaller sizes with growing α. The surface heterogeneity implied in the model, $d(n_D S)/dS$ is shown in C. The shape is a normal heterogeneity type and indicates that our approach is related to the Elovich-type reasoning of Levinson et al (63).

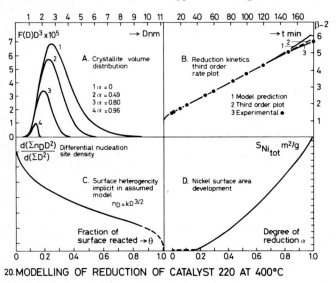

20. MODELLING OF REDUCTION OF CATALYST 220 AT 400°C

We must now introduce two additional complications. Our experiments were started by immersion of the reactor in a preheated salt bath. Heating up is fast but the first minutes are still a blind period: T and zero time are ill-defined. The possibility of badly suppor ted, fast reducing and badly sintering material was already mentioned. These combined problems we solved by arbitrary selection of a point on the reduction curve - 18% - where both are certainly past. The computation was done for the remaining 82 %. The result is shown in B. Up to β^{-2} = 4.5 experimental data, their empirical 3rd order plot and the computed model coincide. The slight irregularity near the β^{-2}-axis + the straight line does not pass through β^{-2}=1 for t=0 - indicates that the reduction up to 18% went too fast though the temperature was too low, indicating some badly supported material. The computed nickel area development finally is shown in D. The smooth curve refers to the well-supported material. For the 18% we put in an arbitrary small area contribution. In this manner we simulated - not necessarily for catalyst 220 - one possible cause for a "toe" in the S_{Ni}/α-curve. The steepening at high α is partly a computation artefact: the assumed continuous size distribution goes to zero size and contains particles of sub-atomic size. A cut-off at a minimum crystallite size might have been used.

An alternative explanation for the "toe" in the S_{Ni}/α-curve lies in the area measurement. At low α the catalyst may well be loaded with water. In the pumping prior to H_2-adsorption this may partially poison the nickel surface, yielding too low adsorption values. This does not explain the more regular A_B-behaviour, which remains surprising.

We did the same computation for 500°C, allowing for the coarser oxide dispersion: D_v=3.0 nm. The fit was equally good, again $D^{3/2}$ in n_D. The remaining T-effect in n_D after allowing for the size effect gave E_a=21 kJ.mol^{-1}. Though the nucleation reaction involved must be complex this may be a true activation energy. As an assumption: nucleation may be equilibrium dissociative adsorption of H_2 on adjacent Ni and O, followed by rate determining H_2O-desorption. In a Langmuir approximation the apparent activation energy $E_a = E_2 - (1-\theta)q_a$ in which E_2 is E_a for desorption of H_2O and q_a is adsorption heat for H_2. For θ small $E_a = E_2 - q_a$, which difference may well be small.

We gave a birds eye view over many part-investigations, often separated by several years, with different workers, apparatus and also catalysts. Transfer of conclusions form one chapter to the next involves risks since different Ni/SiO$_2$-catalysts have much in common but show many differences. We did a last minute verification on catalyst 220. Electron micrographs showed no trace of guhr structure, only wrinkled sheets indicating heavy interaction. Nevertheless x-ray diffraction of the unreduced catalyst showed mainly nickelhydroxide with some silicate. Decomposed at 400°C, quantitative x-ray analysis gave 78% nickel as oxide, 22% as silicate. Mean oxide size 2.2 nm (2.5 assumed in computation). BET-area 1000 m^2/g SiO$_2$ indicates 0.8 nm thick silica sheets, confirming EM-evidence. At α = 73% S_{Ni} was 135 m^2/g Ni, average crystallite size 2.3 nm. The applied model is thus amply confirmed. Cu in Ni 20 ppm, Co + Fe about 1%: in a 1 nm particle 1 impurity atom, in a 5 nm particle 125, which may be relevant for nucleation.

Although with much speculation we hope to have made a case for extreme dispersion to be the main reason for slow reduction of silica supported Ni-catalysts. Regrettably the correlation with EDTA extraction remains unexplained.

6. ACKNOWLEDGEMENT

Thanks are due to Unilever management for permission to publish this work. Many people did the hard work and contributed enormously by their ingenuity and perseverance in meticulous measurement and by provision of constructive ideas. I mention especially W.P. van Beek, J.C.P. Broekhoff, A. de Jonge, Th. J. Osinga F. Pastoor, D. Verzijl and P. van der Vlist. There were many more.

REFERENCES
1 J.W.E. Coenen et al. Proc. 5th Int. Congr. Catalysis, Palm Beach, USA 1972
2 J.J. de Lange and G.H. Visser, Ingenieur 58, 24 (1946)
3 J. Longuet: C.r. 225, 869 (1947)
4 Y. Trambouze and M. Perrin, C.r. 228, 837 (1949)
5 G.C.A. Schuit and L.L. van Reijen: Adv. Catalysis 10, 242
6 J.J.B. van Eyck van Voorthuizen and P. Franzen, Rec. Trav. Chim. 69, 666 (1950)
7 J. Francois-Rosetti and B. Imelik: Bull. soc. chim. Fr. 1957, 1115
8 W.J. Singley and J.T. Carriel: J. Am. Chem. Soc. 75, 778 (1953)
9 J.W.E. Coenen, Thesis Delft 1958
10 J.W.E. Coenen and B.G. Linsen:Phys.and chem.asp. of ads.and cats, Acad.Press (1970)
11 G. Dalmai, C. Leclercq and A, Maubert-Muguet:J.Solid state chem. 16, 129 (1976)
12 V.K. Sharma, Private communication
13 A.A. Andreev and P.W. Selwood: J. Catal. 8, 88 (1967)
14 A.A. Andreev and P.W. Selwood: J. Catal. 8, 375 (1967)
15 B.C. Lippens and J.H. de Boer: J. Catal. 4, 319 (1965)
16 B.G. Linsen in E.A. Flood, ed. The solid-gas interface,M. Dekker,New York 1044 (1967)

17. J. Szekely and J.W. Evans: Metall. Transactions 2, 1699-1710 (1971)
18. J. Szekely, C.I. Lin, H.Y. Sohn: Chem. Eng. Sci 28, 1975-89 (1973)
19. A. Yamaguchi and J. Moriyama: Memoirs Faculty Engineering Kyoto University 28, no 4, 389-403 (1966)
20. K. Kolomasnik, J. Soukoup, V. Zapletal, V. Ruzicka and J. Vacha: Coll. Czech. Chem. Communic. 35, 819-29 (1970)
21. J.W.E. Coenen: Chem. and Ind. in the press
22. H.P. Rooksby: Nature 152, 304 (1943)
23. F. Chiesa and M. Rigaud: Canad. J. of Chem. Eng. 49, 617-20 (1971)
24. A. Roman and B. Delmon: C.r. 269 B, 801-4 (1969)
25. A. Roman and B. Delmon: C.r. 271 B, 77-79 (1970)
26. B. Delmon and A. Roman: J. Chem. Soc.,Farad. Trans 1 (69, pt 5), 941-8 (1973)
27. R. Fréty, L. Tournayan, H. Charcosset: Ann. Chim. 9, 341-55 (1974)
28. M. Foex: Bull Soc. Chim. Fr. 1952, 373-9
29. G. Nury: C.r. 234, 946-8 (1952)
30. B. Delmon: Bull. Soc. Chim. Fr. 1961, 590-7
31. W. Verhoeven and B. Delmon: Bull Soc. Chim. Fr. 1966. 3065-73
32. T. Kurosawa, R. Hasegawa and T. Yagibashi: Transact. Japan. Inst. Metals 13, 265-71 (1972)
33. A.N. Kuznetsov: Russ. J. Phys. Chem. 34, 15-18 (1960)
34. H. Charcosset, G. Dalmai, R. Fréty, C. Leclercq: C.r. 264C, 151-4 (1967)
35. R. Fréty: Ann. Chim. 1969 t4, 453-74
36. Y. Iida and K. Shimada: Bull. Chem. Soc. Japan 33, 1194-6 (1960)
37. J. Bandrowski, C.R. Bickling, K.H. Yang and O.A. Hougen, Chem. Eng. Sci. 17, 379-90 (1962)
38. R. Fréty, H. Charcosset, Y. Trambouze: C.r. Congrès National des Soc. Sav. Section Sciences (1966), 2, 91-101 (1967)
39. H. Charcosset, R. Fréty, Y. Trambouze, M. Prettre: Proc. 6th Int. Symp. React. Solids 171-9 (1968)
40. H. Charcosset, R. Fréty, P. Grange, Y. Trambouze: C.r. 267 C, 1746-8 (1968)
41. H. Charcosset, R. Fréty, A. Soldat, Y. Trambouze: J. Catalysis 22, 204-12 (1971)
42. A. Roman and B. Delmon: C.r. 273 C, 94-7 (1971)
43. R. Fréty, H. Charcosset, P. Turlier, Y. Trambouze: C.r. 264 C, 1451-4 (1967)
44. B. Delmon and M.F. Pouchot: Bull. Soc. Chim. Fr. 1966, 2677-82
45. H. Charcosset, R. Fréty, P. Grange, G. Labbé, A. Soldat and Y. Trambouze: J. Chim. Phys. 68, 49-55 (1971)
46. H. Charcosset, P. Grange, Y. Trambouze: C.r. 273C; 1298-1309 (1971)
47. J. Szekely & J.W. Evans: Metall. Transactions 2, 1691-8 (1971)
48. Y. Yamashine and T. Nagamatsuya, J. Phys. Chem. 70, 3572-5 (1966)
49. T.D. Roy and K.P. Abraham: Phys. Chem. Process. Metall. Richardson Conf. Pap. (1973), 1974, 85-93
50. J.R. Tomlinson, L. Domash,R.G. Hay and C.W. Montgomery, J.A.C.S. 77, 909-10 (1955)
51. K. Nakada, M. Tanaka, S. Aizawa, M. Kasuga: Asahi Garasu Kogyo Gijutsu Shoreikai Kenkyu Hokoku 25, 221-32 (1974)
52. G. Parravano: JACS 1952, 1194-8
53. K. Hauffe & A. Rahmel: Z. Physik. Chem. (Frankfurt) 1954, 104-128
54. G.A. Martin, A. Renouprez, G. Dalmai-Imelik and B. Imelik J. Chim. Phys. 67, 1147-60 (1970)
55. R.H. Tien & E.T. Turkdogan: Metallurgical Transaction 3, 2039-48 (1972)
56. J. François-Rosetti, M. Th. Charton, B. Imelik, Bull.Soc.Chim.Fr. (1957) 614-5
57. A. Roman and B. Delmon: C.r. 273 C, 1310-13 (1971)
58. A. Roman and B. Delmon: J. Catalysis 30, 333-42 (1973)
59. E.J. Nowak: J. Phys. Chem. 73, 3790-4 (1969)
60. V.A. Telipko, G.A. Skorokhod, V.M. Vlasenko and A.G. Fefer: Kinetika i. Kataliz 11, 759-64 (1970)
61. V. Zapletal, K. Kolomasnik, J. Soukup, V. Ruzicka, J. Tikalova: Coll. Czech. Chem. Communic. 37, 2544-8 (1972)
62. M. van den Tempel: Coll. C.N.R.S. 938, Physicochimie des composés amphiphiles Bordeau (1978)
63. G.S. Levinson: A.C.S. Chicago meeting, Div. Petr. Chem. 47-55 (1967)

DISCUSSION

A. BARANSKI : Your description of kinetics of reduction is made from a "solid state" point of view. Do you plan to extend this description taking into account the effect of partial pressures of hydrogen and water. There is literature evidence on the retarding effect of water on the reduction of oxides.

J.W.E. COENEN : It would be very satisfying to do what you suggest. The problem is that the situation is very complex. It appears that low water concentrations already have an effect on nucleation. It will be very difficult to define with sufficient accuracy the prevailing water concentration, which may be different in different parts of the catalyst grain and will change with time. The methods you have further developped in your paper may help to bring the problem closer to a solution.

H. CHARCOSSET : What is your hypothesis concerning the nucleation sites on the surface of the NiO particles, with respect to their reduction by hydrogen ? When studying the reduction of unsupported NiO, we only could establish unambiguously that the corners and edges do not constitute preferential areas for the nucleation of the Ni particles. But we could not determine the exact nature of the nucleation centres i.e., cracks, impurity centres or defects (i.e. Ni^{3+}).

J.W.E. COENEN : I only can speculate on this question. Nucleation is clearly the formation of enough metallic nickel to dissociate hydrogen. Thereafter propagation runs smoothly as your own work has demostrated and -for our small oxide crystallites - very soon complete reduction is reached.
On the surface of each oxide crystal we are looking for sites which are energetically more favourable than the other parts of the surface to provide this spark. Inspired by many literature indications that the nucleation rate is sensitive to the oxide used (its purity and pretreatment) and that black NiO reduces faster than green NiO, I feel that lattice distortion and impurities may well enhance the chance to find this favourable site in the surface of a crystal. The chance of finding such faults should be less in small crystals, which is a justification for our assumption that the surface nucleation rate is size dependent.

R. MONTARNAL : You have presented a strong inhibition of the reduction by water vapor of oxidized nickel supported on silica. On the other hand the drastic increase of sintering rate by water vapor, of nickel metal deposited on carrier, is well known. Do you think that some analogy can exist between the two phenomena, or, on the contrary, that they rather proceed by different mechanisms.

J.W.E. COENEN : The influence of water on the sintering rate of supported metal is outside my experience. We did observe that the presence of water vapour during reduction definitely repels the reduction but at the same time that exposure to water vapour during or prior to reduction at high temperature tends to result in more finely divided nickel. I do not see a direct relation between these two observations.

This effect on the reduction rate may partly be due to poisoning of nucleation and to a shift of the surface equilibria in an unfavourable direction. The hydrothermal treatment may result in the formation of difficultly reducible silicates. This last phenomenon we invoke tentatively to explain that the dispersion of nickel after the "wet" reduction is improved. Alternatively it may be that some of the hydroxide which originally was badly attached to the support, adheres better after the treatment and sinters less on reduction.

J.T. RICHARDSON : Do you have any evidence other than benzene conversion data confirming that higher H_2 flow rates result in larger crystallites ? Could the reduction precompound (oxide or silicate) be a factor ?

J.W.E. COENEN : The paper shows only benzene hydrogenation activity to demonstrate the finer nickel dispersion after wet reduction. Also nickel surface area from hydrogen adsorption shows the same trend. In the paper we propose the idea that the wet conditions at high temperature give a kind of hydrothermal treatment which may well induce some additional silicate formation. One should not take this too strictly, however. The observed effects certainly demonstrate some structural reorganization in the catalyst. This may be additional silicate formation or better attachment to the support of some badly supported hydroxide.

M.V. TWIGG : With regard to the flow rate of hydrogen and the effect of the partial pressure of water on the reduction rate, are the data obtained using an inert carrier gas in quantitative agreement with the results of reductions using various flow rates of hydrogen alone ?

J.W.E. COENEN : I have no definite information on this point. Since in the practical process dilution of hydrogen is unattractive we neglected to consider it in laboratory studies. My impression is that for the reduction mechanism the ratio p_{H_2O}/p_{H_2} governs the situation. Inert gas would not only dilute the water but also the hydrogen and the effect would not be the same as additional hydrogen.

INTERACTION OF NICKEL IONS WITH SILICA SUPPORTS DURING DEPOSITION-PRECIPITATION

L.A.M. HERMANS and J.W. GEUS
Department of Inorganic Chemistry[*], *University of Utrecht, The Netherlands*

ABSTRACT

The conditions to be fulfilled to precipitate catalytically active components exclusively onto the surface of a suspended carrier are dealt with. It is argued that local supersaturations must be kept strongly limited during the precipitation. Precipitation from a solution kept as homogeneous as possible leads to the above deposition onto the support, provided the support interacts sufficiently strongly with the precipitating compound.

The effect of suspended silica on the precipitation of nickel ions was investigated by recording the pH-value and the light-transmission. It was established that the specific surface area of the support, the temperature and the mode of addition of hydroxyl ions affect the reaction of the nickel ions. At 25°C adsorption of nickel ions changes over smoothly into growth of nickel hydroxide that contains an amount of silicate depending on the reactivity of the suspended silica. At 90°C nucleation of nickel hydrosilicate proceeds abruptly and the silica reacts appreciably to nickel hydrosilicate. The elementary reactions revealed are confirmed by investigation of the loaded carriers in the electron microscope and determination of their texture by nitrogen sorption at 77 K.

1. INTRODUCTION

1.1. *Production of supported catalysts*

A high activity of a solid catalyst generally calls for a large specific surface area and, hence, for a finely divided solid. Since most catalytically active materials sinter, however, rapidly at the conditions of the catalytic reaction, the active component has generally to be applied on a highly porous, thermostable support, such as silica or alumina. The support, that is often not active itself, dilutes the catalyst. The support is used most efficiently, when the active material is distributed densely and uniformly over the surface of the support. A rapid transport of reactants and reactions products through the porous catalyst together with a low pressure drop is also a prerequisite for an active solid

[*] *Croesestraat 77A, Utrecht, The Netherlands.*

catalyst (refs. 1-2). As supports of the required porous structure can be produced presently, deposition of the active material in a dense and uniform state must be done without affecting markedly the porous structure of the support.

Carrier materials produced in a separate process are generally loaded with the active component by impregnation and drying or by precipitation of the active material. A dense coverage of the surface of the support by the active component asks for a high degree of loading of the carrier. At a high degree of loading impregnation and drying do not lead to the required uniform distribution of the active material over the support (refs. 1-2). A high degree of loading together with a uniform distribution of the active component can be achieved by starting from a compound in which the constituents of the active component and the support are mixed on an atomic scale. An instance is the ammonia synthesis catalyst, where the catalytically active iron is atomically mixed with aluminium ions. Reduction brings about the formation of the active metal and the alumina support. The atomically dispersion, that was obtained by melting with the ammonia catalyst, can also be attained by coprecipitation (refs. 3-4). Instances are copper-zinc oxide catalysts and magnesium-nickel oxalates. It is, however, not possible to control effectively the porous structure of coprecipitated catalysts. Coprecipitation of the active material and the support is, furthermore, not generally possible. The metal ions in coprecipitated catalysts, finally, can require rather elevated temperatures to be completely reduced. When the reactor cannot be used at these high temperatures, *in situ* reduction of the catalyst is not possible. Elevated reduction temperatures lead, furthermore, to sintering of the metal particles.

Since precipitation of the active material or its precursor onto the surface of the support offers most promise, we will consider the precipitation of solids from a solution more closely.

1.2. *Theoretical Background*

In figure 1 the phase diagram of a solid in equilibrium with its solution is represented. When at a constant temperature the concentration of the homogeneous solution is raised as indicated by (1), the solubility curve is reached first. Crossing the solubility curve does not lead generally to formation of a precipitate, but a metastable state results. Only if the concentration of the solubility curve is exceeded by a critical amount, nuclei of the precipitate are spontaneously generated. Growth of the nuclei brings about the transition to the equilibrium state. The concentrations where nuclei start to develop spontaneously in homogeneous solution, are indicated by the supersolubility curve.

The fact that no precipitate grows when the solubility curve is crossed is due to the considerable surface energy of very small particles of the precipitate. At the concentration of the supersolubility curve the larger decrease in free energy compensates for the surface energy of the nuclei, which removes the nucleation barrier. When the concentration is raised homogeneously according to the path (1) of figure 1, spontaneous nucleation starts as the concentration of the supersolubility curve is touched. A limited number of stable nuclei develop that subsequently grow rapidly. The growth leads to the decrease of the concentration of the solution indicated in figure 1 (ref. 5). When more precipitant is generated homogeneously in the solution the particles of the precipitate grow and the concentration of the solubility curve is approached.

Homogeneously raising the concentration to the value of the supersolubility curve leads to a limited number of nuclei. If the solution is maintained homogeneously, the concentration remains between the solubility and supersolubility curve, where no new nuclei can develop. As a result, precipitation from a homogeneous solution leads to a relatively small number of large particles of the precipitate. Pouring a precipitant into the solution, on the other hand, gives rise to an inhomogeneous solution. The concentration will be locally increased far beyond that of the supersolubility curve, as indicated by path (2) in figure 1. The locally high degree of supersaturation gives rise to formation of a large number of nuclei. Before the concentration has decreased by homogenizing the solution, the nuclei have grown sufficiently to be stable at the lower concentration. Working with an inhomogeneous solution hence results in a relatively large number of small particles (refs. 6-7-8).

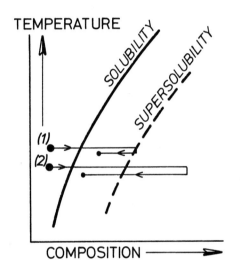

Fig. 1. Phase diagram.

The above shows that pouring a precipitant into a suspension of a porous carrier leads to precipitation of the active material where the precipitant enters the suspension. The precipitate hence will not develop uniformly over the surface of the support. Precipitation from a homogeneous solution proceeds equally in the pores of the support and in the bulk of the solution, it hence can eliminate the difficulties inherent in the inhomogeneous addition of the precipitant. Working with a homogeneous solution, however, generally results in large particles of the precipitate, whereas an active supported catalyst system asks for very small particles

1.3. *Interaction with the Support.*

Since using of a homogeneous solution is not a sufficient condition to precipitate very small particles onto the surface of a suspended carrier, a second condition must be met as well, viz. a sufficiently strong interaction of the precipitating compound with the support. The interaction must decrease the nucleation barrier so far that nucleation at the surface of the support can proceed at a concentration between the solubility and the supersolubility curve. Then the precipitate can nucleate at the surface of the support, whereas nucleation in the bulk of the solution is prevented. Nucleation at the surface must, furthermore, proceed rather rapidly, to avoid growth of a small number of nuclei to large particles of the precipitate adhering to the support. The above has demonstrated that the interaction with the support cannot be too weak if a dense uniform distribution of the precipitate over the support is to be obtained. A strong interaction, on the other hand, can lead to reaction of a substantial fraction of the finely divided support to a compound with the precipitating metal ions. Reaction of the support brings, consequently, about the adverse effects of coprecipitated catalysts.

Interaction confined to the surface of the support and reaction of the support can be studied very well with nickel ions precipitating in the presence of suspended silica. Geus has shown that the hydrolysis of urea can be used to precipitate nickel ions from a homogeneous solution onto silica (ref. 9). The resulting catalysts displayed the desired dense uniform distribution of nickel particles and thus a high thermostability. This author also precipitated nickel in the presence of suspended silica by injection of a hydroxyl ions containing liquid below the level of the vigorously agitated suspension (ref. 9). The injection method can be used at room temperature and at more elevated temperatures, whereas the hydrolysis of urea proceeds with a marked rate only above about 70°C. To study the effect of the temperature we investigated injections into silica suspensions kept at 25°C and at 90°C. Nickel hydroxide dissolves into an excess of ammonia owing to complex formation. Ammonia was therefore used as the precipitant to be injected. As nucleation can proceed in a limited concentration range only, nucleation at the tip of the injection tube will be minimized. The precipitation vessel was, moreover, specially designed so as to avoid local concentration differences as much as possible. To evaluate the degree of homogeneity obtained by the injection method, we compared injection into a suspension kept at 90°C with hydrolysis of urea at 90°C.

Continuous measurement of the pH-value and the light-transmission of the suspension was used to investigate the precipitation process. The interaction with the suspended silica was established by comparing curves recorded with a solution of nickel nitrate alone, with suspended silica alone, and with a suspension of silica in nickel nitrate. The light-transmission indicated formation of a precipitate (nickel nitrate alone) or coagulation of the suspended silica. Comparison of curves recorded on dosewise addition of hydroxyl ions with curves measured during continuous addition of hydroxyl ions yielded information on the rate of consumption of hydroxyl ions and, hence, on the rate of nucleation and growth of precipitates.

As dealt with above the nuclei grow as more hydroxyl ions are made available and, consequently, become more stable. Ageing of the fresh precipitate can, furthermore, raise the stability of the precipitate by rearrangement of the ions in the precipitate and by ion exchange with the solution and with the support. We therefore measured the pH-calues at which precipitates were rapidly established as well as the pH-values after ageing precipitates for 24 or 48 hours in the mother liquor.

Attack of the silica support by precipitating nickel ions can be expected to depend on the temperature and on the particle size of the silica. We therefore investigated silica of a surface area of 200 m^2 per g and of 380 m^2 per g. Substantial reaction of the silica was indicated by the pH-curves. To confirm the interpretation of the pH-curves we investigated the loaded carriers in the electron microscope and by measurement of the nitrogen sorption at 77 K.

2. EXPERIMENTAL

2.1. *Apparatus*

Both injection of ammonia and hydrolysis of urea are used to increase homogeneously the hydroxyl ion concentration of solutions and suspensions. For an extensive description of the apparatus, we refer to (ref. 10) and to (ref. 11).

2.2. *Materials*

As a silica support Aerosil (Degussa, West Germany) grades 200 V and 380 V were used. The surface areas were determined as 200 and 380 m^2/g respectively.

2.3. *Procedures*

2.3.1. Dosewise addition of dilute sodium hydroxide

The titration experiments with dilute sodium hydroxide were done in the small vessel filled with a suspension containing 7.6 g Aerosil 380 V in a solution of 37 g $Ni(NO_3)_2 \cdot 6H_2O$ p.a. per 1 water. After adjustment of the pH-value at a value between 2 and 3 by addition of a few drops of nitric acid and establishment of the required temperature, small doses of sodium hydroxide were injected by means of the Autoburet. After addition of a dose the next dose was added when the pH-value had remained constant for at least 15 min. Nitrogen was bubbled through the suspension to prevent interference by carbon dioxide from the air.

2.3.2. Measurement of pH-values of aged precipitates.

These experiments were also done in the small vessel. The precipitate was obtained from a solution of 37 g of $Ni(NO_3)_2 \cdot 6H_2O$ per 1 water. The suspended amount of silica varied from 0.76 to 7.6 g per 1 of the solution depending on the silica-to-nickel ratio to be established in the precipitate. After adjustment of the temperature and addition of a few drops of nitric acid, the precipitation was done by injection of 0.5 N ammonia at a flow-rate of 2.8 ml/min. within 30 min.

To increase the rate of ageing a small amount of nitric acid was injected subsequently. The ageing time used was 24 h and 48 h at 90°C and at 25°C, respectively.

2.3.3. Measurement of pH-values during continuous precipitation.

The small vessel filled with the suspension containing 7.6 g aerosil 380 V in a solution of 37 g of $Ni(NO_3)_2 \cdot 6H_2O$ per l water was used. After adjustment of the temperature and addition of a few drops of nitric acid, the Autoburet was used to inject 0.5 N ammonia at a flow-rate of 2.8 ml/min. In the experiments with urea, 23 g urea per l of the suspension was added. The hydrolysis of urea was carried out at 90°C.

In all the above experiments the Vitatron UC 200 S colorimeter was used to record continuously the transmission of light of a wavelength of about 780 nm.

2.3.4. Preparation of samples for electron microscopy and nitrogen sorption.

These samples were prepared in the large vessel. After addition of 8.5 g aerosil per liter and/or 41 g $Ni(NO_3)_2 \cdot 6H_2O$ p.a. per liter a few drops of nitric acid were added and the temperature was adjusted. Next 26 g urea per liter was added or 1 N ammonia was injected at a rate of 1.4 ml per minute. When after about 20 hours the pH had increased to about 8 at 25°C and to about 6.5 at 90°C the suspension was filtered. In the presence of silica the precipitation of nickel ions was quantitativ as was evident from the clear filtrate. The residu was subsequently washed two times with distilled water to remove nitrate, whereafter drying was performed at the preparation temperature (in vacuo at 25°C and in air at 90°C).

2.3.5. Electron microscopy and nitrogen sorption.

Transmission electron microscopy was performed on ultra thin sections of the dried precipitates embedded in methylmetacrylate (ref. 12). Nitrogen sorption at 77 K was measured on samples outgassed at 120°C in vacuum with a Carlo Erba Sorptomatic type 1810 equipped with an accurate ($\pm$ 0.5 torr) pressure transducer. The t-plot was used to analyze and interpret complete adsorption-desorption isotherms and the pore-size distribution was calculated according to the method of Broekhoff (ref. 13).

3. RESULTS AND DISCUSSION

3.1. *Dosewise addition of dilute sodiumhydroxide.*

Titration curves, in which the pH-value is plotted as a function of the amount of alkali added, can provide conclusive information about the interaction of precipitating ions with the surface of suspended silica (refs. 14-15). In a first series of experiments a sodium hydroxide solution was injected dosewise into solutions or suspensions kept at 25°C or at 90°C. Measurements were done with pure water, a nickelnitrate solution and with suspensions of silica in pure water and in a nickel nitrate solution. With water and with silica suspended in water the pH-value assumed rapidly an equilibrium value. When nickel ions were present, the time needed to establish a constant pH-value depended on the total amount of sodium hydroxide added. As long as precipitation of nickel ions or coagulation of silica (ref. 16) was not apparent, a constant pH-value was observed rapidly, but the time to reach a constant value rose steeply after the onset of precipitation or coagulation. As a consequence the observed constant pH-values, recorded after the onset of precipitation or coagulation, do not correspond to real equilibrium (compare 2.3.1). Essential for measurement of these phenomena is the method of addition of sodium hydroxide described above, which prevents as much as possible, a temporary, locally high supersaturation. Inhomogeneous addition of sodium hydroxide leads to premature formation of precipitate. Redissolution of the precipitate, according to the equilibrium at the final pH-value of the homogenized solution, proceeds very slowly at pH-values near to that corresponding to the composition of the solubility curve.

Figure 2 shows the (stable according to 2.3.1) pH-values recorded after dosewise addition of sodium hydroxide at 25° (left) and at 90°C (right). At 25°C the silica suspension in water displayed a less rapid rise in pH-value than pure water, which is due to reaction of hydroxyl ions with hydrogen ions set free by dissociation of surface hydroxyl groups:

$$\equiv Si - OH + H_2O \rightleftarrows$$
$$\rightleftarrows \equiv SiO^- + H_3O^+$$

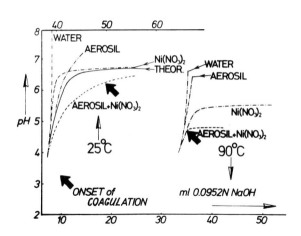

Fig. 2. Titration curves.

The precipitation of nickel hydroxide was at 25°C observed to start at a pH-value of 6.7 in the absence of silica. The reaction of hydroxyl ions to nickel hydroxide caused the pH-value to increase only slightly on further addition of sodium hydroxide. The onset of precipitation being displayed at a pH-value of 6.7 agrees well with the solubility product of nickel hydroxide, which was found by Feitknecht (ref. 17) to lie between $10^{-14.7}$ and $10^{-17.2}$ at room temperature. It can be seen from figure 2 that before the onset of precipitation the pH-value of the nickel nitrate solution increased markedly less rapidly than that of pure water. According to Burkov (ref. 18) this is due to the formation of (polynuclear) hydroxyl complexes. If addition of hydroxyl ions does not lead to interaction of nickel ions with the suspended silica, a pH-curve would be measured that corresponds to the added amounts of hydroxyl ions consumed, with respect to pure water, by nickel ions and silica separately. This calculated pH-curve has been included in the curves measured at 25°C. Figure 2 shows that the curve recorded at 25°C with a silica suspension in nickel nitrate solution lies appreciably below this theoretical curve. The silica consequently strongly reacted with the nickel ions. The pH-curves indicate that nickel ions were adsorbed markedly by the silica from a pH-value of about 4.2 onward. The amount of (hydrolized) nickel ions, exchanged for protons from the surface hydroxyl groups, increased smoothly with the pH-value. At a pH-level of about 6.3 adsorption of nickel ions onto the silica had decreased the electrostatic charge so far, that a slight coagulation of the silica was evident from the decrease in the light transmission. At 25°C adsorption of nickel ions on the silica surface smoothly changes over into precipitation on the surface. Nucleation and growth of hydroxide at the silica surface brings about a drastic coagulation of the silica, as was evident from a strong decrease in transmission. (compare ref. 16, figure 6). The hydroxide precipitated onto the silica surface is less soluble than pure nickel hydroxide as is evident from the lower pH-value where precipitation onto silica starts. The interaction of (hydrolyzed) nickel ions and silica is completely different at 90°C. Without silica the onset of precipitation is found at a pH-value of 5.5. The shift from 6.7 at 25° to 5.5 at 90°C is mainly due to the change in the dissociation constant of water from 10^{-14} at 25°C to about $10^{-12.4}$ at 90°C. The pronounced difference with the curve recorded with pure water also before the start of the precipitation points to a tendency to form (polynuclear) hydroxyl complexes which is larger at 90° than at 25°C (ref. 19). The pH-curve measured at 90°C with a suspension of silica in a nickel nitrate solution coincides with the curve recorded for nickel nitrate alone, to branch abruptly at a pH-value of 4.8, where the light transmission indicated the onset of coagulation. The 90°C curves of figure 2 therefore show that nickel ions precipitated onto the silica surface without a previous adsorption of nickel ions onto the silica, in contrast to the behaviour at 25°C. That the nickel ions interacting at 90°C with silica are much less soluble than pure nickel hydroxide can be concluded from the pH-value at which precipitation with the silica proceeded, viz. 4.8. This value lies appreciably below that of 5.8 at which pure nickel hydroxide started to precipitate at 90°C.

3.2. *Aged precipitates.*

If (hydrolized) nickel ions are adsorbed preferentially on the silica surface, a monolayer of nickel ions should be taken up before precipitation of nickel hydroxide onto the layer of adsorbed hydrolized nickel ions can proceed. Above we established that adsorbed hydrolized nickel ions are less soluble than those in bulk nickel hydroxide. Together with the known surface area of the silica used in this work, this enabled us to demonstrate the preferential adsorption and precipitation of a monolayer of nickel ions experimentally. To this end we used the pH equilibration measurements that also showed the reason of the different behaviour at 90°C. As dealt with above, the equilibrium pH-value was measured of suspensions of relatively rapidly precipitated nickel hydroxide. The hydroxide was precipitated both in the presence and in the absence of silica. Two silica grades of a different specific surface area were used. During ageing of the precipitates the pH-value can change owing to: (i) exchange of occluded and/or adsorbed nitrate ions for hydroxyl ions (refs. 20-21), (ii) Ostwald ripening which leads to larger particles of lower solubility (ref. 5) and (iii) increase in the silicate content of solid species precipitated in the presence of silica. Reaction with the silica proceeded especially at 90°C (ref. 22); it resulted in a less soluble solid and thus in a lower equilibrium pH-value.

The results are represented in figure 3, where the equilibrium pH-value is plotted as a function of the silica-to-nickel ratio. We have calculated the silica-to-nickel ratios corresponding to a monolayer of nickel ions using the nickel spacings of nickel hydroxide and nickel hydrosilicate (refs. 23-24). For the two grades of silica used the monolayer ratios are indicated in figure 3. The nickel hydroxides precipitated without silica ($SiO_2/Ni = 0$) showed equilibrium pH-values of 7 at 25°C and of 4.8 at 90°C. The pH-value at 25°C being slightly higher than in fig. 2 is due to the fact that in the experiments of figure 3 ammonia was used as a precipitant. At 25°C an appreciable concentration of ammonium ions increased the solubility of nickel ions owing to complex formation. Ageing of the precipitate at 90°C, where the ammonium concentration is low at a pH-level of about 5, led to the equilibrium pH-value being markedly lower than that of the freshly nucleating nickel hydroxide of figure 2.

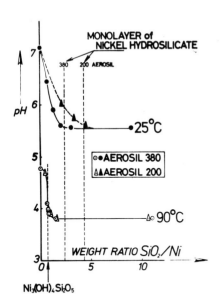

Fig. 3. pH-values for aged precipitates.

The curves in figure 3 measured at 25°C nicely show that after adsorption of a monolayer of nickel ions the equilibrium pH-value gradually shifted to the pH-value of pure nickel hydroxide. With the silica Aerosil 200, that has a smaller surface area, the shift started at a higher silica-to-nickel ratio than with the aerosil 380, the surface area of which is larger. The gradual increase in the solubility of nickel ions with a decreasing silica-to-nickel ratio demonstrates that the interaction with the silica surface continuously decreases. As was indicated already in figure 2 hydrolized nickel ions appear from figure 3 to react completely differently with silica at 90°C. When the amount of nickel corresponding to monolayers was exceeded no different equilibrium pH-values were exhibited. The pH-value measured with the different silica grades, furthermore, coincided. The equilibrium pH-value started to rise only when the silica-to-nickel ratio of nickel antigorite (ref. 24) was approached. This ratio is also indicated in figure 3. The results of figure 3 thus demonstrate that at 90°C the reaction between hydrolized nickel ions and silica is not confined to the silica surface. The small silica particles react to nickel hydrosilicate to a considerably extent. The different grades of silica investigated did not contain silica particles of a size sufficiently differe to lead to a markedly different extent of reaction.

3.3. *Measurements of pH during continuous precipitation.*

Nickel-on-silica catalysts are produced by deposition-precipitation by making available continuously hydroxyl ions. The next three figures therefore show the change in pH-value during continuous injection or generation of hydroxyl ions. In figure 4 curves measured at 25°C are represented. The onset of precipitation was established from the light transmission. As said above a smooth transition from adsorption of nickel ions to growth of nickel hydroxide proceeds at 25°C. This is apparent from the pH-curve of figure 4 recorded with suspended silica. The curve does not display a discontinuity indicating the onset of precipitation. When nickel hydroxide is precipitated without silica being present, an overshoot of the pH-value can be expected. After the pH-value of the supersolubility curve has been attained, a sudden nucleation and growth can proceed. The rate of accommodation of hydroxyl ions into the growing hydroxide generally exceeds the rate at which the hydroxyl ions are being made available. Nickel hydroxide, however, exhibits a low rate of growth (refs. 20, 25) presumably owing to its layer-like lattice structure. As a result the growth rate remained limited in the experiment of figure 4 and the pH-values continued to rise steadily. That an appreciable supersaturation is required to get a marked nucleation and growth rate is evident from the pH-value of about 7.4 where precipitation was established, whereas the equilibrium pH-value of nickel hydroxide precipitated with ammonia is only 7.0.

Figure 5 shows the pH-curves measured on injection of ammonia at 90°C. The curve recorded with nickel nitrate alone again does not exhibit a maximum. The pH-level at which the precipitation proceeded, was again considerably higher than that observed with equilibrated pure hydroxide at 90°C, viz. 6.8 and 4.8, respectively. The interaction of nickel ions with silica being at 90°C strongly different from

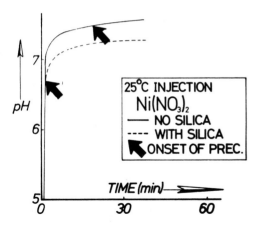
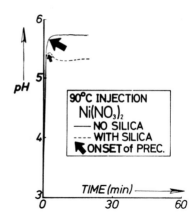

Fig. 4. pH-curves during precipitation   Fig. 5. pH-curves during precipitation

that at 25°C is evident from figure 4 and 5 too. The pH-curve recorded at 90°C with suspended silica displays a maximum, which indicates a nucleation barrier. The data of figure 3 indicated that freshly precipitated nickel hydrosilicate exhibits an equilibrium pH-value of 4.8. When continuous addition of hydroxyl ions had raised the pH-value up to about 5.4, rapid nucleation started all over the large surface area of the silica. The high rate of take up of hydroxyl ions brought about by the sudden nucleation caused the drop in the pH-value, after which the growth of the nickel hydrosilicate proceeded at a lower pH-level.

It is interesting to compare the injection with the completely homogeneous generation of the precipitant. In figure 6 pH-curves are represented recorded during the hydrolysis of urea. The pH-levels at which nickel ions precipitated in the absence of silica (6.8) are equal within the experimental error. The slightly less homogeneous distribution during injection brought about, however, that hydroxide crystallites were discerned in the light transmission at a lower pH-value than with the hydrolysis of urea. With suspended silica the pH-curve obtained with the reaction of urea does not differ markedly from that of figure 5. The decomposition of urea leads also to formation of carbonate ions. Owing to this the solid precipitated in the absence of silica contained about 13 wt. % of carbon dioxide. In contrast to the precipitate obtained without suspended silica, the nickel compound precipitated with silica only contained 1 wt. % of carbon dioxide. This again reflects the large interaction between silica and the precipitating nickel ions; reaction with silicate ions prevents the take up of carbonate ions in the precipitating solid.

3.4. *Electron microscopy and nitrogen sorption.*

The above pH-measurements indicated that at 25°C nickel ions are first adsorbed onto the silica surface to precipitate on further addition of hydroxyl ions as a nickel hydroxide layer, bound to surface silicate ions. At 90°C, on the other hand, nucleation of nickel hydrosilicate sets in abruptly when the pH-value is continuously raised. The reaction appeared to be not confined to the surface of the silica but a considerable fraction of the silica particles reacted to hydrosilicate. The aerosil silica, used in this work, consists of conglomerates of very small particles (about 15 A) that are very tightly clustered together to non-porous larger particles. The conglomerates and the non-porous particles, which are generally intimately connected, have a globular appearance (ref. 26). Reaction of a substantial fraction of the silica must be clearly apparent in electron micrographs; the globular clusters must be changed into nickel hydrosilicate platelets (ref. 23). We therefore investigated four different preparations, all having a silica-to-nickel ratio of 1.0, in the electron microscope.

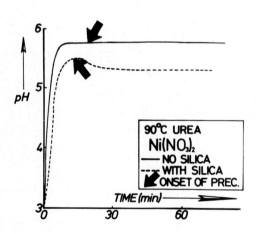

Fig. 6. pH-curves during precipitation

The above picture is substantiated by figure 7a which represents a micrograph of aerosil 380 loaded with nickel ions by injection of ammonia at 90°C. It can be seen that the original globules have completely disappeared. The silica has reacted to relatively well crystallized flat and bended (ref. 23) platelets. Precipitation by hydrolysis of urea leads to slightly less well crystallized platelets as can be seen in figure 7b. This figure shows a micrograph of aerosil 380 (SiO_2/Ni = 1.0) loaded by hydrolysis of urea at 90°C. The better crystallization resulting from injection of ammonia is presumably due to the fact that near the injection point the pH-value was slightly higher during the addition of ammonia, which led to a more easy reaction of the silica (ref. 27).

At 25°C the extent of reaction of the silica should be appreciably smaller. Figure 7c shows a micrograph of aerosil 380 loaded by injection of ammonia at 25°C. Again platelets, but now being much thinner, can be seen. The platelets are situated at a greyish background that reflects the original silica structure. With aerosil

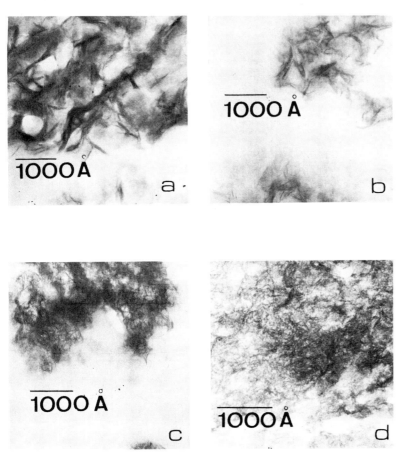

Fig. 7: electron micrographs. a: injection 90°C ; b: urea 90°C ; c: injection 25°C, 380 V ; d: injection 25°C, 200 V.

the platelets are less well developed as can be seen in figure 7d that shows a micrograph of aerosil 200 onto which nickel ions had been precipitated by injection of ammonia at 25°C. It is difficult to establish whether the platelets present in the preparations of figure 7c and 7d consist of pure nickel hydroxide or contain an appreciable fraction of hydrosilicate. Nickel hydroxide crystallizes with the brucite structure that contains flat layers of nickel ions separated by two layers of hydroxyl ions (ref. 24). Though the flat layers can be slightly curved, they are not likely to follow the locally strong curvatures of the silica globules. The platelets might therefore consist of pure nickel hydroxide. Since the globules of aerosil 380 have surfaces more strongly curved than those of aerosil 200, nickel hydroxide platelets can adhere better to the globules of aerosil 200.

The data of figure 3 show, however, that the equilibrium pH-value of pure nickel hydroxide is not exhibited at a silica-to-nickel ratio of unity. Platelets of pure nickel hydroxide protruding from the silica particles are hence unlikely. The platelets must therefore contain some silica, though much less than the nickel species precipitated at 90°C. Owing to the larger curvature of the particles of aerosil 380 the silica will be more reactive. The platelets will therefore contain more silica than the nickel precipitate obtained with aerosil 200 and will protrude further from the silica particles. Preparations in which the silica has been loaded by deposition precipitation at 25°C nevertheless still retain a large fraction of the silica unreacted. This can be inferred from infrared experiments carried out by Joziasse (ref. 28), who observed that silica samples loaded at 25°C still displayed the characteristic silica lattice vibrations. Silica loaded at 90°C did not exhibit silica lattice vibrations. When the sample was reduced the nickel ions migrated out of the hydrosilicate to form small nickel particles adhering to the remaining silica platelets (ref. 10). The removal of the nickel ions led to the reappearance of the silica lattice vibrations that remained after reoxidation of the metallic nickel particles to nickel oxide.

Though electron micrographs can provide very detailed information, electron-microscopic results must be corroborated by results obtained on macroscopic samples. To this end we used the adsorption and capillary condensation of nitrogen at 77 K. The experimental data were processed according to Broekhoff (ref. 13). The resulting data are very well suited to confirm the conclusions drawn from the above pH-measurements and investigations in the electronmicroscope.

TABLE 1

Adsorption and Capillary Condensation of Nitrogen at 77 A.
Silica-to-nickel ratio 1.0

sample	S_{BET} m^2/gcat	S_{BET} m^2/gSiO	S_t m^2/gcat	V_m ml/gcat	V_p ml/gcat	S_{cum} m^2/gcat	S_m m^2/gcat	$\bar{d}_m$ A
aerosil380V	-	380	-	-	-	-	-	-
I42(25)	522	1305	431	0.043	1.515	103	328	18
I42(90)	198	495	121	0.043	0.845	56	65	21
U42	256	663	201	0.029	1.294	185	-	-
aerosil200V	-	200	-	-	-	-	-	-
I22(25)	323	808	242	0.043	0.180	-	242	15
I22(90)	264	660	190	0.037	1.071	94	96	26
U22	245	613	186	0.027	1.155	111	75	33

I injection of ammonia; U hydrolysis of urea at 90°C.
42 aerosil 380; 22 aerosil 200
(25) 25°C ; (90) 90°C injection temperature.
Samples previously evacuated for 16 h at 120°C to 10^{-5} torr.

In table 1 the results have been collected. The meaning of the symbols is S_{BET} - BET surface area; S_t - surface area as obtained from t-plot (ref. 13); V_m - volume of micropores; V_p - pore volume; S_{cum} - surface area obtained by addition of surface area of pores with width larger than about 30 A; S_m - surface area of micropores ($S_m = S_t - S_{cum}$); $\bar{d}_m$ - mean width of micropores as calculated from S_m and V_m.

Aerosil is a loose packing of silica globules that has thus no fixed pore structure (ref. 26). As has to be expected from the platelets apparent in the electron micrographs the hysteresis loops and pore distribution calculations indicated the presence of slit-shaped pores in all the samples.

In the micrographs it can be seen that the platelets in the samples are generally stacked so as to enclose very narrow pores. These pores are filled with nitrogen at very low relative pressures. This causes the BET-surface areas to be appreciably larger than the surface areas calculated from the t-plots. The overall picture from the micrographs is nicely confirmed. Precipitation onto aerosil by injection of ammonia was observed to lead to relatively thick platelets; as a result the surface area of I42(90) is relatively small. The smaller reactivity of silica by the absence of locally high pH-values (ref. 27) in the hydrolysis of urea, which leads to thinner platelets, is apparent from the larger surface area of U42. The variation in the thickness of the platelets as evident from the surface areas is, however, limited. The specific surface area of extended platelets is proportional to the reciprocal thickness of the platelets. As the surface areas differ by a factor smaller than two, the thickness of the platelets is not likely to vary appreciably more than a factor of two. This indicates that the platelets are better stacked in clusters with a mutually parallel orientation, which is likely to be due to the more rapid dissolution of the silica. The formation of better stacked platelets with injection at 90°C is also reflected from the pore volume of I42(90) being markedly smaller than that of U42. Figure 3 showed that at 90°C the silica almost completely reacts and looses its original structure. This can also be concluded from the fact that both aerosil 200 and aerosil 380 being loaded at 90°C display surface areas of the same order of magnitude. The same trend is shown by the pore volumes. With the more reactive aerosil 380, injection leads to better crystallized platelets, as said above. As can be expected with clusters of platelets an appreciable fraction of the surface area is present in pores of a width smaller than about 30 A.

The conclusions drawn from the micrographs of catalysts loaded at 25°C can also be nicely confirmed by the data of Table 1. The structure of aerosil 200 was largely retained, while a number of thin platelets had grown out of the silica particles. The loose structure of the original aerosil was lost by the coagulation mentioned above. The surface area calculated per g of silica is much larger than that of the original silica owing to the platelets protruding from the silica particles. The platelets are enclosing mainly micropores as evident from the relatively small value of S_t. The micrograph of aerosil 380 onto which nickel ions had been precipitated at 25°C showed very thin platelets that protrude considerably further from the silica particles. The silica globules of the aerosil are kept at larger distances by the larger protruding platelets. This can be seen from the much larger meso-pore volume, while the thin platelets give rise to the large surface area. Since the platelets are rather well separated, they enclose an appreciable surface area in slits more than 30 A in width.

The above findings with respect to the formation of nickel hydrosilica during deposition precipitation and/or the ageing of the precipitates, agree very well with the results published by Signalov (refs.29-30) on the interaction of nickel ions with silica gels.

4. CONCLUSION

This work has demonstrated, we believe, that combining careful characterization of the reaction products with procedures in which deposition-precipitation is carried out under accurately controlled conditions can explain the structure of the resulting catalysts very satisfactorily. Since precipitation from homogeneous solution can be scaled up very easily, catalysts having desired properties can be produced. This work showed, however, also that the elementary processes proceeding during the precipitation of the active components onto the support must be studied in detail. This holds also for the thermal pretreatment of solid catalysts.

REFERENCES

1 G.K.Boreskov in Preparation of Catalysts (B.Delmon, ed.), Elsevier, Amsterdam, 1976, pp.223-50.
2 J. Cervello, E.Hermana, J.Jiminéz, and F.Melo in Preparation of Catalysts (B.Delmon, ed.), Elsevier, Amsterdam, 1976, pp.251-263.
3 K. Morikawa, T.Shirasaki, and M. Okada, Adv. Catal., 20(1969) 98.
4 P. Courty, and C. Marcilly in Preparation of Catalysts (B.Delmon, ed.), Elsevier, Amsterdam, 1976, pp.119-45.
5 A.G. Walton in Dispersion of Powders in Liquids (G.D. Parfitt, ed.),Elsevier, Amsterdam, 1969, pp.122-64.
6 D. Mealor, and A. Towhnshend, Talanta, 13(1969)1069-74.
7 P.F.S. Cartwright, E.J. Newman, and D.W. Wilson, The Analyst, 92 (1967)663-79.
8 W.J.Blaedel, and V.W. Meloche, Elementary Quantitative Analyses, 2nd edn., Harper and Row, New York, 1963, pp.177-84, pp.719-29.
9 J.W. Geus, Dutch Patent Application, 1967, 6705, 259, and Dutch Patent Application, 1968, 6813, 236.
10 A.J. van Dillen, J.W. Geus, L.A.M. Hermans, and J. van der Meijden, Preprints Proceedings Sixth International Congres on Catalysis, London, 1976,B7.
11 A.C. Vermeulen, J.W. Geus, R.J. Stol, and P.L. de Bruyn, J. Coll. Interf.Sci., 51(1975)449.
12 L. Moscou, Leitz.-Mitt. Wiss. u Techn., 11 (1962) 103-5.
13 J.C.P. Broekhoff, and B.G. Linsen in Physical and Chemical Aspects of Adsorbents and Catalysts (B.G. Linsen, ed.), Academic Press, London, 1970, pp.1-62.
14 L.H. Allen, and E. Matijevic, J. Coll. Interf. Sci., 33(1970) 420-429.
15 L.H. Allen, and E. Matijevic, J. Coll. Interf. Sci., 35(1971) 66-75.
16 T.W. Healy, R.O. James, and R. Cooper, Adv. Chem. 79(1968) 62-73.
17 W. Feitknecht, Pure and Appl. Chem., 6(1963) 130-99.
18 K.A. Burkov, L.S. Lilic, and L.G. Sillen, Acta Chem. Scan. 19(1965) 14-30.
19 K.A. Burkov, N.I. Žnevick, and L.S. Lilich, Russ. J. Inorg. Chem. 16 (1971) 926, 927.
20 W. Feitknecht, Koll. Zeitschrift, 136 (1954) 52-66.
21 O. Bagno, and J. Longuet-Escard, J. Chim. Phys. 51 (1954) 434-39.
22 J.J.B. van Eijk van Voorthuijsen, and P. Franzen, Rec. Trav. Chim. 70(1951) 793-812.
23 G. Dalmai-Imelik, C. Leclercq, and A. Maubert-Muguet, J. Solid State Chem., 16(1976) 129-39.
24 J.W.E. Coenen, and B.G. Linsen in Physical and Chemical Aspects of Adsorbents and Catalysts (B.G. Linsen, ed.), Academic Press, London, 1970,pp.473-480.
25 V.M. Chertov, and R.S. Tijutijunnik, Kolloidn. Zhur. 37 (1975) 283-7.
26 D. Barby in Characterization of Powder Surfaces (G.D. Parfitt, ed.) Academic Press, London, 1976, pp.385-93.

27 C. Okkerse in Physical and Chemical Aspects of Adsorbents and Catalysts
   (B.G. Linsen, ed.), Academic Press, London, 1970, pp.215-42.
28 J. Joziasse, thesis, Utrecht, The Netherlands, 1978.
29 I.N. Signalov, and A.P. Dushina, J. Appl. Chem. USSR, 46(1973) 1751-54.
30 I.N. Signalov, G.N. Kuznetsova, Yu. V. Konovalova, T.V. Shakina, A.P. Dushina, and V.B. Aleskovskii, J. Appl. Chem. USSR, 49 (1976) 2436-42.

DISCUSSION

E.MATIJEVIC: In both papers only nickel nitrate was used. I wonder what the effect of other anions would be. Indeed, no metal hydroxide precipitation takes place by direct combination of metal and hydroxide ions. Instead, a series of metal hydrolysis products precede precipitation. The composition of pure complex solutes depends strongly on anions present. Thus, it should be expected that sulfate or acetate salts of nickel would give precipitate characteristics different from those obtained with the corresponding nitrate salt.

J.W. GEUS : Without suspended silica the precipitation of nickel hydroxide is affected by the anions present in the solution. It has been observed that the structure of the precipitate depends strongly on the presence of carbonate ions in the solution. Carbonate ions drastically suppress the growth of nickel hydroxide crystallites (to be published).
The hydrolysis of urea leads to formation of carbonate ions. The precipitation obtained using the hydrolysis of urea hence contains about 13 wt.% of carbon dioxide as mentioned on p.11. With suspended silica the situation is completely different. The low solubitility of nickel hydrosilicate brings about that the loaded carrier did not contain carbon dioxide. The precipitation of nickel with suspended silica is not markedly influenced by the anions in the solution.

D.J.C. YATES : I realize that the particle sizes of the urea-precipitated catalysts are rather small, as shown by your electron-microscope photographs. Can you be sure that the catalyst after precipitation is nickel silicate as distinct from, say, nickel hydroxide ?

J.W. GEUS : When the deposition-precipitation was carried out at 90°C, we observed virtually complete reaction of the silica support we used, as dealt with in our paper. This conclusion was based on :

(i) the pH-value of equilibrated suspensions as a function of the silica-to-nickel ratio represented in Fig. 3;
(ii) electron micrographs which do not show the silica particles of the support with the carrier loaded at 90°C;
(iii) the infra-red evidence obtained by Joziasse; with the loaded carrier he did not observe the characteristic lattice vibrations of silica. Reduction led to decomposition of the hydrosilicate and hence to reappearance of the lattice vibrations of silica. The lattice vibrations remained on reoxidation.
When the precipitation was done at 30°C, reaction of the support to nickel hydrosilicate is much less extensive. This can be concluded from the pH-values of Fig. 3, the electron micrographs of Fig. 7 and the infra-red results mentioned on p. 14. The results of Fig. 3 show that the silica content of the nickel precipitate decreases gradually with the silica-to-nickel ratio.
It is difficult to get evidence from X-ray or electron diffraction. The diffraction patterns of platelets of nickel hydroxide and nickel hydrosilicate are difficult to distinguish.

J.W.E. COENEN : I refer to the question of Dr. Yates on the X-ray evidence for hydrosilicate formation and answer of Dr. Geus who found it difficult to get this information. I would like to say that first igniting the material facilitates this detection : nickel hydroxide decomposes quantitatively to oxide at 400°C, but hydrosilicate does not.

J.W. GEUS : Whereas pure nickel hydroxide decomposes already at temperatures below 200°C, nickel hydrosilicates dehydrate at more elevated temperatures. Van Eijk, van Voorthuijsen and Franzen (1951) found two different nickel hydrosilicates, one being dehydrated at a considerably lower temperature than the other. Nickel hydrosilicate decomposing at the higher temperature was assigned as nickel montmorillonite and the one decomposing at lower temperature as nickel antigorite. We wonder whether keeping the catalyst at 400°C leads to dehydration of nickel hydroxide and not of nickel hydrosilicate. The gradual drop in the silica content of th precipitate (Fig. 3) should lead to a continuously changing dehydration temperature. Therefore we are investigating presently the differently prepared nickel-on-silica catalysts by thermogravimetry and differential thermal analysis.

CRYSTALLITE SIZE DISTRIBUTIONS AND STABILITIES OF HOMOGENEOUSLY DEPOSITED Ni/SiO$_2$ CATALYSTS

J. T. RICHARDSON, R. J. DUBUS, J. G. CRUMP, P. DESAI, U. OSTERWALDER and T. S. CALE
Department of Chemical Engineering, University of Houston, Houston, Texas USA 77004

ABSTRACT

Nickel on silica catalysts were precipitated with urea. After reduction the crystallite size distribution (CDS) of the nickel was determined from magnetic measurements. This technique gives very narrow CSD's and is much more reproducible than conventional impregnation methods. Nickel loading is controlled by solution concentration and precipitation time but higher values lead to broader CSD's, lower BET surface areas, and constricted pores. The amount of reduced nickel but not the CSD is a function of hydrogen flow rate and time. Increasing reduction temperatures produce broader CSD's. Calcination or passivation has no effect on the CSD's.

The CSD is very stable at 673 K but shows broadening for sintering times up to 100 hours. Above 723 K, small crystallites disappear and the CSD reaches a limiting log-normal shape, independent of initial distribution or nickel concentration. The decline of metal surface area has an order of 10 and an activation energy of 200 kJ/mole. At temperatures approaching 873 K, the sintered distributions are bi-modal.

INTRODUCTION

Nickel catalysts are most effectively prepared through the optimal combination of high dispersion and metal loading (ref. 1). Small crystallite sizes ensure high specific metal areas but increasing nickel concentrations result in agglomeration. Common impregnation and precipitation procedures lead to inhomogeneous crystallite sizes except for the dilute samples (<5%). Increasingly broad and even bimodal crystallite size distributions (CSD) are found as the nickel increases. This not only changes the activity and selectivity patterns for demanding-type reactions but may also affect thermal stability, since resistance to sintering depends to some extent on the initial CSD (ref. 2).

Van Dillen and co-workers developed a method for the homogeneous precipitation - deposition of nickel hydroxide on silica in aqueous suspension (ref. 3).

Slow decomposition of urea in water is a controlled source of hydroxyl ions. In the presence of nickel ions, these hydroxyls precipitate nickel slowly and homogeneously throughout the suspension. Furthermore, nickel reacts with silica upon precipitation to form a hydrosilicate which is resistent to coalescence. Catalysts prepared in this manner exhibit a uniform dispersion of very small nickel particles. We have used this method together with CSD measurements to investigate (a) preparational parameters (precipitation time, solution concentrations and initial pH, (b) reduction parameters (time, temperature and flow rate), (c) calcination (d) oxygen passivation, and (e) sintering.

EXPERIMENTAL
Catalyst Preparation

Urea was added to a suspension of Cab-O-Sil HS5 (SiO_2) in nickel nitrate solution after the solution reached 363 K. In one series of experiments, the concentrations of nickel nitrate, urea and the support were 0.14M, 0.42M and 7.6 kg/m^3 respectively; in another, 0.28M, 0.84M and 10kg/m^3. The pH-time curves were the same as those reported by Van Dillen et al. (ref. 3). Increasing amounts of total nickel were obtained by precipitating for longer times. The precipitate was filtered, washed and dried at 393 K. Total nickel content on each sample was measured with a colorimetric method described by Coenen and Linsen (ref. 4). Selected samples were calcined in air at various temperatures and times.

Reduction procedures

The catalysts were reduced in the same cell used for magnetic measurements (Fig. 1).

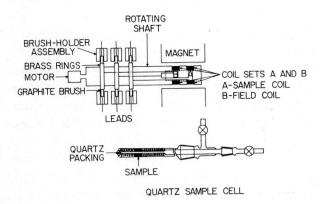

Figure 1. Experimental sample cell and rotating magnetometer

The outer 12mm quartz tube contained one centimeter of catalyst. Reducing and purge gases passed through the inner capillary tube. Each sample (2 to 8×10^{-4} kg) was charged and the cell connected to a gas system. High purity grade hydrogen was passed through the bed at 25 to 200 cm^3/min and the temperature raised to reduction temperatures at eight degrees per minute. After reduction, the sample was outgassed in argon or helium for one hour at 25 degrees above the reduction temperature and cooled to ambient with the inert gas still flowing. The cell was transferred to the magnetometer for magnetization measurements and then to a Micromeretics Pore Size Analyzer for nitrogen BET surface area, pore size distribution and hydrogen nickel surface area measurements.

Magnetic Measurements and Calculations

Details of the measurement of magnetization using the rotating coil magnetometer shown in Fig. 1 are given elsewhere (ref. 5). Magnetization curves were measured at 298 K for each catalyst after reduction and sintering for various times and temperatures. This included the change of magnetization upon removal of adsorbed hydrogen (ref. 6). The last step was determination of the saturation magnetization and degree of reduction after excessive sintering in helium at 1073 K.

Crystallite size distributions were calculated from magnetization data using developed procedures (ref. 7). Model calculations confirmed that this technique is sufficiently sensitive to describe changes in distributions resulting from these treatments.

RESULTS AND DISCUSSION

Homogeneous precipitation - deposition

Fig. 2 shows a comparison between samples prepared by urea precipitation and impregnation with nickel nitrate. Both samples contained about the same amount of total nickel and were reduced at identical conditions (15 hours at 673 K) yet the precipitation CSD is very narrow and small compared to the result of the impregnation. The precipitation procedure is also extremely reproducible. For example, two separate preparations (with different operators) resulted in almost identical CSD's and reductions. Impregnation, however, was very difficult to repeat. Even specimens from the same preparation were different.

Effect of precipitation time

The effect of precipitation time is to increase the total amount of nickel deposited, as shown in Fig. 3.

Changes in sample pore structure as a result of increasing precipitation time are represented in the nitrogen isotherms of Fig. 4.

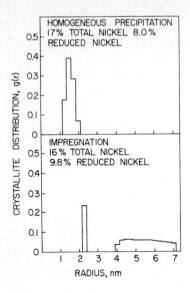

Figure 2. Effect of method of deposition

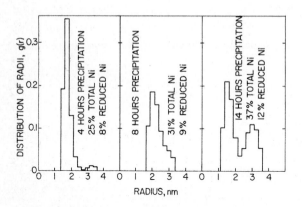

Figure 3. Effect of precipitation time

The hysteresis trends indicate pore neck constrictions that become more severe as precipitation continues. Loss of total surface area from 2.91×10^5 m^2/kg to 1.66×10^5 m^2/kg also occurs, with the percentage of surface in pores below 10 nm decreasing from 90 to 76. The extent of reduction (29-35%) is approximately the same. The increased constrictions in the pore may be responsible for the bi-modal trend in the CSD. Since growth of crystallite in the narrow necks are constrained against further growth, whereas those in the larger body of the pores are not. From these and subsequent results, we conclude that the nickel concentration is much more important in determining the CSD than is the amount of reduced metal. This implies that crystallite growth during reduction is a consequence of nucleation and not sintering, at least up to 673 K.

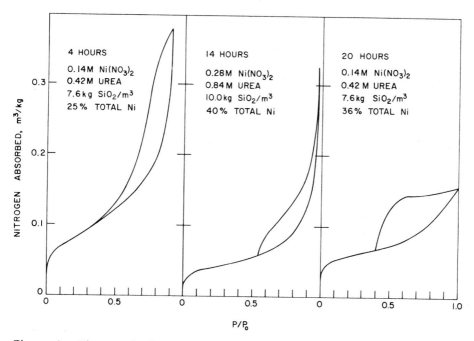

Figure 4. Nitrogen isotherms for increasing precipitation times

Effect of solution concentrations

Increasing concentrations of urea and nickel nitrate increase nickel deposition rates with decreased dispersion. The effect of precipitation time on the CSD is much less for the higher concentrations, possibly because the amount of nickel deposited reaches a steady value for the longer times. One difference is significant. The extent of reduction is higher and increases with time although the total nickel content does not change significantly.

Effect of initial pH

This variable was explored in only one experiment. Two samples were prepared with 4 hours precipitation from 0.4M urea solution but sufficient nitric acid was added to one of them, lowering the initial pH from 4 to 2.5. The result is a slightly lower CSD with less total nickel for the sample with lowered pH. However, the extent of reduction increases, showing some effect on the deposition mechanism.

Effect of reduction time

Figure 5 shows the effect of time on reduction at 673 K. The CSD is remarkably uniform but the extent of reduction increases. Sintering does not appear to involve the reduction in any way. Initially the reduction is fast so that about 30% of the nickel is reduced at five hours. Thereafter the process is slower. This may indicate rapid reduction of the outer layer of nickel hydrosilicate followed by slower reduction deeper in the particle.

Effect of reduction temperature

Reduction at temperatures higher than 673 K result in broader CSD. Many of the samples have bi-modal distributions. These effects may come from accelerated sintering during reduction or from reduction of separate precursers which are harder to reduce at lower temperatures yet result in larger crystallites.

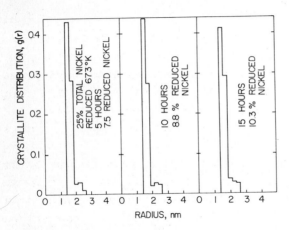

Figure 5. Effect of reduction time

Effect of hydrogen flowrate

Fig. 6 demonstrates the dependence of the extent of reduction on hydrogen gas velocity. The CSD are again unchanged. Particle Reynolds numbers at these conditions are extremely small (.01 - .05) so that the flow is almost in the Stokes region, yet the mass transfer coefficient for hydrogen is six orders of magnitude higher than the fastest reduction rate. The reduction process may be controlled by water removal from the pores.

Varying the hydrogen flow rate is a very sensitive and reproducible method of controlling the amount of reduced nickel.

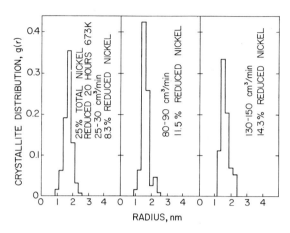

Figure 6. Effect of hydrogen flow rate

Effect of calcination

Calcination in air prior to reduction has little effect on the resulting CSD. Temperatures from 623 K to 723 K were tried with similar results. This implies that the reduction precursor, presumably nickel hydrosilicate, is not drastically altered by the air treatment. This is in contrast to other systems, such as alumina, where calcination promotes support compound formation.

Effect of passivation

Reduced nickel catalysts are often passivated by treatment in dilute oxygen to protect the nickel during subsequent handling operations. A sample was reduced, cleaned and measured in the usual manner. A 2% mixture of air in helium was passed over the catalyst at 298 K for one hour. Air concentration was increased slowly to one atmosphere and the cell exposed to the laboratory atmosphere overnight. Subsequent reduction and measurement resulted in a CSD almost identical to that before passivation.

The following observations were made about the pyrophoric nature of dispersed nickel. After a reduced catalyst has been cleaned in argon at 25° above the reduction temperature, the magnetization curve increases about thirty percent. Adsorption of hydrogen at room temperature returns the curve to its initial value, indicating that the surface is saturated with hydrogen after reduction. If the catalyst is now exposed to air, rapid oxidation and heating of the catalyst follows. Dispersion of the nickel is destroyed by the exothermic heat of oxidation. However, exposure of the catalyst to air after cleaning produces no temperature increase and the magnetization curve shows a decrease consistent with the adsorption of a monolayer of oxygen (ref. 6). Further reduction, cleaning and measurement gives a CSD almost identical with the non-oxidized sample.

These results confirm the observation of Popowicz et al. (ref. 8) that the adsorbed hydrogen and not the nickel initiates the dangerous pyrophoric oxidation of the nickel.

Sintering Effects

Sintering results at 673 and 773 K and are given in Fig. 7. This preparation is very stable at 673 K. The mean radius is almost constant at 1.5 nm, but the distribution first broadens and then moves toward higher radii. Finally, after 150 hours, small crystallites disappear. At 773 K, the shift of

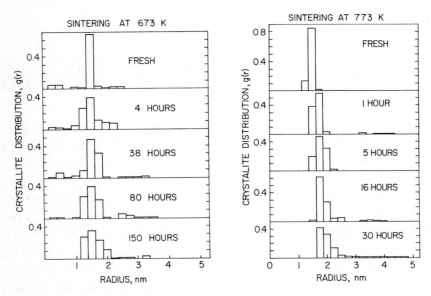

Figure 7. Crystallite size distributions for sintering at 673 and 773 K.

the mean radius toward higher values is so pronounced that all crystallites below 1.5 nm are removed. Sintering is further illustrated in the results of Table 1 and Fig. 8.

TABLE 1

Total and nickel surface areas for sintered catalysts

Catalyst	Surface Area m^2/g (sample or Ni) $\times 10^{-5}$			
	S_{BET}	S_{CSD}	S_H	$S_{\Delta M}$
4 hrs precipitation reduced 15 hrs, 673 K	2.67	1.86	1.82	1.69
sintered 30 hrs, 823 K	2.20	1.24	0.84	0.94
sintered 16 hrs, 1073 K	1.75	<.30	0.15	-

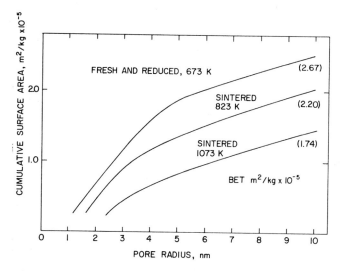

Figure 8. Pore size distributions for sintered catalysts

There is a moderate loss of total surface area upon sintering at 823 K, due mostly to the collapse of pores smaller than about 1.5 nm. The values of S_{CSD} were calculated from the CSDs, S_H came from hydrogen adsorption measurements and $S_{\Delta M}$ were estimated from the change of magnetization after cleaning using procedures to be published elsewhere (ref. 9). For the freshly reduced catalyst, the results are in satisfactory agreement, indicating internal consistency of procedures and emphasizing the availability of metal surface. After sintering at 823 K, S_H and $S_{\Delta M}$ are in agreement but are less than the calculated surface. This can only mean that some surface is inaccessible to hydrogen due to either trapping of small crystallites in collapsed pores or to the surface bonding with the support. The CSD's show a similar trend as those at 773 K in Fig. 7 with the disappearance of crystallites smaller than 1.5 nm.

Other conclusions are as follows: (1) At all temperatures the final CSD is independent of the initial CSD, (2) calculated surface areas, S_{CSD}, fit the kinetic expression $-\frac{ds}{dt} = ks^n$ with n = 10 and an activation energy of 200 kJ/mole, (3) sintering rates are independent of reduced nickel concentrations up to 17 wt% Ni, and (4) for temperatures above 873 K, bi-modal CSD's result.

These results are most consistent with the model of crystallite migration (ref. 10) but the effect of changes in support pore texture during sintering must be reconciled before definite conclusions can be reached.

ACKNOWLEDGEMENTS

The authors thank the Robert A. Welch Foundation and the National Science Foundation for support of this work.

REFERENCES

1. J. W. Geus, Sintering and Catalysis, Plenum Press, New York, 1975, p. 29.
2. P. C. Flynn and S. E. Wanke, J. Catal., 34(1974)290.
3. J. A. Van Dillen, J. W. Geus, L. A. M. Hermans and J. Van der Mejden, Proceedings of the Sixth International Congress on Catalysis, London, July 1976, The Chemical Society, London, 1977, p. 677.
4. J. W. G. Coenen and B. F. Linsen, Physical and Chemical Aspects of Adsorbents and Catalysts, Academic Press, New York, 1970, p. 472.
5. J. T. Richardson and P. Desai, to be published.
6. J. A. Dalmon, G. A. Martin, and B. Imelik, Surf. Sci. 41(1974)587.
7. J. T. Richardson and P. Desai, J. Catal. 42(1976)294.
8. M. Popowitz, W. Celler, and E. Treszczanowicz, Int. Chem. Eng. 6(1966)63.
9. J. T. Richardson and T. Cale, to be published.

DISCUSSION

S.P.S. ANDREW : One of the conclusions drawn is that the results on sintering are most consistent with the model of crystallite migration. However, one of the observations is that sintering rates are independent of reduced nickel concentrations. This does not seem consistent with the migration model which is a second order process.

J.T. RICHARDSON : Two samples with identical crystallite size distributions but differing by a factor of two in reduced nickel concentration, were found to sinter with the same kinetics. You are indeed correct that this is inconsistent with the particle migration model. Perhaps this reflects the influence of support texture and existing theories must be modified. However, a factor of two differences in nickel is not expected to give a large difference according to current models. This may very well be within our experimental error.

H. CHARCOSSET : What do you think of the stoichiometry of oxygen adsorption on Ni at room temperature after cleaning procedures ? Are one or two oxygen atoms chemisorbed per exposed Ni atom ? Have the authors to admit that about two O atoms are chemisorbed per surface nickel atom ?

J.T. RICHARDSON : We have assumed one oxygen atom per nickel at the surface but have no evidence for this. The stoichiometry is not important to our conclusion that one layer of nickel is affected by oxygen during our passivation procedure.

J.W.E. COENEN : Do you have quantitative information on the depth of oxidation in passivation ? In our experience in passivation, the depth of oxidation is about 2 nickel atom layers.

J.T. RICHARDSON : The loss of magnetization after passivation corresponds to the "demetallization" of one layer of nickel atoms. We conclude that only one layer of nickel is oxidized and that the phenomenon is equivalent to chemisorption.

B. DELMON : Concerning Fig. 3 (at 14 hrs), the authors suggest that the two categories of crystallites might correspond to two different precursors. Another explanation is that longer reaction times bring about a recrystallization of a single precursor, with the larger crystallites growing at the expense of the smaller ones. Fig. 3 indeed shows that, even after only 4 hrs., there is already a family of larger crytallites : these might be the ones growing, later on, at the expense of the smaller ones.
Concerning Fig. 2, the two families of crystallite sizes observed with the catalysts impregnated with nickel nitrate are very different. As nickel nitrate impregnation is a widespread method for preparing nickel catalysts, it would be interesting to know the origin of these families. Have you evidences that the larger crystallites might come from nitrate deposited at, or having migrated to, the mouth of the pores during the drying process ? The smaller crystallites could correspond to deposits inside the pores.

J.T. RICHARDSON : The example was taken from a number of samples showing similar heterogeneity. We have no explanation for the variation in distributions but suspect that uneven and difficult to control drying procedures are responsible.

L.L. MURELL : What is your opinion of the structure of the unreduced nickel portion of a nickel on silica catalyst ? Is for $Ni(NO_3)_2$ impregnated onto silica the silicate layer only a monolayer thick ? If so, why is there such difficulty in getting complete reduction of this phase as shown in your work ? If the nickel silicate layer is greater than monolayer thickness then difficulties in complete reduction might be expected.

J.T. RICHARDSON : Examination of many preparations suggests that $Ni(OH)_2$ is easily reduced but nickel hydrosilicate is not. The nickel silicate layer is more than one layer thick and even consumes most of the particle. The deeper this layer, the more difficult the reduction.

THE PREPARATION AND PRETREATMENT OF COPRECIPITATED NICKEL-ALUMINA CATALYSTS FOR METHANATION AT HIGH TEMPERATURES

E.C. KRUISSINK[a], L.E. ALZAMORA[b], S. ORR[b], E.B.M. DOESBURG[a], L.L. VAN REIJEN[a], J.R.H. ROSS[b] and G. VAN VEEN[b]

[a]Laboratory of Inorganic and Physical Chemistry, University of Technology, Delft (Netherlands) and [b]School of Chemistry, University of Bradford, Bradford BD7 1DP (U.K.)

ABSTRACT

The preparation and properties of a series of coprecipitated nickel-alumina catalysts have been examined. For $2 \leq Ni/Al \leq 3$, the precipitate has a layer structure of the type previously described by Feitknecht and by Longuet-Escard. Our results indicate that the nickel and aluminium ions occupy interchangeable positions in the "brucite-type" layers of the structure, while various anions ($CO_3^=$, NO_3^- or Cl^-, depending on the method of preparation) are found in the "interlayer" in addition to water of crystallisation. On calcination, poorly crystallised nickel oxide, which probably contains dissolved Al^{3+} ions, is formed; the crystallinity of this phase depends strongly on the ions present in the precipitate. Phase separation of $NiAl_2O_4$ occurs at temperatures around 1000 K. The metallic surface areas and the methanation activities of the reduced samples depend directly on their degree of reduction; for identical calcination and reduction conditions, it is found that samples originating from the carbonate-containing precipitates have higher activities under methanation conditions than the nitrates or chlorides. The presence of sodium has a detrimental effect on the properties of the catalysts.

1. INTRODUCTION

Because of rising demand for natural gas, there has recently been an increasing interest in the methanation of carbon monoxide rich gases. The methanation reaction

$CO + 3H_2 \rightarrow CH_4 + H_2O$

is highly exothermic and large temperature rises will occur under typical reaction conditions (refs. 1, 2); hence, a good methanation catalyst must have activity at the low temperatures of the entry to the catalyst bed, but it must also have high stability under the hydrothermal conditions of the reactor exit. The methanation reaction is also utilised in the "Adam and Eve" process under development in West-Germany for the transportation of heat from a high temperature nuclear reactor (ref. 2); in this, the helium coolant of the reactor supplies the energy for the steam reforming of the methane, and the energy is regained with high efficiency by subsequently carrying out the methanation reaction.

Thermal efficiency requires a high temperature at the reactor exit and hence again excellent catalyst stability.

Currently, the production of synthetic natural gas (SNG) is most commonly achieved by the steam reforming of naphthas at low temperatures using a coprecipitated Ni-Al_2O_3 catalyst with high Ni content (refs. 3, 4). Preliminary results of an examination of the preparation and activities of such catalysts showed that the properties of the final catalyst depend markedly on the parameters which were changed during the preparation and pretreatment of the samples (ref. 5). The aim of our present work is to gain a greater understanding of the inter-relationships between the method of preparation of such catalysts, their structures and their activities and stabilities for the methanation reaction with the hope that we may be able to design a stable methanation catalyst for possible use in the Adam and Eve project.

This paper presents some of our main conclusions to date; some preliminary results have been published elsewhere (refs. 6, 7). We have found that within a definite range of Ni/Al ratios, nickel-aluminium hydroxy compounds are formed on precipitation and, depending on the conditions of preparation, that these may contain carbonate, nitrate or chloride ions. On calcination, decomposition of the precipitate occurs in two stages: (a) loss of water of crystallisation, and (b) loss of anions and decomposition of the hydroxide structure; the oxidic form of the catalyst probably consists of nickel oxide with dissolved aluminium ions, together with amorphous alumina. The final stage of the preparation is reduction in hydrogen; metallic nickel is formed in a highly dispersed state whose high stability at elevated temperatures must be due to its interaction with the alumina.

We have examined the effect of the following variables on the properties of the final catalyst:
(a) the Ni/Al ratio;
(b) the pH of precipitation;
(c) the type of metal salt used;
(d) temperature of reduction.

The precipitates have been characterised by means of X-ray diffraction, by thermal techniques (D.S.C., D.T.A., T.G.A.) and by chemical analysis; the calcined and reduced materials have been examined by X-ray line broadening, by magnetic susceptibility measurements and by chemisorption of hydrogen; the activities (and selectivities) of the catalysts for the CO + H_2 reaction were measured in a low pressure recirculation reactor and their activities and stabilities were measured in an atmospheric-pressure flow reactor.

2. EXPERIMENTAL

2.1 Catalyst Preparation

Coprecipitation was carried out using two solutions: one contained nickel and aluminium nitrates with a total concentration of 0.9 mol dm^{-3} while the other contained Na_2CO_3 and/or NaOH. 175 cm^3 of each solution was added simultaneously with constant stirring to 150 cm^3 of distilled water; the temperature was maintained at 353 K throughout and the pH was also kept constant at a desired value by adjusting the rate of addition of alkali. In another set of preparations, a solution of Na_2CO_3 was added to a solution of the nickel and aluminium nitrates in such a way that the pH of the solution increased gradually from either one to five or one to seven. After precipitation, part of the product was filtered on a glass filter and dried at 353 K; the remainder was hydrothermally treated in the mother liquid in a teflon tube in an autoclave at 423 K and 5 atm for two days before filtration and drying.

The dried samples were calcined (decomposed) in air, either in a quartz tube closed at one end or in a system in which the sample was maintained in porcelain boats; temperatures ranged from 450°C to 1000°C. Several were also decomposed in a high temperature X-ray camera (see below) or using thermal analysis techniques (D.S.C., D.T.A. or T.G.A.). The calcined materials were reduced in H_2, either in the catalytic apparatus described below, or in a flow of hydrogen in the system used for the calcination step; several samples were also reduced in the high temperature X-ray camera using T.G.A.

2.2 Catalyst Characterisation

X-ray measurements were carried out by means of a Guinier-de Wolff camera and a Guinier-Lenné high temperature camera, both manufactured by Enraf-Nonius (Delft).

Some samples were analysed chemically for nickel, aluminium, sodium, carbonate and/or nitrate. Nickel, aluminium and sodium were determined by atomic absorption spectroscopy, carbonate by dissolving the precipitate in phosphoric acid and titrating the CO_2 evolved, and nitrate by reduction to NH_3 which is then titrated (Devarda method).

Chemisorption and physical adsorption measurements were made volumetrically using a vacuum apparatus capable of background pressures of better than 10^{-5} N m^{-2}. The sample vessel consists of a quartz tube separated from the remainder of the system by a trap maintained at 78 K.

Activity measurements were made in two systems, both of which utilize G.L.C. analysis of reactants and products. The first of these (ref. 6) is a low-pressure recirculation system, operating at about 3 kN m^{-2}, in which a background pressure of $\sim 10^{-4}$ N m^{-2} can be attained; water produced during the reaction is trapped out at 193 K. The second (ref. 7) is an atmospheric pressure flow reactor of conventional design in which the water is removed by a condenser prior to admitting samples of the reactants and products to the G.L.C. In both cases, the samples are reduced in situ prior to commencing the reaction.

2.3 Materials

All chemicals were of Pro-analysis or Analar quality. Gases were supplied by the British Oxygen Company; for the low pressure reactor, they were obtained in sealed ampoules (Grade X) and for the flow reactor, a standard mixture of high purity CO and H_2 in the ratio 1:3 was used.

3. RESULTS AND DISCUSSION

3.1 Structure and Composition of the Precipitate

Coprecipitation of nickel and aluminium gives rise to badly crystallised hydroxy compounds of which the crystallinity is much improved by the application of hydrothermal treatment. This enables better characterisation of the samples to be made, while it has little or no effect on the crystallinities of the calcined and reduced samples, or on the activities of the latter. The hydroxy compounds appear to belong to a large class of hydroxide layer structures made up of layers of the brucite type (as found in $Mg(OH)_2$) separated by disordered interlayers containing anions and water molecules. Important unresolved problems concerning these compounds for the Ni/Al system are (a) the relative positions of the nickel and aluminium, and (b) their exact composition, especially with respect to the presence of anions other than hydroxyl ions. According to Feitknecht (ref. 8), nickel occupies octahedral positions within the brucite layer while the aluminium is situated somewhere in the interlayer. In contrast to these conclusions, Longuet-Escard (ref. 9) assumes that both elements are statistically distributed about octahedral positions within the brucite layer. Both authors mention only the inclusion of hydroxyl ions.

We will summarise here only the main conclusions of our investigation of the properties of the Ni-Al system. The majority of these results have been obtained from hydrothermally aged samples, as the X-ray data for these exhibit more sharply defined spacings, but similar, less quantitative results have been obtained for the unaged samples used for catalyst preparation.

As long as the Ni/Al ratio lies between two and three, a single-phase layer compound is obtained. Outside this range, the excess Ni or Al forms a separate phase, $Ni(OH)_2$ or $Al(OH)_3$ respectively. Measurements of the dependence of the cell constants on the Al content together with magnetic measurements indicate that the Al ions substitute for Ni within the brucite layer (ref. 10). It was also found that the interlayer spacing depended markedly on the pH of precipitation. Table 1 shows selected results for both aged and unaged samples.

Table 1. Effect of Preparation Conditions on the Layer Spacings and Chemical Compositions of the Precipitates.

$\frac{Ni}{Ni+Al}$ (%)	Precipitating Agent(s)	pH of precipitation	Layer Spacing/Å	NO_3/ wt %	CO_3/ wt %
66[a]	$NaOH/Na_2CO_3$	10	7.58	<0.2	7.9
66[a]	$NaOH/Na_2CO_3$	5	8.92	18.3	0.24
50[b]	Na_2CO_3	7	~7.5	0.88	8.6
50[b]	NaOH	7	~9.0	13.8	---

a, hydrothermally aged; b, unaged

The results show that the layer spacing is high at low pH's (e.g. pH ~ 5) but decreases at higher pH (e.g. pH ~ 10) when the precipitating conditions are otherwise identical; when the pH was kept at 7 and different precipitants were used, it was found that the lower spacing was obtained with Na_2CO_3 as precipitant. Chemical analysis of the precipitates, also reported in Table 1, shows that the higher layer spacing corresponds to situations when NO_3^- ions are incorporated in the precipitate (low pH or in the absence of $CO_3^=$) while the lower layer spacing corresponds to incorporation of $CO_3^=$ ions. The equilibria

$$H_2CO_3 \underset{\leftarrow}{\overset{1}{\rightarrow}} H^+ + HCO_3^- \underset{\leftarrow}{\overset{2}{\rightarrow}} 2H^+ + CO_3^=$$

have values of pK_1 of 6.4 and of pK_2 of 10.3. Hence, the concentration of HCO_3^- and $CO_3^=$ will be low at pH values much less than 6. Apparently, a large amount of nitrate ions is only included if carbonate concentration is low; otherwise carbonate ion is preferentially incorporated. If the precipitation is carried out from a solution of the chlorides at low pH, chloride ions are incorporated in the structure rather than nitrate ions. Similar results were obtained for other samples; see the data of Table 3 discussed below.

It is interesting to compare our conclusion that anions other than hydroxyl ions are incorporated in the structure with the work of Merlin et al (ref. 11), who claimed to have prepared a well defined compound $NiO \cdot Al_2O_3 \cdot 8 \, H_2O$. In their preparation method, nickel and aluminium nitrates are dissolved in excess concentrated ammonia and the ammonia is removed by heating on a water bath until the pH reaches 6-7; a precipitate having an X-ray pattern similar to that of our preparations was obtained. Samples prepared by us following the same procedure also showed considerable carbonate and/or nitrate contents, depending on the exact preparation conditions.

Chemical analysis of the precipitates prepared in this work has shown that it is often difficult to remove sodium ions from the precipitates, particularly for samples prepared at pH = 10; the presence of sodium ions has a marked effect on the activities of the final catalysts (see section 3.5).

3.2 Calcination

Thermal decomposition of the precipitates takes place in two stages, the first corresponding to loss of water of crystallisation and the second to loss of CO_2 (or NO/NO_2 or HCl) and the decomposition of the hydroxide layers. For example, T.G., D.T.A. and D.S.C. results for the hydroxy carbonates all show that the first decomposition occurs at about 520 K and the second at about 625 K; typical D.S.C. results were shown in ref. 5. Results obtained using the high temperature X-ray camera showed that the layer structure disappeared during the second decomposition, when only badly crystallised NiO was obtained. At higher temperatures of around 1170 K, formation of $NiAl_2O_4$ was observed together with a recrystallisation of the NiO; strong sharp reflections of both phases were found after calcination at 1273 K.

In the preparation of samples for activity measurements, calcination was generally carried out at 723 K in air. In agreement with the results of the X-ray camera, only NiO is observable. The cell constant of the highly dispersed NiO is somewhat too small, which suggests that some Al ions have dissolved in it, resulting in the formation of a phase which can be represented as $NiAl_{2\delta}O_{1+3\delta}$. There may in addition be present some X-ray amorphous alumina phase.

The crystallinity of the NiO formed depends strongly on the anion present in the precipitate. The crystallinity increases in the order carbonate < nitrate < chloride, as is illustrated by the results of Table 2, for three samples with mole fraction Ni/Ni+Al = 0.50 (see page 7).

3.3 Reduction

The extent of the reduction process at any temperature was found to depend on the conditions under which the reduction was carried out. For example, in the low pressure recirculation reactor, approximately 50% reduction was achieved at 873 K for most samples, calcined at 723 K (see Table 3, discussed below), whereas in the TGA system, in which a flow of hydrogen was used, almost 100% reduction

Table 2. Relation between anion incorporated in the precipitate, particle sizes of resulting NiO and Ni, and Ni surface area.

Sample	Anion	Particle size* of NiO in calcined material	Particle size* of Ni in reduced material	Ni surface area ($m^2 g^{-1}$)	Specific activity C atoms $s^{-1} m^{-2} 10^{17}$
A_6	$CO_3^=$	4.0 nm	6.2 nm	49	0.137
B_2	NO_3^-	5.5 nm	8.5 nm	28	0.115
C_1	Cl^-	>50 nm	>50 nm	21	0.063

*calculated from X-ray line broadening data

was measured. The crystallinity of the nickel in the reduced material appeared to be directly related to that of the calcined samples (see Table 2); in consequence, the smallest nickel crystals are obtained from the carbonate samples. A sample calcined at 1273 K was more difficult to reduce than those calcined at lower temperatures and was found to contain $NiAl_2O_4$ after reduction at 873 K.

3.4 Surface Areas

Hydrogen adsorption isotherms were obtained for various samples after reduction over a range of temperatures. The isotherms behaved according to the Langmuir isotherm for dissociative adsorption and nickel surface areas were calculated from plots of the linearised form of the equation. Fig. 1a shows the dependence on reduction temperature (T_r) of the metallic areas of two samples of different nickel contents, one derived from a nitrate precipitate and the other from a carbonate precipitate; also shown are the corresponding degrees of reduction. Fig. 1b shows the surface areas as a function of degree of reduction. The areas rise almost linearly with degree of reduction up to about 80% reduction, when the area decreases once more; in both cases, this corresponds to a temperature of about 1000 K and it is reasonable to assume that sintering of the nickel crystallites has occurred at this temperature. Other results showed that there was a close correspondence between the nickel particle size calculated from X-ray results and the nickel surface area (ref. 10). These results will be discussed further below (section 3.5) in connection with the activity data.

Table 2 shows the results of metallic area (H_2 adsorption) measurements for samples with the same Ni/Al ratios (1) but prepared from precipitates containing the different anions. The sample prepared from the carbonate-containing precipitate has the highest metallic area. The relationship between particle size and metallic area is demonstrated.

Table 3. The dependence of methanation activity on the method of catalyst preparation and composition.

Sample	Ni/Ni+Al	pH of Precip.	Layer Spacing (Å)	Anion Incorporated	Na content (wt %)	% Reduction at 873 K	Activity Atoms of C g^{-1}s^{-1}.10^{17}
A_1	0.85	10	7.9	$CO_3^=$	?	48	5.6
A_2	0.75	10	7.7	$CO_3^=$, 8.2%	0.09	71	9.4
A_3	0.75	7	7.7	$CO_3^=$	0.05	64	6.1
A_4	0.72	7*	7.7	$CO_3^=$	?	53	5.3
A_5	0.50	10	7.5	$CO_3^=$, 8.4%	0.07	55	5.8
A_6	0.50	7	7.5	$CO_3^=$, 8.6%	0.05	45	6.7
A_7	0.25	10	7.5	$CO_3^=$	?	71	5.2
B_1	0.75	6†	8.4	NO_3^-, 13.8%	?	55	3.3
B_2	0.50	7†	9.0	NO_3^-, 13.8%	0.21	46	3.3
B_3	0.37	5*	8.9	NO_3^-, 16.4%	?	46	2.0
C_1	0.50	7†	7.7	Cl^-	0.05	45	1.3
D_1	0.75	10	7.7	$CO_3^=$	0.37	49	2.5
D_2	0.66	10	7.7	$CO_3^=$	0.70	50	2.5
D_3	0.40	10	7.5	$CO_3^=$	1.13	66	2.2

* Prepared at increasing pH, to the limit given

† Precipitated with NaOH

3.5 Activity, Selectivity and Stability

The hydrogenation of CO was examined in the low pressure reactor; preliminary results have been described elsewhere (ref. 6). It was found, especially at low degree of reduction, that hydrocarbons other than methane were formed; Fig. 2a shows the effect of temperature of reduction (T_r) on the activity and selectivity (methane/total carbon atoms reacted) for the two samples for which the surface area results were given in Fig. 1. Fig. 2b shows the relationship between activity and surface area, from which it can be seen that the activity increases with surface area, although there are probably differences in the specific activity for the two catalysts examined.

Table 3 shows activity data for a number of samples prepared in different ways and with different Ni/Al ratios. Series A shows the influence of different Ni/Al ratios for precipitates containing carbonate ions. The activity is not greatly affected by this ratio. For samples prepared with increasing pH, only the final pH appears to be important; compare sample A4 with A3. Series B was prepared from nitrate containing precipitates; the activities are considerably lower than those of Series A. This effect appears to be due to sintering during the calcination process in the presence of the oxides of nitrogen which are evolved. It seems that the effect may be minimised if the evolved gases are able to escape easily during calcination. This depends on the rate of heating to the calcination temperature, and on the texture of the powder.

The chloride containing sample C_1 had a low activity, which is in agreement with the low metal area measured. Also, its specific activity (see Table 2) is almost a factor two lower than that of the catalysts of Series A or B. This might be due to the presence of chloride on the metal surface.

The D series shows that all samples with high sodium contents have low activities. The same poisoning effect of sodium has been observed by Rostrup-Nielsen for steam reforming catalysts (ref. 12). The stabilities of several of the samples shown in Table 3 were examined in the atmospheric pressure flow reactor. Preliminary results will be published elsewhere (ref. 7). They confirm the order of activity $CO_3 > NO_3 > Cl$ and also show that the most of the samples lose less than 20% of their activity over two weeks operation at 773 K. Tests of longer duration and under more severe conditions are planned.

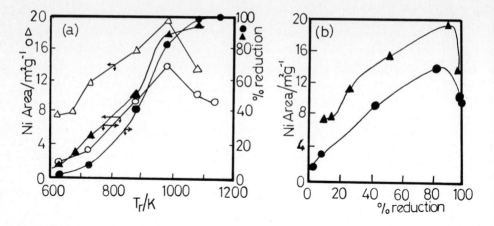

Fig. 1. (a) The dependence of Ni surface area (open symbols) and % reduction (closed symbols) on the temperature of reduction for catalysts A4 (△,▲) and B3 (○,●). (b) The dependence of Ni area on the % reduction (A4, ▲; B3, ●).

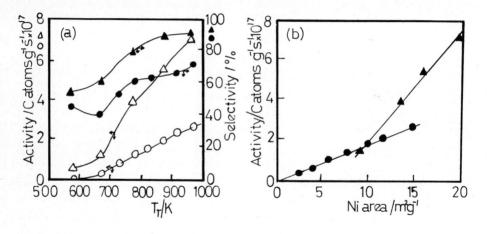

Fig. 2. (a) The dependence of activity for hydrogenation of CO under standard conditions (open symbols) and selectivity for methane formation (closed symbols) as a function of temperature of reduction for catalysts A4 (△,▲) and B3 (○,●). (b) The activity as a function of surface area (A4, ▲; B3, ●).

ACKNOWLEDGEMENTS

The authors acknowledge with thanks an award from NATO under the Science Programme (Grant No. 1085). E.C.K. thanks the Z.W.O. for a Research Fellowship, L.E.A. thanks the British Council for a Technical Assistance Award, S.O. thanks the S.R.C. for a Research Studentship, and G.v.V. thanks the Royal Society for a Visiting Fellowship. They also wish to thank mr. Y. Timmerman at Delft for experimental assistence.

REFERENCES

1 G.A. Mills and F.W. Steffgen, Catalysis Revs., 8 (1973) 159.
2 B. Höhlein, Jül. Report No. 1433, KFA Jülich GmbH, 1977.
3 R.G. Cockerham, G. Percival and T.A. Yarwood, Inst. Gas. Eng. J. (1965), 109.
4 J.R.H. Ross, in M.W. Roberts and J.M. Thomas (Eds.), "Surface and Defect Properties of Solids", Vol. IV, Specialist Periodical Reports, Chemical Society, London, 1975, p. 34.
5 T. Beecroft, A.W. Miller and J.R.H. Ross, J. Catal., 40 (1975) 281.
6 E.B.M. Doesburg, S. Orr, J.R.H. Ross and L.L. van Reijen, J. Chem. Soc., Chem. Comm., (1977) 734.
7 G. van Veen, E.C. Kruissink, J.R.H. Ross and L.L. van Reijen, submitted for publication in Rn. Kinet. Catal. Lett.
8 W. Feitknecht, Helv. Chim. Acta, 25 (1942) 555.
9 J. Longuet-Escard, J. Chim. Phys., 47 (1950) 238.
10 L.E. Alzamora, E.C. Kruissink, J.R.H. Ross and L.L. van Reijen, unpublished results.
11 A. Merlin, B. Imelik and S.J. Teichner, C.R. Acad. Sci., 238 (1954) 353.
12 J.R. Rostrup-Nielsen, Steam Reforming Catalysts, Teknisk Forlag A/S, Copenhagen (1975) 102, 122.

DISCUSSION

D.I. BRADSHAW : X-ray diffraction and infrared spectroscopy carried out by us on similar co-precipitated $Ni-Al_2O_3$ catalysts indicate that vacancies exist in the "brucite" layer with aluminium bonded to carbonate ions in the interlayer. Carbonate bands were observed at 1325 and 1550 cm^{-1}. Would the authors care to comment on this alternative structure for the precipitate.

E.C. KRUISSINK : Serna et al. have shown similar infrared spectra for samples of Mg-Al coprecipitates in which a sodium aluminium hydrate carbonate structure (dawsonite) has separated. In these, the carbonate and Al^{3+} ions are in close proximity. It is possible that the samples you mentioned contain a similar phase. Infrared spectra obained with our samples do not contain the band at 1550 cm^{-1} with the exception of one sample in which also dawsonite was found by X-ray examination. In this particular case a substantial excess of Na_2CO_3 during precipitation led to further reaction (formation of

dawsonite) upon subsequent hydrothermal treatment. In view of these data we do not believe that Al^{3+} and CO_3^{2-} ions are in close proximity in the remainder of our samples.
Serna, C.J., White, J.L. and Hem, S.L. J. Pharm. Sci., **64**, 468(1975); **67**, 324(1978).

J.W. GEUS : Though generally anions strongly affect the precipitation process, the effect of anions with silica is much smaller than with alumina because of the acid character of silica. We accordingly observed that the decomposition of urea leads without suspended silica to a carbonate-containing precipitate, whereas with suspended silica the nickel precipitate did not contain markedly carbonate. Studying the precipitation of Al(III) ions without other metal cations being present, we found consumption of OH^- ions at pH values much lower than at which nickel ions react with hydroxyl ions. Did you investigate whether precipitation of Al(III) ions induces precipitation of nickel ions at a lower pH level than at which nickel ions alone precipitate from a homogeneous solution ?

E.C. KRUISSINK : We carried out one experiment in which Ni(II) and Al(III) ions were precipitated from a nitrate solution by urea hydrolysis at 90°C. Formation of a nickel containing phase was observed at a pH value of 4.5, which is lower than the value 5.5, reported by Hermans et al. for the precipitation of nickel ions only from a homogeneous solution at 90°C, in the same range of nickel concentration. Our concentrations were : Ni(II) 0.48 M, Al(III) 0.17 M, and initial urea concentration 3.6 M.
Hermans, L.A.M. and Geus, J.W. , this symposium.

K. KOCHLOEFL : We found that not only the precipitation agent but also the Ni/Al ratio plays an important role in the thermal stability of coprecipitated $Ni-Al_2O_3$ catalysts.

P.G. MENON : Why do you need such high concentrations of Ni, as high as 50-70 wt% ? Most of the Ni ultimately ends up in large crystallites of average size 140 Å. Will not a catalyst with less Ni, but better dispersed and hence with smaller crystallite size, be equally suited, if not better for the methanation reaction ?

E.C. KRUISSINK : Our answer to this point is speculative as we have not yet examined in any detail the effect of nickel content

on stability. We have reason to believe that the high stability
of the reduced catalyst results from the well defined structure of
the precipitate : the aluminium and nickel are intimately mixed in
the brucite-like layer and remain associated with one another
during calcination and reduction. For Ni/Al>2, separate alumina
phases result and these may be detrimental to the life of the
catalyst, blanketing the active components. Catalysts with
Ni/Al≈3 are preferred for industrial steam reforming applications;
however, this may be because the separate phases of alumina present
at low nickel contents encourage undesirable coking of the catalysts
and this objection is not of such importance under methanation
conditions.

MATIJEVIC : At pH 10 aluminum hydroxide should not coprecipitate
as it is soluble as aluminate and as such it would react with
nickel hydroxide. However, in the presence of carbonate basic
sparingly soluble complexes could account for the coprecipitation.
Indeed, in table 3 no data in the presence of nitrate at pH 10 are
given. Obviously, the question arises can nickel hydroxide
coprecipitate with nickel nitrate solution at this pH.

E.C. KRUISSINK : Indeed, at pH=10 no precipitate was observed in
the absence of carbonate ions, from an aluminium nitrate solution.
However, solubility of aluminium hydroxide as aluminate $Al(OH)_4^-$ is
$2.0 \cdot 10^{-3}$ mole/l at pH=10, according to data of Sillén (1964). This
is small compared with the Al concentration of our starting solution
which is about 0.2 mole/l. Therefore we think $Al(OH)_3$ may be
dissolved in a colloidal state.
The situation is totally different when both Ni(II) and Al(III) ions
are present. We carried out a precipitation at pH=10, from a
nickel and aluminium nitrate solution, rigorously excluding CO_2.
The precipitating agent was sodium hydroxide. Under these conditions
Al precipitates in a nickel-aluminium hydroxy-nitrate, crystallizing
in a hydrotalcite-like structure. The same is known for the Mg-Al
system (Miyata, S. 1975).
Sillén, L.G. "Stability constants of metal-ion complexes".
Chem. Soc. London Spec. Pub. 17, 1964.
Miyata, S. Clays and Clay Minerals 23, 369-375(1975).

P.E.H. NIELSEN : Could you elaborate, i.e. give data on the observed

$NiAl_{2\delta}O_{1+3\delta}$ phase, such as crystal size and lattice parameters.

E.C. KRUISSINK : Crystal size of the $NiAl_{2\delta}O_{1+3\delta}$ phase depends on the anion present in the starting material, and may vary from 40 Å to > 500 Å (see our paper).
The well-crystallized samples show d-values equal to those of pure NiO. For the samples consisting of small NiO particles the following d-values were measured : 2.402 ($\pm$ 5); 2.076 ($\pm$ 5); 1.470 ($\pm$ 5) for respectively the 101, 012 and 110-104 NiO reflections (indexed on a hexagonal basis). The corresponding d-values of pure NiO are 2.412, 2.088 and 1.477-1.476.

A. OZAKI : $Ni(NO_3)_2$ usually gives a better catalyst than does $NiSO_4$ when it is precipitated on SiO_2. This effect seems to come from NO_3 ion incorporated in the precipitate. In view of this I would like to ask your idea on the role of carbonate ion although you are working on Ni/Al_2O_3.

L.L. VAN REIJEN : The role of carbonate ion in our preparations is to prevent the incorporation of nitrate or chloride ions. As we have shown in our paper, incorporation of these ions may cause sintering of nickel oxide during the calcination stage.
Probably also incorporation of sulphate ions is detrimental. P.B. Wells and coworkers have recently shown that Ni/SiO_2 samples prepared by impregnation with sulphate are contaminated by sulphur.

D.L. TRIMM : What would you relate selectivity to ? Do you feel that it reflects crystallite size, residual anion, pore size or some other factor ?

E.C. KRUISSINK : We believe that the selectivity is associated with the degree of reduction of the catalyst, low degrees of reduction favouring the formation of higher hydrocarbons. There has recently been renewed interest in the mechanisms of the methanation and Fischer-Tropsch (FT) reactions; a number of authors favour the participation of surface carbon, formed by dissociative adsorption of CO, both in the methanation [1] and FT reactions [2], whereas for the latter, van Barneveld and Ponec [3] have suggested a mechanism involving the insertion of molecular CO into metal-carbon bonds. Our most recent results tend to favour the van Barneveld and

Ponec mechanism : at low degrees of reduction, the presence of a proportion of unreduced nickel ions on the surface (or, alternatively, sites where nickel is associated with the alumina) discourage dissociative adsorption of CO and hence favour the FT reaction, whereas high degrees of reduction cause the formation of larger nickel crystallites on which dissociative adsorption, followed by methanation, occurs. However, it is not necessary to invoke dissociative adsorption to explain methane formation: more extensive hydrogen adsorption may occur on the larger metal crystallites, encouraging methane formation rather than the FT reaction.

1. P.R. Wentreck, B.J. Wood and H. Wise, J. Catal., 43, 363(1976).
2. R.W. Joyner, J. Catal., 50, 176(1977).
3. W.A.A. van Barneveld and V. Ponec, J. Catal. 51, 426(1978).

J.L. WHITE : The crystalline phase of the material described by Kruissink et al. is the synthetic equivalent of minerals of the pyroaurite-söjrenite-hydrotalcite group. The structure of the synthetic material was worked out by Gastuche, Brown and Mortland in 1967. They showed Mg/Al ratios of 3:1 and 5:1 gave stable structures. The Na^+ observed by the authors is probably due to adsorption on amorphous aluminum hydroxide precipitated at a moderate to high pH. Pertinent references include : G. Brown and M.C. Gastuche (1967), Clay Miner. 7, 193; M.C. Gastuche, G. Brown and M.M. Mortland (1967), Clay Miner. 7, 177; B. Kobo, S. Miyata, T. Kumura and T. Shimada (1969), Yakuzai Gaku 29, 55; C.J. Serna, J.L. White and S.L. Hem (1978), J. Pharm. Sci. 67, 324.

D.J.C. YATES : What factors led to your choice of sodium carbonate and sodium hydroxide as precipitants in view of the fact that sodium is a well-known poison for nickel catalysts ?

J.R.H. ROSS : Initially we have adopted commercial practice, in which it is considered that the sodium can be washed out of sufficiently low levels. Our experience agrees with this conclusion for most conditions of preparation.

CONTROLLED CATALYST DISTRIBUTION ON SUPPORTS BY CO-IMPREGNATION

E. R. BECKER* and T. A. NUTTALL**
Chemical Engineering Research Group - CSIR, Pretoria, South Africa

ABSTRACT

The performance of catalytic reactions, catalyst durability and reaction selectivity can be controlled by appropriate catalyst distribution on supports. Burried catalyst layers on supports enhance rates of reactions which obey negative order kinetics and extend catalyst life when reactions are accompanied by pore mouth impurity poisoning.

The preparation of catalysts with active subsurface layers was studied via the impregnation of alumina support spheres with hexachloroplatinic acid in the presence of citric acid. The influence of the impregnation process variables on platinum distribution were determined. Subsurface layers of platinum on a support particle with controlled depth and layer width were produced. Impregnation time, severity of reduction and the correct choice of an immobilization and drying sequence were found to be the critical variables in the process. The impregnation times measured 5 minutes. Mild reduction of the wet particles in hydrazine vapor succeeded in preserving concentrated, narrow platinum bands inside the support. The depth of the platinum layers was reproducibly controlled by the citric acid concentration in the impregnating solution.

INTRODUCTION

Catalysts with bands of active ingredients concentrated near the support particle exterior are well known and have been in commercial use for some time, e.g., reforming catalyst. The eggshell catalyst distribution minimizes the diffusion path of the molecules which serves to maximize the effective utilization of the catalyst material on the support. In the absence of poisoning or selectivity considerations, the eggshell catalyst distribution is optimal for catalytic reactions with positive order rate dependence on reactant concentrations. The eggshell catalyst is not the most suitable for reactions which have negative order rate dependence on reactant concentrations or for reactions which are accompanied by pore mouth poisoning.

---

*to whom correspondence should be addressed; at: Air Products and Chemicals, Inc. P.O. Box 538, Allentown, PA 18105 U.S.A.
**currently with Dept. of Chem. Eng., Loughborough University, Loughborough, U.K.

Burried layer catalysts with active ingredient in an egg white or egg yolk position were calculated to be superior to eggshell catalysts or uniform catalysts by Becker and Wei (1), (2). The systems they studied were CO oxidation over platinum and first order reactions with impurity poison deposition. The calculations showed that oxidation rates of CO could be enhanced by several factors in catalyst particles with an egg white distribution of active ingredient. Experimental work in the General Motors Research Laboratories (3) have demonstrated the substantial advantages of automotive catalyst performance and catalyst durability by controlled subsurface platinum and paladium layers on an alumina support. Other notable contributions to the concept of controlled distributions on supports and their effect on reaction rate, selectivity and durability are those of Corbett, et al. (4), Shadman Yadzi, et al. (5), Cervello, et al. (6), Michalko (7) and Roth, et al. (8).

Maatman and Prater (9), (10) have explored the effects of competitive species in a platinum solution to achieve uniformly distributed platinum on alumina catalyst supports. Michalko (7) suggested the use of co-impregnation to achieve interior layers of platinum on alumina for automotive oxidation catalysts.

This research is an effort to determine the feasibility of manufacturing catalysts with accurately controlled distribution of active ingredients in support particles. An understanding of the underlying impregnation processes is sought to determine the influence of the processing variables on the resultant catalyst distributions.

CATALYST IMPREGNATION

The preparation of a catalyst by impregnation of a support consists of several distinct processing steps. The wetting of the support with impregnating solution containing the active specie is followed by drying and immobilization of the active ingredient, not necessarily in that order. In the process of co-impregnation, one or more chemical species are added to the solution to modify the deposition characteristics of the active ingredient in order to achieve a desired distribution effect within the catalyst support particle.

The experiments carried out in this research consisted of wetting a dry alumina support with platinum containing solution. Solution is transported into the pores by capillary suction. The characteristic time governing the capillary suction in a cylindrical pore may be characterized by: (10)

$$t = 2\eta x^2 / \sigma r$$

where t is the penetration time in seconds, σ the surface tension of the solution, r the mean pore radius, η the dynamic viscosity of the liquid, and x the effective capillary length. For $r = 10^{-6}$m, $\eta = 10^{-3}$Ns/m^2, $\sigma = 72 \times 10^{-3}$N/m and $x = 10^{-3}$m, a penetration time of 0.03 seconds is predicted. Liquid flow into the pores is accompanied by adsorption of the active ingredients on the pore walls. The rate of adsorption and desorption of the active species and their relative amounts determines the shape of the distribution curve. If left to equilibrate, the adsorbed species will redistribute inside the pores. Only rapid immobilization, by reduction in the case of platinum chloride, can preserve the distributions achieved during the impregnating step.

EXPERIMENTAL

Spherical alumina support pellets from Universal Oil Products (UOP) measuring 3.2 mm and 1.6 mm in diameter were chosen for impregnation. Three-gram batches of either pellet size were dried and heated at 250°C in air for 8 hours preceding impregnation.

Seven impregnation solutions were prepared. An aliquot of 10 ml, well in excess of the catalyst pore volume, contained a fixed amount of platinum. The solutions contained citric acid varying from zero to 37 g/l. Compositions of the solutions, numbered one to seven, are given in Table 1.

TABLE 1
Compositions of impregnating solutions: Platinum concentration = 5.8 g/l.

Solution Number	1	2	3	4	5	6	7
Citric Acid Conc. g/litre	0	0.925	1.85	3.70	9.25	18.5	37.0
Citric Acid Conc. $\frac{\text{Mass of Citric Acid}}{\text{Mass of Alumina}}$ %	0	0.31	0.62	1.23	3.10	6.20	12.3

Seven batches of each size of dry pellets were immersed and stirred in solutions 1 to 7. After a measured contact time the solution was filtered off. The pellets were washed briefly with distilled water to remove interparticle solution. The wet pellets were transferred to a fluidized bed tube for reduction. Alternatively, the wet pellets were dried in an unreduced state. Impregnation times varied from 2 to 30 minutes using solution 4 to determine the effect of contact time on internal platinum distribution. In all other experiments an impregnation time of 5 minutes was chosen.

The absorbed platinum chloride was reduced in a hydrazine containing nitrogen gas stream at ambient temperatures. The color of the pellets changed from a light yellow to black indicating the completion of the reduction. Drying of the reduced pellets at 90°C completed the catalyst preparation.

The platinum distribution within the support bead was measured in a scanning electron microprobe (EMP). The plastic mounted pellets were sectioned equatorially.

This cross section was analyzed by the scanning EMP. Three scans were taken across the section as indicated in Figure 1 by a typical scanning trace. The signal intensity was assumed to be proportional to the platinum concentration.

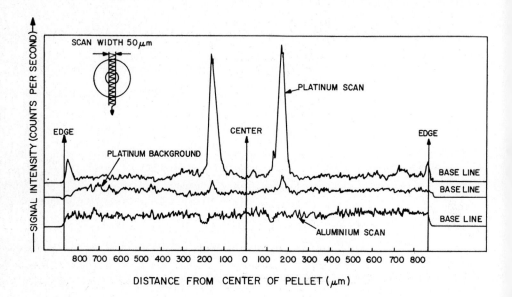

FIGURE 1: Electron microprobe profile of an impregnated alumina sphere

RESULTS

SUPPORT PROPERTIES

The properties of the two aluminas used in this study are listed in Table 2.

TABLE 2

Properties of 3449-11 A & B UOP alumina pellets

Catalyst Support	Support Diameter 10^{-3} m	Modal pore radius μm	Porosity m^3/kg	Surface Area m^2/kg	Average bulk density kg/m^3
3449-11B	1,6	$4,7 \times 10^{-2}$	$2,35 \times 10^{-3}$	$1,73 \times 10^5$	295
3449-11A	3,2	$1,7 \times 10^{-2}$	$1,36 \times 10^{-3}$	$1,49 \times 10^5$	385

PLATINUM DISTRIBUTION

The photograph of the sectioned 1.6 mm alumina spheres, taken under a microscope and shown in Figure 2, clearly shows the platinum bands against the alumina background. The relative sharpness of these rings is believed to depend primarily on the elapsed time between impregnation and reduction.

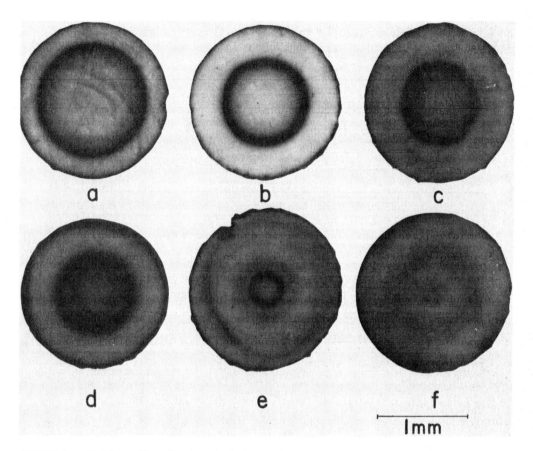

FIGURE 2: Platinum distribution in 1.6 mm diameter alumina spheres: a: solution 1 b: Soln 2 c: Soln 3 d: Soln 4 e: Soln 5 f: Soln 6.

EFFECT OF CITRIC ACID CONCENTRATION: A comparison of the normalized EMP concentration profiles shows that the platinum band is deposited deeper towards the center of the pellet with increasing citric acid concentration. Within the errors of measurement, the depth of platinum penetration is a linear function of citric acid concentration (11). Figure 3 shows the relative profiles for the 3.2 mm alumina "spheres". Since these pellets are more elipsoid in shape than the 1.6 mm pellets, the radii (or edge to elipse-focus paths) were scanned. A more complete description may be found elsewhere (11).

For citric acid concentrations exceeding 3.1%, no distinct layer of platinum is observed in the smaller pellets. Figure 2 shows a diffuse double ring of a grey shade in contrast to the black platinum ring in Figures 2 (a)-(c). The

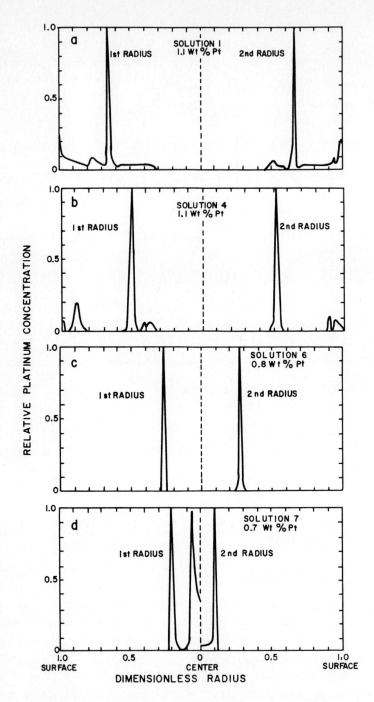

FIGURE 3: Normalized radial EMP platinum profiles in 3.2 mm diameter alumina beads.

width of the platinum band varied from 50 µm in 3.2 mm diameter particles to 430 µm in the 1.6 mm alumina supports.

The reproducibility of the distribution is shown in Figure 4 and corresponds to a maximum variation of ± 10% of the radius of the support. For example, the depth of the catalyst layer in Figure 4 measures 0.53 mm with a standard deviation of 0.07 mm.

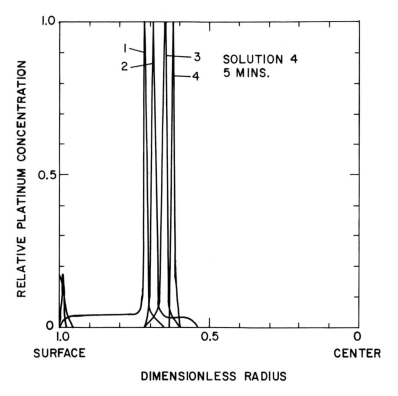

FIGURE 4: Experimental variation of platinum distribution in a 3.2 mm diameter alumina bead.

EFFECT OF IMPREGNATION TIME: Experiments using different impregnation times showed that platinum subsurface layers were already evident after the first two minutes. Pellets impregnated longer than 10 minutes revealed a general grey background (instead of white), suggesting a broader distribution of platinum throughout the support.

The EMP analysis confirmed the broader distribution and is shown in Figure 5.

EFFECT OF REDUCTION AND DRYING: When drying of the impregnated pellets preceded reduction, a "washout" peak of platinum was observed. This washout peak was

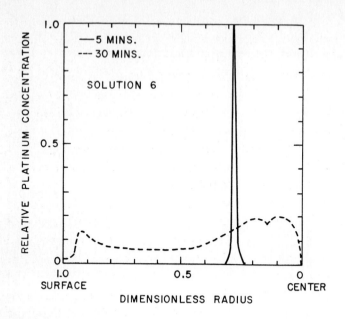

FIGURE 5: Effect of impregnation time on platinum distribution in 3.2 mm diameter alumina beads.

displaced from the original peak toward the exterior of the support bead giving rise to a bimodal platinum distribution of wide and varying shape.

A wet reduction procedure was chosen to avoid the redistribution of platinum during the drying process. This technique, described above, was successful in preserving what was thought to be the original catalyst distribution. The severity of the hydrazine reduction had to be controlled to avoid the dislocation of reduced platinum particles from the surface. Reduction of the wet supports in a 5% hydrazine solution resulted in loss of platinum from the support and redistribution of platinum inside the support beads. Further details of the reduction effects on distribution may be found elsewhere (11).

GENERAL DISTRIBUTION CHARACTERISTICS: The noticeable features exhibited by the platinum profiles obtained with the co-impregnation method are:
- One sharp peak of platinum per radius i.e. concentration of platinum in a small volume of the support (an exception is shown in Figure 3d).
- There was invariably some platinum present on the particle exterior.
- The regions of low platinum concentration were essentially platinum-free in most cases.

DISCUSSION

A catalyst impregnation technique has been demonstrated to achieve controlled distribution of platinum within an alumina support. Subsurface layers as thin as 50 μm and as deep as 0.8 of the bead radius can be obtained.

The depth of a platinum subsurface layer can be predetermined to an accuracy of ± 10% of the pellet radius by adjustment of the citric acid concentration.

Short impregnation times of 5 minutes followed by rapid, but mild reduction in hydrazine vapor were found to give the best results.

Platinum bands in the 3.2 mm diameter supports were thinner but less uniform than those in the 1.6 mm diameter supports. The role of the support characteristic could not be clearly determined from these experiments.

The results of this exploratory research have shown that it is possible to control active ingredient distribution within a support to a degree not heretofore published. The shape of the platinum distributions and the time required to form subsurface layers suggests a mechanism of selective adsorption coupled with capillary suction. A chromatographic separation of platinum hexachloride and citrate ions is envisaged. This was in part confirmed by similar distributions measured even when the pellets were prewetted. To gain a better understanding of the impregnation mechanism, the surface chemistry and the ion-transport mode needs to be understood.

The ability to accurately control internal catalyst distribution on supports adds a valuable tool to the resources of the catalytic reaction engineer and should result in catalysts with better performance characteristics as has been demonstrated by Summers and Hegedus (3) for automotive oxidation catalysts.

ACKNOWLEDGEMENTS

The authors are indebted to Mr. W. G. B. Mandersloot for supporting the research. The analytical contribution of J. Thirlwall and Dr. Colborn is gratefully acknowledged. Alumina supports were kindly supplied by Universal Oil Products and the Houdry Division of Air Products and Chemicals, Inc. U.S.A.

REFERENCES

1. E. R. Becker and J. Wei, J. Cat., 46 (1977) 365-371.
2. E. R. Becker and J. Wei, J. Cat., 46 (1977) 372-381.
3. J. C. Summers and L. L. Hegedus, J. Cat., 51 (1978) 185-192.
4. W. E. Corbett and D. Luss, Chem. Eng. Sci., 29 (1974) 1473-1783.
5. F. Shadman-Yadzi and E. E. Petersen, Chem. Eng. Sci., 27 (1972) 227-237.
6. J. Cervello, J. F. Garcia de la Banda, E. Hermana and J. F. Jimenez, Chem. Ing. Tech., 48 (1976) 520-525.
7. E. Michalko, U. S. Patent 32 59 589 (1966).
8. J. F. Roth and T. E. Reichard, Jnl. of the Res. Inst. Cat., Hokkaido Univ. 20 (1972) 85-94.
9. R. W. Maatman and C. D. Prater, Ind. Eng. Chem., 49 (1957) 253-257.
10. R. W. Maatman, Ind. Eng. Chem., 51 (1959) 913-914.
11. T. A. Nuttall, CSIR Report CENG 182, Catalyst with subsurface active layers prepared by co-impregnation, (1977), CSIR, Box 395, Pretoria, South Africa.

DISCUSSION

K. KOCHLOEFL: Do you think that e.g. formaldehyde can be used instead of hydrazine?

E.R. BECKER: Yes, other reducing compounds will give similar effects, however the conditions of treatment may be quite different from those for hydrazine.

G.H. VAN DEN BERG: What is the explanation for the phenomenon described, viz. the subsurface deposition of platinum occurring at zero citric acid concentration.

E.R. BECKER: I have no good explanation for this phenomenon. However, it is unique to the UOP alumina support spheres. The homogeneity and the preparation method of these unusually light alumina supports was not determined. Other aluminas, e.g. Rhône Poulenc alumina spheres, did not show this behaviour and the penetration distance of platinum "rings" is zero in the absence of citric acid.

E.A. IRVINE: Are there any conditions (i.e. time length, pH) at which platinum comes out of solution? If so, what effect does this have on the platinum dispersion, and on factors such as reproducibility?

E.R. BECKER: During impregnation there is no precipitation of platinum from solution. In these experiments the pH of the solution was between 3 and 5. The precipitation is observed during reduction of the wet pellets. Under strong reducing conditions black platinum sludge is formed and redistributes within the support. The dispersion of platinum was not measured in the catalysts, as they have very high local concentrations of platinum, up to 20 wt%. Rapid precipitation from solution also results in poor reproducibility of platinum distribution within the support particles.

D.E. WEBSTER: Some years ago we observed a similar "tree-ring" effect with $Pd(NO_3)_2/HNO_3$ solutions on alumina pellets. More recently, this has not been observed on the same supplier support, suggesting a strong dependence on the alumina base. Chloroplatinic acid reacts with citric acid at circa 90°C. Does this reaction

occur under your conditions ?
What do you see as the mechanism of the time - dependence of the platinum distribution ?

E.R. BECKER : 1. Our experiments would confirm your observation that the alumina base does have an important role in this process of co-impregnation.
2. We can rule out any reaction of chloroplatinic acid with citric acid at the temperatures of the impregnation experiments, viz. 25°C.
3. We envisage a chromatographic separation of the citrate ions from the platinum chloride ions. The interaction of platinum ions with the alumina surface is thought to be chemisorption and physical adsorption of precipitate. Redistribution occurs due to the mobility of the platinum ions coupled with the high local concentration gradients. The characteristic times of the impregnation processes cover reaction times (adsorption) to diffusion times (redistribution). I feel the key to successful distribution lies in accurate time control of all steps.

D.C. TRIMM : Presumably the mechanism must involve diffusion - either of citric acid or the platinum salt. Did you attempt to model the phenomenon for dried or prewetted supports in terms of diffusion effects ? If so, which do you consider more important in determining the width of the platinum ring ?

E.R. BECKER : Certainly the prewetted pellet impregnations involve diffusion coupled with chemical reaction. Diffusion plays a secondary role in the dry pellet experiments. Dr. Hegedus, in the following presentation shows a model for prewetted pellets. The modeling of dry pellet impregnation has not been attempted to the best of my knowledge.

M. MARTAN : Did you find any difference between the UOP and Houdry alumina ? Did you try to correlate the amount of Pt adsorbed with the basic sites/acid sites ?

E.R. BECKER : Yes, there was a difference in impregnation characteristics from one support to another. Houdry supports show less penetration of platinum for an equivalent amount of citric acid. We did not correlate adsorbed platinum with acid or basic sites.

We found that the amount of platinum on the support was very close to the amount contained in the volume of solution adsorbed.

H. CHARCOSSET : What about the particle size or the percent dispersion of platinum after reduction ?

E.R. BECKER : We did not measure platinum particle size or dispersion as the platinum levels in our experimental catalysts were very high, viz. up to 20 w/w. These concentrations are not comparable to industrial catalysts such as CO combustion catalysts which would have maximum platinum levels of 1 w/w platinum.

R. POISSON : Does your method work with other complexing ions of alumina like F^- or EDTA ? If yes, do you know the mechanism ?

E.R. BECKER : Dr. Hegedus shows in the following paper that F^- is indeed suitable as a competing ion in palladium impregnation. The mechanism is related to the transport of ionic species into the catalyst support and their relative reactivities with the surface.

R.P. SIEG : Did you find an effect on Pt profile of impregnation time prior to reduction ?

E.R. BECKER : Yes. Times greater than 5 minutes caused broadening of profile.

MULTICOMPONENT CHROMATOGRAPHIC PROCESSES DURING THE IMPREGNATION OF ALUMINA
PELLETS WITH NOBLE METALS

L. L. HEGEDUS, T. S. CHOU, J. C. SUMMERS, and N. M. POTTER
General Motors Research Laboratories, Warren, Michigan, 48090 USA

ABSTRACT
A mathematical model of a competitive, multicomponent diffusion-adsorption process is presented, and applied to the impregnation of porous γ-alumina pellets by a Rh complex, in the presence of HF as a site blocking agent which drives the Rh below the surface of the pellets. The model correctly predicts the measured Rh peak penetration depths and Rh uptakes from solution.

INTRODUCTION
Due to its technical importance and scientific interest, the literature of catalyst impregnation has been rapidly growing. Especially interesting are problems related to nonuniform catalyst impregnation, that is, when an activity profile exists along the radius of the porous catalyst pellets.

Both the experimental [e.g., Briggs, et al. (ref. 1), Hoekstra (ref. 2), Whitman and Leyman (ref. 3), Retallick (ref. 4), Roth and Reichard (ref. 5), Summers and Hegedus (ref. 6)] and theoretical [e.g., Kasaoka and Sakata (ref. 7), Minhas and Carberry (ref. 8), Shadman and Petersen (ref. 9), Corbett and Luss (ref. 10), Wei and Becker (ref. 11), Smith (ref. 12), Villadsen (ref. 13), Becker and Wei (ref. 14,15), Cervello et al. (ref. 16), Hegedus and Summers (ref. 17)] literature recognize the importance of nonuniform impregnation profiles in affecting the activity, selectivity, or durability of catalysts. Intrapellet impregnation profiles usually become most important when the main reactions or the poisoning reactions (or both) are significantly influenced by intrapellet diffusion resistances.

The preparation of nonuniformly impregnated catalysts has also received a great deal of attention in the literature. Maatman and Prater (ref. 18) provided an early analysis of the roles of adsorption and solvent exclusion in determining the activity profiles; Maatman (ref. 19) showed how Pt can be uniformly distributed by the addition of salts or acids to the impregnating solution; Michalko (ref. 20) showed the use of site blocking agents (various organic acids) to achieve an impregnated section at the pellet's edge, below the pellet's edge, or in a core at its center; and finally, Chen and Anderson (ref. 21,22) discussed various impregnation techniques leading to nonuniform Cr profiles in γ-alumina spheres.

The nonuniform impregnation profiles usually arise due to a complex interaction of intrapellet flow, diffusion, adsorption, and reaction phenomena, depending on the particular system at hand and on the method of impregnation process employed. A comprehensive treatment of heterogeneous sorption-diffusion problems, as they apply to catalyst impregnation by a single component, was provided by Weisz et al. (ref. 23,24,25), including both irreversible and reversible processes. Harriott (ref. 26) analyzed an intrapore precipitation technique and found good agreement between a diffusion model's predictions and experiments with Ag-alumina catalysts. Cervello et al. (ref. 27) compared a diffusion-reaction model with measured Ni distributions in a NiO-alumina catalyst. In another work, Cervello et al. (ref. 28) provided an experimental and theoretical analysis of the variables which influence the distribution of the active components in porous pellets.

The effects of intrapellet flow during catalyst impregnation were first analyzed by Maatman and Prater (ref. 18), and then in more detail by Vincent and Merrill (ref. 29).

The purpose of this work is to provide a theoretical and experimental analysis of multicomponent catalyst impregnation. Such processes are important when the catalyst is either simultaneously impregnated by two or more active components which may compete with each other for the sites of the support, or when site blocking agents are used to manipulate the intrapellet profiles of one or more active components.

First we will outline a somewhat general theoretical description of the multicomponent diffusion-adsorption process, and then we will illustrate the applicability of the model to Rh-alumina catalysts prepared by the use of HF as a site blocking agent, so that the Rh is deposited below the catalyst pellet's surface.

THEORETICAL ANALYSIS

We will consider the simultaneous diffusion of n components (catalytic metals and site blocking agents), and their reaction (or adsorption) on the sites of the catalyst support. The support is prewetted, so that its pores are initially filled with the solvent.

The impregnation process can be described by a set of transient differential equations. For the liquid phase,

$$D_i \nabla^2 l_i - A\rho_p \frac{\partial s_i}{\partial t} = \epsilon \frac{\partial l_i}{\partial t}, \quad i = 1 \ldots n. \tag{1}$$

In formulating the solid phase conservation equations, we have to specify the mechanism by which the impregnating species interact with the support's surface. As we will see in the experimental part of this paper, different species have different saturation concentrations over the same support surface. Consequently, we use a Langmuir adsorption scheme where the various species are allowed to have different saturation concentrations. A physical interpretation of this mechanism implies species of differing molecular cross sections, covering differing number of support sites upon saturation, but anchored only at one site to the surface.

With the above mechanism, the solid phase conservation equations for all species i become

$$A\rho_p \frac{\partial s_i}{\partial t} = A\rho_p [k_{+,i} l_i (1 - \frac{N_A}{N_{s,i}} s_i - \frac{N_A}{N_{tot}} \sum_{\substack{j=1 \\ j \neq i}}^{n} s_j) - \frac{k_{+,i}}{K_i} \frac{N_A}{N_{s,i}} s_i],$$

$$i = 1 \ldots n. \tag{2}$$

N_{tot} is conveniently defined as the highest saturation concentration of all the species which participate in the process; in our experiments, the saturation concentration of the site blocking agent HF was found to satisfy this choice.

The boundary conditions are in part a function of the way the experiments are conducted, so they need some discussion. The solvent-saturated pellets (their number is N_p) are placed into a solution of volume V and stirred. Due to the finite volume of the impregnating solution, the concentration of species i in the surrounding liquid will change with time.

The following equations represent the initial and boundary conditions:

$$l_i(r,0) = 0, \quad i = 1 \ldots n. \tag{3}$$

$$s_i(r,0) = 0, \quad i = 1 \ldots n. \tag{4}$$

$$D_i \frac{\partial l_i}{\partial r}(R,t) = k_{m,i} (l_i(\infty,t) - l_i(R,t)), \quad i = 1 \ldots n. \tag{5}$$

$$\frac{\partial l_i}{\partial r}(0,t) = 0, \quad i = 1 \ldots n. \tag{6}$$

$$V \frac{\partial l_i(\infty,t)}{\partial t} = N_p 4\pi R^2 D_i \frac{\partial l_i(R,t)}{\partial r}, \quad i = 1 \ldots n. \tag{7}$$

$$l_i(\infty,0) = l_{io}, \quad i = 1 \ldots n. \tag{8}$$

where Equation (7) describes the conservation of species i in the bulk impregnating solution which surrounds the catalyst pellets.

Equation (2) can be used to determine the adsorption equilibrium isotherm of species i. At equilibrium, and for a single-component system:

$$\frac{\partial s_i}{\partial t} = 0, \text{ and so, from Equation (2),} \qquad (9)$$

$$\frac{1}{s_i(\infty)} = \frac{N_A}{N_{s,i}} + \frac{N_A}{N_{s,i} K_i} \frac{1}{l_i(\infty,\infty)} \qquad (10)$$

In Equation (10), $s_i(\infty)$ and $l_i(\infty,\infty)$ are, of course, the solid and liquid concentrations at equilibrium. Figures 2 and 5 illustrate how $N_{s,i}$ and K_i can be determined using equation (10).

Equations (1) to (8) were solved numerically on an IBM 370 computer, using the PDEPACK routine of Madsen and Sincovec (30). The results, together with the numerical values of the parameters involved, will be discussed later on.

EXPERIMENTAL PART

A commercial γ-alumina support was used, either in the form of spherical pellets or as a powder. The support's properties are shown in Table 1 below.

TABLE 1
Properties of the γ-alumina support employed

Pore volume	0.629 cm^3/g
Pore size distribution	Bimodal
Surface area (BET), A	97×10^4 cm^2/g (powder)
	89×10^4 cm^2/g (pellets)
Particle diameter, 2R	0.365±0.018 cm (spherical pellets)
	149-177 μm (powder)
Pellet density, ρ_p	1.09 g/cm^3
Pellet void fraction, ε	0.69

Before each experiment, the pores of the support were filled with a HCL solution at pH = 2.7. (This pH was chosen because the Rh solution which we used in the impregnation was also adjusted to pH = 2.7, using HCl).

At the beginning of each experiment, the wet support was suddenly brought into contact with a known volume of impregnating solution in the reaction vessel. The reactions were carried out in sealable, stirred containers which were thermostatically controlled at 31±0.5°C. Typical experiments used 4 to 20 g alumina (dry basis), and 100 to 250 cm^3 of impregnating solution (in excess of the pore volume).

Liquid samples of 0.5 cm^3 volume were drawn at predetermined intervals. HF was analyzed by a solid-state fluoride electrode, while Rh was determined by atomic absorption.

Impregnation profiles in the catalyst pellets were determined by ion microprobe mass analysis.

RESULTS AND DISCUSSION

a. <u>Rh-Alumina</u>

The interactions of noble metal complexes with alumina surfaces are often complex, and little investigated. Recent papers which deal with the subject were written by Santacesaria, et al. (ref. 31, 32), Summers and Ausen (ref. 33) and by Spek and Scholten (ref. 34). Instead of elucidating the detailed mechanism of the chemical interactions, our purpose is to develop a semiquantitative interpretation of the process, suitable to generate input parameters for our mathematical model.

The kinetics of the adsorption of our Rh complex on the alumina surface was determined in Rh uptake measurements as a function of time, on a powdered support so that diffusion effects are minimized. Figure 1 shows the data: the initial slopes were used to determine the forward rate constant $k_{+,Rh}$, while the long-time asymptotes ($l_{Rh}(\infty,\infty)$) served to evaluate the equilibrium constant K_{Rh} and the saturation concentration $N_{s,Rh}$.

The linearized plot of the equilibrium isotherm is shown in Figure 2, from which K_{Rh} and $N_{s,Rh}$ were determined. It is interesting to note that the saturation concentration of Rh does not reach a monolayer over the alumina (1.5×10^{14} molecules/cm^2 BET vs. approximately 10^{15} for a monolayer), partly due to the fact that not all the sites may attract Rh, and partly that one Rh complex molecule may cover several sites, even if we assume that it is anchored to only one of them.

The effective diffusivity of the Rh complex in the porous alumina pellets was determined by comparing the results of Rh uptake measurements on pellets with numerical diffusion-adsorption calculations [Equations (1) to (8)] in which all the parameters, with the exception of D_{Rh}, were evaluated from the powder experiments discussed above. As Figure 3 shows, $D_{Rh} = 3 \times 10^{-6}$ cm^2/s appears to be a reasonable approximation.

b. <u>HF-Alumina</u>

The kinetic and equilibrium parameters of this system were determined by techniques similar to those employed for Rh. Figure 4 shows the HF uptake

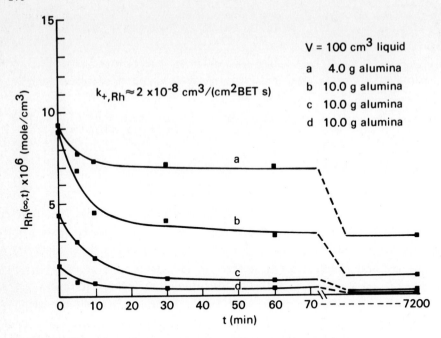

Figure 1. Rh uptake experiments on powdered alumina.

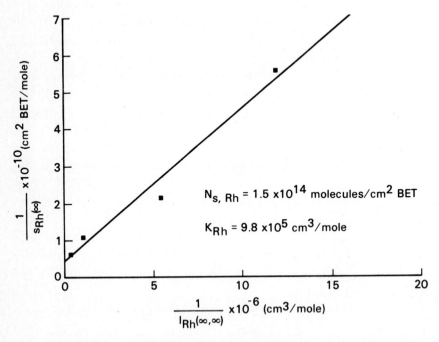

Figure 2. Linearized plot of the adsorption isotherm for Rh.

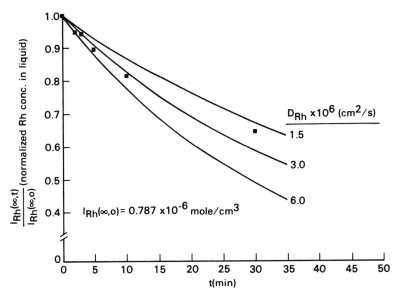

Figure 3. Spherical catalyst pellets: comparison of theory with experiment to determine the effective diffusivity of the Rh complex.

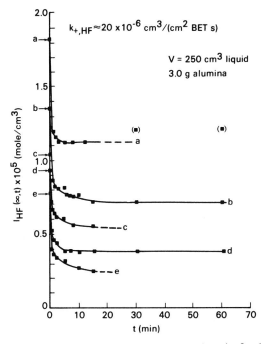

Figure 4. HF uptake experiments on powdered alumina.

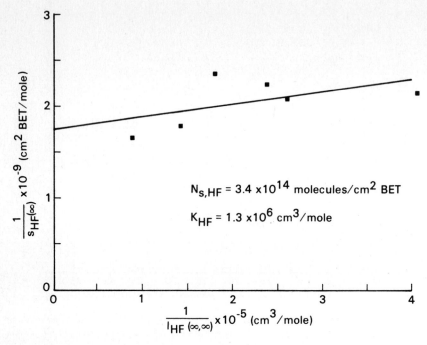

Figure 5. Linearized plot of the adsorption isotherm for HF.

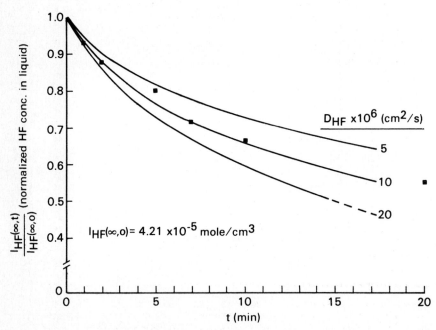

Figure 6. Spherical catalyst pellets: comparison of theory with experiment to determine the effective diffusivity of HF.

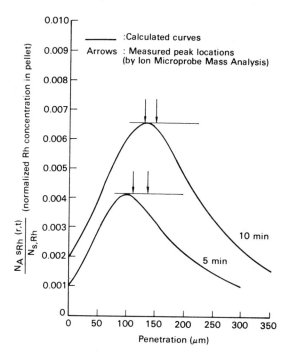

Figure 7. Rh distribution in the catalyst pellets in an experiment with HF + Rh: comparison of theory with measurements.

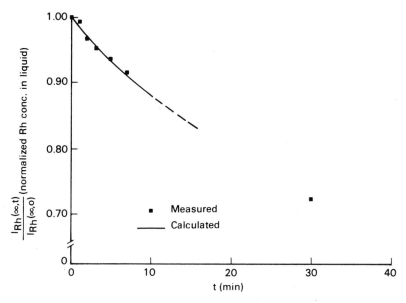

Figure 8. Rh uptake by porous catalyst pellets in an experiment with HF + Rh: comparison of theory with measurements.

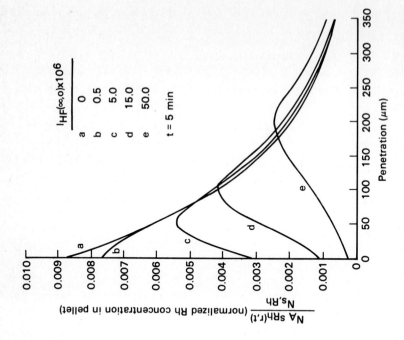

Figure 10. Effect of the initial HF concentration on the distribution of Rh in the pellets.

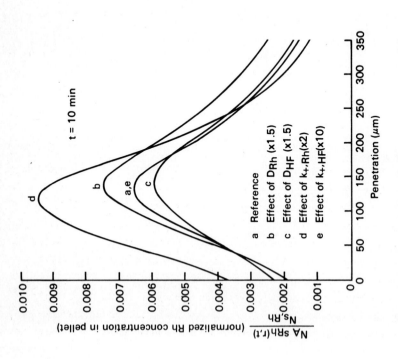

Figure 9. Effects of various parameters on the subsurface Rh peak (to illustrate parametric sensitivity).

experiments on powdered alumina. The initial slopes are very steep, resulting in a considerable uncertainty in determining $k_{+,HF}$. However, as we will see later on, $k_{+,HF}$ is so large that the uptake of HF in a catalyst pellet is essentially diffusion controlled. Consequently, both the indirectly determined effective diffusivity of HF and its impregnation profile are a weak function of $k_{+,HF}$.

Figure 5 displays the linearized adsorption isotherm for HF: notable is the approximately 2.3 times larger saturation concentration of HF than that of Rh.

Similarly to the case of Rh, the effective diffusivity of HF in the catalyst pellets was determined from a comparison of numerical calculations with HF uptake results on spherical pellets. The resulting effective diffusivity of HF (Figure 6) is approximately 3.3 times larger than that of the more bulky Rh complex: a reasonable finding.

c. **Rh-HF-Alumina**

We determined the parameters from independent experiments; let us now compare the predictions of our mathematical model with an experiment in which both Rh and HF are employed. The parameters employed in the corresponding computer runs are listed in Table 2.

TABLE 2
Parameter values for the Rh-HF impregnation of spherical alumina pellets

	Rh	HF
$l(\infty,0)$	0.412×10^{-6}	15.0×10^{-6}
k_+	2×10^{-8}	2000×10^{-8}
K	9.8×10^{5}	13×10^{5}
D	3×10^{-6}	10×10^{-6}
k_m	0.003	0.003
N_s	1.5×10^{14}	3.4×10^{14}

$V = 100$ cm^3, $N_p = 380$, $2R = 0.365$ cm, $\rho_p = 1.0$ g/cm^3

Figure 7 shows the computed Rh profiles in the pellets, at 5 and 10 minutes after the experiment began. The computed peak locations are in reasonable agreement with ion microprobe mass analysis measurements.

Beyond the location of the Rh peak, we are also interested in seeing if our model correctly predicts the amount of Rh which was taken up by the pellets. Figure 8 compares the measured and computed bulk liquid Rh concentrations as a function of time. Again, the model's prediction is quite reasonable.

To test the effects of inaccuracies in the numerical values of D and k_+, computer calculations were carried out to see how perturbations in them may affect the location and magnitude of the Rh peak in the pellets. As Figure 9 shows, relatively large perturbations in the above parameters resulted in modest changes in the peak locations.

Our model can be used to simulate a variety of impregnation experiments, with the purpose of modifying the location of the Rh peak in the pellets. Of the many possible variations, we illustrate only one here: the effect of initial HF concentration on the Rh distribution at t = 5 minutes (Figure 10). As the results indicate, the Rh peak can be flexibly positioned to its desired location by an appropriate choice of the initial HF concentration.

ACKNOWLEDGMENTS

The impregnation experiments and part of the analyses were carried out by S. A. Ausen.

NOMENCLATURE

A (cm^2/g)	BET surface area of the alumina
D_i (cm^2/s)	Effective diffusivity of species i in the catalyst pellets
$k_{+,i}$ $(\frac{cm^3}{cm^2\ BET\ s})$	Forward rate constant for species i
$k_{m,i}$ (cm/s)	Mass transfer coefficient for species i
K_i $(cm^3/mole)$	Equilibrium constant for species i
l_i $(mole/cm^3)$	Liquid phase concentration of species i
N_A (molecules/mole)	Avogadro's number
N_p	Number of pellets in beaker
$N_{s,i}$ $(molecules/cm^2\ BET)$	Saturation concentration of species i over the alumina surface
N_{tot} $(molecules/cm^2\ BET)$	Total number of sites per unit surface (see text)
r (cm)	Pellet radial coordinate
R (cm)	Average pellet radius
s_i $(mole/cm^2\ BET)$	Solid phase concentration of species i
t (s)	Time
ε	Pellet void fraction
ρ_p (g/cm^3)	Pellet density

REFERENCES

1   W. S. Briggs, W. A. Stover, and D. S. Henderson, U.S. Patent No. 3 288 558, November 29, 1966.
2   J. Hoekstra, U.S. Patent No. 3 360 330, December 26, 1967.
3   R. H. Whitman and E. Leyman, U.S. Patents No. 3 819 533, June 25, 1974, and 3 956 459, May 11, 1976.
4   W. B. Retallick, U.S. Patent No. 3 901 821, August 26, 1975.
5   J. F. Roth and T. E. Reichard, J. Res. Inst. Catalysis, Hokkaido Univ., 20(2)(1972)85.
6   J. C. Summers and L. L. Hegedus, J. Catalysis, 51 (1978) 185.

7   S. Kasaoka and Y. Sakata, J. Chem. Eng. of Japan, 1(2)(1968)138.
8   S. Minhas and J. J. Carberry, J. Catalysis, 14(1969)270.
9   F. Shadman-Yazdi and E. E. Petersen, Chem. Eng. Sci., 27(1972)227.
10 W. E. Corbett and D. Luss, Chem. Eng. Sci., 29(1974)1473.
11 J. Wei and E. R. Becker, Adv. in Chemistry Series, 143(1975)116.
12 T. G. Smith, Ind. Eng. Chem. Process Des. Dev., 15(3)(1976)388.
13 J. Villadsen, Chem. Eng. Sci., 31(1976)1212.
14 E. R. Becker and J. Wei, J. Catalysis, 46(1977)365.
15 E. R. Becker and J. Wei, J. Catalysis, 46(1977)372.
16 J. Cervello, J. F. J. Melendo, and E. Hermana, Chem. Eng. Sci., 32(1977)155.
17 L. L. Hegedus and J. C. Summers, J. Catalysis, 48(1977)345.
18 R. W. Maatman and C. D. Prater, Ind. Eng. Chem., 49(2)(1957)253.
19 R. W. Maatman, Ind. Eng. Chem., 51(8)(1959)913.
20 E. Michalko, U.S. Patents No. 3 259 454 and 3 259 589, both July 5, 1966.
21 H. C. Chen and R. B. Anderson, Ind. Eng. Chem. Prod. Res. Dev., 12(2)(1973)122.
22 H. C. Chen and R. B. Anderson, J. Catalysis, 43(1976)200.
23 P. B. Weisz, Trans. Faraday Soc., 63(1967)1801.
24 P. B. Weisz and J. S. Hicks, Trans. Faraday Soc., 63(1967)1807.
25 P. B. Weisz and H. Zollinger, Trans. Faraday Soc., 63(1967)1815.
26 P. Harriott, J. Catalysis, 14(1969)43.
27 J. Cervello, E. Hermana, J. F. Jimenez, F. Melo, in B. Delmon, P. A. Jacobs, G. Poncelet (editors): "Preparation of Catalysts," Elsevier, Amsterdam, (1976)251.
28 J. Cervello, J. F. Garcia de la Banda, E. Hermana, and J. F. Jimenez, Chem. -Ing.-Tech. 48(6)(1976)520.
29 R. C. Vincent and R. P. Merrill, J. Catalysis, 35(1974)206.
30 N. K. Madsen and R. F. Sincovec: "PDEPACK: Partial Differential Equations Package," Scientific Computing Consulting Services, 531 Zircon Way, Livermore, California.
31 E. Santacesaria and S. Carra, I. Adami, Ind. Eng. Chem. Prod. Res. Dev. 16(1)(1977)41.
32 E. Santacesaria, D. Gelosa, and S. Carra, Ind. Eng. Chem. Prod. Res. Dev.16(1)(1977)45.
33 J. C. Summers and S. A. Ausen, J. Catalysis, in press, 1978.
34 T. G. Spek and J. J. F. Scholten, J. Molec Catalysis, 3(1977/78)81.

DISCUSSION

V. FENELONOV : What is the difference between experimental and calculated data in the distribution along the pellet radius ?

L.L. HEGEDUS : Figure 7 shows computed Rh distribution profiles in the catalyst pellets at two selected times (5 min. and 10 min.). Only the penetrations at the peaks (indicated by arrows) could be compared with ion microprobe mass analysis data, because the ion microprobe instrument was not calibrated for absolute Rh concentrations. Such a calibration would be difficult due to the low Rh levels and due to the possible non-linearities involved.

L.L. MURRELL : Do your isotherms of HF and HCl adsorption take into account the fact that Al_2O_3 in aqueous solution is quite reactive toward acidic solutions such as oxalic and hydrochloric acid ?

L.L. HEGEDUS : Our catalyst supports were pretreated in an HCl solution (pH = 2.7) before the impregnation experiments. The isotherms for HF and Rh(III) were measured in the presence of HCl in the solutions (initial pH = 2.7) and should not be extrapolated to other HCl concentrations. While the chemistry of the problem is probably more complex than what our model implies, it seems that the simplifications we employed are permissible to predict the penetration of Rh into the pellets and the uptake of Rh from solution.

M.V. TWIGG : Rhodium(III) + chloride ions in aqueous solution form a number of complex ions. Their interconversion is not rapid, and their ratio at equilibrium is concentration dependent. In connection with the theory described which assumes but one type of rhodium ion, what are the rhodium species present in your impregnation solutions ? Were any interesting time or concentration effects observed ?

L.L. HEGEDUS : We are indeed aware of the fact that aqueous $RhCl_3$ solutions contain a number of partially hydrated species in equilibrium. Our model stipulates that their reactivity and diffusive properties are similar, so that they can be lumped into one hypothetical species. The good agreement of our computations with the experimental data indicates the validity of this assumption for modeling purposes. However, we decided to explore the adsorptive properties of various aquo-chloro-rhodium(III) species in further experiments.

J.W. HIGHTOWER : It appears from your Fig. 10 that all the deposition penetration curves at various HF levels blend into the zero HF curve. The total Rh content must then be decreased by the presence of HF. Rather than the Rh being "pushed inside" by the HF, the Rh adsorption sites near the external surface are simply poisoned by the HF. Would you agree with this interpretation ?

L.L. HEGEDUS : Figure 10 displays the computed Rh profiles at a fixed time elapsed (5 min.). Indeed, the presence of HF delays the accumulation of Rh in the pellets, since they compete for the same sites on the alumina surface.

FACTORS CONTROLLING THE RETENTION OF CHLORINE IN PLATINUM REFORMING CATALYSTS

S. SIVASANKER, A.V. RAMASWAMY and P. RATNASAMY

Indian Institute of Petroleum, Dehradun 248 005, India

ABSTRACT

The influence of various preparative and operational parameters on the uptake and retention of chlorine in Pt-alumina and Pt-Sn-alumina is reported. Catalysts based on eta-alumina adsorb and retain more Cl^- than those based on the gamma form. It is shown that Cl^- ion-exchanges for OH^- on the alumina surface. During the simultaneous adsorption of HCl and H_2PtCl_6, while Cl^- ions do not affect the uptake of Pt significantly on eta-alumina, there is a decrease in the case of gamma-alumina as the concentration of Cl^- increases. Apart from the type of alumina used (eta- or gamma-) the temperature of calcination/reduction and the water content of the air/H_2 used are found to be the two major factors which control the retention of chlorine during the activation of these catalysts. The nature of chlorine held on alumina is discussed.

INTRODUCTION

Chlorine has a major influence on the performance of naphtha reforming catalysts, such as Pt-alumina and Pt-Sn-alumina. It is incorporated into the catalyst formulation during the preparative stage by impregnation from a solution containing chloride ions (ref. 1). After impregnation with chlorine and platinum, the catalyst is dried at around 383 K and then calcined at around 773 K. Before use, the catalyst is reduced in a stream of hydrogen to convert the ionic platinum to the metallic state. During the above activation and reduction processes, the amount of chlorine retained on the catalyst is likely to change. No systematic study of the uptake of chlorine during the impregnation process and its retention during the subsequent calcination and reduction has so far been reported. For instance, are there differences between eta and gamma aluminas in their ability to adsorb and strongly retain chloride ions ? How is the incorporation of Pt affected by Cl^- ions present in the same solution ? During the activation and reduction of the catalyst, how does the purity of the gases (especially its moisture content) affect the loss, if any, of chlorine from the catalyst ? Does tin have any influence on the uptake and retention chloride ? The present study is an attempt to answer some of the questions.

EXPERIMENTAL

Two modifications of alumina, the gamma and the eta forms were used in this study. The gamma alumina (S_{BET} = 190 m^2/g) was obtained from Labofina and had been prepared from boehmite. The eta alumina (S_{BET} = 200 m^2/g) was prepared by hydrolysis of aluminium isopropoxide following the procedure of Yoldas (ref. 2). Water loss experiments indicated that this alumina contained more than 80% of the eta form.

The Sn- and Pt-alumina was prepared by impregnating the support with solutions of $SnCl_2$ and H_2PtCl_6, respectively, in the presence of small amounts of HCl to yield catalysts containing 0.4 wt.% of the metal. The impregnations were carried out at 293 K. The impregnated materials were dried at 383 K and then calcined at 673 K in air for 1.8 x 10^4 s. In the case of Sn-Pt-Al_2O_3, the support was first impregnated with Sn and calcined at 673 K prior to Pt impregnation.

Adsorption isotherms were obtained by equilibrating 5 g amounts of catalysts with 100 ml volumes of HCl (without or with added H_2PtCl_6 to give 0.4% Pt on the catalyst) for 8.28 x 10^4 s. Chloride ion uptake was estimated by titration of both the supernatant liquid and the catalyst. The supernatant liquid was also analysed for Al and Pt ions by emission spectroscopy and colorimetric methods, respectively. The dispersion of Pt was determined by the method of Benson and Boudart (ref. 3).

Extractions of the catalysts for chloride and Pt estimations were made with six 15 ml aliquots of NH_4OH (1N) for each sample. The extracts were mixed and analysed either for Pt by a colorimetric procedure (ref. 4) or for chloride ions by titration against $AgNO_3$ after acidification with HNO_3 (6N) (ref. 5,6). This extraction procedure was sufficient to remove all the chlorine from the catalyst in whatever form. For instance, in the case of 0.88% Cl-gamma-alumina sample, the first, second and third extractions removed 91.4, 7.7 and 0.9% respectively of the total chlorine present in the sample.

RESULTS AND DISCUSSION

The uptake of chloride ions - equilibria and rates

Adsorption isotherms of HCl on eta and gamma aluminas. The adsorp

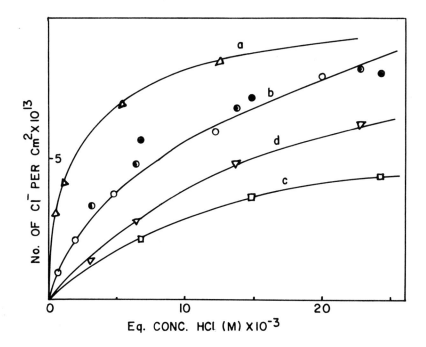

Fig. 1. Adsorption isotherms of HCl at 293 K on a) eta alumina -△-, b) gamma alumina -O-, c) Cl-gamma-alumina -□-, and d) Sn-gamma-alumina -▽-. Total Cl contents for Cl-gamma-alumina -●- and Sn-gamma-alumina -◐- are also shown.

tion isotherm of Cl^- ion on eta and gamma aluminas are shown in Fig. 1, curves a and b, respectively. The number of chloride ions adsorbed per cm^2 of alumina surface is higher on eta than on gamma alumina. The values range between 10^{13} and $10^{14}/cm^2$. It may be recalled that the concentration of surface hydrolysis on both eta and gamma aluminas is about $10^{15}/cm^2$ (ref. 7). The adsorption of mineral acids and aqueous electrolytes on alumina surfaces has been studied by Jacimovic et al. (ref. 8) and Ahmed (ref. 9). In such solutions, the occurrence of charged (±) or neutral surfaces at the oxide-solution interface is due to the formation of metal-aquo complexes as shown below:

$$[-Al(H_2O)_2OH]^+ \rightleftharpoons [-Al(H_2O)(OH)_2]^o \begin{matrix} \nearrow [-Al(H_2O)_2(Cl)]^+ \\ B \\ \searrow [-Al(H_2O)(OH)O]^- \end{matrix} \quad (1)$$

A

In species A, the anions (Cl^-, in our case) do not replace the surface hydroxyl groups but stay as counter-ions outside the primary hydration shell of the surface. In species B, the anions replace the OH⁻ ions and are thus attached directly to the metal cation. This latter case is analogous with the ion exchange of these OH⁻ ions. Since our adsorption isotherms were carried out at pH values lower than that corresponding to the zero point charge of alumina, the surface of alumina is expected to bear a net positive charge (species A or B) due to proton addition to the neutral aqueous complex.

In order to distinguish between the species A and B, the kinetics of adsorption of Cl⁻ ions on both eta and gamma aluminas was followed and the results are shown in Fig. 2 A. The rates of adsorption on eta alumina are higher than on gamma alumina. If the adsorption of Cl⁻ ions is essentially an ion exchange process (exchanging for surface OH⁻ ions), then the rate equation for exchange controlled by film diffusion in the shallow-bed method (ref. 10) should fit the observed kinetic data. In the rate question,

$$1 - Y/Y_\infty = \exp(-3Dt/RK\Delta r) \qquad (2)$$

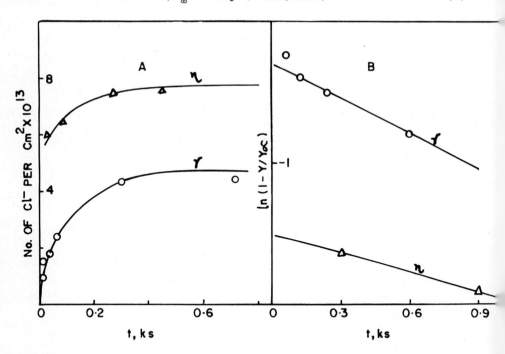

Fig. 2. Kinetics of adsorption of HCl on eta and gamma aluminas :
A. Plot of chloride ion uptake vs. time. B. Plot of data according to equation (2).
Y and Y_∞ are the amounts of Cl⁻ adsorbed at time t and after an infinite time interval, D is the diffusion coefficient in the boundary film, K is the equilibrium concentration of Cl⁻ in alumina

divided by the concentration of Cl⁻ in the solution, R is the radius of the alumina particles and Δr is the thickness of the boundary film. A linear plot of $\ln(1-Y/Y_\infty)$ vs. t shown in Fig. 2 B indicates that Cl⁻ ions ion-exchange with surface OH⁻ ions and are directly linked to the surface Al^{3+} cations.

Adsorption isotherms of HCl on Cl-gamma- and Sn-gamma-aluminas. Curves c and d of Fig. 1 illustrate the influence of preimpregnated HCl and $SnCl_2$ on the subsequent uptake of Cl⁻ from aqueous HCl. Even though both reduce the further uptake of Cl⁻ ions from aqueous HCl, the <u>total</u> amount of Cl⁻ retained in the alumina (i.e., the amount pre-impregnated plus that subsequently taken up) in equilibrium with aqueous HCl of comparable concentrations is similar to that on gamma alumina. This is shown in Fig. 1 where the data for the total Cl⁻ content on Cl-gamma- and Sn-gamma-alumina samples also fall on curve b, the isotherm data for the pure gamma alumina sample. Thus the capacity of the alumina surface for the retention of Cl⁻ ions in aqueous media depends mainly on its inherent structural features. Multiple impregnation techniques or pre-impregnation with tin chloride does not alter the total chlorine content of the surface. This picture of specific adsorption of Cl⁻ further supports the conclusions of the preceding section that Cl⁻ ion-exchanges with specific OH⁻ groups in the primary coordination shell of the surface Al^{3+} ions.

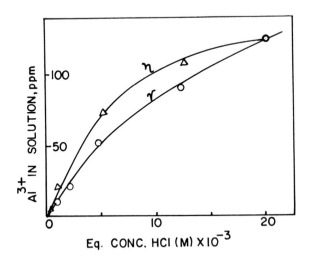

Fig. 3. Solubility of eta and gamma alumina in aqueous HCl.

<u>Solubility effects</u>. At the low concentrations of HCl employed in the present investigation, the solubility of Al_2O_3 was very low.

Jacimovic et al. (ref. 8) had reported negligible solubility for solutions whose pH was above 2.5. The solubility of eta and gamma alumina in aqueous HCl of different concentrations is illustrated in Fig. 3. At all values of the concentration of HCl, the eta form is more soluble than the gamma form.

Simultaneous adsorption of HCl and H_2PtCl_6 on eta and gamma aluminas. The isotherms for the simultaneous adsorption of chloride and platinum ions on eta and gamma aluminas are shown in Fig. 4 A and B. In these experiments enough H_2PtCl_6 was taken to give 0.4% wt. of Pt in the final Pt-alumina catalyst. Eta-alumina adsorbs and retains more Cl^- than the gamma sample (Fig. 4 A) at a given concentration of Cl^- in the aqueous solution. An interesting difference in the ability of adsorption and retention of platinum in the presence of Cl^- is evident from Fig. 4 B. While Cl^- ions do not affect significantly the uptake of Pt on eta alumina, there is a decrease in the case of the gamma alumina as the concentration of Cl^- increases.

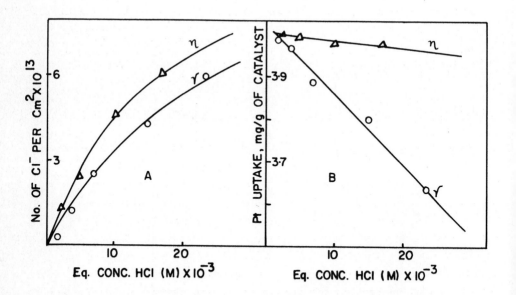

Fig. 4. Isotherms for simultaneous adsorption of chloride ions (A) and Pt ions (B) on eta and gamma aluminas.

TABLE 1

Influence of various calcination and reduction procedures on the chlorine content of the catalysts

Catalyst	%Cl after drying at 383 K	%Cl after calcination in dry air at 823 K	%Cl after calcination in static atm. contg. 15000 ppm of water at 673 K	823 K	%Cl after reduction at 773 K in dry H_2	H_2 contg. 1500 ppm water
γ-alumina	0.26	0.25	-	-	-	-
"	0.55	0.51	-	-	-	-
"	0.71	-	0.55	0.15	0.65	0.61
"	1.12	0.87	-	-	-	-
"	1.53	0.90	-	-	1.26	0.88
η-alumina	1.20	0.98	0.85	0.52	-	-
Sn-γ-alumina	(0.79)	-	0.60	0.20	-	-
Pt-γ-alumina	(0.70)	-	0.52	0.22	-	-
Pt-γ-alumina	(0.54)	-	0.52	0.22	0.51	0.48
Pt-Sn-γ-alumina	(0.71)	-	0.70	0.22	0.70	0.72
Pt-η-alumina	(0.57)	-	0.56	-	0.56	0.54

Note : a) The values are based on analysis, excepting those given in parentheses, which are based on theoretical values. b) Metal loadings are 0.4% by wt. for both Pt and Sn. c) All drying, calcination and reduction were done for 1.8×10^4 s. d) Flow rate of hydrogen was 1.7×10^{-6} m^3 s^{-1} during reduction.

Influence of calcination and reduction conditions on the retention of chlorine

Table 1 illustrates some of our results. In addition, the retention of chlorine during the reaction of n-heptane was measured. A mixture of H_2 and n-heptane (mole ratio, 2.7) was passed through the catalyst bed at 773 K for 1.44×10^4 s. The feed rate of n-heptane was 5.6×10^9 mole s^{-1} kg^{-1}. The catalysts were held in H_2 at 773 K for 3.6×10^3 s prior to introduction of n-heptane. Care was taken to dry the n-heptane feed thoroughly. The moisture content of the H_2 stream was controlled by passing through a NaOH solution. The concentration of Cl^- ion in the catalyst was estimated at the end of the run. In the case of the 0.4 wt.% Pt-gamma-alumina catalyst, there was no loss of chlorine in a H_2 + n-heptane atmosphere if water was absent. The Cl^- content decreased from 0.54 to 0.49% by wt. if the feed stream contained 1500 and 25,000 ppm of water. Experiments in which toluene was replaced for n-heptane did not reveal any influence of the type of hydrocarbon on the loss of Cl^- from the catalyst.

Temperature of calcination/reduction and the water content of the air/H_2 used during the process are the two major factors which control the retention of chlorine during the activation of the Pt-alumina or Pt-Sn-alumina catalysts. The following points may be noted :

1) A decrease in the chlorine content with increasing temperature of calcination is observed in the case of eta and gamma alumina supports. In the case of 0.4 wt.% Pt-gamma-alumina also, an increase from 673 to 823 K of the calcination temperature at constant H_2O content decreased the chlorine content from 0.52 to 0.22 % by wt.

2) The amount of chlorine retained in eta alumina at a calcination temperature of 823 K decreases from 0.98 to 0.52 % by wt. when the H_2O level is increased from essentially dry conditions to 15,000 ppm.

3) During reduction in hydrogen, the presence of H_2O leads to a loss of chlorine only if the chlorine level is above 0.7 % by wt. Below this value, there is no loss of chlorine in the presence of H_2O (at least up to a level of 1500 ppm of H_2O). Chlorine is lost from alumina and Pt-alumina in the form of HCl as well as $AlCl_3$. Our studies indicate that from pure alumina, chlorine is lost predominantly as $AlCl_3$ in a stream of dry H_2. In moist H_2 (1500 ppm of H_2O) evolution of HCl occurs.

TABLE 2

Concentrations of extractable Pt complex after calcination in air

Catalyst[a]	Temperature of calcination, K[b]	Percentage of Pt extracted
Pt-γ-alumina	383	8.8
Pt-γ-alumina	673	2.0
Pt-η-alumina	383	2.2
Pt-η-alumina	673	3.2
Sn-Pt-γ-alumina	383	24.5
Sn-Pt-γ-alumina	673	2.7

[a] The catalysts contained 0.4 % Pt by wt. and 0.4 % Sn by wt.
[b] All calcinations were carried out in a static atmosphere of air containing 15,000 ppm of H_2O. Duration was 1.8×10^4 s.

4) Retention of Cl^- ion during the activation process is generally more on Pt-eta-alumina than on Pt-gamma-alumina.

5) While Pt dispersion values were insensitive to the presence of H_2O (upto 1500 ppm) during reduction in H_2 at 773 K in the case of

Pt-gamma-alumina and Pt-Sn-gamma-alumina, there is a drastic reduction from 93 to 59 % in the case of Pt-eta-alumina. In Pt-alumina, in addition to the chlorine held by the support, part of the chlorine will also be associated with Pt. The variation in the concentrations of the platinum extracted by hot NH_4OH (1N) after calcination in air is shown in Table 2. Incorporation of tin increases the concentration of extractable Pt in the catalyst dried at 383 K. Calcination at 673 K in moist air leads to a drastic diminution due to its conversion into different species.

The nature of chlorine held by alumina

As noted earlier, the impregnation of Cl^- on Al_2O_3 at low concentrations is essentially an ion-exchange process with surface hydroxyl groups being replaced by Cl^-. What are the surface hydroxyls that are most liable to exchange for Cl^-? A recent model for the surface hydroxyls of alumina (ref. 7) classifies them into 5 groups depending on their net charge. Type I a and I b hydroxyl groups bear negative charges of -0.25 and -0.5, respectively and will, hence, undergo ion-exchange with Cl^- more easily than others. Of the two, type Ia forms part of the 'x-site', the site postulated to be active in reactions of hydrocarbons, like double-bond and cis-trans isomerisation (ref. 7). In order to accomodate the larger Cl^- (diameter, 3.62 Å) in the place of OH^- (diameter, 2.80 Å) some type IIa hydroxyl groups which are adjacent to both the original type Ia hydroxyl group as well as the 'x-site' have to be removed (on dehydration at higher temperatures) thus increasing the coordinative unsaturation and hence the strength of the acid sites on the alumina surface (see Fig. 12 of ref. 7). This enhanced acid strength is manifested in the ability of chlorided aluminas to catalyse such reactions as the methyl shift and other skeletal isomerisation reactions of hydrocarbons which the relatively weaker acid sites of pure alumina are unable to catalyse (ref. 11). The larger concentration of such OH^- groups on eta alumina and consequently its higher uptake of Cl^- ions are understandable in view of the lower packing density and higher concentration of stacking faults in its oxygen lattice. The retention of chlorine is also higher on the eta form. Under certain conditions the presence of tin enhances the retention of chlorine on the catalyst.

CONCLUSIONS

At low concentrations of chloride ions (below 1% by wt.), the uptake of Cl^- ions on gamma and eta aluminas is controlled by the concentration of surface OH⁻ groups carrying a net negative charge. The larger concentration of such OH⁻ group on eta alumina and consequently its higher uptake of Cl⁻ ions are understandable in view of the lower packing density and higher concentration of stacking faults in its oxygen lattice. The retention of chlorine is also higher on the eta form. Under certain conditions the presence of tin enhances the retention of chlorine on the catalyst.

ACKNOWLEDGEMENT

We thank Labofina (Belgium) for the supply of the gamma alumina sample. We are grateful to our colleagues in the Analytical Physics section for Al estimations, and to Miss. S. S. Vishnoi for Pt estimations. We thank Drs. I.B. Gulati and K.K. Bhattacharyya for encouragement and support.

REFERENCES

1. R.W. Maatman, Ind. Eng. Chem., 51 (1959) 913.
2. B.E. Yoldas, J. Appl. Chem. Biotechnol. 23 (1973) 803.
3. J.E. Benson and M. Boudart, J. Catal., 4 (1965) 704.
4. K. Kodama, Methods of Quantitative Inorganic Analysis, Interscience, New York, 1963, p. 241.
5. A.I. Vogel, A Text-book of Quantitative Inorganic Analysis, 3rd Ed., ELBS and Longman, London, 1969, p. 267.
6. S. Sivasanker and L.M. Yeddanapalli, Curr. Sci., 41 (1972) 878.
7. Knözinger and P. Ratnasamy, Catal. Rev., 17 (1978) 31.
8. Lj. Jacimovic, J. Stevovic and S. Veljkovic, J. Phys. Chem., 76 (1972) 3625.
9. S.M. Ahmed, J. Phys. Chem., 73 (1969) 3546.
10. W. Rieman and H.F. Walton, Ion-exchange in Analytical Chemistry, Pergamon, New York, 1970, p. 55.
11. B.H. Davis, J. Catal.,23 (1971) 355.

DISCUSSION :

R. POISSON : Did you take the precaution to analyse the properties of bayerite in your precursor ? I mean by ATG to take 100% of Al_2O_3 into account.

P. RATNASAMY : Yes, the precursor hydrates were analysed by thermogravimetry and found to contain more than 90% of the trihydrate.

J.W. GEUS : Some people have mentioned that they have troubles in getting reliable estimates of free Pt surface area by using the oxygen-hydrogen (or reversibly) titration. They are preferring therefore the adsorption of hydrogen onto the previously cleaned

Pt surface. Do you know whether the authors have had difficulties with this technique ?

P. RATNASAMY : In spite of the extensive work reported in this area, there is still no satisfactory method to measure accurately the dispersion of Pt in typical reforming catalysts. Both the H_2 chemisorption and H_2-O_2 (or O_2-H_2) titrations have their own drawbacks. Factors like cleanliness of the Pt surface, spill-over of H_2 and/or O_2, variation in the stoichiometric values with metal crystallite size, etc., preclude an unambiguous preference of one method over the other. In our laboratory, we have obtained almost identical values of metal dispersions by both the H_2-O_2 or the reverse O_2-H_2 titrations using a conventional volumetric set-up. We find that this method gives satisfactory and reproducible values of metal dispersions for samples not differing significantly in metal loadings, dispersion levels, nature of support, etc..

H. BREMER : 1) Chlorine in practice is not only added during preparation but also during e.g. reforming reaction in the form of CCl_4 and $CHCl_3$ etc... Is its action similar to that of HCl ? 2) To what extent dispersion is influenced by added chlorine ?

P. RATNASAMY : 1) We believe that the chloro compounds added during the reaction get first converted into HCl and hence their action will probably be similar to that of HCl added during the preparation stage. In fact, chlorine compounds are added during the reforming reaction mainly to compensate for the loss of chlorine from the catalyst under the reaction conditions.
2) Chlorine added during the reaction is not expected to alter the dispersion of the metal. However, chlorine added during the rejuvenation of the used catalyst reportedly helps in dispersing the Pt better.

P.G. MENON : With reference to H. Bremer's questions we have found recently (1) that chlorination with CCl_4 affects both Al_2O_3 and Pt on the $Pt-Al_2O_3$ catalyst. In addition to enhancing the acidity of Al_2O_3, chloriding drastically suppresses the hydrogenolysis activity of Pt. The chloride retained by the catalyst under reaction conditions still depends on the temperature and moisture content in the feedstock and the recycle gas, as emphasized in this paper. Since chlorine adds onto Pt as well, H_2-chemisorption measurements on the chlorided catalyst will show a lower value (1). But it does not mean any change in Pt dispersion, it only shows that part of the Pt is covered by Cl and hence inaccessible to H_2.

(1). P.G. Menon, R.P. De Pauw and G.F. Froment. Ind. Eng. Chem. Prod. Res. Dev. (in press)

B. DELMON : Concerning Fig. 2, could you comment on the kinetic features of the Cl^- uptake. Do you attribute them to diffusional kinetics, activated exchange or other causes ?

P. RATNASAMY : The kinetic data were fitted to rate equations derived from various models of adsorption and exchange processes. As mentioned in the text, the rate equation for ion-exchange controlled by diffusion of the chloride ions through the boundary film was found to fit, mathematically, most appropriately our kinetic data.

PREPARATION OF ALUMINA OR SILICA SUPPORTED PLATINUM-RUTHENIUM BIMETALLIC CATALYSTS

G. BLANCHARD[1], H. CHARCOSSET[1], M.T. CHENEBAUX[2] and M. PRIMET[1]

[1] Institut de Recherches sur la Catalyse du C.N.R.S., 79, bd du 11 novembre 1918 69626 Villeurbanne Cédex France

[2] Institut Français du Pétrole, 1 et 4, avenue de Bois Préau, 92506 Rueil-Malmaison, France

ABSTRACT

The purpose of the study was to prepare well dispersed (Pt, Ru) bimetallic cluster particles supported by γ-Al_2O_3. A (Pt, Ru)/SiO_2 alloy catalyst, poorly dispersed, and hence suitable for X-Ray Diffraction analysis, was also investigated for comparison. UV visible spectroscopy, electron microprobe analysis, Temperature Programmed Reduction and Titration, electron microscopy and Infrared Spectroscopy of chemisorbed CO were used as complementary methods. We arrive to the conclusion that the difficulty in preparing the above (Pt, Ru)/Al_2O_3 clusters arises mainly from a microscopic heterogeneity of the catalyst at the end of the impregnation step. Segregation of Pt and Ru may further result from an oxidizing heat treatment which gives rise to considerable sintering of Ru.

INTRODUCTION

The present study deals with the alumina (1) or silica supported platinum-ruthenium bimetallic catalysts. Their reduction by hydrogen was studied, by Temperature Programmed Reduction (TPR) (2), to determine if all of the Pt and Ru were reduced to the metallic state and to check the possible synergetic effects in the reduction of the Pt and Ru species. The degree of interaction between Pt^o and Ru^o and their degree of dispersion were studied altogether, by Temperature Programmed Titration (TPT) of previously chemisorbed oxygen, by hydrogen. That method was found very useful to show the presence if any, of pure Ru particles in a bimetallic catalyst. Electron microprobe analysis, U.V. visible absorption spectroscopy, X-Ray diffraction analysis, electron microscopy and Infrared Spectroscopy of chemisorbed CO were also used. The main conclusion is that the preparation of homogeneous (Pt, Ru) bimetallic particles, highly dispersed on γ-Al_2O_3, needs for further work.

EXPERIMENTAL

1. Supports and reagents

They were : $-Al_2O_3$ Rhône Progil GFS400 (pellets of $\sim$ 1.5 mm in diameter, 15 mm in length ; $S \sim 200\ m^2/g$; broad pore size distribution with a maximum at $\bar{r}_p = 60-70$ Å, total pore volume $\sim 0.7\ cm^3/g$).

$-SiO_2$ Aerosil Degussa (powder, $S \sim 200\ m^2/g$, non porous).

-chloroplatinic acid solution (Merck) and chlororuthenic acid solution (Comptoir Lyon Allemand).

2. Preparation of the catalysts

2.1. Impregnation

It was proceeded to total (co) impregnation of SiO_2 with H_2PtCl_6, H_2RuCl_6 to have a number of (Pt+Ru) atoms equivalent to 10 wt % Pt. Drying was carried out under reduced pressure ($\sim 10^{-2}$ Torr) from 25 to 50°C. The nominal composition of the bimetallic catalyst was 50 at % Ru. The Al_2O_3 pellets were calcined in a dry air flow 2 hrs at 500°C, before impregnation by water at first and subsequently by the active species solution.

Pt/Al_2O_3 ($\sim$ 2 wt % Pt) : 2.4×10^{-4} mole HCl/gAl_2O_3 were found necessary to add in the H_2PtCl_6 solution, to get macroscopic homogeneous distribution of Pt inside the Al_2O_3 pellets. 1 hr agitation was succeeded by water washing and by drying in a dry air flow from 25 to 110°C.

Ru/Al_2O_3 ($\sim$ 1 wt % Ru) : The amount of HCl ($\sim 1.5 \times 10^{-3}$ mole/gAl_2O_3) to be added in the H_2RuCl_6 solution was about ten times as much as used for the Pt/Al_2O_3 catalyst 10 hrs agitation were followed by water washing and drying, as for Pt/Al_2O_3

$(Pt+Ru)/Al_2O_3$ ($\sim$ 1 wt % Pt, 0.5 wt % Ru) : It was not found possible to obtain a homogeneous distribution of Pt, Ru inside the Al_2O_3 pellets, when using a coimpregnation procedure. The bimetallic catalyst was therefore prepared as follows : a) impregnation by the H_2RuCl_6, HCl solution ($1.5\ 10^{-3}$ mole HCl/gAl_2O_3), b) water washing to eliminate the excess Cl^- ions, c) impregnation by the H_2PtCl_6 solution, without further addition of HCl, d) washing and drying as above.

2.2. Reduction

While the silica supported catalysts were reduced directly by hydrogen, two modes of reduction were used for the alumina supported catalysts : a) direct reduction by a hydrogen flow of the dried catalysts, 2 hrs at 500°C, b) reduction by hydrogen as in a) but following 2 hrs pre-calcination in a dry air flow at 500°C. From literature data (3) the first mode of reduction should result in a better dispersion of Ru in Ru/Al_2O_3, while the second mode is closer to the usually applied industrial conditions of reduction.

3. U.V. visible absorption spectroscopy, electron microprobe analysis and X-Ray diffraction data for the unreduced Al_2O_3 supported catalysts

3.1. U.V. visible absorption spectroscopy

The results are reported in Fig. 1. The spectra related to Ru are for : - the chlororuthenic acid impregnation solution (atomic overall Cl/Ru ratio ∿ 30)(Spect 1) - the solid ammonium hexachlororuthenate(IV) (Fluka,puriss)(Spect 2) -the non hydrolyzed orange Ru/Al_2O_3 catalyst (Spect 3) and the hydrolyzed and dried green-dark Ru/Al_2O_3 catalyst (Spect 4). The chemistry of the Ru chloride solutions is highly complex (4) (5) and dimeric species are not excluded (6). Even the spectrum (2) for the solid NH_4 chlororuthenate may not be ascribed unambiguously to $RuCl_6^{--}$, according to the data in (4) (5) and some oxygenated ligands are possibly present.

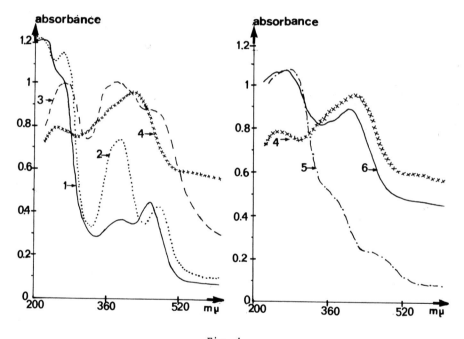

Fig. 1

There is a strong analogy between the non-hydrolyzed Ru/Al_2O_3 (Spect 3) and the NH_4 chlororuthenate (Spect 2) which means that Ru is in the same state in the two solids. Hydroxychlorospecies of Ru^{4+} like $Ru(OH)_2Cl_4^{2-}$ or the dimeric species $Cl_5RuORuCl_5^{4-}$ seem to be the most probable. Comparison of Spect 4 to Spect 3 suggests that the hydrolysis of Ru/Al_2O_3 gives rise to Ru^{4+} species with an increased ratio of oxygen containing ligands to Cl ligands.

Spect 5 for the dried Pt/Al_2O_3 catalyst was identical to that for the H_2PtCl_6, HCl solution, of atomic ratio Cl/Pt ∿ 7.5. Comparison of Spect 6 for the bimetallic dried catalyst to Spect 4 and 5 for the Ru and Pt monometallic catalysts respectively shows strong evidence of additivity. The bimetallic catalyst behaves at least in a first approximation as a mixture of the two monometallic catalysts.

3.2. Electron microprobe analysis

Fig. 2 reports the Ru, Pt and Cl profiles in the bimetallic catalyst either dried at 110°C (left) or further calcined in air at 500°C (right) (Subsequent reduction by H_2 did not change the profiles). It is shown that the procedure followed in 2 gives a reasonable macroscopic homogeneity of the Ru, Pt, Cl distributions inside the Al_2O_3 pellets.

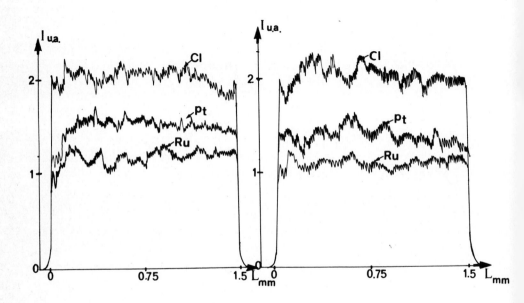

Fig. 2

3.3. X-Ray diffraction analysis

It was not possible to observe any diffraction line for Pt, Ru or (Pt, Ru) species also after calcination in air of the Al_2O_3 supported catalysts.

4. Temperature Programmed Reduction of the catalysts

TPR consisted to follow the variation in the thermal conductivity of a 3 % H_2 in N_2 mixture flowing through the catalyst bed at a heating rate $\sim$ 10°C/min. A liquid nitrogen trap was included between the catalyst and the catharometer to which the H_2-N_2 mixture was directed through an automatized valve (carrier gas : N_2).

Fig. 3 reports the reduction curves for the silica supported catalysts. The reduction curve for Pt/SiO_2 shows strong analogy with recent data of Jenkins et al.(7) The reduction of Pt in the bimetallic catalyst is strongly inhibited presumably

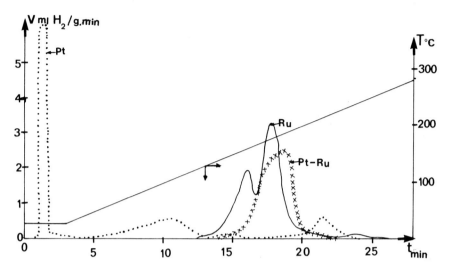

Fig. 3

because of the cocrystallization of the Pt and Ru chloride compounds during the drying step. The overall H_2 consumption was about 95 % the value calculated for the reduction of Pt^{4+}, Ru^{4+} to Pt^o, Ru^o, in the three catalysts.

Fig. 4 reports the reduction curves for the Al_2O_3 supported catalysts, dried at 110°C. The overall H_2 consumption at 500°C (balance between the H_2 consumption due to reduction and the H_2 evolution due to thermal desorption) corresponds to 95 % reduction of Pt^{4+} to Pt^o in Pt/Al_2O_3, but only to ∼ 55 % reduction of Ru^{4+} to Ru^o in Ru/Al_2O_3. The uncomplete reduction of Ru in Ru/Al_2O_3 at 500°C is corroborated by a further H_2 consumption near 700°C which corresponds to a 15 % increase (absolute value) in the degree of reduction. The bimetallic catalyst does not show any H_2 consumption in this temperature range, which suggests a ∼ 100 % reduction of Ru^{4+} to Ru^o at 500°C ; accordingly, the overall H_2 consumption at 500°C corresponds to a nearly complete reduction of Pt and Ru. When a hydrogen flow (p = 1.2 atm) was used to reduce Ru/Al_2O_3 2 hrs at 500°C, the subsequent TPR experiment with the 3 % H_2 in N_2 mixture did not show any H_2 consumption near 700°C, hence the % reduction of Ru at 500°C was certainly at least ∼ 70 %.

Fig. 5 reports the TPR curves for the Al_2O_3 supported catalysts, calcined 2 hrs in dry air flow at 500°C. The reduction curve of Pt/Al_2O_3 is not modified significantly following precalcination in air, while the reduction of Ru/Al_2O_3 occurs then at a very low temperature. The % reduction of Ru/Al_2O_3 at 500°C in the conditions of the TPR is about 80 and the reduction wave at ∼ 700°C corresponds to a further increase of ∼ 15 % (absolute value) in the degree of reduction. In the bimetallic catalyst, the main reduction wave of Ru is shifted to a ∼ 60°C greater temperature

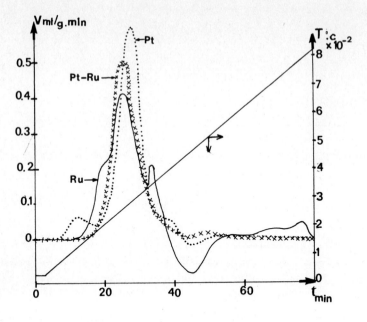

Fig. 4

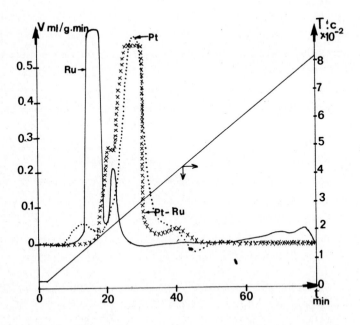

Fig. 5

compared to the monometallic catalyst. The reason for this is not clear, we suppose that the first amounts of H_2O issue of the reduction of Pt oxide species could inhibit the reduction of RuO_2. There is no further H_2 consumption at $\sim 700°C$ and the % reduction of (Pt^{4+}, Ru^{4+}) to (Pt^0, Ru^0) is in fact ~ 100 at 500°C from the H_2 consumption at that temperature.

5. X-Ray Diffraction and electron microscopy data over the reduced catalysts

The (Pt+Ru)/SiO_2 catalyst reduced 2 hrs in H_2 flow at 700°C showed X-Ray diffraction patterns characteristic of the cfc structure of Pt (Pt/SiO_2) or hcp structure of Ru (Ru/SiO_2). The bimetallic catalyst showed a cfc structure with a lattice parameter a = 3.88 Å. From the X-Ray diffraction data for bulk Pt-Ru alloys (8), that value of a corresponds to a ~ 40 at % Ru in Pt solid solution (to compare to ~ 49 at % Ru, the composition derived from chemical analysis). The relatively easy formation of SiO_2 supported Pt-Ru alloys has already been reported (9) and also the preparation of unsupported Pt-Ru alloy powders (10). The metallic particles in the above catalysts were rather coarse ($\phi \sim 300$ Å in Pt/SiO_2, 150 to 300 Å in Ru/SiO_2, and very broad particle size distribution in (Pt+Ru)/SiO_2).

The electron microscopy data for the Al_2O_3 supported catalysts are summarized in Table I. In agreement with (3)(11) precalcination in air of Ru/Al_2O_3 before reduction leads to a considerable decrease in the % dispersion of Ru.

TABLE I

Catalyst (approximate composition)	Calcined in air at 500°C before reduction	Type of particle size distribution (very strongly bimodal or not)	Mean ϕ (Å)
2%Pt/Al_2O_3	Not	Not	15
2%Pt/Al_2O_3	Yes	Not	15
1%Ru/Al_2O_3	Not	Not	< 10
1%Ru/Al_2O_3	Yes	Not	1000
1%Pt, 0.5%Ru/Al_2O_3	Not	Not	20
1%Pt, 0.5%Ru/Al_2O_3	Yes	Yes	(broad distribution)

The precalcined bimetallic catalyst looked strongly like a mixture of the precalcined Pt and Ru monometallic catalysts. No information could be drawn about the existence of Pt and Ru as a mixture of Pt and of Ru clusters or as (Pt, Ru) bimetallic clusters, in the bimetallic catalyst reduced without precalcination.

6. Temperature Programmed Titration by hydrogen of the oxygen previously chemisorbed at 25°C on the reduced catalysts

The method derives from previous work about the (Pt+Re)/Al_2O_3 (12) and (Pt+Ir)/Al_2O_3 (13) catalysts. Strong evidence was found there that clustering Re, Ir respectively, with Pt, results in a strong increase in the reducibility of the oxygen chemisorbed by the Re (or Ir) exposed atoms (the technique was microthermogravimetry and the O_2-H_2 titrations were carried out at $\sim 25°C$).

The hereunder reported experiments were carried out as follows. The samples were reduced 2 hrs in H_2 flow at 700°C (SiO_2 support) or 500°C (Al_2O_3 support), then outgassed 2 hrs in He flow at 500°C, before cooling in He down to 25°C. Differential frontal analysis with a 1 % O_2 in He mixture allowed then to chemisorb oxygen. The sample was then purged 1 hr by Ar at 25°C, before the H_2 titration with a 1 % H_2 in Ar mixture. The H_2 titration was carried out 20 min at the room temperature at first and then at increasing temperature (10°C/min) up to 200°C. The oven was finally removed, allowing the catalyst to cool to room temperature under the H_2/Ar flow.

Fig. 6 reports the H_2-TPT curves for the Pt, Ru, and (Pt+Ru) silica supported catalysts. The titration is complete at 25°C over Pt and only very little if any H_2 consumption or evolution occurs during the subsequent heating and cooling. On the other hand, the titration is only slow at 25°C over Ru and it takes place (largely) during the heating, the small H_2 consumption peak during cooling is approximately equal to the H_2 desorption wave during the plateau at 200°C. The bimetallic catalyst shows considerable deviation from additive behaviour between Pt and Ru studied separately ; most of the oxygen chemisorbed on Ru is titrated at the room temperature and only a small titration wave is observed when increasing the temperature.

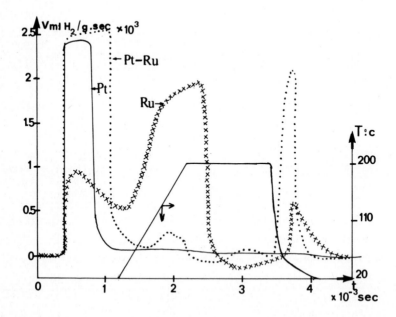

Fig. 6

It could be due to a small % of unalloyed ruthenium, the presence of which was suggested by comparison of the experimental lattice parameter to the value calculated if all of the Ru would be alloyed with Pt. The large H_2 consumption wave during cooling, compared to the monometallic catalysts, would suggest a decrease in the mean hydrogen metal bond strength due to alloying, but other phenomena like oxido-redox processes could also be involved.

Fig. 7 reports the H_2-TPT curves for the Al_2O_3 supported catalysts, reduced directly by H_2. It is shown that : a) the Pt_sO reduction takes place at the room temperature, like over Pt/SiO_2, b) none of the reduction of the oxygen chemisorbed on Ru/Al_2O_3 takes place at 25°C, at the difference of Ru/SiO_2. Particle size effects (Ru/Al_2O_3 is much better dispersed than Ru/SiO_2), probably also support effects, are involved, c) the TPT curve for the bimetallic catalyst shows considerable evidence of additivity between Pt and Ru studied separately ; note that the base line is practically recovered at t ∿ 1000 sec, that is between the two reduction waves.

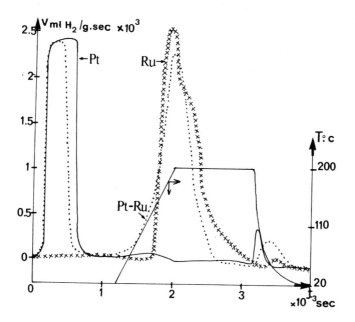

Fig. 7

The room temperature wave is more than a half the value for Pt/Al_2O_3 which suggests that Pt is somewhat better dispersed in the bimetallic catalyst. The high temperature wave is also more than a half the value for Ru/Al_2O_3 which is accounted for by the increase in the % reduction of Ru.

Fig. 8 reports the results for the Pt, Ru and (Pt+Ru) alumina supported catalysts calcined in air before reduction. The curve for Ru/Al_2O_3 has only a very small area, as expected from the electron microscopy data. On TPT curve for (Pt+Ru)/Al$_2$O$_3$ note that : a) the H$_2$ consumption when t ~ 2000 sec is approximately a half the value for Ru/Al$_2$O$_3$, b) the H$_2$ consumption when t ~ 1000 sec is clearly above the value for the two monometallic catalysts, which suggests that a small part of the Ru could be in interaction with Pt.

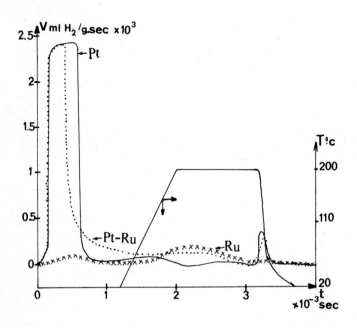

Fig. 8

7. <u>Infrared spectroscopy of CO chemisorbed on the reduced Al$_2$O$_3$ supported catalyst</u>

This technique was found useful in determining the state of Pt and Re in (Pt+Re)/Al$_2$O$_3$ catalysts (12). The Ru/Al$_2$O$_3$ (directly reduced form) and (Pt+Ru)/Al$_2$O$_3$ (reduced either directly or after calcination) catalysts were cooled down in H$_2$ from 500 to 25°C. H$_2$ was then evacuated at RT and the chemisorbed H displaced by CO (p = 50 Torr) The spectra over the two bimetallic catalysts, following evacuation of gaseous CO, are reported in Fig. 9 A (a : direct reduction ; b : reduction after calcination). The ratio of the optical densities of a to b is 1.7, which supports the above conclusion of a larger number of (Pt+Ru) atoms exposed, after direct reduction than after reduction following calcination. The spectra A do not show any evidence of a band at 2140 cm^{-1} which was observed over the directly reduced form of Ru/Al$_2$O$_3$, corresponding to CO adsorbed on oxidized species of Ru. This is in agreement with the above conclusion of a more complete reduction of Ru in the bimetallic catalysts than in the directly reduced form of the monometallic Ru catalyst. The spectra A

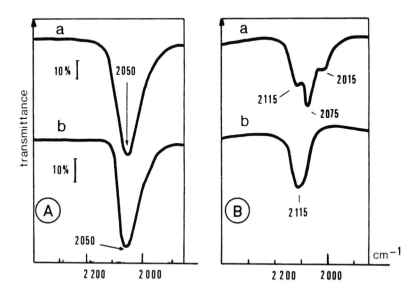

Fig. 9

finally suggests overlapping of the CO species adsorbed on Pt and Ru respectively, a phenomenon which will be discussed separately in as much as the interaction of CO with Ru monometallic catalysts is rather complex (14). The low $\nu(CO)$ frequencies (2050 cm^{-1}) compared to the value for CO adsorbed on bare platinum (2075 cm^{-1}) mentioned in (12) is ascribed to the effect of adsorbed water in the present experiments; this effect of H_2O on the $\nu(CO)$ frequencies for CO adsorbed on Pt has been reported previously (15).

After recording the spectra A, the catalysts were submitted to oxidation by O_2 at 150°C in order to burn out the chemisorbed CO. They were then cooled in O_2 at RT, O_2 was evacuated and CO was chemisorbed again. CO was here interacting with a oxygen covered metal surface and giving rise to the spectra B. The main point is that the precalcined catalyst (Spect b) shows a pure Pt catalyst behaviour (the band at 2115 cm^{-1} corresponds to Pt$\underset{CO}{\overset{O}{\diagdown}}$ species (12)) while the directly reduced form (Spect a) shows a like additive behaviour between Pt and well dispersed Ru monometallic catalysts (the bands at 2075 and 2015 cm^{-1} were observed on Ru/Al$_2$O$_3$ in the same conditions). This again does not conflict with the above conclusion that the directly reduced (Pt+Ru)/Al_2O_3 consists in a mixture of small Pt and Ru particles while the precalcined catalyst would mainly consist in a mixture of well dispersed Pt and coarse Ru particles.

DISCUSSION AND CONCLUSION

There is a considerable evidence from several techniques, in particular the hydrogen programmed titration of chemisorbed oxygen that we did not succeed in preparing well dispersed and homogeneous in composition (Pt,Ru) bimetallic cluster particles on the γ Al_2O_3 support. Referring to our previous studies on the (Pt+Re or Ir)/Al_2O_3 catalysts (12)(13), it follows that the (Pt,Ru) pair of elements behaves more like the (Pt,Ir) pair rather than the (Pt,Re) pair. Nevertheless Ru is more sensitive than Ir to sintering due to heat treatment in an oxidizing atmosphere.

The heterogeneity of the precalcined (Pt+Ru)/Al_2O_3 catalyst could be connected to the segregation of most of the Ru as large RuO_2 particles during the calcination ; the Pt oxide or oxychloride species do not co-agglomerate with Ru as (Pt+Ru) mixed oxide particles.

The heterogeneity of the directly reduced (Pt+Ru)/Al_2O_3 catalyst may not be ascribed to segregation during the reduction step since there are only minute differences in the temperature of reduction of the two active elements species (see Fig. 4). Therefore, we conclude that a microscopic heterogeneity in the repartition of Pt and Ru on the Al_2O_3 surface preexists before the reduction. Improvement in the homogeneity of the bimetallic phase needs, therefore, further studies in order to obtain a more homogeneous microscopic distribution of Pt and Ru at the end of the impregnation procedure.

The quantitative analysis of the Temperature Programmed Titration data in terms of concentrations in exposed Pt and Ru atoms may be done on the basis of 1 0 chemisorbed per Pt_s (12) and of 2 0 chemisorbed per Ru_s (16). The results were compatible with the electron microscopy data in Table I and are not reported here for a purpose of concision. Finally, let us note that the TPT technique is not confined to platinum based bimetallic catalysts. It may be useful every time the difference in reducibility of the oxygen chemisorbed on the two component elements (at least one of which but even the two may activate hydrogen) is sufficiently significant. It is especially useful to discriminate between separate particles of the two elements and bimetallic cluster particles, a task which is well known to be not evident especially over well dispersed catalysts (17). Of course, the complementary use of other techniques, when it is possible, is very profitable ; it is the reason why we make a pluri-technique approach of our problem.

Note : After the paper being finished, we learnt from the supplier that the "ammonium hexachlororuthenat" used for Fig. 1, Spect 2 was in fact ammonium -oxo- decachlorodiruthenat (IV), $(NH_4)_4$ (Ru_2OCl_{10}), in agreement with the presence of oxygen containing dimeric species considered above.

ACKNOWLEDGMENTS

The authors kindly acknowledge Mrs. M.T. Gimenez, C. Leclercq and H. Praliaud for help in the X-Ray Diffraction, electron microscopy, U.V. spectroscopy work respectively.

REFERENCES
1. G. Martino, J. Miquel and P. Duhaut, Fr. Patent 2.234.924, 24 Janv, 1975, (CA80, 20080 p).
2. S.D. Robertson, B.D. McNicol, J.H. De Baas and S.C. Kloet, J. Catal. 37 (1975) 424-31.
3. R.A. Dalla Betta, J. Catal. 34 (1974) 57-60.
4. Chr.K. Jørgensen, Acta Chem. Scand. 10 (1956) 518-34.
5. Chr.K. Jørgensen, Molecular Phys. 2 (1959) 309-332.
6. I.V. Prokoficva and N.V. Fedorenko, J. Inorg. Khim 13 (1968) 1348-53.
7. J.W. Jenkins, B.D. McNicol and S.D. Robertson, Chem. Tech. (1977) 316-20.
8. N.W. Ageev and V.G. Kuznetsov, Izv. Akad. Nauk SSR 1(1937) 753.
9. M.F. Brown and R.D. Gonzalez, J. Catal. 48 (1977) 292-301.
10. S. Engels, Nguyen-Phuong-Khuê and M. Wilde, Z. Chem. 16 (1976) 455-57.
11. C.A. Clausen III and M.L. Good, J. Catal. 38 (1975) 92-100.
12. C. Bolivar, H. Charcosset, R. Fréty, M. Primet, L. Tournayan, C. Betizeau, G. Leclercq and R. Maurel, J. Catal. 45 (1976) 163-78.
13. L. Tournayan, J. Barbier, H. Charcosset, R. Fréty, C. Leclercq, G. Leclercq and P. Turlier, Thermochimica Acta, in press.
14. R.A. Dalla Betta, J. Phys. Chem., 80 (1975) 2519-25.
    M.F. Brown and R.D. Gonzalez, J. Phys. Chem. 80 (1976) 1731-35.
    A.A. Davydov and A.T. Bell, J. Catal., 49 (1977) 332-44.
15. M. Primet, J.M. Basset, M.V. Mathieu and M. Prettre, J. Catal. 29 (1973) 213-23.
16. H. Kubicka, Reaction Kinetics and Catal. Letters, 5 (1976) 223-28.
17. J.H. Sinfelt, Accounts of Chemical Research 10 (1977) 15-20.

DISCUSSION

N. PERNICONE : Concerning your TPR data, you have performed a quantitative measurement of the amount of H_2 consumed in the various reduction steps. You know that the response of thermal conductivity cells is very sensitive to ambient conditions (temperature, pressure, gas flow rate, etc.). Have you estimated the accuracy of your measurements and which expedients have you adopted to improve it ?

G. BLANCHARD : The accuracy of our measurements is estimated to be $\pm$ 5% for the volume of hydrogen consumed in the range of magnitude of 2 cm^3 NTP. The most useful expedient to improve the accuracy has been the use of an electronic gas flow regulator. Of course further verification has been carried out, that is : (i) absence of any base line drift during blank experiments, (ii) proportionality of the response signal of the detector to the gas concentrations in the concentration range considered. Nevertheless a drawback of the TPR experiments should be not omitted : the reduction is carried out under a much lower H_2 pressure than the atmospheric one. This disadvantage may be overcome to some extent by carrying out the reduction under 1 atm. pressure at first, before drawing the TPR curve which is then relevant of the reduction of the unreduced part of the active species in the normal conditions of 1 atm. pressure of hydrogen.

C.J. WRIGHT : Am I correct in thinking that none of the evidence you put forward could unequivocably differentiate between small particles of Pt/Ru alloy - and platinum and ruthenium particles in close proximity.

G. BLANCHARD : We think to be able to differentiate unequivocably between two situations where the first one should correspond to small alloy particles and the second one to a mixture of small particles of pure Pt and of pure Ru. We do not think that a situation where pure Pt and pure Ru particles would be in close contact with each other is likely to occur. With respect to the sintering of monometallic catalysts it is accepted that as soon as two small metallic particles come in close contact, they co-agglomerate into a single particle at rather low temperatures. It should be considered that our catalysts were heat treated in hydrogen at 500°C; interdiffusion of very small Ru and Pt particles would be very rapid if these particles would come in close contact. Therefore in our catalysts the Pt and the Ru particles are most likely to be separated by some positive, but unknown distance.

R. MILES : I would like to make some comments with regard to the technique of temperature-programmed reduction (TPR). Firstly, McNicol et al. have observed that, as in the case of a SiO_2 support, alloy formation is readily achieved in bimetallic Pt-Ru mixtures supported on graphitic carbon, as shown by TPR. Secondly, they have found that the ease of reducibility as determined by the temperature of the maximum in the hydrogen depletion rate, is largely governed by such factors as the support interaction, oxidation state and crystallite size of the metal catalyst. Furthermore, he has shown that more information concerning the detailed nature of the platinum component in such systems can be gained if the thermal scan is commenced at low temperature, e.g. about 77 K. Have you investigated the TPR behaviour of your catalyst at low temperature ?

H. CHARCOSSET : We are well aware of the TPR studies of McNicol et al. performed on various metallic catalysts (Ref. 2 cited as example in the present paper). We have not yet investigated our catalysts in the subambient temperature region.

PREPARATION OF CATALYSTS BY ADSORPTION OF METAL COMPLEXES ON
MINERAL OXIDES*

J.P. BRUNELLE

Procatalyse, Centre de Recherches Rhône-Poulenc, Aubervilliers, France

ABSTRACT

The preparation of dispersed metal supported catalysts by adsorption of metal complexes on oxides has led us to analyze the phenomena occurring at the interface oxide-solution. This analysis is based on simple principles such as surface polarization of oxides versus pH and adsorption of counterions by electrostatic attraction.

The three most important parameters which seem to rule the adsorption phenomena are: isoelectric point of the oxide, pH of the aqueous solution, and nature of the metal complex.

This simplified approach toward adsorption phenomena is in agreement with the results in literature concerning the fixation on alumina or silica carriers of chlorometallic and amine complexes of metals belonging to 7a, 8 and 1b groups.

An extension of this analysis to other mineral oxides is proposed.

---

1. IMPORTANCE OF METALLIC CATALYSTS

The importance of metal-based catalysts in our economic system need certainly no longer be demonstrated. They are involved in such different fields as oil refining, automobile, petrochemical and fine chemical industries. A list of the principal processes operating with metal-based catalysts will be found in table 1.

The properties of these catalysts in general and their activity in particular are closely related to the state of dispersion of the active elements. This explains why three-fourths of the processes listed in table 1 use catalytic systems in which the active phase consists in very small crystallites of about ten angströms dispersed on the surface of a support. It is what we usually call dispersed metal catalysts.

The metals used in these catalysts belong generally to groups 7a, 8 and 1b of the periodic table, very often such as platinum and palladium. Four main supports are used in their preparation: alumina, silica-alumina, active carbon and molecular sieves.

---

* Published in Pure Appl. Chem., 50, 9-10 (1978), p. 1211.

TABLE 1
Metal catalysts used in oil refining, automobile, petrochemical and fine chemical industries.

	PROCESS	TYPE OF CATALYST	
		METAL	CARRIER
DISPERSE METALLIC CATALYSTS	• Auto-exhaust gases Post-combustion		Alumina or
	Oxydation catalysts	Pt+Pd	Alumina coated
	Three-way catalysts	Pt+Rh	Cordierite
	• Selective hydrogenation of :		
	Olefins streams in ethylene plants	Pd	Alumina
	Pyrolysis gasoline in ethylene plants	Pd	
	• Catalytic Reforming	Pt (+Re,Ir,Au..)	Alumina + Cl
	• Hydrocracking	Pd	Y zeolite based
	• Isomerization of :		
	Paraffins	Pt	Alumina + Cl or mordenite
	Xylenes	Pt	Alumina + Cl or Alumina-Silica
	• Dismutation of Toluene	Cu, Ni	Zeolite
	• Fine organical chemistry	Pt,Rh,Pd,Ru, Ni Raney	Charcoal
	• Off-gas Treatments	Pt,Pd	Alumina
	• Fuel Cells	Pt,Pd	Charcoal
NO DISPERSE Metallic Catalysts	• Ammonia oxidation	Pt+Rh	-
	• Selective oxidation of ethylene in ethylene oxide	Ag	α Alumina
	• Selective oxidation of methanol in formol	Ag	Carborundum

    It is to be noted that the Raney nickel, which is still used nowaday in several liquid phase hydrogenation processes, constitutes a particul case of a dispersed metallic catalyst which is non supported.

    If we consider the tonnage of catalyst involved, the most important processes are catalytic reforming, hydrogenation of different petroche mical streams, and above all, automobile post-combustion, which repre- sents the greatest turnover for the industry.

In fact, only three important processes use metal catalysts that do not require a high dispersion of the active element: 1)ammonia oxidation to make nitric acid on a platinum-rhodium wire catalyst, 2) methanol selective oxidation into formol and 3) ethylene selective oxidation into ethylene oxide. Silver-based catalysts are used in the latter two processes.

2. METHOD OF PREPARATION OF DISPERSED METAL CATALYSTS

Several studies (1-6) have shown the interest of using the adsorption or exchange of complexes with some surface sites of mineral supports to obtain highly loaded metal surfaces. It should be apparent that :
-The reduction of metal ions adsorbed on the surface of the support must lead initially to an atomic deposit of the metal.The whole problem consists in preserving a dispersion of the metallic deposit as close as possible to its initial state, thus avoiding too great a crystallization of the metal during the catalyst activation.
-On the other hand, if the impregnation is made in condition where metallic precursors do not fix on the surface of the support,these precursors are going to deposit according to a process of crystallization, precipitation or decomposition during the drying step. The size of the crystallites depends on a great number of parameters such as support texture,precursor solubility and drying velocity. In this case, crystallites smaller than 50 Å are rarely obtained.

For instance, figure 1 enables the comparison of the mean crystallite size of two series of platinum-silica catalysts with variable percentage of platinum and prepared :
-either by a conventional method of impregnation without exchange from a chloroplatinic acid solution.
-or by tetramine-platinum II cation-exchange with ammonia-polarized surface sites of silica.

The mean crystallite size of catalysts prepared by the conventional method varies between 60 and 170 Å, while the cation exchange method leads to a mean platinum crystallite size of about 10 to 20 Å, for a platinum content varying between 0.4 and 5.5 wt%.

3. ADSORPTION OF ION COMPLEXES ON THE SURFACE OF MINERAL OXIDES

Although several studies have shown the practical interest of using adsorption and exchange of metal complexes on mineral oxides, few studies in fact have been carried out on this subject. Some of them deal with the adsorption of platinum (1,2,4,6,13), palladium (3,5,11,13), cobalt, nickel, or copper (13,14) cationic complexes on the surface of silica or alumina supports. Some others treat the adsorption of anionic complexes of platinum (7-10, 12, 13), palladium (5, 11) or other precious metals such as gold, rhodium, ruthenium, iridium (12,13) on alumina.

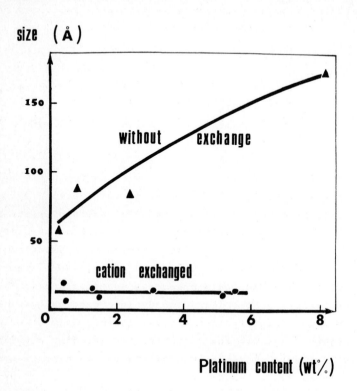

Fig. 1. Variation of the mean crystallite size vs metal content of Pt-SiO$_2$ catalysts. Specific surface area of silica support = 260 m^2/g. Activation of catalysts : drying in air at 120°C for 16 hr and reduction in hydrogen at 410°C for 6 hr.

Our purpose is not to describe the results obtained from these different studies but to treat generally the adsorption of metal complexes and the exchange phenomena on the surface of mineral oxides. In using simple principles such as surface polarization of an oxide as function of pH and adsorption of complex ions by electrostatic attraction, we shall try to extricate the most important thermodynamic parameters which seem to control the adsorption phenomena, and to suggest the basic principles for the choice of these parameters.

On the other hand, the problem of fixation of metallic hydrolyzable cations (15-41) will not be treated. This problem, though it has often been studied, does not yet seem to be clearly explained as the conclusions are sometimes contradictory. The most common mechanism is hydrolytic adsorption of hydrolyzable cations. This mechanism, which certainly deserves to be carefully studied, involves a hydrolysis and not an adsorption of the cation on the oxide.

We shall also restrict our analysis to the phenomena which occur during impregnation at the interface oxide-solution without taking into account the other phenomena which may take place afterwards during activation (reaction of the complex with the support, partial or total decomposition of the complex ...).

Finally, we will ignore the kinetic aspect of these phenomena, even if it is significant.

3.1. ISOELECTRIC POINT OF AN OXIDE

A particle of a mineral oxide in suspension in an aqueous solution tends to polarize and to be electrically charged. Most oxides are amphoteric. Thus, the nature and importance of this charge are a function of pH of the solution surrounding the particle.

For example, in an acid medium, the particle is positively charged. The principle of electroneutrality implies the presence of a layer of ions with opposite charge near this particle, the two electric charges compensating each other. If we consider a schema in agreement with the GOUY theory, counteranions will thus be located around the particle in a thin diffuse layer, as shown in figure 2.

Schematically, the equation of the surface polarization may be written :

$$S-OH + H^+A^- \rightleftarrows S-OH_2^+A^-$$

where S-OH represents a surface adsorption site and H^+A^- a mineral acid.

In a basic medium, the reverse is true. The particle is negatively charged and is surrounded by compensating cations. The equation of the surface polarization may be written as follows :

$$S-OH + B^+OH^- \rightleftarrows S-O^-B^+ + H_2O$$

where B^+OH^- represents a base.

One conceives easily that between these two cases, a given value of pH exists at which the overall charge of the particle is zero. This value, which is a characteristic of the oxide, corresponds to its zero point of charge (Z.P.C.) or its isoelectric point (I.E.P.S.).

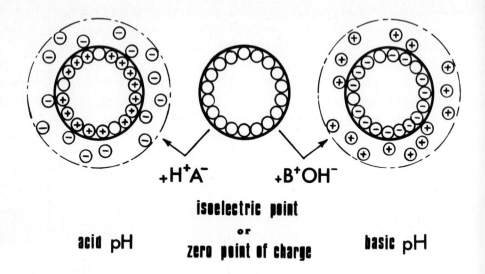

Fig. 2. Schematic representation of the surface polarization of an oxide particule as a function of the solution pH.

Let us now examine two methods allowing us to estimate the nature importance of the surface polarization of an oxide as a function of p of the solution in which it is dipped.

3.1.1. Electrophoresis

Electrophoresis techniques enable us to measure the velocity of a charged particle in suspension placed in an electrical field. The ele trophoretic velocity is proportional to the potential difference exis ting between the opposites areas of the double layer (Zeta potential) For this reason it is possible to determine experimentally the sign a the importance of the polarization of a given oxide particle.

In order to illustrate this method, we have shown in figure 3 four curves of Zeta potential versus pH corresponding to four different pr ducts: a silica gel, a neodyme hydroxide gel, a titania gel and a gibbsite gel (42).

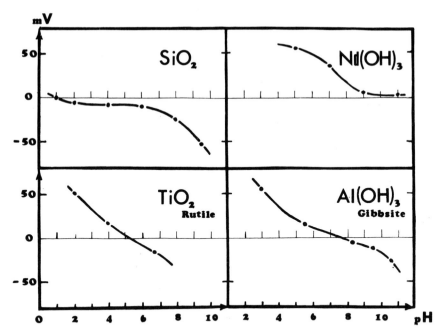

Fig. 3. Various types of Zeta potential curves vs pH obtained by electrophoresis.

The following informations can be drawn out of these curves :
-The I.E.P.S. of silica, which corresponds to the cross-over point of Zeta potential curve and pH-axis, is very low, around pH 1. This result indicates the acid and non-amphoteric type of this oxide. A negative polarization of its surface occurs at pH higher than 1 (negative zeta potential) but becomes significant only above pH 7. In other words, silica may only adsorb cations, and this phenomenon is important only above pH 7.
-Neodyme hydroxide has a quite opposite behaviour compared to that of silica. Its I.E.P.S. is around pH 12, which is in agreement with the basic type of this hydroxide. Its surface polarizes positively below pH 12, which makes possible the adsorption of anions on its surface.
-Titania and gibbsite gels correspond typically to amphoteric compounds which results in an inversion of the polarization sign when passing from acid to basic medium. Their I.E.P.S. are respectively equal to 5.5 and 7.5. It is then possible to adsorb cations or anions if the pH of the solution is higher or lower than the I.E.P.S. of these two amphoteric gels.

3.1.2. Neutralization at constant pH

The second method which permits to follow the polarization of an oxide surface consists in measuring the capacity of adsorption of the oxides at constant pH (2).

The procedure is as follows: an oxide is dipped into an aqueous solution, which pH is maintained constant by using ammonia or monochloracetic acid. The basic or acid quantities delivered to maintain the pH at a given value are recorded as a function of time. These quantities corrected with the blank experiment, correspond to the quantities neutralized by the support. If the support does not dissolve, they also correspond to the quantities of ions adsorbed on the support surface.

Fig. 4 shows the amounts of ammonium or monochloracetate anions adsorbed on gamma alumina and silica as a function of pH. These results, obtained according to the method described above, lead to the same conclusions as those of electrophoresis :

-Alumina, due to its amphoteric properties, adsorbs ammonium cations and monochloroacetate anions at pH respectively higher and lower than 8. The particular point at pH 8 correlates well with the I.E.P.S. of gibbsite in figure 3 and with the several I.E.P.S. measurements of alumina found in literature (43).

-Silica adsorbs ammonium at pH higher than 6 but does not adsorb monochloroacetate anions in the pH range investigated.

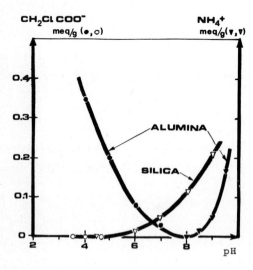

Fig. 4. Ammonium and monochloracetate adsorption on γ-alumina or silica supports as a function of pH

(●,▼) : γ alumina of 190 m^2/g

(○,∇) : Silica of 160 m^2/g

(●,○) : 5 g of support + 100 ml of a 0.02 M solution of CH_2ClCOO Na; addition of a 0.1 M solution of $CH_2ClCOOH$

(▼,∇) : 5 g of support + 100 ml of a 0.02 M solution of NH_4NO_3; addition of a 0.1 M solution of ammonia

In summary, a particle of oxide dipping in a solution at a pH lower than its I.E.P.S. tends to polarize positively and to adsorb compensating anions. On the contrary, the same particle dipping in a solution at a pH higher than the I.E.P.S. gets a negative surface charge which is compensated by adsorbed cations. Thus, three parameters seem to be important: the isoelectric point of the oxide, the pH of the impregnating solution and the charge of the ion to be adsorbed.

3.2. ANIONIC AND CATIONIC METAL COMPLEXES

Our practical aim is to prepare supported metal catalysts by adsorption. We are now going to specify the form in which these metal complexes may occur.

We can use two well-known families of complexes concerning a number of metals from group 7a, 8 and 1b.

-The family of chlorometallic complexes $(MCl_n)^{x-}$ shown in table 2.

TABLE 2
Anionic complexes of metals of 7a, 8 and 1b groups.

Mn MnO_4^-	Fe	Co	Ni	Cu
Tc	Ru	Rh $RhCl_6^{3-}$	Pd $PdCl_4^{2-}$	Ag
Re ReO_4^-	Os $OsCl_6^{2-}$	Ir $IrCl_6^{2-}$	Pt $PtCl_6^{2-}$	Au $AuCl_4^-$

In this case, the metal is in the form of an anionic complex in which the coordination sphere is constituted by four or six chlorine atoms.

We have, for instance, tetrachloroaurate $(AuCl_4)^-$, hexachloroplatinate $(PtCl_6)^{2-}$, hexachloroiridate $(IrCl_6)^{2-}$, tetrachloropalladate $(PdCl_4)^{2-}$, hexachlorhodate $(RhCl_6)^{3-}$ and hexachloroosmate $(OsCl_6)^{2-}$ anions. To these anions, we can add peroxide anions such as perrhenate $(ReO_4)^-$ and permanganate $(MnO_4)^-$ anions in which the coordination sphere is constituted by four oxygen atoms instead of chlorine atoms.

-The family of amine complexes $|M(NH_3)_n|^{x+}$ shown in table 3.

TABLE 3
Cationic complexes of metals of 8 and 1b groups.

Mn	Fe	Co $Co(NH_3)_x^{2+}$	Ni $Ni(NH_3)_x^{2+}$	Cu $Cu(NH_3)_x^{2+}$
Tc	Ru $Ru(NH_3)_5Cl^{2+}$	Rh $Rh(NH_3)_5Cl^{2+}$	Pd $Pd(NH_3)_4^{2+}$	Ag $Ag(NH_3)_x^{2+}$
Re	Os	Ir $Ir(NH_3)_5Cl^{2+}$	Pt $Pt(NH_3)_4^{2+}$	Au

In this family, the metal is in the form of a cation coordinated to several amine or ammonia groups.

Among the most often used complexes, we can indicate the chloropent amine complexes of ruthenium, rhodium and iridium with valence III, th tetramine complexes of palladium and platinum with valence II and ami complexes of silver, copper, nickel and cobalt.

3.3. EXAMPLES

To illustrate what has been said, we can now examine on the basis o the results given in literature, what are the possibilities of adsorption between several platinum complexes and silica or alumina.

For silica, two conditions are required: a solution pH higher than and preferentially about six and the use of a metal cationic precurso

These two conditions are not met with a chloroplatinic acid soluti: this explains why chloroplatinic acid does not adsorb on silica (2,4 With a tetramine platinum II chloride solution, we have a cationic co plex but the pH of the solution is not high enough. Thus, the adsorpt is very small (2,4). With a tetramine platinum hydroxide, both condit are fulfilled and in this case platinum adsorption (2,4) effectivel occurs. Likewise, the use of a solution containing tetramine platinum chloride and ammonia leads to platinum adsorption (1,2,4,6).

For alumina, two conditions are also required: either an anionic precursor solution with a pH lower than about 8, or a cationic precursor solution with a pH higher than about 8.

In the cases of chloroplatinic acid solution, tetramine platinum hydroxide solution or tetramine platinum chloride ammonia solution, both conditions are fulfilled and consequently platinum adsorption occurs. (1,2,4,7-10,13). On the other hand, if we impregnate alumina with a sodium chloroplatinate solution, we have only very little adsorption because the pH of the solution is close to the isoelectric point of alumina. As for the tetramine platinum chloride solution, it does not lead to cation adsorption because the pH condition is not met.

Thus, the results in literature concerning the adsorption of platinum complexes on oxides such as alumina and silica are consistent with the simplified schema presented above. We think that this schema remains valid for other oxides and other metal complexes (3,5,11-13).

3.4. EXCHANGE OF ADSORBED SPECIES

Until now, we have considered the case of an oxide particle in contact with a solution containing one salt. We shall now consider the case where the impregnating solution contains, for instance, two mineral acids HA and HB.

If the oxide is basic or amphoteric, its surface polarizes positively and is surrounded by two types of counterions: A^- and B^-.

Schematically, the three following reactions take place at the interface oxide-solution :

$$M-OH + H^+A^- \rightleftarrows (M-OH_2^+ A^-) \qquad (1)$$
$$M-OH + H^+B^- \rightleftarrows (M-OH_2^+ B^-) \qquad (2)$$
$$(M-OH_2^+ A^-) + B^- \rightleftarrows (M-OH_2^+ B^-) + A^- \qquad (3)$$

The first and the second reactions correspond to the polarization of a surface site S-OH by HA or HB. The equilibria are displaced towards the right side.

On the contrary, reaction (3) corresponds to a true exchange of A^- and B^- between the oxide surface and the solution.

To illustrate these various considerations, we can take the practical example of the impregnation of alumina by hydrochloric acid and chloroplatinic acid solutions.

According to (3) the equilibrium of exchange can be written as follows :

$$2(Cl)_{ads} + PtCl_6^{2-} \rightleftarrows (PtCl_6)_{ads} + 2Cl^-$$

If the solution is diluted and assuming homogeneity of the adsorption sites on alumina, we can express the equilibrium constant as follows :

$$K = \left[\frac{Cl^-}{Cl_{ads}}\right]^2 \cdot \frac{1}{Kd}$$

where Kd is the distribution coefficient of platinum between the solution and alumina.

Now, if we take the logarithmic expression of the equilibrium constant, we should obtain a linear relationship between log K and $\left[\log(Cl^-) - \log Cl_{ads}\right]$, the value of the slope giving the valence of the metal anion.

This is what we observe experimentally, when gamma alumina is impregnated with a hydrochloric solution of chloroplatinic acid, chloroiridic acid or chlorauric acid (figure 5). In the range studied, straight lines are obtained with slopes of ∼ 2 for platinum and iridium and ∼ 1 for gold.

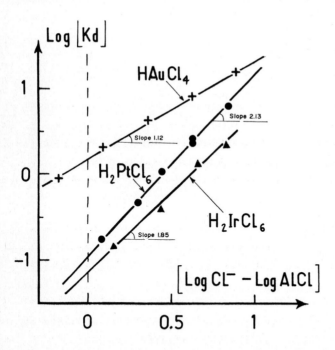

Fig. 5. Anionic exchange equilibrium - For details, see Ref. (13).

Similar results are obtained when silica is impregnated with an ammonium buffer solution of platinum tetramine chloride, palladium tetramine chloride or amine copper salts (figure 6).

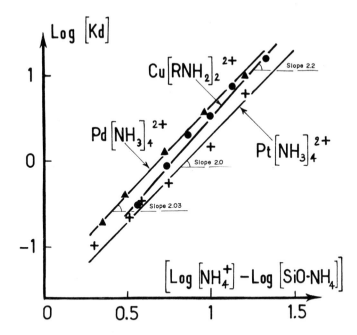

Fig. 6. Cationic exchange equilibrium - For details, see Ref. (13).

From a practical point of view, the reversibility of adsorption is very important because it permits to obtain a redistribution of the adsorbed complex ions on the polarized sites with a competitive ion and thus, to obtain an homogeneous deposit of the metal ions on the support.

3.5. ISOELECTRIC POINT OF OXIDES

We have seen that, given the isoelectric point of an oxide, we may foresee the adsorption capability (anionic or cationic) of this oxide and roughly what the pH range (acidic or basic) of the impregnating solution will be.

The isoelectric point of a large number of oxides and hydroxides is known and available in literature. For this purpose, we can refer to the synthesis article published in 1965 by PARKS (43).

A number of values are given in table 4. They allow us to rank approximately the oxides in three categories :
- The first one corresponds to the oxides that are acidic, such as Sb_2O_3, WO_3 or SiO_2. Their I.E.P.S. are very low and we may suppose that they are only able to serve as adsorbents or exchangers of metal complex cations.

-The second one corresponds to the oxides that are basic, such as La_2O_3 or MgO. Their I.E.P.S. are very high, generally greater than 10. We think that these oxides are able to adsorb essentially metal complexes
-Finally, the third and most numerous class, includes the oxides that ar amphoteric such as TiO_2, Cr_2O_3 or Al_2O_3.

TABLE 4
Isoelectric points of various oxides.

Oxide	I.E.P.S.	Adsorption
Sb_2O_5	<0.4	CATIONS
WO_3 Hydrous	<0.5	
SiO_2 Hydrous	1.0-2.0	
U_3O_8	~4	CATIONS or ANIONS
MnO_2	3.9-4.5	
SnO_2	~5.5	
TiO_2 Rutile Anatase	~6	
UO_2	5.7-6.7	
γFe_2O_3	6.5-6.9	
ZrO_2 Hydrous	~6.7	
CeO_2 Hydrous	~6.75	
Cr_2O_3 Hydrous	6.5-7.5	
$\alpha, \gamma Al_2O_3$	7.0-9.0	
Y_2O_3 Hydrous	~8.9	ANIONS
αFe_2O_3	8.4-9.0	
ZnO	8.7-9.7	
La_2O_3 Hydrous	~10.4	
MgO	12.1-12.7	

Thus, with these values, we can hope to predict roughly the behaviou of an oxide when it is impregnated by a solution of a metal complex salt.

3.6. LIMITS AND SECONDARY PHENOMENA

In practice, the preparation of metal catalysts by complex adsorpti may be limited or complicated by many factors or secondary phenomena. Three of them are discussed below :

3.6.1. Texture of the oxide

We have seen that an important adsorption capacity requires the choice of adequate pH conditions. However, this condition is necessary, but not sufficient in itself. Thus, if the oxide in operation has a low surface area, that is to say, a small number of adsorption sites, its adsorption capacity will be necessarily low.

For instance, we cannot hope to fix more than 0.2 wt% of platinum on an alpha alumina of 10 m^2/g from a chloroplatinic acid solution. In the same conditions, a gamma alumina of 200 m^2/g fixes about 3 wt% of platinum.

3.6.2. Stability of metal complexes

The formulas of several chlorometal and amine complexes have been listed in table 2 and table 3 to illustrate the possibilities of metal anion and cation adsorption.

The stability of these complexes is not always excellent.

For instance, in an aqueous medium, some chlorometal anions are subject to aquation, hydrolysis, reduction (if the metal has several valences), and even polymerization reactions (45).

These reactions lead to different complexes whose formulas are given in table 5.

TABLE 5
Aquation, hydrolysis, reduction and polymerization of chlorometal anions $(MCl_n)^{x-}$

- **Chloro Aquo Complexes**

$$\left(M(H_2O)_m Cl_{n-m}\right)^{(x-m)-}$$

- **Chlorohydroxy Complexes**

$$\left(M(OH)_m Cl_{n-m}\right)^{x-}$$

- **Chloro Aquohydroxy Complexes**

$$\left(M(H_2O)_m (OH)_{m'} Cl_{n-m-m'}\right)^{(x-m)-}$$

- **Reduced Chloro Aquohydroxy Complexes**

$$\left(M(H_2O)_m (OH)_{m'} Cl_{n-m-m'}\right)^{(x'-m)-}$$

- **Polymeric Species**

$$\left(M_p (H_2O)_m Cl_n\right)^{(n-p(n-x))-}$$

The kinetics of these reactions depends on a number of parameters such as pH, chlorine concentration or temperature.

Concerning the adsorption of these anions on oxides, the reactions of aquation, hydrolysis, reduction and polymerization modify the charge of the coordination sphere of the anion, and this modification may have a significant influence on the thermodynamics of the adsorption equilibri

In this case, there are as many exchange equilibria as there are species in solution.

It is therefore important to know the metal species in operation during impregnation, and, if possible, to maintain conditions where only one metal species exists.

3.6.3. Solubility of the oxide

The apparent solubility of mineral oxides or hydroxides depends on their acid-base character (figure 7).

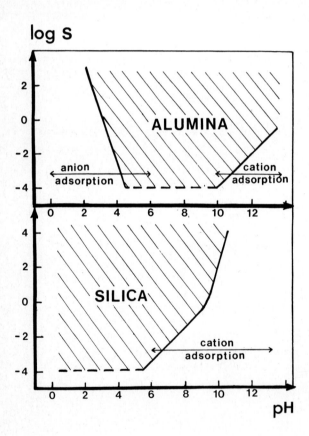

Fig. 7. Variation of the solubility of alumina and silica versus pH.

For instance, the solubility of an acid oxide or hydroxide, such as silica, increases very rapidly in a basic medium. On the other hand, a basic oxide or hydroxide dissolves better in acid medium. For an amphoteric oxide, the solubility increases as well for acid pH as for basic pH.

Anionic or cationic exchanges on basic or acid oxides often imply using solutions that are respectively acid or basic. Consequently, the support may dissolve if the apparent solubility is high at the pH of the impregnation solution. In this case, it involves solubilization of ions containing the specific metal of the support.

This phenomenon of dissolution has been referred to many times in literature for an alumina support that has been brought into contact with a solution containing a strong acid such as HCl or H_2PtCl_6 (8,9, 10,15,46).

Figure 8 shows for example that when gamma alumina is in contact with a hydrochloric acid solution containing much more chlorine than that which corresponds to the adsorption capacity of the support, then a part of the hydrochloric acid is adsorbed on the alumina, and the excess HCl is neutralized by surface dissolution of alumina into aluminium chloride (13).

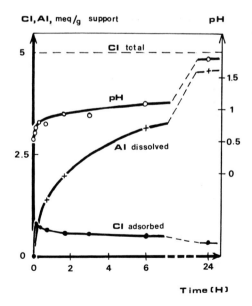

Fig. 8. Chlorine adsorption and aluminium dissolution for the HCl-γ-Al_2O_3 system.

4. CONCLUSIONS

In analyzing the phenomena occurring when an oxide particle is put into an aqueous solution, it is possible to bring out the importance of a specific property of the oxide: its isoelectric point (I.E.P.S.) or zero point of charge (Z.P.C.).

Indeed, an oxide in contact with a solution whose pH is below its I.E.P.S. tends to polarize positively and be surrounded by compensating anions. On the other hand, the same oxide in a solution with a pH above its I.E.P.S. tends to polarize negatively and be surrounded by compensating cations.

With this simplified schema, the pH range that has to be used and the kind of metal complexes (anions or cations) that must be selected can be estimated. Thus, the I.E.P.S. of an oxide and the pH of the impregnating solution seem to be the most important parameters that regulate adsorption of metal complexes. However, knowing only the I.E.P.S. of an oxide is not enough. It must be kept in mind, for instance, that the capacity of cationic adsorption of silica becomes significant above pH 6, which is five pH units from its isoelectric point.

Consequently, in order to know with a good precision the adsorption capacity of an oxide at a given pH, it is preferable to know the evolution of the polarization as a function of pH. This can be done, for example, either by measuring the electrophoretic velocity, or by neutralization experiments at constant pH.

Notwithstanding the complication raised by secondary parameters such as small surface area of the oxide, instability of the metal complexes or solubilization of the oxide, we think that all these considerations which seem to be valid in the adsorption of metal complexes on alumina or silica, are also valid for other oxides.

Furthermore, other possibilities may be considered, such as the doping of a catalytic active oxide by surface adsorption of a foreign metal ion, or even the selective depositing of a metal on one of the constituents of an oxide mixture. This is to be confirmed, however, through further original experiments.

ACKNOWLEDGEMENTS

We thank Pr. R. MONTARNAL (I.F.P.) and Drs. M. MICHEL and R. POISS (RHONE-POULENC) for helpful discussions.

REFERENCES

1. V.S. Boronin, V.S. Nikulina and O.M. Poltorak, Russ.J.Phys.Chem., 37 (1963) 626-629.
2. H.A. Benesis, R.M. Curtis and H.P. Studer, J. Catal., 10 (1968) 328-335.
3. E. Echigoya, L. Furuoya and K. Morikawa, J. Chem.Soc. Japan,Ind. Chem., Sect. 71 (1968) 1768-1773.
4. T.A. Dorling, B.W.J. Lynch and R.L. Moss, J. Catal., 20 (1971) 190-201.
5. B. Samanos, P. Boutry and R. Montarnal, C.R. Acad. Sci. Série C, 274 (1972) 575-578.
6. J.P. Brunelle, A. Sugier and J.F. Lepage, J. Catal., 43 (1976) 273-291.
7. R.W. Maatman and C.D. Prater, Ind. Eng. Chem., 49 (1957) 253-257.
8. R.W. Maatman, Ind. Eng. Chem., 51 (1959) 913-914.
9. R.W. Maatman, P. Mahaffy, P. Hoekstra and C. Addink, J. Catal., 23 (1971) 105-117.
10. E. Santacesaria, S. Carra and I. Adami, Ind. Eng. Chem., 16 (1977) 41-47.
11. F. Bozon-Verduraz, A. Omar, J. Escard and B. Pontvianne, J. Catal., 53 (1978) 126-134.
12. J.C. Summers and S.A. Ausen, J. Catal., 52 (1978) 445-452.
13. J.P. Brunelle and A. Sugier, C.R. Acad. Sci. Série C, 276 (1973) 1545-1548.
14. R.L. Burwell, R.G. Pearson, G.L. Haller, P.B. Tjok and S.P. Chock, Inorg. Chem., 4 (1965) 1123-1128.
15. F. Umland and W. Fischer, Naturwissenschaften, 40 (1953) 439-440.
16. H. Schäfer and W. Neugebauer, Naturwissenschaften, 38 (1951) 561.
17. P.W. Jacobs and F.C. Tompkins, Trans. Farad. Soc., 41 (1945) 388-394.
18. L. Sacconi, Disc. Farad. Soc., 7 (1949) 173-179.
19. W. Naddack and E. Bankmann, Z. Elektrochemie, 58 (1954) 725-731.
20. F. Umland, Z. Elektrochemie, 60 (1956) 701-721.
21. W. Fischer and A. Kulling, Z. Elektrochemie, 60 (1956) 681-688.
22. K.C. Williams, J.L. Daniel, W.J. Thomson, R.I. Kaplan and R.W. Maatman, J. Phys. Chem., 69 (1965) 250-253.
23. J. Stanton and R.W. Maatman, J. Colloid. Sci., 18 (1963) 878-885.
24. C.R.A. Clauss and K. Weiss, J. Colloid. Interface Sci., 61 (1977) 577.
25. D. Briggs and Y.M. Bosworth, J. Colloid. Interface Sci., 59 (1977) 194-196.
26. J.E. Duval and M.J. Kurbatov, J.Phys.Chem., 56 (1952) 982-984.
27. M.H. Kurbatov, G.B. Wood and J.D. Kurbatov, J. Phys. Colloid.Chem., 55 (1951) 1170-1182.
28. E.A. Forbes, A.M. Posner and J.P. Quirk, J. Colloid. Interface Sci., 49 (1974) 403-409.

29. P.H. Tewari and W. Lee, J. Colloid. Interface Sci., 52 (1975) 77-78.
30. R.O. James and T.W. Healy, J. Colloid. Interface Sci., 40 (1972) 42-81.
31. L.H. Allen and E. Matijevic, J. Colloid. Interface Sci., 33 (1970) 420-429, 35 (1971) 66-76.
32. E. Matijevic, F.J. Mangravite and E.A. Cassell, J. Colloid Interface Sci., 35 (1971) 560-568.
33. H. Tominaga, M. Kaneko and Y. Ono, J. Catal., 50 (1977) 400-406.
34. P.W. Schindler, B. Fürsf, R. Dick and P.U. Wolf, J. Colloid Interface Sci., 55 (1976) 469-475.
35. J.P. Bonsack, J. Colloid Interface Sci., 44 (1973) 430-442.
36. H. Hohl and W. Stumm, J. Colloid Interface Sci., 55 (1976) 281-288.
37. C. Huang and W. Stumm, J. Colloid Interface Sci., 43 (1973) 409-420.
38. M.G. Mac Naughton and R.O. James, J. Colloid Interface Sci., 47 (1974) 431-440.
39. G.R. Wiese and T.W. Healy, J. Colloid Interface Sci., 51 (1975) 434-442.
40. J. Stanton and R.W. Maatman, J. Colloid. Sci., 18 (1963) 132-146.
41. A. Kozawa, J. Elec. Soc., 106 (1959) 552-556.
42. M. Michel, personal communication.
43. G.A. Parks, Chem. Reviews, 65 (1965) 177-198.
44. J.F. Lepage, Catalyse de contact, Technip, Paris (1978).
45. W.P. Griffith, The Chemistry of the rarer platinum metals Interscience Publishers, London (1967).
46. J.M. Vergnaud, B. Rey-Coquais, B. Buathier and R. Neyron, Bull. Soc. Chim. Fr., (1968) 3881-3885.

DISCUSSION

MATIJEVIC : The picture given by Dr. Brunelle is greatly oversimplified. Assuming that electrostatic attraction is solely responsible for adsorption is not correct (i.e. that anionic adsorption is going to take place below I.E.P. and cationic adsorption above I.E.P.). Indeed, opposite charges will attract solute species into the interfacial layer, but it will not make them adsorb. For the latter to happen a reaction has to take place, such as surface precipitation, condensation, complexation, etc. As a matter of fact uncharged solutes may show higher adsorption density on the same substrate than the oppositely charged species of similar composition.

J.P. BRUNELLE : First of all, we must remind that our topic is not the "fixation" of any solute on any substrate. In front of the complexity of interaction phenomena between solutes and carriers, we have excluded for example the case of the fixation of hydrolyzable cations on oxides. In the same manner, we have not considered the case of fixation of complexing or precipitating anions such as oxalate, citrate, selenate or phosphate on alumina.

In fact, we have limited our speech to the adsorption phenomena of chlorometallic complexes and amine complexes of groups 7a, 8 and 1b metals on mineral oxides such as alumina or silica, and this, for two reasons: it is the field which seems the best known (or rather the least unknown) and it is of great interest for the catalysis searchers. In this very case, we have attempted to present a simplified analysis of the parameters which govern the adsorption phenomena.

This analysis, though simplified, has been till now confirmed experimentally, and further work will show us the limits of our presentation.

L. RIEKERT : Comment : Because of the many chemical and physical interactions taking place in these systems any exact quantative description will be helpful, although different phenomena may be lumped in such a description. In this respect the contribution by Dr. Brunelle appears welcome and essential. Perhaps it would be advisable to distinguish exactly in terms of experimentally measured quantities ... the terms "adsorption", "fixation", "ion-exchange" and "reaction with the carrier".

J.P. BRUNELLE : Fixation and interaction are general terms which may include many types of phenomena such as precipitation, exchange, adsorption, neutralization, complexation (or specific adsorption)... In a schematic manner, the polarization in an acidic medium of a surface hydroxyl group may proceed by the two following ways :

(1) $\quad M-OH + H^+A^- \rightleftharpoons M-OH_2^+A^-$

or

(2) $\quad M-OH + H^+A^- \rightleftharpoons M^+A^- + H_2O$

In the same way, the polarization in a basic medium of a superficial hydroxyl group may be written :

(3) $\quad M-OH + B^+OH^- \rightleftharpoons M(OH)_2^- B^+$

or

(4) $\quad M-OH + B^+OH^- \rightleftharpoons MO^-B^+ + H_2O$

In our opinion, equations (1) and (3) correspond to <u>adsorption</u> processes. Equations (2) and (4) correspond to <u>neutralization</u> processes.

Exchange reactions correspond in fact to the following cases :

(5) $\quad M^+A_1^- + A_2^- \rightleftharpoons M^+A_2^- + A_1^-$

$\quad$ (or $MOH_2^+ A_1^-$) $\qquad$ (or $MOH_2^+ A_2^-$)

(6) $\quad MO^-B_1^+ + B_2^+ \rightleftharpoons MO^-B_2^+ + B_1^+$

For example, polarization of alumina in hydrochloric acid with chloride fixation may occur via an adsorption process or a neutralization process. On the other hand, fixation of chloroplatinate anions on chlorinated alumina proceeds via an exchange reaction between adsorbed chloride anions and chloroplatinic anions of the solution.

V. FATTORE : The average crystallite dimensions of platinum on a carrier are dependent not only on the impregnation technique used (see your first figure). It is possible to vary and to obtain very low and similar crystallite size starting from different impregnation methods just chosing the proper controlled drying and calcining conditions like, for instance, oxidizing or reducing atmosphere, fast or slow rate of drying.

J.P. BRUNELLE : We agree with you when you say that the choice of optimal activation treatment may also lead to small crystallites. But, one may also recognize that it is generally easier to obtain a great dispersion of a metal on a carrier when an adsorption or exchange procedure is used.

ANALYSIS OF STEPS OF IMPREGNATION AND DRYING IN PREPARATION OF SUPPORTED CATALYSTS

V.B. Fenelonov, A.V. Neimark, L.I. Kheifets, A.A. Samakhov
Institute of Catalysis, Novosibirsk, U S S R

ABSTRACT

The steps of impregnation and drying and their influence upon the uniformity of distribution of catalyst components on supports have been analyzed. The component adsorption during capillary impregnation of mono- and bi-dispersed supports as well as component redistribution on drying have been examined. Several parameters determining the distribution of precipitated component within the catalyst particles have been proposed.

INTRODUCTION

Distribution of the active components of supported catalysts, introduced into the support from solution, may depend significantly upon the conditions of impregnation and drying (ref. 1-6). The final distribution of components adsorbed from solution may be determined by sorption-diffusion mechanism before drying (adsorption catalysts) (ref. 2-4). If the extent of adsorption of components is negligibly small as compared to their total concentration, then distribution depends primarily on drying conditions (impregnated catalysts) (ref. 5,7-9). The best known works in this field (partially reviewed by J.R.Anderson, ref. 1) offer extensive consideration on the details of precipitation of adsorbed components. However, only limited studies have been carried out on the role of drying and the combined effects of capillary transfer and adsorption during drying. The objective of this paper is to systematize all known factors effecting the distribution of an active component during impregnation and drying.

RESULTS

1. Analysis of equilibrium conditions

The ratios between adsorbed and dissolved component in

the impregnated support under equilibrium conditions is determined by the adsorption isotherm, equilibrium concentration of solution C_{eq}, and support structural characteristics (pore volume V_Σ and surface area S). For the general case, during the contact of m grams of the support with a solution volume V_o, at initial concentration C_o, the following equations are true:

$$a = V_o \frac{C_o - C_{eq}}{m} + C_{eq} V_\Sigma (f_v - f_o) \qquad (1)$$

$$b = a + C_{eq} V_\Sigma (1 - f_v) \qquad (2)$$

$$C_{eq} = \frac{C_o (1 - am/V_o C_o)}{1 - (f_v - f_o) \frac{V_\Sigma m}{V_o}} \qquad (3)$$

where a and b are values of adsorption and total component concentration in one gram of the supporting substance; f_o is the fraction of pore volume filled with pure solvent before contact with the solution; f_v is the fraction of the pore volume inaccessible to the component. Introduction of parameter f_v permits the cases of "negative" adsorption and molecular-sieve effects (when $f_v > 0$) to be taken into account. As follows from equation (3), the decrease in the component concentration after impregnation does not necessarily point to adsorption and may be observed, for example, at $a=0$ and $f_o > 0$. At $f_v > 0$ the increase in C_p is possible as well. Parameter P is the ratio between the adsorbed and the dissolved component in the support volume:

$$P = \frac{a}{b-a} = \frac{a}{C_{eq} V_\Sigma (1-f_v)} = \frac{K_v}{V_\Sigma (1-f_v)} = \frac{K_s S(1-f_s)}{V_\Sigma (1-f_v)} \qquad (4)$$

where $K_v = \frac{a}{C_o}$ and $K_s = \frac{a}{C_{eq} S}$ are characteristics of the adsorption isotherm (in the linear region of the isotherm, K_v and K_s are values of the Henry constant); f_s is the fraction of the surface inaccessible to the component.

At $P \gg 1.0$ the component is primarily in an adsorbed form ("adsorption" catalyst) and at $P \ll 1.0$ adsorption is negligibly small ("impregnated" catalysts). Figure 1 shows the change in parameter P during adsorption of H_2PtCl_6 on three typical supports (ref. 4). For these cases adsorption isotherms are described by the Langmuir equation. In the Henry region of the isotherm, parameter P is constant but decreases subsequently down to the value corresponding to the saturated solution (C_s).

2. Impregnation

The concentration and distribution of an active component depends

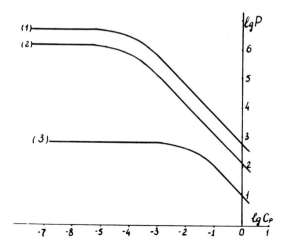

Fig. 1. Plot of parameter P <u>versus</u> equilibrium concentration C_{eq} in adsorption of H_2PtCl_6 from aqueous solutions on activated carbon (1), active alumina (2) and alumo-silicate (3) (data taken from ref. 3).

significantly upon the conditions and duration of impregnation. If the initial support is saturated with a pure solvent ($f_o=1.0$), then impregnation reduces to the diffusion of the component into the support pores, accompanied by adsorption. Hereafter this method will be called "diffusional impregnation". The characteristic time of diffusional impregnation (ref. 1,2) is found from the equation:

$$t_d = R^2(1+P)/D \qquad (5)$$

where R is the grain radius and D is an effective coefficient of the component diffusion in the pore medium. The uniform distribution is reached only at $t > t_d$. Concentration profiles during diffusional impregnation have been calculated by several investigators (ref.2,3,6). If $f_o=0$ then the impregnation process may be divided into two periods: capillary impregnation t_{cap}, and diffusional impregnation t_d. In the case of capillary impregnation the component may be transferred by both convection and diffusion. The duration of capillary impregnation is determined by the viscosity of the solution μ, the value of surface tension σ, the wetting angle θ and structural characteristics of the support. For the model of a monodispersed grain with capillary length R and radius r, the following equation is valid:

$$t_{cap.} = \frac{2R^2 \mu}{r \sigma \cos\theta} \qquad (6)$$

During the limited time of impregnation, the component distribution is determined by parameter α:

$$\alpha = \frac{t_d}{t_{cap.}} = \frac{r \sigma \cos\theta}{2 \mu D}(1 + P) \qquad (7)$$

At $\alpha \leqslant 1.0$ the component, together with a solution, moves inside the grain. At $\alpha \gg 1.0$ distribution depends on dissusion and for calculation the ratios analogous to those cited in ref. 2, 3 may be used. Peculiarities of the impregnation of bi-dispersed structures may be illustrated by the model of the support grain of radius R_o, consisting of aggregates of radius R_2, which, in turn, consist of primary species of radius R_1. When the specific pore volume between aggregates (the secondary structure) equals V_2 and pore volume in the aggregates (the primary structure) equals V_1, the ratio between the time required for grain impregnation into pores of the secondary structure, t_o, and aggregate impregnation time, t_1, is found by the following equation:

$$\alpha_2 = \frac{t_1}{t_o} = \frac{R_o^3}{R_o^2 R_1} \cdot \frac{V_2/V_1}{1 + V_1 \rho} \qquad (8)$$

Here ρ is the true support density. Equation (8) is valid for both capillary and diffusion impregnation. In the case of diffusional impregnation, we find

$$\alpha_2 = \left(\frac{R_2}{R_o}\right)^2 \frac{1 + \frac{K_s S}{V_1}}{1 + \frac{K_s S_2}{V_2}} \approx \left(\frac{R_2}{R_o}\right)^2 \frac{S}{S_2} \frac{V_2}{V_1} \qquad (9)$$

In this equation S is the support surface area and S_2 is the surface area of aggregates. $S_2 = \frac{6}{R_2 \rho}(1 + V_1 \rho)$. Here and hereafter we shall assume $f_v = 0$. According to eq. 9, at $\alpha \gg 1.0$ $\alpha_2 \gg 1.0$ and with limited time of impregnation, we may anticipate relatively uniform distribution of the component on the external surface of the aggregates. A non-uniform distribution will be found along the radii of the aggregates. At $\alpha_2 \ll 1.0$ a non-uniform distribution is possible along the radii of both grains and aggregates.

3. <u>Drying</u>.

An analysis of the mechanisms of solution transfer and redistribution of components in the grain during drying is reported by several authors (ref. 7-9).

When the rate of evaporation from the grain surface, j_n^o, is significantly less than the characteristic rate of solution transfer in the grain volume under the influence of capillary forces, $j^o_{cap.}$, we may designate as "slow" drying. Under these circumstances the following equation is applicable (ref. 11,12):

$$\alpha_3 = \frac{j^o_{cap.}}{j^o_n} = \frac{\varepsilon \sigma \cos\theta \, \bar{r}}{4 \mu D_1 \zeta} (1 + Nu_d)^{-1} \left(\frac{\rho_L}{\rho_v}\right) \Delta \gg 1.0 \qquad (10)$$

Here ρ_L and ρ_v represent the densities of solvent liquid and vapor, respectively. Nu_d is the Nusselt criterion, ε is porosity, $\bar{r}$ is mean pore size, D_1 is the molecular diffusion coefficient of vapor and ζ is the tortuosity factor. Also Δ is the parameter characterizing pore distribution along the radii (for uniform distribution $\Delta = \frac{r_{max} - r_{min}}{\bar{r}}$). If, in addition to equation 10, the following condition is fulfilled

$$\alpha_4 = \frac{D_1 \zeta (1 + Nu_d)}{D \varepsilon} \frac{\rho_v}{\rho_L} \ll 1.0 \qquad (11)$$

then diffusional gradients in the grain must be absent. A schematic representation of the movement of the liquid evaporation boundary at $\alpha_3 \gg 1$ in the section of an arbitrary porous body is shown in Fig. 2. This movement is due to the removal of liquid from

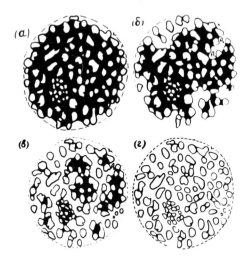

Fig. 2. Scheme of liquid movement during evaporation of a porous body: a - step I, , ઠ-ઠ -step II - formation of domains, ૨ - step III - liquid phase is present only in micropores and adsorption films.

larger pores at the evaporation boundary. In this case liquid is partly evaporated directly and partly moves to pores of smaller diameter at the evaporation boundary. Such occurs because of capillary pressure gradients. This type of transfer is possible only between the portions of solid bound by liquid. The extent of binding by the liquid decreases with evaporation with evaporation. At the beginning the grain is completely bound. However, as evaporation proceeds, isolated domains are gradually formed (step II). Their numbers increase and their size decreases. At the last step (III) liquid is found only in largely isolated domains of minimum size connected only with the adsorption film.

When $P \ll 1.0$, (impregnated catalysts), the solvent volume V_s at the concentration of saturation C_s is

$$V_s = C_o V_\Sigma / C_s \tag{12}$$

If value V_s is reached at steps I or II then an appreciable portion of an active component should be precipitated in accordance with the redistribution mechanism. As a result, precipitation occurs at the mouth of the limited number of pores of small diameter coming to the external surfaces of either grains or macro-pores. This redistribution is caused by both the volume transfer of the solvent effected by capillary forces and diffusional transfer of the component toward fairly large particles formed at the mouth of small pores. At $\alpha_4 < 1.0$, growth of these particles prevents the supersaturation of a solvent and the formation of new particles at the boundary of shifting meniscus. As a result one may expect the formation of a rather coarsely-dispersed deposit of the component (ref. 9). A finely-dispersed deposit is formed only in the domains of minimal size at step III of drying at $V_s < V_h$. According to ref. 7,8, V_h corresponds to the critical liquid concentration. Its value is similar to the adsorption value at the lowest point of the hysteresis loop on the adsorption isotherm of solvent vapors on the support.

Condition for the formation of only a finely-dispersed deposit is the following (ref. 7,8):

$$H = \frac{C_o V_\Sigma}{C_s V_h} < 1.0 \tag{13}$$

At $H > 1$ the portion of coarsely-dispersed phase is $B = 1 - H^{-1}$. In the regime of fast drying (or "moving evaporation front" (ref. 11-12)) at $\alpha_3 \ll 1.0$, the redistribution of solvent under the influence of capillary forces may also be neglected. The rate of drying falls with deepening of the evaporation front and the corresponding increase in the resistance of the vapor diffusion to the grain external surface. The fall in the value of parameter with time at $t > 0$ is described by the equation:

$$\alpha_3(t) = \frac{[R_0 - R(t)] \sigma \cos\Theta \; \bar{r} \Delta}{R(t) \; 4 \; D, \; M} \left(\frac{\rho_l}{\rho_v}\right) \tag{14}$$

where $R(t)$ is the position of the boundary of evaporation zone. As follows from eq. (14), there is a position of the evaporation zone boundary R^* such that $\alpha_3 = 1.0$ and condition for the above regime is not fulfilled. If $(R^*/R_0)^3 \ll 1.0$ then the amount of residual liquid may be neglected. Moreover we may consider that drying of the whole grain proceeds in the regime of moving evaporation front. With this regime we may anticipate a uniform distribution of the component in the volume of the support grain beginning with depth R_H, corresponding to the concentration C_s of a saturated solution. The depth of zone R_H may be estimated from the balance equation, for example, for a spherical mono-dispersed grain with radius R_0:

$$\Delta R = (R_0 - R_H) = R_0 \left[1 - (C_0/C_s)^{1/3}\right] \tag{15}$$

4. Influence of the component adsorption on distribution during drying

Consider some peculiarities of drying of adsorption catalysts ($P > 1$) when the component concentrating results in additional sorption. After the concentration of a saturated solution C_s is reached, precipitation of the component starts at the boundary vapor - liquid. Introduce the value of the relative moisture content $u = V_i/V_\Sigma$ at a given moment of drying and the known dependence of the specific area at the boundary solution(support)-vapor ($S = S_0 \eta(u)$). Then we may write the following equation for the dependence of the solution concentration C on the specific moisture content u:

$$\left[u + \frac{S}{V} \eta(u) \frac{d\,f(c)}{d\,c}\right] \frac{dc}{du} + c(u) = 0 \tag{16}$$

In the Henry region the following analytical solution may be obtained:

$$C = C_0 \; e^{\int_u^1 [u + P_s \eta(u)]^{-1} du} \tag{17}$$

where $P_s = \dfrac{a_s}{C_o V_\Sigma}$, a_s is the value of sorption in a saturated solution at C_s. Functions u, $\eta(u)$ in eqs. 16, 17 may be calculated using models of the support structure.

In the specific case of "slow" drying regime, distribution of the solution in the support volume is close to equilibrium. Down to low solution concentrations, corresponding to the monolayer coverage, the entire support surface is covered with a solution, i.e. $\eta(u) = 1.0$. Solution (17) may be simplified:

$$\frac{a}{b} = \frac{P'_s}{1+P'_s}\left(\frac{C_s}{C_o} - 1\right) \tag{18}$$

$P'_s = P_s \dfrac{C_o}{C_s} = \dfrac{a_s}{C_s V_\Sigma}$. A more rigid (as compared to $P \ll 1$) condition for adsorption to be neglected follows from eq. (18):

$$P_s = \frac{a_s}{C_o V_\Sigma} \ll 1.0 \tag{19}$$

i.e., the ratio between the maximum possible amount of adsorbed component and the total amount of inserted component must be low.

In a "fast" regime of drying at $\alpha_3 \ll 1.0$ we may assume for monodispersed systems that $\eta(u) = u$. In this case

$$\frac{a}{b} = 1 - \frac{[C_o/C_s]\, P'_s}{(1 + P'_s)} \tag{20}$$

and condition for the sorption to be neglected will be presented as:

$$P'_s \ln \frac{C_s}{C_o} \ll 1.0 \tag{21}$$

If conditions reversible to eq.(19) and (21) are fulfilled, then the whole component is bound upon drying. Furthermore, at $\alpha_3 \ll 1.0$ the surface concentration increases with movement along the grain radius while at $\alpha_3 \gg 1.0$ the component is concentrated at the mouth of small pores coming either to the external grain surface or to the surface of macropores.

Finally, the condition for preparation of a finely-dispersed deposit, resulting from the component precipitation after the solution saturation, transforms to

i)
$$H = \frac{1}{u_{cr}}\left(\frac{C_o}{C_s}\right)^{(1+P'_s)} < 1.0 \tag{22}$$

(provided that $\eta(u) = u$) , and

ii) $H = \dfrac{1}{u_{cr}} \left[\dfrac{C_o}{C_s} - P_s'(1 - \dfrac{C_o}{C_s}) \right] < 1.0$ \hfill (23)

(provided that $\eta(u) = 1.0$).
In both cases $u_{cr} = V_h/V_\Sigma$. As follows from eqs. (22) and (23), the portion of a finely-dispersed deposit increases with P_s'.

CONCLUSIONS

The step of impregnation contributes significantly to the distribution of an active component for the catalyst of the adsorption type at $P \gg 1.0$. In this case the uniformity of distribution depends on impregnation time. To obtain the uniform distribution (termed as distribution of the first type), the time of impregnation must be longer than the characteristic time, t_d. The maximum amount of the inserted component is determined by sorption isotherm. At shorter periods of time of impregnation the distribution is characterized by the increased concentration of the component near the external surface of the grain (distribution of the second type). In the case of impregnated catalysts ($P \ll 1.0$), after the step of impregnation (provided that time of impregnation exceeds t_{cap}), an equilibrium distribution is established which changes significantly at the step of drying. The catalysts of the first type are obtained if $\alpha_3 > 1.0$, $\alpha_4 \ll 1.0$, $H < 1.0$ ($b \le C_s V_h$ may be introduced) or if $\alpha_3 \ll 1.0$, $\alpha_v \gg 1.0$ and any H ($b \le C_s V_\Sigma$ may be introduced). The catalysts of the second type are obtained if $\alpha_3 \gg 1.0$ and $\alpha_4 < 1.0$, $H > 1$, especially if $\alpha_4 \ll 1.0$. Condition $\alpha_3 \ll 1.0$, $\alpha_4 \ll 1.0$ leads to the catalysts of the third type characterized by high concentration at the grain center. In the intermediate cases, boundaries between these principal types of distribution become less distinct.

This analysis should be considered only as a first approximation. It does not take into account possible changes in the support structure during impregnation, nor does it consider the details of interaction between active component and support. The distribution of an active component may also change during heating.

Actually the significant process factors which determine distribution are more complicated. We believe, however, that our analysis may be useful for comparative estimation of the effects of principal parameters on the distribution of an active component.

It may also provide assistance to understand the physico-chemical details of the processes involved.

REFERENCES

1. J.R. Anderson, Structure of Metallic Catalysts, Acad. Press, London - N.Y., 1975.
2. P.B. Weisz, Trans. Faraday Soc., 63 (1967) 1801, 1807.
3. P. Harriott, J. Catalysis 14 (1969) 43.
4. R.W. Maatman and C.D. Prater, Ind. Eng. chem., 49 (1957) 253.
5. R.W. Maatman, Ind. Eng. chem., 51 (1959) 913.
6. R.C. Vincent and P.P. Merrill, J. Cat., 35 (1974) 206.
7. A.V. Neimark, V.B. Fenelonov and L.I. Heifets, React. Kinet. Catal. Lett., 5 (1976) 67.
8. V.B. Fenelonov, Kinetika i Kataliz, 16 (1975) 732; 17 (1976) 1582.
9. G.K. Boreskov, Proc. 1st Int. Symp. Catalysts Preparation, Brussel, 1976, p. 223.
10. V.B. Fenelonov in sb. Adsorption and Porosity, "Nauka", Moscow, 1976, p. 55.
11. A.V. Neimark and L.I. Heifets, Doklady Akademii Nauk SSSR, 228 (1976) 135.
12. A.V. Neimark, L.M. Pis'men, V.E. Babenko and L.I. Heifets, Teor. Osnovy Khimicheskoi Tekhnologii, 9 (1975) 369.

Note added in proof:

It can be added that in the case of "adsorption" catalysts, introduction of adsorbed competitive substances allows to regulate the distribution of the main component within wide interval.

This regulation is based on the chromatographic effects resulting in the displacement of weakly adsorbed components by strongly adsorbed ones. In contrast with the close analogous method - paper chromatography - component motion is associated with the migration from the external surface into the grain and can be accompanied by the additional concentrating of the displaced component.

For example, simultaneous impregnation of Al_2O_3 catalysts with solutions H_2PtCl_6, and $H_2C_2O_4$ or H_2SO_4 permits the preparation of Pt/Al_2O_3 catalysts with an active component distributed either in the central zones or in the annular form inside the grain. In this case the competitive agent is adsorbed more strongly than H_2PtCl_6. Simultaneous impregnation of Al_2O_3 with solutions of H_2PtCl_6 and CH_3COOH, $HCOOH$, HNO_3 or HCl aids in the distribution of Pt at the external boundary of the grain. These competitive agents are adsorbed in a weaker form than H_2PtCl_6.

Variations in the concentration of components, their types, duration of their impregnation as well as application of successive treatments by a competitive agent or an active component significantly enlarge the possibilities of regulating the distribution of an active component.

DISCUSSION

R. GALIASSO : In my opinion, your model for support structures is oversimplified : 1) How do you take into account that during the drying processes the α_3 function changes due to the increasing influence of diffusional gradients, mainly during the period the rate decreases ? 2) Could you comment on the form of the η function, which appears in equations 16 and 17 for bidispersed catalysts ?

V. FENELONOV : 1) Our analysis should be considered only as a first approximation which is necessary for a physico-chemical analysis of processes, responsible for formation of structures on catalyst supports. We have taken into account that the value of the parameter α_3 may be increased during the fast drying, as a result of a displacement of the evaporation zone in the depth of the pellet and an increase in the vapor diffusion resistance (in the text "fall in the value α_3" must be replaced by "increase in the value α_3"). In this case, in the central part of the pellet, conditions may be produced similar to those obtained during slow drying.
2) During the slow drying regime a bidispersed structure is obtained : at first liquid is removed from the large pores between aggregates and only afterwards from the narrow pores in the aggregates. So we may represent the drying processes in this simplified way: first of all processes of removal of liquid from the secondary structure between aggregates should be considered. The aggregates are viewed as non porous material. Then the independent processes of removal of liquid from the interior structure of the aggregates has to be taken into account. This approach permits the use of our equations for bidispersed structures too.

H. CHARCOSSET : What do you know about the particle size of platinum in the catalysts prepared from different ion competitors (last slide in the communication) ?

V.B. FENELONOV : Generally speaking, the dispersion of the active component depends on its surface concentration. It may be defined as the amount of component per unity of support surface and as the number of component crystals per unit of support surface. It is connected to sintering and to other related processes of redistribution. So, in our experiments the crystal size is about 10-15 Å in the case of a uniform distribution; if the component is concentrated in a narrow zone of the pellet, the crystal size is increased.

SOME MECHANISTIC CORRELATIONS BETWEEN IMPREGNATION AND ACTIVATION OPERATIONS FOR THE PREPARATION OF HIGH-SELECTIVITY SUPPORTED METAL CATALYSTS

H. SHINGU[x] and T. INUI
Department of Hydrocarbon Chemistry, Faculty of Engineering, Kyoto University, Kyoto, 606 Japan [x] (38 Higashikishimoto, Shimogamo, Sakyo, Kyoto, 606 Japan)

SUMMARY

From reaction-engineering analysis of catalyst performances, it is concluded that the essential structural requirements for the high-selectivity (i.e., demonstrative of 100% purity of catalytic function) supported metal catalysts, particularly, for the oxidation catalysts on silver and copper bases, comprise three classes of uniformity principles concerning the distribution, the size or the microsurface structure, and the microscopic environment of catalyst materials in structural matrices of the catalyst system, which are respectively related to (1) the porous support, (2) the dispersed metal particles or agglomerates, and (3) the situation of supporting the latter.

According to these uniformity principles, the selection of suitable supports having the prerequisite uniform pore and surface structure and the correct impregnation and activation procedures for the preparation of high-selectivity catalysts are discussed, and some demonstrative examples are illustrated together with interpretations of mechanistic correlations between the operations.

1. INTRODUCTION

For the establishment of scientific bases for the preparation of heterogeneous catalysts it is imperative that adequate reaction engineering studies should first be made to ensure the clarification of the engineering objectives, particularly, the essential structural requirements for a genuinely heterogeneous catalyst which may be definable, by analogy to enzyme systems, as a chemical "reacting-apparatus".

These structural requirements could be concluded from surface-kinetic or certain mechanistic studies of the <u>real working state</u> of the catalyst, i.e., from reaction engineering analysis of the catalyst performances, namely, the selectivity, stability, and activity. As a result, these three fundamentals of catalyst performance are found to be, functionally as well as structurally, in close connection with each other from the reacting-apparatus viewpoint.

The present paper shows firstly, that with the supported metal catalysts, for some selective oxidations, the structural requirements for an ideal selectivity are eventually almost the same as those for high stability and activity (except in the last case, one additional scalar parameter, namely, the bulk concentration of active sites in structural matrices of the catalyst system). The essential requirements may be summarized :

(1) Uniform distribution of the in-pore surface in the structural matrices of a uniform macro- and/or micro-pore system of the support.

(2) Uniform distribution of reduced metal particles or agglomerates of suitable uniform size, eventually supported in dispersion through the in-pore surface.

(3) Uniform physical and chemical situation of supporting the dispersed metal particles or agglomerates on the in-pore surface which renders them capable of satisfactory reproduction and/or reversibility of the redox transformation of the individual metal particle or the microscopic surface layer of the metal agglomerates under the reaction conditions.

In view of the uniformity principles which occur in all the above items, especially item (1), it is evident that the non-uniformit of the pore structure of the conventional preformed supports precludes the application of various measures for uniform impregnation. Therefore, it has been necessary to select or eventually to synthesize suitable supports having the prerequisite uniform pore and/or surface structure for the investigation of impregnation and activation operations for the purpose of preparing high-selectivity metal catalysts.

It has also been found that even with supports of sufficient uniformity the ordinary impregnation and activation procedures failed to produce a final uniform distribution of impregnants and dispersed metal support particles at both macro- and micro-porous levels. In particular, certain mechanistic correlations between impregnation and activation processes, depending upon the chemical and physical nature of the impregnating compounds, should be taken into consideration.

Thus, with bimodal supports which possess a three-dimensionally uniform, isotropic and continuous, pore structure, the macroporous uniformity control could be effected by filling the macroporous space or surface uniformly with a commensurate quantity of impregnant liquid or solution of appropriate concentration using wetting or spraying techniques, provided that the impregnants could be deposited in an amorphous or a homogeneous state (such as glassy) on evaporation and/or activation.

If the crystalline impregnant deposits are to melt prior to the activation process, such as thermal decomposition, the above control may be neglected, and a correct procedure with the activation process under microporous uniformity control becomes the main subject.

The microporous uniformity control is essentially related to the activation process, but it affects the impregnation procedure in that it also corresponds to the formation of the final microsurface structure or the active sites. The microporous surface structure of the support as such does not always meet the requirements of the above items (2) and (3), and certain chemical activators and/or structural promoters should be used as necessary according to the activation mechanism, to modify the activation as well as impregnation procedures, so that the usual control of the adsorption and/or ion-exchange effects in micropores, much less of the chromatographic effects in multiporous systems, may seem to be almost immaterial in the present context.

All these controls over catalyst preparation parameters have been effected together with the corresponding systematic reaction-engineering analyses of catalyst performance data obtained by the proper use of one-dimensional isothermal integral reactors in fixed-bed flow systems for laboratory as well as pilot tests, which at the same time have systematically revealed the essential structural requirements for high-selectivity catalysts and the underlying uniformity principles.

It is therefore considered as appropriate to present at the outset some description of fundamental concepts and definitions, from both macro- and micro-kinetic aspects, of the experimental foundation of chemical reaction engineering in heterogeneous catalysis.

In the course of the systematic presentation of the principles of macro- and micro-kinetic controls of catalyst performances, some examples of the preparation of high-selectivity catalysts may be illustrated which are indicative of 100% purity of catalytic function, on copper or silver bases for vapour phase oxidations.

2. FUNDAMENTALS OF REACTION-ENGINEERING ANALYSIS OF CATALYST PERFORMANCES

In view of the developments of chemical engineering in recent years it is surprising that, until now in the description of rate processes the fundamental difference between homogeneous and heterogeneous catalysis has been overlooked or even deliberately neglected. The classical chemical kinetics (which deal essentially with mathematical treatment of ideally homogeneous reactions on macroscopic or thermodynamic basis and may therefore, be renamed, after A. Skrabal (ref. 1) as homogenkinetics) have always formed the basis for describing the rate process of heterogeneous catalysis, leading to various expressions for the homogenkinetic approximation to heterogeneous kinetics$^+$ such as the Langmuir-Hinshelwood (LH) or Hougen-Watson (HW) treatment in combination with the semi-empirical pseudo-equilibrium assumption about the gas-solid interactions, namely, the Langmuir adsorption isotherm. The HW treatment could cause further speculation about the detailed surface mechanism by applying the transition-state (TS) theory which is an empirical, statistical-thermodynamic, theory and is applicable only to homogenkinetics (ref. 2 & 3). Consequently these versions of homogenkinetic approximation could have afforded at best correlation data of limited applicability which could not generally be rationalized, especially from the microscopic point of view. Since the kinetics of heterogeneous catalysis must include the essential part of the catalyst surface, certain molecular, microscopic description of the active, i.e., reacting surface is required in order to characterize the heterogeneous reaction system or the surface phase kinetically. But, owing to the lack of physical knowledge of the active surface or the working state of the catalyst, we could only define the catalyst surface in a chemical, phenomenological way. According to such chemical definition, the concept of the active surface may be substantiated in quantity and quality by assuming that a macroscopic description of the catalyst surface may represent a uniform reaction phase which should be microscopically analogous to that of homogeneous system, so that a kinetic expression such as the bulk concentration of active sites or centers within the extent, i.e., the volume, of the catalyst system may be assumed as rational (ref. 4) or condition that the uniformity of the catalyst surface and the preclu-

---

$^+$In contrast to homogenkinetics, we may here introduce the term heterogenkinetics as a theoretical representation for heterogeneous kinetics, which characterizes in general terms the microscopic restrictions due to heterogeneity in chemical kinetics.

sion of diffusional (mass and heat transfer) problems, both external and internal to the catalyst system, are ideally assured. Obviously in this chemical approach to the homogenkinetic approximation, there can be no reason for assuming a two-dimensional, macroscopic, unit extent of the reaction system, and the concept of the active surface or sites merges with that of the adsorption surface sites, as is implicitly indicated by the LH or HW treatment, where the microscopic characterization of the heterogeneous surface or site is deliberately neglected and the physical description of the extent and nature of the surface is replaced with a chemical, empirical one by use of such an intensive parameter as the limiting (saturation) capacity for adsorption or the specified rate for surface reaction in concentration units.[+]

Thus, the neglect of heterogenkinetic aspects of the surface process has obscured the kinetic distinction between homogeneous and heterogeneous catalysis in that the surface active site should represent an activated state associated with a certain unit surface ensemble in which microscopically, many complex elementary-stochastic processes are involved. As a consequence, the heterogenkinetic features of the elementary surface processes, including adsorption and desorption as necessary steps in the reaction, should be confirmed phenomenologically in the chain nature of each and all processes involved, as well as in the mechanistic complexity of the definition of the active site in comparison with the molecular simplicity of that inherent to homogeneous catalysis. There is already much experimental

---

[+]The conventional representation of the specific surface rate or intrinsic activity of the catalyst on the basis of a two-dimensional unit extent of the catalyst surface (e.g., g-mole cm^{-2} sec^{-1}) should be considered as physically ill-defined, since the extent of the adsorption surface or site bears, microscopically, no direct relation to the intensive parameter of the active surface or sites, although the BET surface area may be regarded as a certain measure of the adsorption capacity of unit mass of the catalyst.

Also, the term turnover number or frequency (e.g., molecules $site^{-1}$ sec^{-1}) may be regarded as of no exact physical meaning, because the active site, as a kinetic term, should represent the chemical reaction center, macroscopically, in terms of intensive parameter (i.e., concentration in the catalyst system) or, microscopically, in reference to a certain activated state of the catalyst surface. Its quantity and quality could only be determined experimentally through rate measurements, so that the evaluation of turnover number would depend solely upon the mechanistic definition of the active site itself. If the active sites are assumed to be equivalent to or, at least, proportional in quantity to the adsorption sites, the mechanism of the surface reaction should correspondingly become too simple and one-sided to be realistic for heterogeneous surface catalysis, except for the extremes such as quasi-homogeneous or "immobilized" homogeneous catalysis.

evidence for these heterogenkinetic characteristics of heterogeneous catalysis : the "sigmoid kinetics" which is indicative of the prevalence of a certain set of elementary chain processes has been observed for a number of surface processes such as adsorption (including not only irreversible chemisorption but also reversible physi- and chemisorption); desorption (the first observation of autocatalysis in thermal decomposition of solid silver oxide by G.N. Lewis (ref. 5) may be regarded as the prototype), surface oxidation and reduction (of metals and oxides uniformly dispersed on supports (ref. 6), and numerous solid-catalysed reactions (especially in relation to the transient behaviours of retardation or inhibition (ref. 7). Also, the commonly observed low value of saturation pressure for chemisorptions of Langmuir-type with low heat of adsorption (ref. 8) and the chemical relaxation measurements at solid catalyst surfaces (ref. 9) may be considered as indicative of the extent of mechanistic complexity of the adsorption and the active sites, respectively.

The inconsistencies due to the neglect of heterogenkinetics seem to have been reflected in the reaction engineering of heterogeneous catalysis, so that, in contrast to the well defined and elaborated macrokinetics, the microkinetics has been almost disregarded without proper definition[+] since van Krevelen (1957; ref. 10) misdefined it as identical to "classical molecular" chemical kinetics. It has been realized (ref. 11),however, that for reaction engineering contro of catalyst performances it is most important to make a clear distinction between macro- and micro-kinetic factors in designing the experimental reactors and operating conditions, as well as in evaluating the observed data, and that, as a result, the essential

---

[+]Microkinetics may be defined as a branch of applied or engineering kinetics which, complementary to macrokinetics, is characterized as encompassing the regime of molecular-microscopic description of rate phenomena relevant to chemical engineering. For the reaction engineering of heterogeneous catalysis, in particular, microkinetics is to describe molecular transport phenomena, as well as microscopic mass and energy transfer with or without surface chemical transformations, in the realm of catalyst surface structure in order to accomplish the reaction engineering analysis of catalyst performance or catalytic functioning of the catalyst system (ref. 11). In short, microkinetics, dealing with engineering aspects of heterogenkinetics, may play on phenomenological basis a most important role in treating the engineering problems of the catalyst performance control concerning the heterogeneous surface processes which are inclusive of mass and heat transfer in the surface environment or in the micropores, under exclusion of the macrokinetic mass and heat transfer limitations.

part of the catalyst performance controls consists of the analysis
and control of microkinetic rather than macrokinetic factors. In
this way, considering that microkinetic factors may represent the
crucial heterogenkinetic characteristics of the active surface
site or reaction center, the kinetic distinction between homogeneous
and heterogeneous catalysis may further be delineated, and it follows
from reaction engineering experimentation that fundamental concepts
and definitions concerning the kinetic description of heterogeneous
catalysis should be re-examined and, if need, be revised.

2.1 <u>Integral representation of the rate is only valid for heterogeneous catalysis</u>

Firstly, from the foregoing heterogenkinetic arguments, it is
evident that the differential representation of the rate such as
specific rate or catalytic constant, i.e., moles converted per unit
concentration of catalyst per unit time, in homogeneous catalysis
cannot apply in heterogeneous cases, although formally a similar
differential rate expression may be assumed through homogenkinetic
approximation in the macrokinetic regime of heterogeneous catalysis.
Furthermore, it is commonplace in experimental practice for basic
rate information to consist of integral or average rates (ref. 12),
but the differential treatment of the integral rate data obtained
in heterogeneous catalysis is at variance with that obtained in
homogeneous systems in that for the former case there is an irreducible
discontinuity between the time dimensions for macro- and microkinetic regimes due to the complexity of heterogeneous surface
structure, whereas for the latter the microkinetic time dimension
is assumed to be included in the time dimension of molecular-kinetic
continuity. Practically, discontinuity limits may correspond to the
limit extent of the catalyst system which is almost equal to the
catalyst particle size of less than 1 mm diameter for a fluidized or
suspensoid catalytic reactor of complete-mixing batch type or the
porous catalyst pellet diameter of less than 6-8mm[+] for a fixed-bed
tubular reactor of steady flow type. Therefore, it may be concluded
that it is more correct for heterogeneous catalysis not to make use

---

[+]Naturally, these limiting figures may be modified with certain
factors depending on the porous structure of the catalyst pellets,
as well as on the reaction, especially fluid flow, conditions. As
for larger particle or pellet diameters, the figures should be
multiplied with a macrokinetic correction factor such as Thiele
modulus or "effectiveness" factor.

of the differential representation of the rate, on the analogy of
homogenkinetics, such as specific rate (constant), initial rate,
and other differential terms, especially <u>differential reactor</u>,
but to refer to integral terms such as the <u>specific activity</u> (for
specific rate) as expressed by the integral or average rates per
unit extent of the catalyst system (space-time expressions are not
only practically but scientifically valid also for heterogeneous
catalysis : <u>Space-Time-Yield</u>, STY, and <u>Space-Time-Conversion</u>, STC,
in moles per liter catalyst space per hour) the quantity of which ca
be accurately evaluated directly from measurements with a practical
<u>isothermal internal reactor</u> of either batch or flow type.

2.2 <u>Contact time vs, true reaction time problem in heterogeneous
     catalysis</u>

Now that the integral rate character of the microkinetic regime
of heterogeneous catalysis is acknowledged as above, it is possible
to explain that the <u>contact</u> time, t_c, in heterogeneous catalysis,
defined as the reciprocal of <u>space velocity</u>, SV, and corrected with
respect to temperature and pressure, may become linear measure for
the true time, as well as the <u>residence time</u> in homogeneous catalysi
with the condition that the macrokinetic factors are well controlled
so that diffusion to and from the external boundary of the catalyst
system does not influence the reaction rate or the specific activity
This condition can be fulfilled experimentally, for example, by
efficient stirring (>600-800 r.p.m.) in suspensoid-catalytic batch
reactors or by selecting flow conditions in fixed-bed reactors packe
with catalyst pellets of smaller size (<6mm) as shown in Fig. 1.
What is remarkable is that the essential point of the macrokinetic
control of contact time is the production of an efficient steady
flow condition with turbulence at the catalyst boundary and that,
therefore, there is a definite possibility of modifying the specific
activity by influencing microkinetic factors, through particular
macrokinetic control (ref. 13).

2.3 <u>The stationary-reacting state of the catalyst system represents
     the real working-state of catalyst surface</u>

Knowing that the above-mentioned condition of macrokinetic contro
of contact time corresponds to providing the homogenkinetic
approximation in the macrokinetic regime of heterogeneous catalysis
it becomes conceivable that, under these macrokinetic controls with
microkinetic factors remaining practically unchanged, the catalytic
reactions, as represented collectively by the specific activity, are

proceeding within the external boundary of the catalyst system, integrally, in quasi-stationary state, and microscopically, in a diffusional steady state which may be contrasted phenomenologically with the <u>homogenkinetic steady state</u>. The basic conditions for providing this category of homogenkinetic approximation in the microkinetic regime consist firstly of uniformity and stability, i.e., the reproducibility of the catalytic function of the active catalyst surface or sites and then of a proper microkinetic control over reaction engineering parameters such as temperature, pressure, reactant compositions including diluents and especially inhibitors, in order to obtain sets of reproducible data for specific activities. Naturally, these microkinetic controls relate precisely to the whole subject except that of macrokinetic controls of the reaction engineering analysis of catalyst performance, as well as, in another version, to that of catalyst evaluation, so that the <u>stationary-reacting</u> state of the catalyst system as a unit integral reactor should uniquely represent the real working state of the catalyst surface (ref. 14).

Experimentally, this state of the real catalyst system or bed can best be realized in a fixed-bed flow system by the use of an <u>isothermal one-dimensional integral reactor</u> under proper macrokinetic control and be most comprehensively investigated with respect to the axial profiles of temperature and reaction figures as shown in Fig. 1. With this type of tubular reactor which is applicable not only to laboratory but also to industrial practice, the isothermicity, in macrokinetic measures, is ensured by applying efficient external cooling (by use of a molten salt bath with high speed stirring) to maintain the wall temperature of the reactor tube (of smaller diameter <20 mm) almost equal to the constant bath temperature throughout the catalyst-bed, and may be confirmed by the observed linear proportionality between the average temperature rise, ΔT, in the catalyst-bed and the integral conversion (of oxygen), $\Delta(O_2\text{-STC})$, in the range of temperature rises up to as large as 20-50°C : $\Delta T_{av.} = \text{const.} \times \Delta(O_2\text{-STC})$.

2.4 <u>The Linear Reciprocal Space Time Relationship (LRSTR) presents basic correlation data for microkinetic analysis of catalyst performance and also for a posteriori design of catalytic reactors</u>

That the homogenkinetic approximations both in macro-and microkinetic regimes may be practically established in a continuous flow reaction with the isothermal one-dimensional reactor, can be domonstrated conclusively by examining the linear correlation between the reciprocals of integral rates, STY or STC, and space velocity, SV, that is, LPSTR (ref. 11 & 15) :

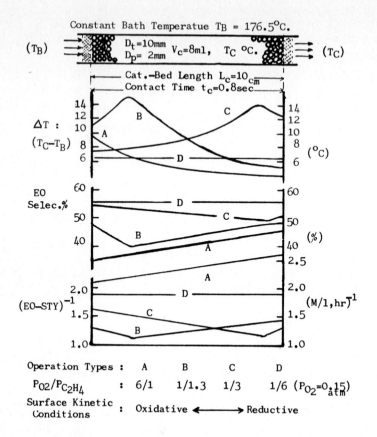

Fig. 1. The real working state of the catalyst surface as represented by the <u>stationary-reacting state</u> of the catalyst bed in an <u>isothermal one-dimensional integral reactor</u>:
Axial profiles for catalyst bed temperature, T_c, selectivity of ethylene oxide formation, EO-Selec.%, and reciprocal space-time-yield of ethylene oxide, $(EO-STY)^{-1}$, in vapour phase oxidation, under ordinary pressure, of ethylene-oxygen mixture with silver catalyst supported on preformed spherical alpha alumina supports of 2mm in diameter.

$$(STY)_{t_c}^{-1} = (STY)_0^{-1} + \alpha t_c \quad (1),$$

where $t_c = 3600 \times (SV)^{-1}$, contact time in sec, and α is a coefficient which, as the gradient for reciprocal integral rates, comprises a homogenkinetic factor for stoichiometric deceleration with reaction time and certain retardation ($\alpha > 0$) or acceleration ($\alpha < 0$) factors specific to the catalytic surface processes, and accordingly may be referred to as either <u>retardation</u> or <u>acceleration coefficient</u>. To confirm the applicability of homogenkinetic approximation in macrokinetic regime, an alternative expression for LRSTR is preferred for comparison with homogeneous reactions, which is also suitable for treating the integral rates of suspensoid-catalytic reactions in a complete-mixing batch reactor : this is the linear relationship between reciprocals of fractional conversion, $x=p'/p_o$, or product partial pressure, p', and reaction time, t, which for a homogeneous first-order reaction with the initial rate, $r_o = k_o p_o$, can be expressed as

$$1/x = 1/k_o t + 1/2 \quad \text{or} \quad 1/p' = 1/r_o t + 1/2p_o \qquad (2).$$

Eq. (2) proved a good approximation in the range of conversions up to 60%.

Similarly, for heterogeneous catalysis for which homogenkinetic approximations in both regimes are applicable, so that HW treatment of double-sided mechanism may be assumed with the specific rate formula : $r = kp/(1+Kp+K'p')^2$ and $r_o = kp_o/(1+Kp_o)^2$, the linear reciprocal relation can be expressed with about the same degree of approximation as Eq.2 as

$$1/p' = 1/r_o t + (1/2p_o) \times (2(1+K'p_o)/(1+Kp_o) - 1) \qquad (3).$$

Comparing Eqs 1 & 3 and noting that $t_c = t$, p's are expressed in atm, and the adsorption terms may be approximated as $(1+K'p_o)/(1+Kp_o) \approx K'/K$, we may obtain

$$160.6 x\alpha = (1/2p_o)(2K'/K - 1) = 1/2p_o + (1/p_o)(K'/K - 1) \quad (4).$$

The last term of Eq.4 signifies, or at least explains qualitatively that the particular retardation or acceleration of the integral rate, i.e., specific activity, with respect to individual catalyst particles in the stationary-reacting state, as observed in the axial profiles in Fig. 1, is caused substantially by the microkinetic concurrence of adsorption and desorption of certain molecular species relevant to reaction components, especially the reaction product. In this respect, it is stressed that the retardation or inhibition due to adsorbates is the most important factor for microkinetic control of catalyst performance, expecially stability and selectivity, and that without the control over inhibitory factors it is impossible to establish the stationary-reacting state or the real working state of the catalyst.

In this way, LRSTR not only provides basic correlation data for microkinetic analysis and control of catalyst performance, revealing the heterogenkinetic characteristics of surface processes involved in the stationary-reacting state of the catalyst system which may therefore, be referred to as "reacting-apparatus", but it also presents in its final form, with correction due to macrokinetic control, a practical a posteriori design equation.

3. PERFORMANCE ANALYSIS FEED BACK TO CONTROLS OF CATALYST PREPARATION

The co-ordination of catalyst preparation and reaction engineering in developing both the catalyst structure and performance with feedbacks to both controls of catalyst preparation and reaction parameters may be illutstrated in the scheme :

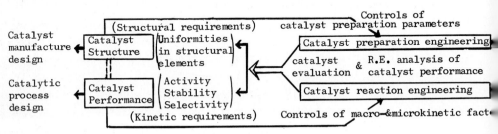

It can be seen that the three fundamentals of catalyst performance as well as structural requirements represented by the uniformity principles for each of structural elements, are functionally as well as structurally interrelated. Originally, certain controls over uniformities in structure relative to activity and stability are prerequisite for the establishment of LRSTR. In the course of further development of the catalyst performance, namely, in a sequential cycl process : Activity ⟶ Stability ⟶ Selectivity, further refinements in structural uniformity principles are followed up, and at a certai stage of the development where a knowledge of the structural element postulating the realization of 100% purity of catalytic function would necessitate the use of preformed uniform-porous supports to effect activity and stability improvements, the impregnation and activation operations become mandatory as a method for High Selectiv (HS) catalyst preparation, and the controls over mechanistic correla between both operations also become important in relation to the formation, distribution, and stabilization of the active sites on the surface matrices of the selected supports. Scientific bases for such controls of catalyst preparation parameters will be illustrated in the following examples :

    3.1 HS SILVER CATALYSTS FOR EPOXIDATION OF ETHYLENE AND PROPYLENE
    3.2 HS BLACK SILVER OR COPPER CATALYSTS FOR COMPLETE OXIDATION.

3.1 HS SILVER CATALYSTS FOR EXPOXIDATION OF ETHYLENE AND PROPYLENE

3.1.1 ESSENTIAL STRUCTURAL REQUIREMENTS

ACTIVE SITES : Pure metal particle of 10 to 20 nm diameter for ethylene oxide formation and of 5 to 6 nm diameter for propylene oxide formation.

SITUATION OF SUPPORTING :

Physically, sufficient seclusion of the individual metal particles from each other and heat stability (to above 600 up to 900°C) of the surrounding support materials are required for the complete prevention of thermal migration.

Chemically, complete inertness not only to metallic silver but also to silver oxide at higher temperature (above 400°C) is necessary to prevent chemical migration and deterioration.

SELECTION OF THE POROUS SUPPORT :

The above mentioned STRUCTURAL REQUIREMENTS could have been fulfilled at least as a laboratory curiosity with non-porous supports. But, in order to improve the catalyst activity to an industrially feasible level of several Space Time Yield in mole (epoxide) $1^{-1}h^{-1}$, it was necessary to make use of a porous

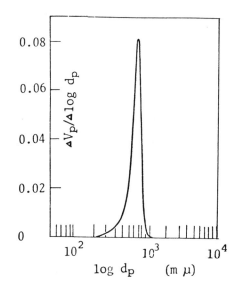

Fig. 2. SEM photomicrogram of a cross-section of the ideally uniform support preformed in spheres of 3.5 mm diameter, showing the isotropic three-dimensionally uniform and continuous pore system of high mechanic strength.

Fig. 3. Differential pore-volume distribution curve for the same support as shown in Fig. 2, showing the pronounced uniformity of pore size of the unimodal support.

support having an ideally uniform pore system of suitable pore diameter. The porous support synthesized for this purpose was an unimodal one preformed in spheres of 3.5 mm diameter, thermally stable up to 1500°C, and having an isotropic three-dimensionally uniform and continuous pore structure, as shown in Figs. 2 & 3.

3.1.2 CONTROLS OF CATALYST PREPARATION PARAMETERS : IMPREGNATION AND ACTIVATION

The use of the above-mentioned porous support precludes the usual controls of the adsorption or ion-exchange effects, as well as the chromatographic effects, in the impregnation operation. However, the STRUCTURAL REQUIREMENTS claims to the necessity of the use of certain structural promoter (specifically, this proved to be effective in the final form as alkaline earth carbonates) in an exactly optimized atomic proportion of the promoter elements (i.e., alkaline earth elements) to silver, so that the impregnation must be conform to the activation process in accordance with the activation or promoting mechanism. First, the activation processes were studied with numerous impregnant compositions, as well as impregnation conditions. It has revealed, according to UNIFORMITY PRINCIPLES, that any kind of crystalline precipitations is prohibitive to a proper activation with optimized impregnant compositions. And finally, a proper procedure for impregnation and activation was established, by using such an impregnation condition that the impregnant compounds, after filling

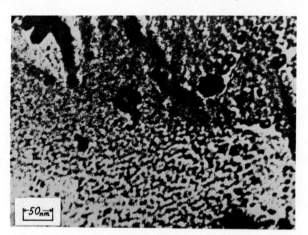

Fig. 4. Electron microgram of the in-pore surface of a HS SILVER CATALYST suitable for propylene oxide formation, showing the uniformly dispersed silver particles of ca. 5 nm diameter as blackened shadow dots isolated by the surrounding materials.

the pore volume with the impregnant liquid, may homogeneously be precipitated in an amorphous state of colloidal homogeneity, so that the activation operation may consequently follow up. The HS SILVER CATALYSTS thus obtained, containing 1-3 wt% Ag, have proved actually to fit the STRUCTURAL REQUIREMENTS, as can be seen from Fig. 4, showing excellent performance figures which are illustrated in Fgs. 5 & 6 along with the results of midget pilot tests.

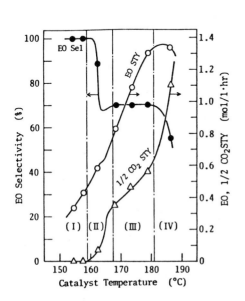
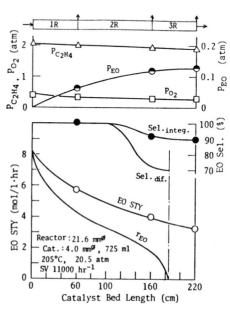

Fig. 5. Temperature jump in selectivity in the stationary-reacting state operations on 100% EO-selectivity basis with an isothermal one-dimensional reactor : P_{total} 16.2, $P_{C_2H_4}$ 1.70, P_{O_2} 0.3 24 (atm); SV=5500 h^{-1}.

Fig. 6. Axial profiles for reactions in a single tube reactor equivalent to the industrial unit reactor operated for 100% EO selectivity, showing EO concentration jump in selectivity at 110 cm catalyst bed length.

Some results of midget pilot tests for pure syntheses of ethylene and propylene oxides are shown in the following table :

Results of Midget Pilot Tests

Olefin	P_Σ (atm)	P_{olefin} (atm)	P_{O_2} (atm)	P_{H_2O} (atm)	T_b (°C)	L (m)	SV (hr^{-1})	Conv. (olefin) (%)	Epoxide STY (mol/l·hr)	Sel (%)
C_2H_4	20.5	2.06	0.41	0	200	3.5	28600	6.2	8.0	100
	20.5	0.90	1.0	0	205	5.0	18000	19.0	5.8	80
	6.6	0.19	0.56	0	240	3.5	9000	63	5.5	70
C_3H_6	20.5	0.70	0.35	0.25	200	3.7	23600	11.8	2.1	71

Reactor I.D. 21.6 mm, Catalyst Dia. 4.0 mm

3.2 HS BLACK SILVER OR COPPER CATALYSTS FOR COMPLETE OXIDATION

This category of High-Selectivity (HS) supported metal catalysts is characterized firstly by its performance standards, i.e., 100% purity of catalytic function for complete oxidation of hydrocarbons and their oxygenated derivatives to the extent of the quantity of oxygen existing in the reaction mixture, with high stability and activity. It may also be characterized by its black colour in general, which is indicative of certain structural resemblance to the old black metal catalysts such as Pt or Pd black, but it differs from these greatly in its structural uniformity due to supporting, hence in its selectivity and stability, especially at higher temperature.

3.2.1 ESSENTIAL STRUCTURAL REQUIREMENTS

ACTIVE SITES : In-pore surface metal layer of less than 10 Å thickness in the microporous region (pore diam. 2 to 5 nm) of the porous support.

SITUATION OF SUPPORTING :

Physically, thermal stability (at above 600 to 800°C) of the microsurface geometry of the microporous support materials is required to exclude in-pore surface migration and sintering.

Chemically, certain chalcogenic affinity to the metal or metal oxide of the microsurface of the support materials is a requisite for a good extension and stability of the very thin metal layer, but it must not be too large.

SELECTION OF THE POROUS SUPPORT :

An unimodal, uniformly microporous support which conforms well to the above REQUIREMENTS can be obtained by calcining (at up to 600 - 800°C) pure and uniform silica gel granules of suitable size. Also, bimodal preformed supports having similar characteristics on silica or silica-alumina basis can be synthesized, which are suitable for large scale applications with high space velocity operations (SV > 300,000 h^{-1}).

3.2.2. CONTROLS OF CATALYST PREPARATION PARAMETERS : IMPREGNATION AND ACTIVATION

The selection of impregnant compounds is important in order to obtain an efficient result with these SELECTED SUPPORTS, which also determines the procedures of both impregnation and activation. The usual proceeding with the operations referring to nitrate ($AgNO_3$ or $Cu(NO_3)_2$) solutions has yielded poor results in respect to supporting efficiency or %dispersion---10 to 15 wt% metal loading

may be regarded as usual. Complexing the nitrate solutions with
NH_3 seems to cause no particular improvements in both impregnation
and activation processes. In general, it can be said that certain
melting of the crystalline impregnant deposits, prior to activation,
e.g., thermal decomposition of impregnant compounds, may profitably
be applied to fit the REQUIREMENTS, provided macroporous uniformity
control over the distribution of the impregnant deposites be
ensured in the impregnation stage. As a specific example, with
cyanide pyridine complexes as impregnant compounds and with
SELECTED SUPPORTS, HS black silver or copper catalysts may be obtained
which show with metal loading of less than 5 wt% very stable metal
surface area of higher than 100-200 $m^2 g^{-1}$ (Ag or Cu).

REFERENCES

1. A. Skrabal, Homogenkinetic, Th. Steinkopff, Dresden, 1941.
2. O.A. Hougen, K.M. Watson, Chemical Process Principles, Part III, J. Wiley, New York, 1947.
3. R.C. Baetzold, C.A. Somorjai, J. Catalysis, $\underline{45}$, 94, 1976.
4. J.A. Christiansen, Adv. Catal., $\underline{5}$, 311, 1953.
5. G.N. Lewis, Z. Physik. Chem., $\underline{52}$, 310, 1905.
6. T. Inui, T. Ueda, M. Suehiro, H. Shingu, J. Chem. Soc. Japan, 1976, (11), 1665.
7. T. Inui, H. Shingu, SHOKUBAI, $\underline{14}$, (4), 195P, 1972.
8. G.C. Bond, Catalysis by Metals, Academic Press, 1962, p. 73.
9. G. Paravano, Catal. Rev. $\underline{3}$, 207, 1969.
10. D.W. van Krevelen, Chemical Reaction Engineering (Ed. K. Rietema), Pergamon Press., 1957, p.8.
11. H. Shingu, T. Okazaki, T. Inui, SHOKUBAI, $\underline{6}$, 32, 1964.
12. R.R. White, S.W. Churchill, A.I. Ch. E. J., $\underline{5}$, 354, 1959.
13. T. Inui, Docter thesis, Kyoto University, Faculty of Engineering, 1971.
14. H. Shingu, T. Inui, 26th Intern. Congress IUPAC Tokyo 1977, ABSTRACTS Session I, p.60.
15. H. Shingu, T. Inui, T. Okazaki, 2nd Petrochemical Symposium, Japan Petrol. Inst., Tokyo, 1971, Nov. 18, Preprints, p. 120.

DISCUSSION

A.V. KRYLOVA : Do you think that selectivity changes may be due to
the existence of discrete adsorbed species at different temperatures?

H. SHINGU : In homogeneous reactions, selectivity changes may be
considered to be due to different reaction intermediates. In heterogeneous catalysis, reaction mechanisms are much more complicated :
the adsorbed species or the adsorbates are not always the only reaction intermediates which determine the selectivity of surface reactions. This remains true even if, at the stationary state, the

homogeneous kinetic approximation is apparently applicable. The selectivity is essentially a kinetic term, and is generally controlled by simultaneous and/or consecutive events. (ref. 14). In the latter case, the change in selectivity may not be ascribed to the change of adsorbate structure. The adsorbate may undergo different surface reactions in competition with product desorption (see in Fig. 5 the second temperature jump in selectivity, at temperatures higher than 180°C. In the former case, most commonly observed are the simultaneous or concurrent surface reactions, due to the non-uniformity of the catalyst surface or active sites. This is clearly seen in Fig. 5 where at the first temperature jump the selectivity for ethylene oxide steadily increases with temperature. The selectivity for CO_2 increases independently at 160-170°C after the development of a transient chain period. This is due to the release of the inhibition predominant at lower temperatures. The possibility that selectivity changes are due to the existence of different adsorbed species, as you mentioned, cannot be excluded for a complex heterogeneous reaction. In the present reaction of olefin oxidation with silver catalysts, this is not the case.

S.S.Y. KRICSFALUSSY : A selectivity of 71% for propylene oxide is a very high value. Could you tell us something about the life time of the catalyst ?

H. SHINGU : This selectivity may also be reached in laboratory tests under sufficiently inhibited reaction conditions but at substantially lower activities.
The catalyst life time is unlimitedly long. At least it is longer than 600 hours of operation. With a semi-continuous laboratory test apparatus completely reproducible results are obtained, provided proper reaction conditions are maintained throughout the tests.
It is to be noted that with HS catalysts within several minutes the stationary-reacting state is reached. These catalysts are very stable and their performance is reproducible.

S.P.S. ANDREW : Your principle of uniformity reminds me of a more ancient version of the principle - namely that inactive catalysts are selective. Are your catalysts pellets operating in a state of diffusional limitation for the ethylene oxidation ?

H. SHINGU : Your remark is very partly correct in this way that inactive catalysts which are sufficiently inhibited or poisoned may retain uniform active sites. However, our HS catalysts are highly active and stable, so that there can be no diffusional limitations with respect to the operating conditions (with high flow rate), as well as to the uniform pore structure of the support.

J.W. GEUS : You mentioned selectivities in the oxidation of ethylene to ethylene oxide of 100%. Sachtler has developed a theory that it is impossible to get higher selectivities than 86% based on an extensive body of experimental evidence. Did I understand you well that you obtained selectivities of 100% ?

H. SHINGU : We obtained selectivities of 100% in laboratory experiments, as well as in pilot tests. They are expressed in stoichiometric molar terms and equal to (100-E)%. E is a definite quantity of inhibitory consumption (oxidation) of reactants per pass. This quantity is inherent to the catalytic reactor and reaction conditions and is estimated usually to be less than 0.1-0.01%. I am aware of those theories which limit the maximal selectivities to 85-86%. They have been forwarded by several authors including Sachtler, and are based on hypothetical mechanisms. It is easy to refute such hypothetical theories by claiming the lack of convincing surface kinetic arguments, but I consider that the experimental evidence for 100% selectivity is the most convincing one.

V.D. YAGODOVSKI : Could you explain why a very narrow metal particle size distribution is necessary for the high activity and selectivity of ethylene oxidation on silver ? How can these data been explained theoretically ?

H. SHINGU : The narrow range of metal particle sizes is derived from an empirical basis as a necessary consequence of the uniformity principle corresponding to the structural requirements for 100% purity of catalytic function. And this requirement also corresponds to the high stability, or the high degree of reproducibility of the catalyst surface, as well as to the highest specific activity, on metal-loading basis, which are attainable at the real working state or the stationary reacting state of the catalyst system. The data for the optimal metal particle size may be explained as a result of the definite (optimal) surface potential, or statistical

distribution of surface disorders, determined by the curvature of the macroscopic surface of the spheroidal, uniformly heat-treated, metal particles. These surface characteristics are particle size dependent as indicated by oxygen adsorption measurements on supported silver catalysts of various particle sizes.
(T. Inui, H. Shingu, Shokubai, 14(1), 49 (1972).

THE IMPREGNATION AND DRYING STEP IN CATALYST MANUFACTURING

G.H. van den Berg and H.Th. Rijnten
Akzo Chemie b.v., Ketjen Catalysts, Research Centre Amsterdam, P.O. Box 15,
Amsterdam, The Netherlands.

SUMMARY

Three metal compound-on-carrier systems, differing in their respective interaction strength and metal loading are discussed. Different impregnation and drying processes have to be applied, to obtain the required properties of the catalysts.

I. INTRODUCTION

Many catalysts in common use today consist of small metal crystallites, dispersed on a high surface area carrier. These metal-on-support catalysts can be produced by impregnating the carrier with a solution of the metal compounds. During impregnation and subsequent drying, the metal species are deposited onto the support. The conditions applied during these steps can influence the final metal distribution through the carrier particle.

Two interdependent factors are decisive for the choice of impregnation technique and drying conditions:
  1. the adsorption strength of the metal compound onto the carrier.
  2. the metal content of the final catalyst.

Two techniques can be applied for <u>impregnation</u> via the liquid phase:

A. At low adsorption strength the impregnation technique chosen will be the "dry" impregnation as indicated e.g. by Berrebi and Bernusset [1]. In this case, the required amount of metal compounds is dissolved in a sufficient volume of water to fill about the total pore volume of the carrier batch (pore volume saturation).

B. At strong adsorption the carrier can be impregnated using an excess amount of solvent and circulating the solution through a bed of carrier particles. The metal compound diffuses from the solution into the carrier particle and is adsorbed then (soaking procedure).

Because the number of strong adsorption sites per unit weight of carrier is limited this "soaked" impregnation technique can be applied for rather low metal contents. For higher metal contents (approx. 5%) generally the dry impregnation technique is applied. If very high metal contents are required, the limited solubility of metal compounds may require two or even three consecutive impregnation steps.

Drying of the impregnated system is generally carried out at temperatures between 80 and 300°C. This step can affect the impregnation result quite drastically in the case of weak adsorption of the metal compound on the carrier. At the area in the carrier particle where evaporation occurs, the metal compound concentration increases. Transport of the solution from the catalyst particle interior towards the evaporation area and continuing evaporation lead to an inhomogeneous deposition of the metal compound through the carrier particle. Even distribution of the metal compound is obtained when diffusion of the metal compound over the carrier pore surface can occur, even during the final stage of the drying process (empty pore system with layers of liquid still present). When the metal compound is adsorbed strongly, a change of the metal distribution on drying is unlikely. In this case, the distrbution is determined by the impregnation procedure.

In this paper we like to discuss three impregnation systems, differing in adsorption strength and metal loading (table 1).
These examples may illustrate the phenomena indicated above.

TABLE 1. Characteristics of the systems under consideration.

	3.1	3.2	3.3
Carrier	Activated carbon	$\gamma\text{-}Al_2O_3$	$\gamma\text{-}Al_2O_3$
Metal compound	$Zn\ (CH_3COO)_2 \cdot 2H_2O$	$CuCl_2$	H_2PtCl_6
Metal content (wt%)	11	5	0.3
Adsorption	weak + strong	weak	strong
Critical step	Impregnation + drying	drying	impregnation
Impregnation procedure	soaked	dry (PV-sat)	soaked

2. EXPERIMENTAL

The equipment used is sketched in figure 1. The dry impregnations have been carried out in a rotating pan, provided with baffles (1A). For the soaked impregnations a glass column has been used (1B). In this case, the solution was circulated through the bed of carrier particles by means of an air-lift. Some physical properties of the carriers we used are given in table 2.

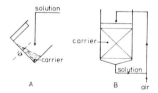

fig.1 TECHNIQUE OF DRY (A) AND SOAKED (B) IMPREGNATION

TABLE 2. Some physical properties of the carriers.

	3.3 Activated carbon	3.2 γ-Al_2O_3	3.3 γ-Al_2O_3
Source	Norit	Akzo Chemie	Akzo Chemie
Type	RKD-3	Ketjen Grade E	Ketjen CK 300
Size	3 mm extrudates	Fluid powder	1.5 mm extrudates
Surface area (m^2/g)	1343	122	180
Pore volume (ml/g) (mercury pyknometer).	1.12	0.39	0.59

Typical impregnation and drying conditions of this work are given in table 3.

TABLE 3. Impregnation and drying conditions

	3.1	3.2	3.3
Carrier (g)	400	5000	50
Water (g)	757	1800	100
Metal compound	$Zn(CH_3COO)_2 \cdot 2H_2O$	$CuCl_2$	H_2PtCl_6
Amount (g)	308	565	0.65
Temperature (°C)	25	25	25
Impregnation time (h)	3	1	3
Drying temp. (°C)	130	120-300	120

3.1. ZINC ACETATE ON ACTIVATED CARBON

A zinc acetate on activated carbon catalyst is applied in the manufacturing of vinylacetate. This catalyst must contain a large amount of zinc acetate in order to ensure a long catalyst life under practical conditions (0.55 gram zinc acetate .$2H_2O$ per gram of carrier). The activated carbon in extruded form has a pore volume of 1.12 ml/g. A saturated aqueous zinc acetate solution of 20°C (density 1.165 g/ml) contains 0.31 gram metal compound per ml. From these data, it can be calculated, that a maximum loading of 0.35 gram of zinc acetate $2H_2O$ per gram of carrier can be obtained in a one step dry impregnation. To reach the desired metal compound loading by this technique, a second dry impregnation must be carried out.

The impregnation might be carried out in one-step, however, using the soaking technique, by exploiting both the sorptive properties of the active carbon and the pore volume capacity of the carrier. The sorption of zinc acetate on active carbon has been demonstrated by measuring the density of the mother liquor during a soaking experiment. The initial density of 1.153 reached an equilibrium value of 1.091 within 3 h. The total uptake of zincacetate has been measured as a function of the zinc acetate concentration in the equilibrium solution (fig. 2). From the concentration in the solution the uptake by the pore volume has been estimated. The difference between both curves represents the adsorbed amount. This figure indicates that an appreciable part of the total zinc acetate loading is originated from adsorption.

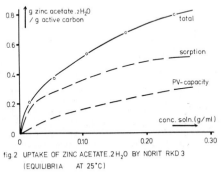

fig. 2 UPTAKE OF ZINC ACETATE.$2H_2O$ BY NORIT RKD 3 (EQUILIBRIA AT 25°C)

To get some insight in the location of the metal compound after careful drying of the impregnated carrier, the carrier and the dried product have been investigated (table 4). All data are related to 1 g of carrier.

TABLE 4. Chemical and physical data of carrier and dried product.

	Norit RKD-3	Dried product
Zinc acetate .$2H_2O$ content (wt%)	-	59.2
Surface area (m^2/g)	1340	410
Total pore volume (ml/g)	1.12	0.79
Pore volume micropores (<3.75 nm) ml/g	0.57	0.20

The mercury penetration curves of both carrier and dried product are given in fig. 3.

The curve given for the carrier indicates two types of pores to be present: macropores, having a radius between 3×10^2 and 2×10^3 nm, and micropores (<3.75 nm), which cannot be penetrated by mercury at the maximum pressure applied. The micropores almost exclusively contribute to the surface area of the carrier.

From table 4 one can see the pore volume of the final product to be 0.33 ml/g lower than the pore volume of the carrier. This volume corresponds with the volume of zinc acetate .$2H_2O$ in the catalyst. The volume of the macropores is hardly affected. Supported by the decrease of the surface area observed, it may be concluded that the metal compound is mainly situated in the micropores of the carrier.

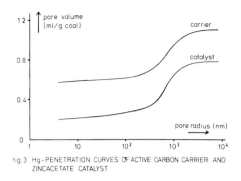

fig.3 Hg-PENETRATION CURVES OF ACTIVE CARBON CARRIER AND ZINCACETATE CATALYST

3.2. COPPER (II) CHLORIDE ON γ-ALUMINA

The required metal loading of this catalyst system is approximately 12 gram $CuCl_2$ per 100 gram of carrier. The pore volume of the alumina carrier amounts to 0.40 ml/g. The solubility of $CuCl_2$ in water is high enough to allow a one-step dry impregnation.

By soaking experiments it has been found, that the adsorption of $CuCl_2$ on γ-alumina is very weak and therefore we might expect a mobility of the metal compound on drying. This has been confirmed experimentally.

Fig. 4a. shows the homogeneous copper distribution over the crosssection of a slowly dried particle as measured by electronmicroprobe (catalyst I). On fast drying an uneven copper distribution is obtained as shown in fig. 4b (catalyst II).

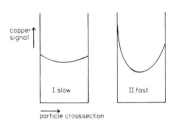

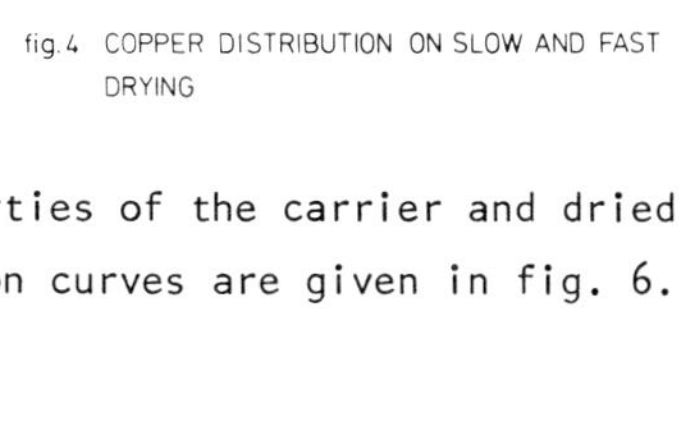

fig.4 COPPER DISTRIBUTION ON SLOW AND FAST DRYING

The physical properties of the carrier and dried catalysts are given in table 5. The mercury penetration curves are given in fig. 6. All data are based on 1 g of carrier.

TABLE 5. Characteristics of carrier and dried catalysts

		Carrier	Catalyst I	Catalyst II
Drying		-	slow	fast
Surface area	(m^2/g)	122	99	105
Total pore volume	(ml/g)	0.40	0.37	0.38
$CuCl_2$ content	(wt %)	-	11.3	11.3

These data show, that the relative decrease of the surface area on impregnating γ-alumina with copper (II) chloride is more severe than the decrease of the pore volume. This indicates, that the smaller pores are preferentially filled with the metal compound. The lowest surface area is found for the slowly dried product.

The fast dried product shows a surface area between those of the carrier and the slowly dried product. The mercury penetration curve of this product is situated between the curves of the carrier and the slowly dried product too.

The data illustrate that upon fast drying, copper (II) chloride migrates towards the larger pores at the outer layer of the particles, giving rise to an uneven copper distribution over the particles.

By exposure to water vapour, 5% of water has been adsorbed on catalyst II. After slowly drying, the copper distribution has become homogeneous and the surface area was decreased to 98 m^2/g.

We may conclude that a weakly adsorbed metal compound is mobile in a carrier particle if a suitable transport medium is present. This phenomenon makes a redistribution of the metal compound over the carrier particle possible.

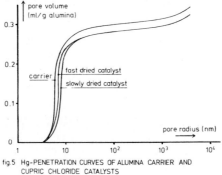

fig.5 Hg-PENETRATION CURVES OF ALUMINA CARRIER AND CUPRIC CHLORIDE CATALYSTS

3.3 PLATINUM ON γ-ALUMINA

An example of strong interaction between metal compound and carrier is the chloroplatinic acid - γ-alumina system, which has been investigated by many institutes. In this paper the behaviour of the Pt compound in solution and the carrier are discussed separately. A model of the competitive adsorption of chloroplatinic acid and other acids is given.

3.3.1. Chloroplatinic acid (CPA)

The behaviour of chloroplatinic acid in solution has been studied by Davidson and Jameson [2]. H_2PtCl_6 is a strong acid which upon dissociation yields two protons per molecule.

From spectroscopic measurements they concluded that the remaining complex may undergo ligand exchange with water molecules:

$$PtCl_6^{2-} + H_2O \rightleftharpoons (PtCl_5H_2O)^- + Cl^-$$

This mono-aquo-chloro-complex can undergo hydrolysis:

$$(PtCl_5H_2O)^- \rightleftarrows (PtCl_5OH)^{2-} + H^+$$

The dissociation constant for this complex has been determined to be pKa = 3.8. This value indicates that $(PtCl_5H_2O)^-$ behaves as a weak acid.

The phenomenon of aquation and hydrolysis of diluted CPA solution can be demonstrated to be time dependent by potentiometric titration with alkali. The titration curves of fresh and aged chloroplatinic acid solutions we found are given in fig. 6

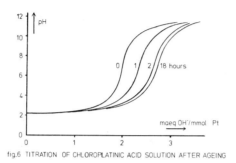

fig.6 TITRATION OF CHLOROPLATINIC ACID SOLUTION AFTER AGEING

The curves indicate, that the aquation reaction does not proceed instantaneously. After 2 h the equilibrium situation is almost reached.

Since impregnation proceeds largely at pH values of approx. 4 and because impregnation of a batch takes several hours in practice (2-4 hrs), it may be assumed, that a substantial part of the platinum is present as $(PtCl_5OH)^{2-}$.

3.3.2 Carrier

The γ-alumina carrier used in these investigations undergoes some interaction with the acid solution. This could be demonstrated during a typical soaking experiment. The pH of the soaking solution increased from 1.2 to a constant level of 4 within 2 h.

At the low initial pH value, an acid attack of the alumina occurs:

$$Al - O - H + H^+ \longrightarrow Al^+ + H_2O$$

The cationic alumina is believed to yield a site for adsorption (ionic interaction) of $PtCl_5OH^{2-}$. This is supported by our observation, that on impregnation of alumina using a K_2PtCl_6 solution without any free acid a homogeneous platinum distribution over the particle and a low Pt impregnation yield are obtained. An equivalent concentration of H_2PtCl_6 does yield a surface layer of platinum under identical conditions. The recently published acid attack model of Santacesaria c.s. is slightly different, but does give the same results [3].

3.3.3 Competitive adsorption of CPA and HCl

When γ-alumina is impregnated with a CPA solution, the protons from chloroplatinic acid attack the alumina surface on penetration of the extrudates by the liquid. The sites created by this attack are situated in the outer layer of the extrudates and serve as strong adsorption sites for $PtCl_5OH^{2-}$.

A homogeneous Pt distribution can be obtained when additional acid is present in the impregnation liquid [4]. This acid creates additional adsorption sites onto the alumina both for $PtCl_5OH^{2-}$ and the acid anions throughout the particle. Due to the adsorption competition between both anions a homogeneous Pt distribution can be obtained.

Impregnation experiments of H_2PtCl_6 with varying amounts of hydrochloric acid have been carried out. The penetration depth and impregnation yield are given in fig. 7.

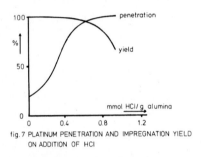

fig. 7 PLATINUM PENETRATION AND IMPREGNATION YIELD ON ADDITION OF HCl

The Pt penetration increases on increasing the HCl concentration in the impregnation solution. A second effect of the competitive adsorption is the decrease of the platinum impregnation yield on increase of the HCl concentration.

The competitive adsorption of $PtCl_5OH^{2-}$ and Cl^- can also be demonstrated by recording the separate adsorption isotherms of HCl and CPA on γ-alumina (fig. 8). The adsorption isotherms have been recorded under the conditions given in table 3.

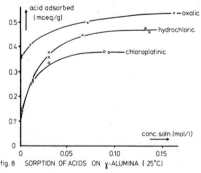

fig. 8 SORPTION OF ACIDS ON γ-ALUMINA (25°C)

The adsorption curves for HCl and CPA indicate that the adsorption strengths of Cl^- and $PtCl_5OH^{2-}$ are similar. Therefore HCl is a good impregnation aid to obtain a homogeneous Pt distribution.

The incomplete platinum penetration obtained without impregnation aid is caused not only by the strong interaction but also by the low amount of platinum applied in our experiments. On addition of CPA to an impregnation solution in sufficient quantity to reach the equilibrium loading (0.38 maeq/g), a homogeneous Pt-distribution will be obtained without impregnation aid.

3.3.4. Competitive adsorption of CPA and oxalic acid

Adsorption phenomena are also exploited in the technique for obtaining a subsurface metal layer, as described in the patent literature [5,6] and recently by Nuttall [7] : Addition of a small quantity of an acid to the impregnation solution which shows a very strong adsorption on alumina as compared to CPA, yields an outer platinum-deficient shell in the carrier particle. The thickness of this layer can be controlled by the added amount of such an acid.

All, at least dibasic, organic acids are claimed to yield this effect (like oxalic, tartaric and citric acid). Aromatic acids containing a hydroxylgroup adjacent to the carboxylic group, like salicylic acid are claimed as well.

The different behaviour of e.g. hydrochloric and oxalic acid on impregnating CPA, is drawn schematically in fig. 9.

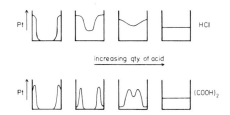

fig.9 PLATINUM DISTRIBUTION IN ALUMINA EXTRUDATES AT FOUR LEVELS OF HYDROCHLORIC AND OXALIC ACID CONCENTRATION

On addition of small quantities of oxalic acid to the CPA solution, the outer layer of the extrudates is attacked by protons and oxalate anions are strongly adsorbed on these sites. The platinum compound is deposited in a subsurface layer, just under the layer of more strongly adsorbed oxalate ions.

The relative thickness of the shell has been determined at varying additions of oxalic acid and citric acid (fig. 10). These results have been compared with the data reported by Nuttall for citric acid. The latter results do not agree with our observations.

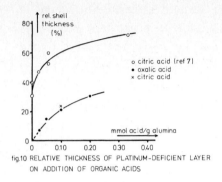
fig.10 RELATIVE THICKNESS OF PLATINUM-DEFICIENT LAYER ON ADDITION OF ORGANIC ACIDS

The adsorption isotherm of oxalic acid on alumina has been determined (fig.8). In the concentration range applied in practice (0.01 mol/l), the curve indeed shows a very strong adsorption as compared to the adsorption of CPA or HCl.

The phenomena observed have to be attributed to chromatographic effects based on the different adsorption strengths and the penetration of the impregnation solution from outside towards the particle centre. This could be demonstrated by separation of the compounds involved, using long extrudates (10 cm), placed in a vertical position. After application of a mixture of CPA and oxalic acid solution, chloroplatinic acid moves under the influence of a 10% HCl solution as mobile phase; oxalic acid hardly migrates.

A second effect, underlining the chromatographic character of this behaviour is its time dependency; we found a broadening of the platinum-rich zone with time. The more strongly adsorbed oxalic acid does not migrate noticeably during a period of 2-20 min.

ACKNOWLEDGEMENT

The authors wish to thank Mr. A.L.L. Südkamp for his valuable contribution in the experiments and discussions.

1. G. Berrebi and Ph. Bernusset, Preparation of catalysts, ed. Delmon, Jacobs and Poncelet, Elsevier, Amsterdam (1976) pag. 24.
2. C.M. Davidson and R.F. Jameson. Trans Faraday Soc. 61 (1965) 2462.
3. E. Santacesaria, S. Carra and I. Adam: Ind. Eng. Chem. Prod. Res. Dev. 16 no. 1 (1977) 41.
4. R.W. Maatman, Ind.Eng. Chem. 51 (1959) 913.
5. U.S. Patent 3.388.077 (1968).
6. Neth. Pat. Appl. 291717 (1963).
7. T.A. Nuttall CSIR report CENG 182. Pretoria 1977.

DISCUSSION

V.B. FENELONOV : Is it possible that in your experiments there exist big temperature gradients, which could be an explanation for the experimental data from the fast drying ? In that case liquid can move in a cold zone of the pellet, particularly in the external surface domains.

G.H. van den BERG : I agree with you that heat transport into the particles is a very important factor in intra-particle metal compound transport. However, measurements of temperature gradients in these particles is impossible. Calculations are difficult to make because reliable data are lacking.

A. OZAKI : What is your definition of fast or slow drying in view of the apparent contradiction with the paper of Kotter and Riekert ?

G.H. van den BERG : I cannot give a quantitative criterion for the difference between fast and slow drying. In our experiments, the fast dried product has been submitted to a forced flow of hot air (large surplus amount).

J.P. BRUNELLE : What kind of mechanism occurs in the fixation of $CuCl_2$ on γ-alumina ? Do you think it is an "hydrolytic adsorption", in other words a mechanism which consists of an anion adsorption and a precipitation of the cation ?

G.H. van den BERG : Apart from its weak nature, we do not know the character of the interaction between copper(II) chloride and γ-alumina.

R. GALIASSO : I would like to point out that we have observed that when γ-Al_2O_3 impregnated with ammonium paramolybdate solution, or nickel nitrate solution is slowly dried, a homogeneous Mo (or Ni) distribution is obtained in the alumina pellets. Fast drying on the other hand, leads to inhomogeneous distribution which is in agreement with your results in the case of the $CuCl_2/\gamma Al_2O_3$ system. It is surprising that you do not have a higher degree of surface reduction in the fastly dried samples compared to slowly dried ones. Would this mean that the macropore-micropore migration during drying is the same under both sets of conditions ?

G.H. van den BERG : 1) Your comment indicates that particle size is not the source of the differences between Maatmans' and our results, obtained on slow and fast drying respectively.
2) Upon fast drying the copper compound in the porous system tends to migrate towards the larger pores. The minor effect on the surface area and pore volume is caused by the rather high metal compound loading.

C.J. WRIGHT and M.V. TWIGG : We are perturbed that the result presented in Figure 1 for the copper distribution produced by fast drying conditions might be an artefact of the experimental technique. The use of the electron probe on a 40μ particle will be influenced by the method of particle sectioning (or sputtering) and by the corrections used to compensate for the escape depth of the X-rays. A further complication will surely arise from the irregular shape of your particles.

G.H. van den BERG : The special sample preparation technique we applied diminishes the risk of migration of the copper species to a large extent.
The penetration depth of the electron beam is 2.5μ at the 40 kV applied; therefore we do not need large corrections.
It is clear that the results obtained cannot be interpreted quantitatively, due to the effects you mentioned and other ones. Nevertheless, the effect of different drying techniques on the copper compound distribution mentioned in the paper has been found to be qualitatively reproducible.

P.G. MENON : There is a good agreement between the results on Pt-γAl_2O_3 in this paper and those in the paper by Sivasanker et al., particularly in the competitive adsorption of HCl and H_2PtCl_6 from a solution mixture (e.g., compare Fig. 7) and 8 of this paper with Fig. 4 of the paper of Sivasanker).

E.A. IRVINE : 1) Have you any comparative data for different types (forms) of aluminas, i.e. is your gamma alumina well defined as such and, if not, is there the possibility that different forms of alumina will give a different picture in terms of the copper distribution ?

G.H. van den BERG : We cannot provide comparative data for other types of alumina than the grade E applied (Table 2).
Due to the different interaction characteristics other aluminas will show a different picture under identical preparation conditions.

L. RIEKERT : a) The distributions of the active component which you found in the case of weak interaction with the carrier for slow as compared to fast drying are at variance with the predictions just made by Fenelonov and with what one would expect from simple qualitative considerations. Is there an explanation ?
b) How have these distributions been obtained experimentally ?

G.H. van den BERG : a) In my opinion the effect found does fit in Fenelonov's model, if the proper values of the various parameters are put into the formulae. Because of lack of sufficiently reliable data, we did not succeed in demonstrating this so far.
b) These distributions have been obtained via an electron microprobe technique.

STUDY OF THE INTERACTION OF Fe_2O_3-MoO_3 WITH SEVERAL SUPPORTS

L. CAIRATI [1], M. CARBUCICCHIO [2], O. RUGGERI [3] and F. TRIFIRO [4]

[1] Euteco - Milano
[2] Istituto di Fisica Università Parma
[3] Istituto di Metallurgia Bologna
[4] Istituto di Tecnologie Chimiche Speciali Università Bologna

ABSTRACT

Fe_2O_3-MoO_3 have been supported on SiO_2 and Al_2O_3 of different surface areas in order to prepare catalysts with high mechanical strength, for the selective oxidation of methanol to formaldehyde. Mössbauer, Raman, X-ray dispersion, electron microscopy and catalytic measurements have been carried out in order to characterize the catalysts. A decreasing activity and selectivity in CH_2O was found both for Fe_2O_3-MoO_3 and MoO_3 catalysts with the increase of surface area of silica. By increasing the surface area of the supports a decrease of the amount of $Fe_2(MoO_4)_3$ was observed with formation of a new compound of Fe(III) different in the case of SiO_2 and Al_2O_3. Formation of surface compounds of the type FeMoSi and FeMoAl has been hypothesized.

1 INTRODUCTION

The oxidation of methanol to formaldehyde is normally carried out in a fixed bed reactor with a catalyst based on Fe_2O_3-MoO_3 (1-2).
In order to obtain the oxidation of methanol in a fluidized bed reactor, supported Fe_2O_3-MoO_3 on SiO_2 and Al_2O_3 has been prepared so as to increase the mechanical strength of the catalyst. In fact unsupported Fe_2O_3-MoO_3 presents very low mechanical strength (3) and therefore it cannot be used in fluidized bed reactors. With a view to optimizing the preparation and to understanding the difference in activity and selectivity observed with supports of different nature and surface area, Raman, Mössbauer, X-ray and electron microscopy measurements have been carried out.

2 EXPERIMENTAL SECTION
2.1. Preparation of catalysts

Silica (Grace) and Al_2O_3 (Saint Gobain 125 m^2/g) with different surface areas have been prepared by calcination at high temperature (1100°C) for different durations.

The grain size of SiO_2 and Al_2O_3 used was 200-270 mesh.

The active components Fe_2O_3 and MoO_3 were supported by impregnation from soluble salts of Fe(III) and Mo(VI) in the presence of citric acid (4). A solution of $(NH_4)_6 Mo_7 O_{24}$ $4H_2O$, citric acid, $Fe(NO_3)_3$ $9H_2O$ was sprayed on the support-grains kept in vacuum under rotation in a rotavapor; after drying at 110°C the catalyst was calcined in a fluidized bed reactor in a flow of air at 360°C for 10 hours; afterwards the catalyst was calcined at 430°C for 16 hours in an oven. The amounts of Fe and Mo as Fe_2O_3 and MoO_3 were in all the catalysts, 2.3% by weight, with a molar ratio Mo/Fe=2. Molybdenum oxide catalysts were prepared by impregnation of the support with the solution of $(NH_4)_6 Mo_7 O_{24}$ $4H_2O$ (as before).

The analyzed samples are labeled in the following way by indicating the kind of support and the surface area between brackets; for example, SiO_2 (100) indicates an SiO_2 supported catalyst with surface area of 100 m^2/g.

2.2. Measurements of catalytic activity

The reactor was made of Pyrex with ∅ 28 mm and 1000 mm long. The reagent was distributed from a porous disc with 10-20 μm perforations. The length of the bed was 450 mm with a flow-rate of 125 N l/h. The concentration of methanol was 5.5% in air. The analysis of products was carried out according to ref. 3.

2.3. Mössbauer, Raman, X-ray and SEM measurements

The Mössbauer absorption spectra were measured by means of a standard spectrometer consisting of a constant acceleration electro-mechanical driving system and a multichannel analyser working in multiscaler time mode. Both the source (25 mCi Co^{57} in Rh matrix) and the absorbers were kept at room temperature.

The powder of catalysts on which Al film was evaporated were observed with a Jeol 15 kV electronic microscope. The chemical nature of the surface was analyzed by X-ray energy dispersion electron microprobe (Montedel). The Raman spectra were recorded by a Cary 83 spectrophotometer.

3 RESULTS AND DISCUSSION

3.1 Measurements of activity

Figs. 1 and 2 show the conversion of methanol, the selectivity to formaldehyde and the yield of CO, CO_2, HCO_2CH_3 vs the silica surface for T=260° and 300°C respectively. For SiO_2 (300) the values of selectivity and activity are only slightly less than those obtained for SiO_2 (100). The table shows the same values obtained at T=300°C for MoO_3-SiO_2 catalysts at two levels of surface area of SiO_2. From Figs. 1 and 2 and the table it is possible to deduce that with increasing surface area of the support a decrease in activity and selectivity occurs. The decrease in activity was not observed with MoO_3-SiO_2 catalysts. The values of selectivity in CH_2O are the same for both groups of catalysts and are independent of temperature.

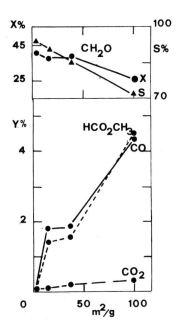
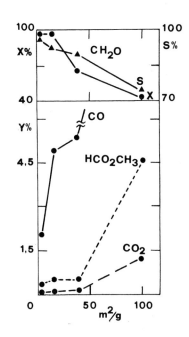

Fig. 1. Activity pattern of Fe_2O_3-MoO_3 on SiO_2 at 260°C.

Fig. 2. Activity pattern of Fe_2O_3-MoO_3 on SiO_2 at 300°C.

Activity pattern of MoO_3 on SiO_2 at 300°C

Catalyst	X%	S%	Yield %		
			CO	CO_2	HCO_2CH_3
7.5% MoO_3 on SiO_2 (2.3)	40.0	90.0	2.3	0.0	1.0
7.5% MoO_3 on SiO_2 (100)	52.0	73.0	7.2	1.4	5.4

3.2. Raman spectra

Raman spectroscopy allows characterization of the supported catalysts as the SiO_2 and the Al_2O_3 are Raman inactive. Therefore it was possible to observe the characteristic bands of molybdenum - oxygen bonds of pure and mixed oxides (5)(6). Fig. 3 records the spectra of Fe_2O_3-MoO_3 on SiO_2 (2,3), Al_2O_3 (0.5) and MoO_3 on SiO_2 (2.3). In the case of samples with higher surface area no more bands were detected.

Fig. 3D is the spectrum of MoO_3 (10%) on SiO_2. Notwithstanding the high amount of MoO_3 present in the catalysts its most intense band was only detectable with the highest gain of the intensity of the instrument. The noise of the instrument covers the other bands.

In Fig. 3C and Fig. 3B it is possible to recognize the bands at 800 and 960 cm^{-1} respectively, as those of the Mo-O-Mo and Mo=O groups of $Fe_2(MoO_4)_3$ (6). In Figs. 3A and 3C the bands at 820 and at 1000 cm^{-1} are respectively those of the Mo-O-Mo and Mo=O groups of MoO_3.

In Fig. 3B the bands of MoO_3 are more difficult to recognize.

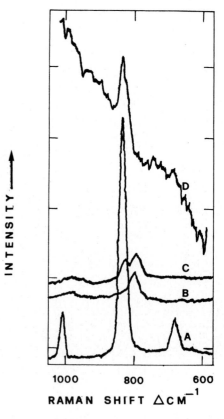

Fig. 3. Raman spectra
  A MoO_3 on SiO_2 (2.3)
  B Fe_2O_3-MoO_3 on Al_2O_3 (0.5)
  C Fe_2O_3-MoO_3 on SiO_2 (2.3)
  D MoO_3 on SiO_2 (100)

3.3 SEM and X-ray analysis measurements

SEM does not show any significant difference in the morphology of the analysed silica-supported Fe_2O_3-MoO_3 catalysts, either before or after their use. Fig. 4 illustrates a characteristic image of these catalysts obtained by means of SEM.

The qualitative analysis carried out by X-ray electronic microprobe on SiO_2-supported Fe_2O_3-MoO_3 catalysts with high surface area allows the supposition that the active component is distributed rather uniformly on the support and does not accumulate on preferential zones. Fig. 5a shows a typical spectrum of this catalyst; it is possible to detect the presence of small quantities of Fe and Mo.

For samples with areas lower than 2.3 m^2/g, where considerable numbers of grains are present, a segregation of hexagonal plates having constant size (8-10 μm side - see Fig. 5b); the said plates are made of a coherent material while the surrounding catalysts have an irregular and incoherent form. The most important datum resulting from these observations has been obtained by X-ray microanalyses; as a result of these we can now assert that the plates are made exclusively of Mo; therefore they can be attributed to MoO_3 (Fig. 5c). Moreover an X-ray map of a plate-containing grain has been produced by preselecting on the microanalyser the energy band of X-ray emission relative to Mo. Fig. 5d shows the characteristic Mo spots, which are particularly concentrated in the plate zones, while **appearing rarefied in the surrounding areas.**

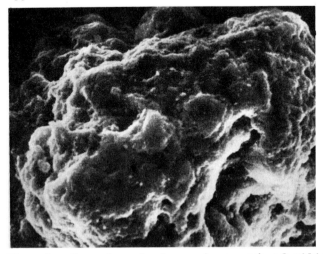

Fig. 4. Scanning electron micrograph of silica supported Fe_2O_3-MoO_3 catalyst X.1300.

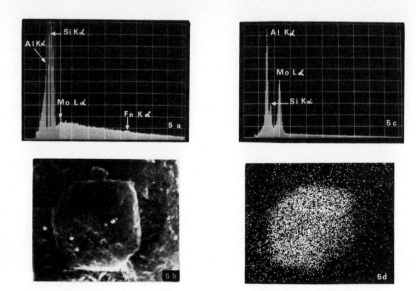

Fig. 5a X-ray chemical analysis at energy dispersion of catalyst SiO_2 without plate.

Fig. 5b Scanning electron micrography of an hexagonal plate on catalyst SiO_2 (2.3)X.300.

Fig. 5c X-ray chemical analysis at energy dispersion of catalyst containing plate.

Fig. 5d X-ray Mo distribution map.

3.4 Mössbauer measurements

The Mössbauer spectra, carried out on samples with surface areas lower than 2.3 m^2/g and on both the SiO_2 and Al_2O_3 supports, display a well-defined single-peak with the same isomer shift and line-width as the one measured for the pure Fe(III)-molybdate (7). Figs. 6a and 7a show the spectra for the samples SiO_2(1) and Al_2O_3(0.5) respectively. These spectra, in agreement with the Raman, are evidence that the supported Fe(III)-molybdate can form for low areas independently of the support.

As the Mössbauer measurements can provide short range information this technique can be utilized for continuously following the modifications of the iron surroundings, independently of the crystalline state of the compound. For 2 m^2/g of sample surface area the Mössbauer spectra display a broadened peak; by increasing the area the broadening increases further. It is worth noticing that for both the silica and alumina supports, the isomer shift remains that typical for Fe^{3+}

285

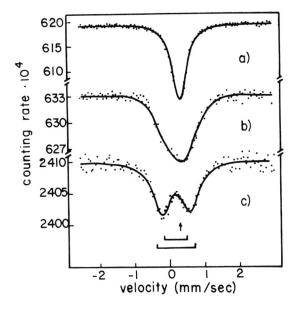

Fig. 6 Room temperature Mössbauer spectra for silica supported catalysts: a) $SiO_2(1)$, b) $SiO_2(100)$ and c) $SiO_2(300)$

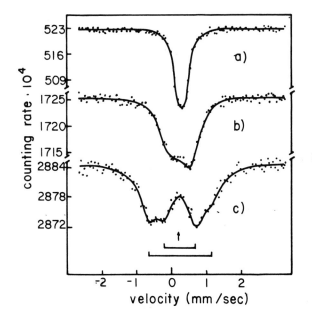

Fig. 7 Room temperature Mössbauer spectra for alumina supported catalysts: a) $Al_2O_3(0,5)$, b) $Al_2O_3(5,4)$ and c) $Al_2O_3(125)$

ions; the line-width increase is linear up to 40 m^2 in the case of SiO$_2$ and up to 10 m^2 for Al$_2$O$_3$ with a greater slope for alumina. Figs. 6b and 7b record the spectra for the SiO$_2$(100) and Al$_2$O$_3$(5.4) samples respectively. As these spectra are peaked corresponding to the line position of the Fe$_2$(MoO$_4$)$_3$, they can be interpreted assuming the super position of at least two spectra; (i) a single-line due to Fe^{3+} ions of Fe(III)-molybdate, (ii) an unresolved spectrum due to Fe^{3+} ions located in distorted sublattices. This second contribution could be due to iron located at the Fe$_2$(MoO$_4$)$_3$ surface in lattice sites which would be highly distorted. However, this assumption implies that the support plays only the role of providing high surface areas and cannot explain the different behaviour for alumina and silica. Therefore, a chemical reaction between the supports and the precipitated complex salt must be assumed.

The observed behaviour of the line-width as a function of the sample surface, suggests that the Fe(III)-molybdate contribution reduces with increasing area while the other contribution to the spectra increases. If this assumption is correct one can reasonably expect to resolve, for very high surface areas, this second contribution, i.e. the one due to iron with distorted surroundings. It must be indicated that, due to the broadening of the lines, the spectra of high area samples need very long counting times in order to reduce sufficiently the statistical error.

Figs. 6c and 7c record the spectra for SiO$_2$(300) and Al$_2$O$_3$(125) respectively. They can be interpreted as a superposition of two quadrupole doublets whose positions are indicated in the figures; the position of the single line of the Fe(III)-molybdate which is no longer present is indicated with an arrow. The spectra of SiO$_2$(300) and Al$_2$O$_3$(125) are clearly different, in agreement with the hypothesized reaction between the precipitate salt and the supports. They could be reasonably interpreted as due to the formation of some of the following compounds :
i) Fe^{3+} substituted alumina or silica,
ii) α-Fe$_2$O$_3$ small particles,
iii) Fe-Mo-Si or Fe-Mo-Al compounds.

In the case of the Al$_2$O$_3$(125) sample, the comparison between the Mössbauer parameters obtained and the data available from the literature suggests (8) that (i) no Fe^{3+} substituted alumina is formed, (ii) growth of α-Fe$_2$O$_3$ small particles cannot be excluded and (iii) Fe-Mo-compound at the support is likely to form. A similar analysis of the SiO$_2$(300) sample indicates (9) that (i) it is possible to exclude the formation of iron silica compounds, while (ii) α-Fe_2O_3 small particles and (iii) Fe-Mo-Si compounds are both likely to form.

It is worth noting that the small α-Fe_2O_3 particles growing should liberate molybdenum oxide. As SEM photography reveals the presence of molybdenum crystallites for low but not for high sample surface area, it is reasonable to assume that the Mössbauer spectra measured for high areas are essentially due to Fe^{3+} ions in Fe-Mo-Si and Fe-Mo-Al mixed compounds.

4 CONCLUSION

As both Raman and Mössbauer measurements revealed the presence of $Fe_2(MoO_4)_3$ and confirmed it for low surface area, we can state that method of preparation used in this paper leads to the formation of the same active species in the absence of support. Therefore all chemical changes observed on high surface area must be attributed to support reactivity. With low surface areas, the only difference with respect to the preparation without support is the segregation of MoO_3 observed by SEM. This indicates that SiO_2, as in the case of low surface area, modifies the catalyst morphology. The absence of bands in Raman spectra and the absence of plates of MoO_3 in high surface area catalysts cannot exclude the formation of MoO_3 and $Fe_2(MoO_4)_3$. The Mössbauer spectra demonstrated the decrease of the amount of $Fe_2(MoO_4)_3$ with increasing surface area of the supports and the formation of a new Fe(III) compound.

The nature of this Fe(III) compound depends on the nature of support. From the catalytic point of view the very low percentage of CO_2 formed may be considered as a sign of absence of Fe_2O_3; in fact it is well known that Fe_2O_3 is particularly active in methanol oxidation, giving CO_2 only as product (10).

The comparison between the activity of the two groups of catalysts investigated, i.e. Fe_2O_3 and MoO_3 on SiO_2, leads to the following considerations: i) for low surface areas those catalysts containing only MoO_3 present lower activity with respect to the mixed oxides, notwithstanding that the amount of MoO_3 is much higher than that present in the mixed oxides. This fact is an indication that iron is a promotor of MoO_3 activity. The formation of $Fe_2(MoO_4)_3$ may explain this promoting effect in agreement with the hypothesis concerning supported catalysts (7); (ii) in the case of high surface area catalysts, those containing only MoO_3 present lower activity with respect to mixed oxides.

Also in this case, a promotor effect of the iron on MoO_3 must be assumed. As Mössbauer measurements revealed the absence of $Fe_2(MoO_4)_3$ the formation of a new surface compound with molybdenum and very probably with the support must be hypothesized.

The different nature of Fe(III) revealed in Al_2O_3 catalysts, confirms the rôle of the support in determining the nature of supported active species.

The decrease in activity with increasing surface area, can be attributed to the formation of this new compound and/or to a diffusion effect that decreases the efficiency of the catalyst (11). The fact that in the case of MoO_3 a slight increase in activity was observed can be an indication that in the case of mixed oxides the diffusion effect cannot play a major rôle in decreasing the activity.

On the contrary, the analogous behaviour in selectivity of the two classes of catalysts confirms that the diffusion phenomena can be contributory.

The over-oxidation of CH_2O can occur as a consecutive reaction within the pores.

REFERENCES

1 M. Dente, I. Pasquon, Chim. Ind. (Milano), 47 (1963) 359.
2 N. Pernicone, J. Less Common Metals, 36 (1974) 289.
3 G. Alessandrini, L. Cairati, P. Forzatti, P.L. Villa and F.Trifiro J. Less Common Metals, 54 (1977) 373-386.
4 C. Marcilly, B. Delmon, Comptes Rendus, 268 (1969) 1795-1805.
5 P.L. Villa, F. Trifiro and I. Pasquon, Reactor Kinetics and Catalysis Letters, 1 (1974) 341-344.
6 I. Pasquon, F. Trifiro and G. Caputo, Chim.Ind. (Milano), 55 (1973) 168-175.
7 M. Carbucicchio and F. Trifiro, J. Catal.,43 (1975) 77-83.
8 Ch. Janot and P. Delcroix, J. de Physique, 35 (1974) 557-561.
9 A.M. van der Knaun, Phys.Stat Sol (0),18 (1973) 215-226.
10 F. Trifiro and I. Pasquon, J. Catal, 12 (1968) 412-416.
11 C.N. Satterfield, Mass Transfer in Heterogeneous Catalysis, M.I.T Press Massachusetts (1970).

DISCUSSION

L. GUCZI : If I understood correctly, you correlated the disappearance of the middle peak to the destruction of the Fe-molybdate compound. I think this is not necessarily so if we assume a large quadrupole

splitting, due to the very large increase in surface area and thus
in dispersion of active components.

M. CARBUCICCHIO : From the analysis of the Mössbauer spectra obtained
for supported catalysts with intermediate areas (up to 120 m^2/g for
SiO_2 and up to 15 m^2/g for Al_2O_3) it is possible to exclude that the
broadening of the lines observed by increasing the support areas
is due to the increase of the quadrupole splitting of only one kind
of iron. The spectra are always due to two contributions : a "single
line" and a "double line" spectrum. By increasing the surface area,
the single line contribution decreases and the double line contribution to the spectra increases while its quadrupole splitting
remains constant. This suggests that no evolution of the iron surroundings occurs. It is a possible hypothesis that this double line
contribution is due to iron ions located at the surface of $Fe_2(MoO_4)_3$.
However the double line contribution is not the same for Al_2O_3 and
SiO_2, as one must expect if the supports are inert. Moreover, also
the smallest quadruplar splitting measured should imply such a high
electric field gradient at the Mössbauer nucleus that, in our opinion,
it is not reasonable to attribute this doublet to iron ions of
$Fe_2(MoO_4)_3$ even located at the surface. This is also supported by the
catalytic measurements : the observed decrease of the catalytic
activity by increasing the catalyst area is in agreement with a
destruction of the active species rather than with the increase of
the active component area.

B. DELMON : There might be much mutual benefit if FeMo catalysts
could be compared to CoMo. In relation to this, I wish to refer to
our recent results (1,2) with the CoMo system in its oxidic form;
we studied this system when supported on γ-Al_2O_3 and SiO_2.
The main feature of the CoMo/γ-Al_2O_3 system is the existence of a
CoMo "bilayer", namely a mixed cobalt-molybdenum oxidic species,
extremely well dispersed on the surface of the support. We do not
know at the moment the stoichiometry of this species. I suggest
this mixed species could be the equivalent of the "distorted" iron
molybdate described by the authors of the present paper. The formation
of the monolayer strongly promotes the dispersion of cobalt oxide.
One could speculate that, if analogies really exist between FeMo/Al_2O_3
and CoMo/Al_2O_3, Fe_2O_3 could segregate when a "bilayer" (as a "distorted" iron molybdate) is formed, as does Co (as cobalt oxide), if

iron is in excess with respect to the stoichiometry of the bilayer. Actually, this speculation seems to be supported by the authors' data. When the species are deposited on SiO_2, in contrast with the case of Al_2O_3, there is less affinity between MoO_3 and the support, and normal molybdates can form. We observe $CoMoO_4$ on SiO_2. If the analogy I suggest really exists, normal $Fe_2(MoO_4)_3$ should form on SiO_2. Have the authors any data supporting this prediction ?
With low surface area catalysts, it is always possible to form the normal molybdate because only a small proportion of the deposited phases can come into strong interaction with the support. As a result, segregation of MoO_3 will occur in your samples. This kind of MoO_3 is representing the excess of the composition (Mo/Fe=2) with respect to the stoichiometry of the normal molybdate (Mo/Fe=3/2). It should be noticed that this segregation effect gives MoO_3, whereas the one caused by the formation of the "bilayer" gives group VIII metal oxide.
(1) B. Delmon, P. Grange, M.A. Apecetche, P.S. Gajardo, F. Delannay C.R. Acad. Sci, in press.
(2) P.S. Gajardo, P. Grange, B. Delmon, submitted for publication.

M. CARBURCICCHIO : The existence of possible analogies between supported Co-Mo and Fe-Mo in the oxidic form, should be further investigated on alumina and on silica. The proposed mechanism of the formation of the "bilayer" can be very useful in order to interpret the nature of the compounds formed in our systems. With respect to your question we can affirm that silica has a much lower reactivity towards $Fe_2(MoO_4)_3$ than alumina. This is in agreement with your prediction. Indeed, in the case of silica, $Fe_2(MoO_4)_3$ is still present at 120 m^2/g while it was found to be absent in the case of alumina having much lower surface areas. However, for very high surface areas and for long calcination times also in the case of the silica support $Fe_2(MoO_4)_3$ is absent. On the supported catalysts with areas lower than these very high values, $Fe_2(MoO_4)_3$ coexists with the products of the reaction between the support and the active species. However, at this moment, we cannot say if $Fe_2(MoO_4)_3$ is in the distorted or in the normal form.

J. SCHEVE : You should strengthen the information from the γ-resonance spectra by drawing in addition spectra made at 77 K. The computed doublets now can only be seen as a shoulder in the original spectra.

M. CARBUCICCHIO : You are right. The Mössbauer spectra at 77 K for catalysts supported on high area silica and alumina confirm the fact that no α-Fe_2O_3, in the form of very small particles, grows as a consequence of the interaction between support and active species. However, the doublets observed at room temperature are not expected to be transformed in a shoulder at 77 K. α-Fe_2O_3 small particles have at room temperature a superparamagnetic behaviour due to the fast relaxation of their magnetic moments : the room temperature paramagnetic doublet transforms in a six-line pattern at liquid nitrogen temperature. In any case the low temperature measurements we performed both on silica and on alumina supported catalysts with high areas (unpublished data), confirmed the information (i.e. the absence of α-Fe_2O_3 small particles) we obtained by comparing our room temperature Mössbauer data both with the ones available in literature and with our catalytic measurements.

S.P.S. ANDREW : Was an analysis made for carbon on the used catalyst ? Could an involatile by-product be the explanation of the different activities of the high and low area catalysts ?

F. TRIFIRO : We do not exclude, as you suggest, that the decrease of the activity with the increase of the surface area of silica can be due to a strong absorption of products of side reactions. However we can exclude that this by-product of carbon is formed since an excess of oxygen is present inside the reactor.

N. PERNICONE : Have you any data about the influence of the iron molybdate concentration on both activity and selectivity of your catalysts ?

F. TRIFIRO : We have data only for SiO_2 with low surface area. We can say that by increasing the amount of the active species, the activity increases but the selectivity decreases. It is necessary to take into account that the contact time in our experiment was very high.

A. KUBASOV : You showed that catalytic activity decreased after the formation of a chemical bond between the catalyst and support. But there are a lot of cases where activity increases. For example, we found out that in the case of boronphosphate on silica the activity increased after chemical interaction between components. So my question is : what is the nature of the interaction in your case ? What is changed, the electronic state, the coordination number or something else ?

F. TRIFIRO : Initially, we were also surprised that the activity declined with an increase of surface area. Afterwards from the analysis of Mössbauer data we revealed that during the preparation of the catalysts not only an increase of surface area of active components occurs, but also an interaction with the support. This strongly changes the nature of the starting catalysts, at least when the electronic properties of iron are considered. However, it is not known whether the coordination of molybdenum, that must be a more important element for the investigated reaction, also changes.

P.R. COURTY : You showed that with silica supported iron molybdate catalysts, the amount of methylformiate formed is of the same order of magnitude as the amount of CO formed. On the other hand, with bulk iron molybdate the amount of methylformiate formed is very low. This suggests that the silica carrier, even with a low surface area, strongly modifies the specific catalytic properties of the active phase. Also, the formation of dimethylether which occurs with the bulk catalyst, is not mentioned for the supported ones. Can you comment on this ?

F. TRIFIRO : We have observed dimethylether and no methylformiate wit pure iron molybdate. We believe that the formation of methylformiate with the supported catalysts could be due to the acid sites of silica rather than to a modification of the iron molybdate. Formic acid formed as intermediate in the production of CO, can react with methanol in the presence of acid sites of SiO_2.

RELATIONSHIP BETWEEN AVERAGE PORE DIAMETER AND SELECTIVITY IN IRON - CHROMIUM - POTASSIUM DEHYDROGENATION CATALYSTS.

Ph. COURTY and J.F. LE PAGE
Institut Français du Pétrole, Rueil-Malmaison, France.
Procatalyse, Paris, France.

ABSTRACT

The dehydrogenation of ethylbenzene into styrene is carried out with bulk catalysts generally composed of iron, chromium and potassium oxides. Beside this dehydrogenation, side reactions producing toluene and benzene are also observed.

As far as the performances of the catalyst are concerned, it has been established that the main step of the catalyst preparation was the final calcination. The selectivity towards styrene production is directly correlated to the specific area which in turn depends on the calcination conditions.

It has also been observed that a 150 hrs running period was necessary to stabilize the catalyst ; during this time, selectivity is progressively improved without change in activity. Surface rebuilding of the crystallites have been invoked to explain this increase of selectivity in the operating conditions.

TRANSFORMATION ANALYSIS

Styrene is essentially obtained by dehydrogenation of ethylbenzene. The range of operating conditions conventionally used is mentioned in table I. The reaction which is highly endothermic is thermodynamically limited and the conversion is increased when increasing the applied temperature.

The feed is much diluted with steam for the following purposes : limitation of the coking rate of the catalyst, decrease of the temperature fall due to the reaction endothermicity and improvement of the conversion at the equilibrium. Industrial reactors are either adiabatic or isotherm and the various reactions occurring during the operation are pointed out in table I ; we can observe that the ethylbenzene dehydrogenation is accompanied by secondary reactions which are detrimental to the selectivity : toluene, benzene but also small quantities of heavy aromatics (diphenyl , naphtalene, anthracene) are produced whereas the light intermediate paraffins are mainly transformed into CO_2.

TABLE I

Industrial Operating Conditions

T °C : 580 - 630 °C

P (bars) abs. : 1 - 5

L.H.S.V.* : 0.3 - 1 h^{-1}

wt H_2O/wt $C_6H_5C_2H_5$: 1.5 - 4

Main Reaction :

$C_6H_5C_2H_5 \longrightarrow C_6H_5CH=CH_2 + H_2 \quad \overrightarrow{\Delta H}^o_{298} = 29$ Kcal.mol^{-1}

Side Reactions :

$C_6H_5C_2H_5 + H_2 \longrightarrow C_6H_5CH_3 + CH_4$

$C_6H_5C_2H_5 + 2 H_2 \longrightarrow C_6H_6 + 2 CH_4$

$C_6H_5C_2H_5 + H_2 \longrightarrow C_6H_6 + C_2H_6$

$CH_4 + H_2O \rightleftarrows CO + 3 H_2$

$CO + H_2O \rightleftarrows CO_2 + H_2$

*For expressing L.H.S.V., only ethylbenzene is considered.

CATALYST DESCRIPTION

Commercial catalysts as well as catalysts being presently studied in research laboratories seem to be prepared from ferric iron oxide as the major active agent (60 - 80 wt %), potassium oxide as a stabilizer (10 - 20 wt %) and chromium oxide as a promoter ; they can also possibly contain various additives as texture or structure stabilizers. Solid state interactions occur between these iron, chromium and potassium oxides, when they are heated at temperatures ranging from 700 to 1000 °C. The mixture of iron and chromium oxides, acting probably as a solid solution, would be the essential active phase, whereas the potassium oxide effect, in presence of steam, would consist in avoiding the building up of coke.

Catalysts A, B, C, D are laboratory catalysts prepared by wet mixing of ferric oxide, potassium carbonate, chromium oxide, binders and texture promoters. The resulting paste is transformed into 5 mm diameter and 5 mm length extrudates ; the product is dried, then calcined for 2 hours at temperatures ranging from 850 to 1000 °C. Figure 1 shows the change of texture characteristics as a function of the calcination temperature. Between 900 and 950 °C, the rapid increase of the average pore diameter indicates a very important sintering effect ; at the same time, the specific surface area decreases, the total pore volume remaining almost constant. Table 2 summarizes textural properties of A, B, C and D catalysts. Catalyst E is an industrial catalyst having the same composition as the above mentioned catalysts.

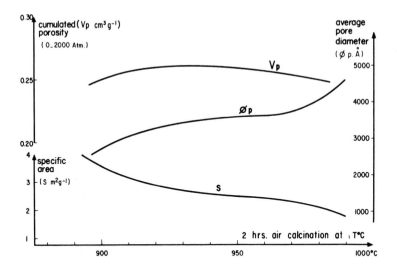

Figure 1

Styrene catalysts (serie A - D)

Textural properties vs. calcination temperature.

TABLE 2 - Textural properties

Catalyst	Calcination Temp. (°C)	Duration (h)	Crystallites Size (scanning microscopy)	$\emptyset_P(\text{Å})$	$S.Sp._{(m^2 g^{-1})}$
A	890	2	2000 - 3000	2400	4
B	920	2	nd	2700	3.2
C	940	2	nd	3200	2.5
D	970	2	4000 - 5000	4800	2.2

EXPERIMENTAL CONDITIONS

The activity of the catalysts is determined in a "Géomécanique" cata-test working at atmospheric pressure ; non stop runs at 2 bars have been performed in the same equipment to study the catalyst stability. Operating conditions are summarized in table 3. Mention has to be made that activity tests are realized with catalysts which have been aged for 150 hrs in the conditions of the stability tests.

TABLE 3 - Test conditions

	Activity	Aging and Stability
P (bars)	1	2
L.H.S.V.* (h^{-1})	0.4	0.4
Wt H_2O/Wt $C_6H_5C_2H_5$	2.2	2.2
T °C	580 - 630	580 - 630

*For expressing L.H.S.V., only ethylbenzene is considered.

Crushed catalysts (5 - 6 mesh) are preheated under nitrogen atmosphere up to 500 °C ; steam is first injected at 500 °C and finally ethylbenzene at 550 °C. Ethylbenzene is purchased from C.D.F. Chimie.

Products are analyzed by G.P.C.

EXPERIMENTAL RESULTS

We shall examine successively the experimental results pertaining to the activity and then to the stabilization of the catalysts.

Catalyst performances will be expressed as follows :

Ethylbenzene conversion (C) in % $= \dfrac{\text{E.B inlet - E.B outlet}}{\text{E.B inlet}} \times 100$

Product (i) Selectivity (Si) in % mol. $= \dfrac{\text{Product (i)}}{\text{E.B inlet - E.B outlet}} \times 100$

Product (i) Yield in % mol. $= \dfrac{\text{Product (i)}}{\text{E.B inlet}}$ with $R_i = C.S_i$
(Ri)

Catalyst activity

Figures 2 to 5 show the results obtained from activity tests, with the catalysts A, B, C, D treated for 150 hours, in the aging conditions. When the calcination temperature of the catalysts is increased, the conversion progressively decreases (fig. 2) as the surface area decreases. A strong effect of the calcination temperature on the styrene yield is observed, as shown in figure 3 ; for instance, the maximum of styrene yield goes from 64 to 71 % when the calcination temperature increases from 890 to 970 °C.

Figure 4 presents the benzene plus toluene yields as a function of the reaction temperature for the various A, B, C, D catalysts ; one can see in the figure that the dealkylating activity of the catalyst decreases as the calcination temperature is increased.

All these experimental results are summarized by the selectivity-conversion curves presented in figure 5 : the higher the temperature of calcination, the higher the catalyst selectivity towards the dehydrogenation.

X-ray diffraction study of the diagrams of the A, B, C, D and E catalysts fails to show any difference. In all of them, a well crystallized Fe_3O_4 phase and a poorly crystallized K_2CrO_4 phase are detected as shown in figure 11. Besides, the $K_2Fe_{22}O_{34}$ phase, previously mentioned [6] and which might be considered as the active species has not been observed.

The increase of the catalysts selectivity towards the dehydrogenation can be correlated with the variation of the pore diameter and, in a better way, with the decrease of the specific surface area, as shown in figures 6 and 7. As the catalysts used in this study are non supported catalysts, it is not surprising that the activity should be proportional to the surface area all the more as iron oxide which contributes for an essential part to the development of this surface is also the active constituent of the catalyst.

297

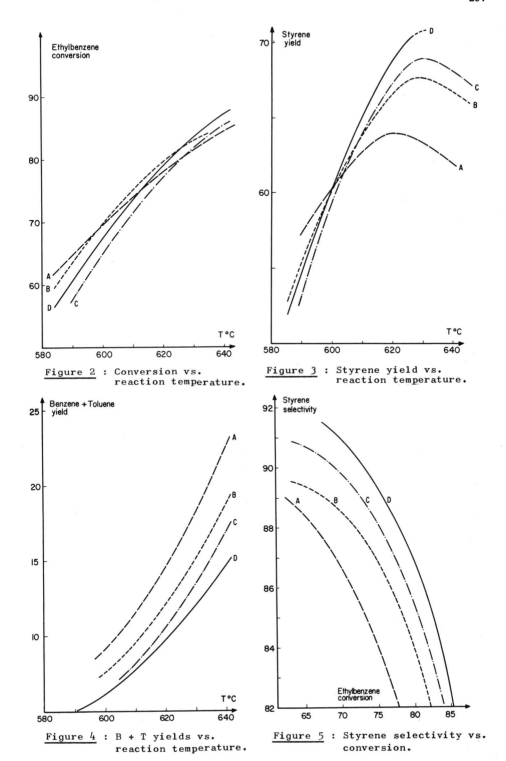

Figure 2 : Conversion vs. reaction temperature.

Figure 3 : Styrene yield vs. reaction temperature.

Figure 4 : B + T yields vs. reaction temperature.

Figure 5 : Styrene selectivity vs. conversion.

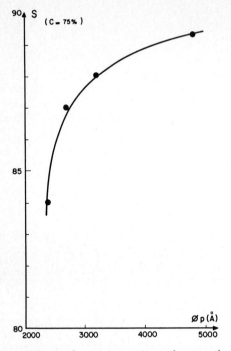

Figure 6 : Selectivity (at 75 %₀ conversion) vs. pore diameter (Å)

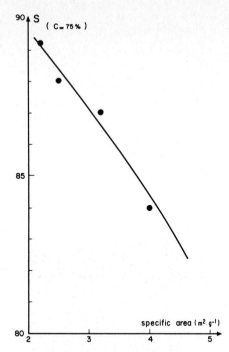

Figure 7 : Selectivity (at 75 % conversion) vs. spec. area ($m^2 g^{-1}$)

The other components, K_2CO_3, CrO_3, K_2CrO_4 offer a negligible surface after being separately calcined at temperatures ranging from 800 to 1000 °C.

The main side reactions which reduce the catalyst selectivity concern the attack of the ethyl radical C-C bonds by steam according to the following reactions :

$$C_6H_5C_2H_5 + 2 H_2O \longrightarrow C_6H_5CH_3 + CO_2 + 3 H_2$$

$$C_6H_5CH_3 + 2 H_2O \longrightarrow C_6H_6 + CO_2 + 3 H_2$$

This is supported by the analysis of the reaction gases which generally contain 10 to 15 % CO_2, about 1 % CO and only 1 to 2 % CH_4 ; toluene to benzene ratio is close to 2.

Steam dealkylation of alkylbenzenes has indeed been recently explained as a selective steam reforming of side chains [7] ; the selectivity of active sites can be correlated with the average size of crystallites. The smallest crystallites corresponding to the highest surfaces exhibit the highest activities ; when the crystallite size is decreased, conventional steam reforming increases to the detriment of C-C bond rupture through an hydrogen attack. Such an observation agrees with the results of Maslianski and Rabinovitch [8] who have observed that, with group VIII metals other than iron, steam dealkylation requests high density of sites for water activation.

Figures 8 and 9 show the electron scanning microphotographs of A and D catalysts, the average pore diameter of which was respectively equal to 2400 Å and 4800 Å. It can be observed that the average crystallites sizes reach about the same value (2000 - 3000 Å for A, 4000 - 5000 Å for D) : catalyst D exhibiting the largest average crystallite size is the most selective, as far as dehydrogenation of ethylbenzene into styrene is concerned (fig. 5).

Figure 8 : Cat. A - Electron scanning microphotograph (G x 20 000).

Figure 9 : Cat. D - Electron scanning microphotograph (G x 20 000).

Catalyst stabilization

A 300 hrs test in very severe conditions has been carried out with a catalyst made at an industrial scale ; this catalyst (E) has the same composition as the catalysts A, B, C, D previously mentioned. Figure 10 shows the evolution of this catalyst performances (conversion, styrene yield and selectivity, benzene and toluene yields) as a function of the running time. It may be observed that, for 200 hrs the selectivity of the catalyst does not cease to increase even though the conversion remains constant : styrene yield varies from 55 to 63 % to the detriment of ligher aromatics yields which decrease from 17 % down to about 10 %.

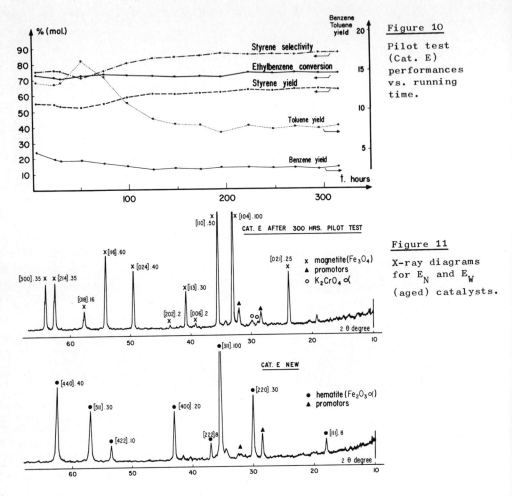

Figure 10

Pilot test (Cat. E) performances vs. running time.

Figure 11

X-ray diagrams for E_N and E_W (aged) catalysts.

Figure 12 : Cat. E_N - Electron scanning microphotograph (G x 20 000)

Figure 13 : Cat. E_W - Electron scanning microphotograph (G x 20 000)

Catalytic characteristics of catalyst E either new (E_N) or after use in the stabilization test E_W have been determined. Texture characteristics (surface area, pores sizes, total pore volume) remain the same for E_N, E_W and even E_W calcined for 2 hours with air at 500 °C as shown in table 4 ; changes of textural characteristics cannot be invoked to explain the observed change in selectivity during the non-stop run.

X-ray diffraction diagrams have also been determined ; as shown in figure 11 the only difference between E_N and E_W lies in a smooth reduction of Fe_2O_3 into Fe_3O_4 during the stabilization period ; crystallite sizes do not change as it is confirmed by the electron scanning microphotographs presented in figs 12 and 13.

TABLE 4 - Catalyst E - Textural properties

	Bulk density (g/ml)	Structural density (g/ml)	Total pore volume (ml/100 g)	Cumulated pore volume (0-2000 bars)	Average pore diameter (Å)	Specific area (m^2/g)
new E_N	1.84	3.86	28.4	23.3	3200	2.7
aged E_W (316 hrs-pilot-plant aged)	1.90	3.86	26.7	26.1	3500	2.9
air-calcined (500 °C/2 hr)	1.96	3.85	25.0	24.1	3200	2.9

CONCLUSION

The stabilization of the catalysts used for the dehydrogenation of ethylbenzene into styrene inside the reactor corresponds to a slow increase of selectivity without change in activity. After a 200 hrs period, the catalyst is stabilized but the selectivity is considerably increased.

Although a reduction of Fe_2O_3 into ferromagnetic Fe_3O_4 goes with the stabilization, this does not justify the observed increase of the selectivity ; as a matter of fact, this reduction is very rapid and requires only a few hours in the operating conditions, to be achieved. As the specific surface area, the pore sizes and the total pore volume do not vary during the stabilization period, only a progressive rebuilding of the catalyst surface can be invoked to explain this phenomenon.

Modifications of the distribution of steps, kinks and terraces on the surface of the active crystallites during this stabilization period might account for this increase of selectivity [9] ; such reconstruction mechanisms are currently mentioned not only with oxide catalysts, but also with metal catalysts [10], [11].

The influence of the calcination temperature on the performances has been therefore studied after a prior aging of the catalysts for 150 hrs reaction conditions. It has been observed that the selectivity increased up to an optimum when the calcination temperature and, consequently, the crystallites sizes increased. For catalysts having the above mentioned composition, this optimum is obtained for an average crystallite size round about 5000 Å ; beyond this value, the selectivity, in iso-conversion conditions, begins to decrease because of the reaction temperature increase, which is necessary to compensate the activity loss, and which relatively favors the detrimental thermal dealkylation.

These results finally show that it is often dangerous to select catalysts taking into account only their initial activity. Moreover in the present study, it has been observed that the catalysts which are the most selective after the stabilization period, have not necessarily the highest stability.

REFERENCES

1 E.H. Lee, Catalysis Review, 1973, $\underline{8}$, (2), P. 285 - 305.
2 Shell Oil U.S.P. 3 360 579, 2 461 147, 2 414 585.
3 Catalyst and Chemicals Inc. U.S.P. 2 971 927.
4 Dow Chemical U.S.P. 3 849 339, 3 839 478, 3 703 593, 3 205 179.
5 Procatalyse. BRIT. P. 1 504 627.
6 K. Shibata, T. Kiyoura, J. Cat., 1969, $\underline{13}$, (1), P. 103 - 105.
7 G.L. Rabinovitch, G.N. Masliansky, L.N. Treiger, Symposium on the mechanisms of hydrocarbon reactions, 5 - 7 june 73, Siofok (Hungary), P. 48, P. 97.
8 D.C. Grenoble, J. Cat., 1978, $\underline{51}$, P. 203 - 211.
9 D.W. Blakely, G.A. Somorjai, J. Cat., 1976, $\underline{42}$, P. 181 - 196.
10 R. Brill and J. Kurzidim, "Structure et propriétés des surfaces des solides" Coll. C.N.R.S., Paris, juillet 1969, P. 99, C.N.R.S. Paris 1970.
11 A.G. Friedlander, Ph. Courty, R. Montarnal, J. Cat. 77, $\underline{48}$, P. 312 - 321.

DISCUSSION

Z. PAAL : The phenomenon of catalyst stabilization reminds me of a similar phenomenon described for potassium-chromia-alumina catalysts (Karansky, B.A., Rozengart, M.I. et al., Kinetika i Kataliz $\underline{4}$, 768(1963)). The stabilization process lasted for about 30 min. and was attributed to the reduction of Cr(VI) to Cr(III) (or eventually to Cr(II)). The latter has been claimed to possess dehydrogenation activity. Reduction proceeded under the effect of the hydrocarbon feed, as shown by the appearance of CO_2. Can this process be excluded here, considering that Fe may hinder reduction ?

P.R. COURTY : According to the above mentioned reference, the stabilization period which corresponds to a reduction of Cr^{6+} into

Cr^{3+} or Cr^{2+} is a short phenomenon (about 30 min.). The stabilization of the styrene catalyst requires more than 150 hrs and cannot be correlated with a simple reduction phenomenon. Anyway, we observe always some CO_2 production (due to steam attack of the side chains) during the whole life of the catalyst.

E.J. NEWSON: Since the particle size of the crushed catalysts used in your experiments was as large as 5-6 mesh (about 3-4 mm diameter), it is just possible that "diffusion-confusion" could equally well explain your Figures 2 and 3. Focus attention on curve A (catalyst pore diam. 2400 Å) and curve D (catalyst pore diam. 4800 Å). From Figure 2, starting at low temperature and showing intrinsic kinetics presumably, it is clear that the conversion over cat A (2400 Å) is higher than over cat B (4800 Å). At high temperatures (640°C) just the opposite is observed. A similar comment applies to Figure 3, the selectivity data.

Do you have the same data for the same batch of catalysts in small particles size (e.g. 0.3 mm diam.) ? If yes, this new data should be used the reply.

P.R. COURTY : The catalyst granulometry (5-6 mesh/3.4-4 mm) for the test procedure was chosen for <u>reducing</u> the intraparticle diffusion limitation and not for supressing it. It is well known that the rate of the dehydrogenation reaction for such catalysts is diffusion limited. The effectiveness factor we evaluated is 1 for an eq. diam. of dp = 0.4 mm and about 0.8 for dp = 3.7 mm. Nevertheless we have selected this granulometry since the smallest granule dimensions used in industrial reactors are dp = 3.2 mm (1/8") and l = 5-6 mm. This corresponds to an eq. diam. of about 3.7 mm (our average granulometry). Studies carried out with smaller particles cannot be directly related to an improvement of industrial catalysts. With respect to your observation, we agree that some diffusion limitation occurs, but the diffusion effect is far from being sufficient to explain the selectivity differences. The conversion <u>vs</u> temperature curves (Fig. 2) are parallel (curves B,C,D) and correspond to catalysts which also exhibit different pore sizes. This suggests that the effectiveness factor should have approximate values. Its value is slightly higher for A. This is supported by a similar study carried out with smaller particles. The same effect is also observed : the selectivity increases with the increase of pore diameter and decrease of the specific area.

J.R.H. ROSS : Under the conditions of the processes under consideration, the reaction $Fe_2O_3 + 3H_2 \rightarrow 2Fe + 3H_2O$ may occur (1). Whether the oxide or the metal is the stable phase depends on the ratio P_{H_2O}/P_{H_2}. Metallic iron is an active steam reforming and hydrogenation/dehydrogenation catalyst.

Have the authors considered the possibility that their activated catalyst may contain some metallic iron ? Could metallic iron have been re-oxidized before the x-ray measurements were made on the used catalysts ?

1. J.R.H. ROSS, in M.W. Roberts and J.M. Thomas (Eds.) "Surface and Defect Properties of Solids" Vol. IV, Specialist Periodical Reports, Chemical Society, London 1975, p. 34.

P.R. COURTY : Under process conditions (T⩾600°C and $P_{H_2}/P_{H_2O} < 0.1$, at the reactor outlet) the more stable iron compound is the magnetic oxide. The formation of metallic iron, from a thermodynamic point of view, seems difficult (see either Chaudron or Ellingham diagrams for the system $Fe_3O_4-Fe-H_2-H_2O$). On the other hand, the formation of metallic iron would be accompanied by an increase of the hydrogenolysis activity. Indeed during the whole catalyst life, we do not observe cracking of the aromatic rings (less than 0.5%). During the stabilization period we do not observe an increase of the methane yield, which would be a result of hydrogenolysis of the side chains on reduced iron. Conversely, toluene and benzene yield decreases in parallel with the CO_2 yield, suggesting that the corresponding reaction is steam dealkylation. Finally, we have never observed metallic iron in our so-called "stabilized catalysts".

M.V. TWIGG : Have the steam reforming dealkylation reactions of toluene and ethylbenzene been studied separately ? If they have, could details be given ?

P.R. COURTY : In the conditions of steam dealkylation reactions (T>600°C; LHSV = 0.4 - 1 h^{-1}; $H_2O/HC \geqslant 10$ mole/mole), the reactivity of ethylbenzene and toluene over the styrene catalyst are very different. With toluene as a feed, steam dealkylation takes place, but the benzene yield is very low (less than 5%). With ethylbenzene as a feed, the main reaction is dehydrogenation; via steam dealkylatio reactions, ethylbenzene demethylation (which produces toluene) and deethylation (which produces benzene) are observed. The most important reaction results in the formation of toluene.

K. KOCHLOEFL : 1) By-products e.g. benzene and toluene in the dehydrogenation of ethlybenzene are formed exclusively by steam dealkylation. This conclusion resulted from our experiments where D_2 was added to the feed (H_2O + ethylbenzene). In this case no benzene or toluene containing deuterium were found.
2) Apart from the processes described in this paper, also K-migration in the particles of the Fe-Cr-K catalyst plays an important role during the activation period.

P.R. COURTY : The K-migration into catalyst particles is a well known phenomenon (1). Although the new catalyst exhibits an uniform radial distribution for chromium and potassium, the used catalyst (20.000 hrs aging) shows the potassium to be largely concentrated near the center of the pellets and some enrichment of chromium near the periphery of the pellets. Since the decrease of the superficial concentration of potassium is accompanied by a decrease in activity (the activity remains constant during the stabilization period), the phenomenon of K-migration alone cannot be invoked to explain the stabilization. On the other hand, the stabilization phenomenon is rapid (200 hrs) with respect to the expected duration of life of the catalyst (20.000 hrs or more).
(1) E.H. LEE, Catalysis Review 8(2), 285-305(1973).

PREPARATION OF HIGHLY-DISPERSED RUTHENIUM ON MAGNESIUM OXIDE SUPPORTS: COMMENTS ON THE ADVANTAGES OF NON-AQUEOUS CATALYST PREPARATIONS

L. L. MURRELL AND D. J. C. YATES
Exxon Research and Engineering Company, Linden, New Jersey 07036

ABSTRACT

The difficulty of preparing highly-dispersed ruthenium catalysts on basic oxide supports is related to hydrolysis of the support by the aqueous phase. Use of anhydrous, aprotic solutions prevents the hydrolysis reactions with the hydroxyl groups of the support, allowing selective adsorption of ruthenium chloride onto the support. An adsorption preparation is not possible from aqueous solution due to the high pH values of the solutions. The preparation of well-dispersed ruthenium on MgO-washcoated monoliths was accomplished by a non-aqueous adsorption preparation. Aqueous solution preparations in contrast lead to ruthenium maldistribution with the metal poorly dispersed on this type of monolith support.

Determination of the metal dispersion for supported ruthenium catalysts is complicated by a number of factors. One of the most important of these factors is multiple adsorption of CO on Ru catalysts. Ruthenium on alumina metal dispersions measured with H_2 chemisorption do not agree with those measured with CO. We have, however, found for many ruthenium catalysts on MgO and on $MgO/MgAl_2O_4$ monoliths that there is good agreement between the metal dispersions obtained by H_2 and CO. We propose that the CO/H_2 ratio is a probe of the metal-support interaction of Ru with MgO.

INTRODUCTION

Ruthenium catalysts are of interest in the catalytic reduction of NO to N_2 without formation of undesirable byproducts, such as NH_3. It has been observed that supported ruthenium is a uniquely selective catalyst for the direct reduction of NO to N_2 (ref. 1). Ruthenium supported on MgO powder has recently been reported to have excellent stability to oxidizing conditions at 900°C (ref. 2). In comparison, ruthenium on alumina readily volatilizes as RuO_3 and RuO_4 under oxidizing conditions at 900°C (ref. 2). Although MgO powder is not a practical support under conditions where reaction of the support with water can occur, such as high temperature automotive exhaust environments, a stable ruthenium catalyst can be prepared by reaction of an alumina washcoated cordierite monolith with excess MgO (refs. 3-4). Therefore, the preparation of highly-dispersed ruthenium catalysts on MgO-washcoated monoliths is of practical importance.

The preparation of supported ruthenium catalysts of high metal dispersion is markedly dependent on the catalyst preparation step where the ruthenium salt is contacted with the support surface. The advantages of aprotic, non-aqueous solutions will be described in detail. Comparison of the ruthenium studies will be made with iridium and platinum supported on MgO. A model of heterogeneous catalyst preparation from non-aqueous solution will be described, and contrasted with, conventional aqueous catalyst preparations.

The carbon monoxide and hydrogen chemisorption values for ruthenium on magnesium oxide provide evidence for a strong metal-support interaction between ruthenium metal and the basic MgO. Comparisons will be made with the chemisorption behavior of H_2 and CO on ruthenium supported on alumina (refs. 5-7).

Recognition of the importance that the support surface chemistry can play in heterogeneous catalyst preparation (ref. 8) is a necessary starting point to obtain highly-dispersed ruthenium on basic oxide supports. A key aspect of the surface chemistry of an oxide in an aqueous solution is the dissociation of the hydroxyl groups (ref. 9). The equilibria is:

$$M-OH_2^+ \rightleftarrows M-OH \rightleftarrows M-O^- + H^+$$

$$\underset{\text{decreasing pH}}{\longleftarrow} \qquad \underset{\text{increasing pH}}{\longrightarrow}$$

This expression indicates that by appropriate adjustment of pH the surface may carry either a positive or a negative charge (ref. 10). Also, the above equilibria implies that a given oxide surface when in an aqueous solution will spontaneously react to give a particular pH value. This has in fact been observed in our work (ref. 11). The situation during heterogeneous catalyst preparation is much more complex, however, than implied by the above equilibria expression. Reaction of the support itself with the aqueous solution can buffer (ref. 12) the pH at a constant value regardless of attempts to externally adjust the solution pH. This buffering capacity of a high surface area oxide support is particularly relevant to the preparation of ruthenium catalysts on alumina washcoated cordierite monoliths, on MgO powder, and on MgO-reacted cordierite monoliths. The implications of this work to other heterogeneous catalyst preparations where highly acidic metal salt solutions are employed will be considered in subsequent discussions.

The advantages of non-aqueous solutions in the preparation of highly-dispersed ruthenium catalysts will be described in detail. These catalysts can be prepared from non-aqueous solution on $MgO/MgAl_2O_4$ cordierite monoliths (refs. 3-4) which have 60% metal dispersion, i.e. 60% of the metal atoms are surface atoms in the supported phase. This is about a factor of five improvement in ruthenium dispersion over the best aqueous catalyst preparation on the analogous monolith.

EXPERIMENTAL

The gas adsorption measurements were carried out as previously described (refs. 13-14). A description of the gas adsorption cell used for the monolith chemisorption studies will be described in detail in a future publication. Ruthenium chloride trihydrate, chloroiridic and chloroplatinic acid were obtained from Engelhard Industries Inc. The MgO powder was Baker reagent grade and was found to have a surface area of 59 m^2/gm. The gamma-Al_2O_3 of 190 m^2/gm was obtained from Engelhard Industries Inc. The alumina washcoated monoliths were obtained from the Corning Glass Works.

Catalyst preparations with monolith supports

Catalyst preparations in this paper are for cordierite monolith cylinders (refs. 15-16) 1" in diameter by 3" long. The monoliths were air dried at 500°C for 16 hrs, then cooled to ambient temperature in a dry atmosphere. The monoliths, while still in the dry box, were put in a tightly fitting glass vessel containing 30 cc of a non-aqueous solution of ruthenium chloride, of a concentration to give 0.03 wt% Ru. This allows an equal volume of solution to be in contact with the interior monolith channels as is in contact with the broken, irregular exterior surface of the monolith. Approximately 10% by weight of the monolith was a tightly adhering alumina washcoat. The area of this washcoat was 120 m^2/gm of alumina, or about 12 m^2/gm of monolith.

The adsorption from the intensely black ruthenium chloride solution required about 30 min from anhydrous acetone or acetonitrile. The rate of adsorption was rapid in the first 5 min and complete adsorption then required a further 25 min, leaving a colorless liquid. Evaporation of the decanted acetone from these cores left no residue. In agreement with these visual observations, X-ray fluorescence examination did not detect any Ru in the decanted liquid after adsorption.

The aqueous adsorption preparation was done in identical fashion, except that an aqueous solution of ruthenium chloride was used. A quasi-incipient wetness impregnation was attempted by flowing an acidified aqueous solution through the monolith channels, the volume being sufficient to just wet the alumina washcoat surface. Similar methods of aqueous preparations were also attempted with $MgO/MgAl_2O_4$ monoliths. The $MgO/MgAl_2O_4$ washcoat had a surface area of 60 m^2 per gm of washcoat or 6 m^2/gm of monolith.

RESULTS AND DISCUSSION

Ruthenium chloride chemistry in aqueous phase

Before discussing the non-aqueous preparations of highly-dispersed ruthenium catalysts, the aqueous chemistry associated with such preparations will be reviewed. The hydrated chloride salts of the second and the third row Group VIII transition metals are standard materials for the preparation of highly-dispersed supported metal catalysts (ref. 17). Aqueous solutions of these metal chlorides are usually highly acidic (refs. 18-19). For example, an aqueous solution of ruthenium chloride of 0.06M concentration has a pH of 1.5. An increase in the pH of a solution, either 0.06 or 0.006M in ruthenium chloride, to 1.7 by the addition of 0.1N NaOH results in quantitative precipitation of ruthenium from solution. When this solution is reacidified to pH 1.5 with 0.1N HCl, the precipitated salt is completely solubilized. It is interesting that allowing a ruthenium chloride solution of pH 1.7 to stand overnight results in an insoluble product, even after reacidification to a pH of less than one. Analogous results are obtained for aqueous solutions of other second and third row Group VIII metal chlorides but, naturally, with different pH and metal concentration dependencies.

The implications of the ready precipitation of ruthenium chloride from aqueous solution with a relatively small increase in the pH is of importance in the preparation of Ru catalysts supported on Al_2O_3 and MgO.

Catalyst support reactivity with acidic metal salt solution

Alumina is one of the most commonly employed heterogeneous catalyst supports. Alumina has been reported recently (ref. 8) to be quite reactive in aqueous solution. This reactivity is due to the amphoteric nature of the hydroxyl groups terminating the lattice of the support (ref. 10). For example, we found that the addition of 5 gm of pure Al_2O_3 to 100 gm of distilled water rapidly changed the pH to 7.8 from 6.0. In addition, titration of the above suspension with aliquots of 0.1N HCl results in the alumina buffering the system to a pH of 4.2 (as long as any undissolved alumina is present). Continuous hydrolysis of the alumina support is responsible for the buffering behavior of the suspension. A similar effect is observed for a silica-alumina cracking catalyst with 14% Al_2O_3 content (ref. 17). Quite different behavior is observed upon titration of aqueous slurries of silica, titania and carbon (refs. 18, 20) with 0.1N HCl.

The implications of the hydrolysis characteristics of alumina and silica-alumina to aqueous heterogeneous catalyst preparations are:

Acidic metal chloride solutions may result in metal salt precipitation when using salts which are insoluble at a pH of 4.2.

Metal salt precipitation must be taken into account, especially for excess aqueous solution preparations in which adsorption from the aqueous phase may be confused with precipitation from solution.

Hydrolysis of alumina rapidly occurs on contact with water, giving a pH of 7.8. This high pH level may lead to even more complete salt precipitation with incipient wetness catalyst preparations. Such preparations necessarily contain relatively small amounts of water.

Alumina containing a 0.2 wt% sodium impurity was found to buffer the slurry to a pH of 8.6, compared to 7.8 for very pure alumina (ref. 11). Therefore the purity of the alumina is also an important factor to consider in preparations where acidic metal salt solutions are used.

Monolith catalyst preparations

Aqueous solutions of ruthenium chloride when contacted with either Al_2O_3 coated monoliths or $MgO/MgAl_2O_4$ coated monoliths gave instantaneous precipitation from solution as observed visually. Prior acidification of the solution with either HCl or HNO_3 failed to prevent this precipitation of the ruthenium due to the buffering action of the support. Gross mal-distribution of ruthenium on the monolith interior channels could not be avoided because of the ruthenium precipitation in these aqueous preparations. With the quasi-incipient wetness preparation, precipitation occurred to such a degree that the metal salt was only deposited on the top third of the monoliths with the solution flowing down the monolith channels.

With no satisfactory method of preparing either a well-dispersed or uniformly distributed Ru monolith catalyst from aqueous solutions, we investigated non-aqueous methods. Use of polar, aprotic, non-aqueous solvents provided a completely satisfactory solution to the problem. Such solutions are defined by Tremillon (ref. 21). Initial studies were with dehydrated monoliths, but subsequent work demonstrated that rehydrated monoliths could also be used satisfactorily. Completely uniform distribution of Ru on the monolith was achieved for adsorption from either acetone or acetonitrile. This adsorption method is clearly a major improvement over earlier aqueous techniques.

Ruthenium dispersion of about 60%, corresponding to 1.8 nm particles, could reproducibly be obtained by our new non-aqueous procedure, see Table 1. The ruthenium dispersion calculated from H_2 chemisorption on a $MgO/MgAl_2O_4$ monolith was 20% less than that calculated from the adsorption of CO. Subsequent work (to be published) on MgO powders indicates that multiple CO chemisorption on surface ruthenium atoms is responsible for the differences in the metal areas derived from H_2 and CO chemisorption. Another advantage of preparations using polar, aprotic solutions is that the usually employed hydrated metal chloride salts can be used (refs. 1,7,19). This avoids the requirement for difficult to handle and expensive metal alkoxides, acetylacetonates, or organometallics in non-aqueous preparations.

The use of protic, non-aqueous solvent to deposit Ru on $MgO/MgAl_2O_4$ coated monoliths was unsatisfactory. Methanol and ethanol were used to dissolve the $RuCl_3 \cdot xH_2O$. Apparently, the protic solutions interact strongly with the dehydrated surface by hydrogen bonding to form charged polar layers (ref. 10) which inhibit uniform adsorption of the ruthenium salt. The final monoliths made by this method always showed incomplete adsorption and very spotty, non-uniform deposition.

The rate of ruthenium chloride adsorption from acetone and the ruthenium dispersion of the catalyst were not influenced by rehydration of the MgO/MgAl$_2$O$_4$ monoliths. Rehydration was accomplished by allowing the calcined monoliths to stand overnight at ambient temperature in an atmosphere saturated with water. Also, acetone-water mixtures can be employed for the ruthenium chloride adsorption if the acetone content is 60 volume %, or greater. Qualitatively, the rate of adsorption is identical for an acetone-water solution containing 60 vol % acetone to that found with anhydrous acetone. Use of acetone-water solutions where the acetone is less than 60% resulted in precipitation of ruthenium and "spotted" adsorption. A mixed aqueous-non-aqueous solution may be of particular advantage for salts of limited solubility in the non-aqueous solvent itself.

The precipitation of full-sized (3.75" x 4.5") MgO/MgAl$_2$O$_4$ washcoated automotive monoliths (ref. 3) presented no problems with ruthenium chloride-acetone solutions. The procedure used was scaled up from that used for the 1" x 3" cores. Excellent distribution of the metal on the large monoliths was demonstrated by Ru analysis, using X-ray fluorescence of a sectioned core.

Hydrogen and carbon monoxide chemisorption studies on the 1" x 3" monoliths show (Table 1) improved Ru dispersion by a factor of five for acetone adsorption preparation compared to an aqueous preparation. Dispersion values calculated from both gases were found to be in fair agreement for these samples. For example, the dispersions for 0.03% Ru on a MgO/MgAl$_2$O$_4$ monolith by H$_2$ and CO were 61 and 74%, respectively. This relatively good agreement between the H$_2$ and CO chemisorptions for well-dispersed ruthenium is not generally found (refs. 5). Apparently a strong metal-support interaction between Ru and MgO significantly retards multiple CO chemisorption, compared with Ru of similar dispersion supported on Al$_2$O$_3$ (refs. 5,7). This point will be discussed in more detail in a subsequent section where comparisons are made between Ru supported on Al$_2$O$_3$ and MgO, both for aqueous and non-aqueous preparations. It is of interest that, regardless of the chemical nature of the support, poorly dispersed Ru does not show anomalous chemisorption behavior (refs. 5,7,13).

Aqueous and non-aqueous preparations on Al$_2$O$_3$ and MgO powders

It has been shown by Benesi, Curtis and Studer (ref. 22) that adsorption catalyst preparations can have advantages over incipient wetness preparations. Both acetone adsorption and aqueous incipient wetness preparations were carried out using RuCl$_3 \cdot$xH$_2$O for Al$_2$O$_3$ and MgO powders. Aqueous adsorption preparations were unsatisfactory, as discussed earlier. The alumina extrudates were crushed and calcined overnight at 500°C in air, then transferred rapidly to a dry box to prevent readsorption of water. An aqueous incipient wetness procedure was used to prepare 0.3% Ru on Al$_2$O$_3$.

Chemisorptions on this sample were performed in the usual manner and are reported in Table 2. As observed in other experiments with alumina supports

TABLE 1

The dispersion of Ru on monolith supports as measured by CO and H_2 chemisorption

Catalyst	Method of Preparation [4,5]	% Disp. by CO [6]	% Disp. by H_2 [7]
0.1% Ru [1]	Aq. F	13	—
0.1% Ru [1]	Aq. O	13	—
0.03% Ru [2]	Ac.	76	—
0.04% Ru [3]	Ac.	80	—
0.04% Ru [3]	Ac.	74	61

(1) The alumina coated monolith (ref. 4) was further coated with magnesium nitrate by a quasi-incipient wetness procedure. After calcination at 600°C (ref. 4), the monolith contained 1.8% MgO.

(2) MgO was added in two steps to the monolith. First, the monolith was immersed in molten magnesium nitrate (90°C), the excess liquid removed. It was dried for 4 hrs at 110°C, then calcined for 4 hrs at 600°C. This was followed by an incipient wetness impregnation as in 1. The monolith after this contained 12% MgO.

(3) Preparation as for 2, but 16% MgO was added.

(4) Aq. F denotes aqueous incipient wetness impregnation, followed by freeze-drying. Aq. O denotes aqueous incipient wetness impregnation, followed by air drying at 110°C. Ac. denotes adsorption from acetone solutions. In all cases ruthenium chloride was used.

(5) After drying or evaporation of acetone, all samples were reduced at 500°C for 2 hrs.

(6) Dispersion values calc. on the assumption of solely linear CO adsorption, i.e. one CO molecule per Ru surface atom.

(7) Dispersion values calc. on the assumption that H_2 dissociates on adsorption and that one H atom adsorbs on one surface Ru atom.

TABLE 2

Chemisorption studies on Ru, Ir and Pt

	Metal Content	Support	Preparation Method [1]	Pre-treatment [2]	% M Disp. by CO	% M Disp. by H_2	CO Disp./ H_2 Disp.
1.	0.3% Ru	Al_2O_3	Aq.	None	79	31	2.5
2.	0.3% Ru	Al_2O_3	Ac.	None	76	32	2.4
3.	0.3% Ru	MgO	Aq.	None	43	38	1.1
4.	0.3% Ru	MgO	Ac.	None	100	—	—
5.	0.3% Ru	MgO	Ac.	A	50	—	—
6.	1.0% Ru	MgO	Ac.	None	18	—	—
7.	1.0% Ru	MgO	Ac.	B	15	—	—
8.	2% Ir	MgO	Ac.	None	—	57	—
9.	2% Ir	MgO	Aq.	None	—	67	—
10.	2% Pt	MgO	Ac.	None	—	19	—

(1) Aq. denotes aqueous incipient wetness impregnation. Ac. denotes an acetone adsorption preparation.

(2) Pretreatment A consisted of calcining catalyst No. 4 in flowing air at 850°C for 16 hrs. B was a calcination of catalyst No. 6 in flowing air at 500°C for 2 hrs. In all cases, the catalysts were reduced at 500°C in flowing H_2 for 2 hrs after the pretreatment.

(refs. 5,7), the H_2 and CO dispersions are not the same, due to the presence of multiple CO chemisorption on surface ruthenium atoms. For a sample with 31% dispersion (based on H_2 chemisorption) a CO/H_2 ratio of 2.5 was found. This dispersion corresponds to a particle size of 2.3 nm, if the particles are spheres or cubes. For similar size Ru particles (2.5 nm) Dalla Betta (ref. 7) reported a CO/H_2 ratio of 2.3. His catalyst contained 5% Ru on Al_2O_3 and was prepared by an aqueous procedure. The analogous anhydrous catalyst was prepared in a dry box using rapid stirring of the acetone slurry to insure uniform adsorption of the ruthenium chloride from solution. The Ru dispersion of this catalyst was essentially the same as that of its aqueous counterpart (cf. samples 1 and 2, Table 2). This is an interesting example where essentially the same metal dispersions result from completely different metal salt addition procedures. The reduction procedures were, of course, identical. The similarity of the dispersions suggest that migration of either the metal complex or of the metal particles occurring at some stage during reduction is the key factor in determining the ultimate metal dispersion.

In contrast, quite substantial differences in Ru dispersions on MgO were obtained when the solvents were changed (lines 3 and 4, Table 2). The MgO powder was precalcined and the ruthenium chloride added by procedures identical to those used with Al_2O_3.

As expected from the results on the monoliths, the acetone preparation on powdered MgO gave much higher ruthenium dispersions than did the aqueous preparation by a factor of 2.4 (samples 3 and 4, Table 2). If one assumes no multiple CO adsorption, the dispersion of sample 4 is 100% or atomic dispersion. Unfortunately, H_2 chemisorption of this atomic dispersion could not be obtained on sample 4, as the H_2 isotherms were steeply sloping. This makes uncertain the monolayer capacity. However, for somewhat less well-dispersed Ru on $MgO/MgAl_2O_4$ monoliths, good agreement in Ru dispersions between H_2 and CO has been found (sample 5, Table 1). A possible explanation is that MgO powder interacts more strongly with the Ru than does the $MgO/MgAl_2O_4$ surface. It is also possible that there is a change in the mechanism or the stoichiometry of H_2 adsorption.

The most surprising result for Ru on MgO was the relatively high dispersion (43%) found for the incipient wetness preparation (sample 3, Table 2). Based on the previous discussion of the aqueous chemistry of ruthenium chloride, there is no doubt that precipitation occurs on contact of the aqueous phase with the MgO. Nevertheless, the ruthenium dispersion is a little higher than that obtained for the analogous catalyst prepared using alumina. This surprisingly high dispersion suggests that the basic oxide surface of MgO has a favorable support interaction with ruthenium (or its complexes) leading to small metal particle sizes after reduction. An intermediate metal chloride or metal chloro-hydrate complex could migrate on the MgO during the reduction process, giving the good Ru dispersion.

This is an example where precipitation of the metal salt during the aqueous impregnation step does not result in thn expected poorly-dispersed metal phase after reduction.

For comparison, 2% Ir and 2% Pt were deposited on the same MgO by the adsorption of the hexachlorides from acetone. After reduction, Table 2 shows that their dispersions were 57% and 19%, as measured by hydrogen chemisorption. A catalyst consisting of 2% Ir on MgO was prepared by aqueous incipient wetness. Surprisingly it was even better dispersed (67%) than the acetone preparation (57%).

Here again we have another instance where a catalyst prepared by a method involving salt precipitation gives an unexpectedly high dispersion. In a fashion similar to Ru, this observation implies a strong metal-support interaction for Ir on MgO.

Major decreases in iridium particle size in an oxidizing atmosphere have been reported recently (ref. 23). The Ir was supported on barium-doped alumina and the mechanism of particle size diminution was postulated to be related to the formation of $BaIrO_3$. Although neither bulk magnesium ruthenates or bulk magnesium iridates have been reported (refs. 2,23), our data shows that MgO surfaces have an unexpectedly beneficial effect on Ru and Ir dispersions.

As the Ru on MgO systems are intended for automotive exhaust purification, it is of interest to study their stability in high temperature oxidizing environments (refs. 2,24). Calcination of the 0.3% Ru catalyst in air at 850°C for 16 hrs gave a sample that was still about 50% dispersed after reduction (samples 4 and 5, Table 2). This was achieved despite the fact that the MgO decreased in area to 30 m^2/gm. Calcination of a 1% Ru sample at 500°C for 2 hrs showed almost no loss of dispersion (samples 6 and 7, Table 2). The retention of good Ru dispersion after high temperature calcination is consistent with the formation of a surface ruthenate (ref. 2). Studies of the effect of calcination on Ru dispersion using a MgO support of invariant surface area will be described elsewhere (ref. 25).

SUMMARY

A combination of low metal dispersion and mal-distribution of Ru coated monoliths prepared from aqueous solutions led us to use non-aqueous, aprotic solutions for these unique systems. Excellent Ru dispersion and uniform distribution was obtained from both acetone or acetonitrile solutions. The agreement between metal areas derived from CO and H_2 chemisorptions for Ru on $MgO/MgAl_2O_4$ monoliths indicates only minor multiple CO chemisorption. This agreement indicates the presence of a strong interaction between Ru and MgO in contrast to the behavior of well-dispersed Ru on Al_2O_3. Studies of similarly prepared Pt and Ir on MgO showed that the dispersions are in the following order Ir >> Pt $\approx$ Ru. With Ir the metal dispersions were the same, for both aqueous and non-aqueous solutions. With MgO a considerable improvement in Ru dispersion was obtained by the use of non-aqueous solvents. The high dispersion of Ru on MgO was retained after a high temperature air calcination.

REFERENCES

1. T. P. Kobylinski and B. W. Taylor, J. Catal., 33 (1974) 376.
2. S. J. Tauster, L. L. Murrell and J. P. DeLuca, J. Catal., 48 (1977) 258.
3. M. W. Pepper, J. P. DeLuca, R. P. Rhodes and L. S. Bernstein, SAE Paper #750684, presented at Fuels and Lubricants Meeting, Houston, Texas, June 3,
4. L. L. Murrell and D. J. C. Yates, U.S. Patent 3,980,589 and J. P. DeLuca, G. B. McVicker and L. L. Murrell, U.S. Patent 3,990,998.
5. R. A. Dalla Betta, J. Catal., 34 (1974) 57.
6. K. C. Taylor, J. Catal., 38 (1975) 299.
7. R. A. Dalla Betta, J. Phys. Chem., 79 (1975) 2519.
8. J. R. Anderson, Structure of Metallic Catalysts, Academic Press, New York, 1975, 33.
9. G. D. Parfitt, Pure and Appl. Chem., 48 (1976) 415.
10. C. B. Amphlett, Inorganic Ion Exchangers, Elsevier, New York, 1964, Chapter 5.
11. Five grams of a given oxide were calcined overnight at 500°C and transferred to a dry box. It was then contacted with 100 cc of distilled water with a pH of 6.0. Reforming grade Al_2O_3 gave a pH of 7.6. An Al_2O_3 sample contaminated with 0.2% Na results in a pH of 8.6. A $BaTiO_3$ support gave a pH of 8.5, silica (Cabosil) 4.8, titania (Caboti) 5.3 and silica-alumina (DA-1 Davison) 4.2.
12. D. D. Perrin and B. Dempsey, Buffers for pH and Metal Ion Control, Chapman and Hall, London, 1974.
13. J. H. Sinfelt and D. J. C. Yates, J. Catal., 8 (1967) 82.
14. D. J. C. Yates and J. H. Sinfelt, J. Catal., 8 (1967) 348.
15. Reference 8, p. 88.
16. J. J. Burton and R. L. Garten (Eds.), Advanced Materials in Catalysis, Academic Press, New York, 1977, Chapter 10.
17. Reference 8, Chapter 4.
18. Reference 8, p. 181.
19. J. Escard, C. Leclerc and J. P. Contour, J. Catal., 29 (1973) 31.
20. The oxides were treated as described in reference 11, except for the carbon (Carbolac) which was dried in flowing helium at 300°C. For silica, titania and carbon the pH values as a function of volume of 0.1N HCl added were identical to those obtained by addition of the same volumes of 0.1N HCl to distilled water alone.
21. B. Tremillon, Chemistry in Non-aqueous Solvents, D. Reidel, Pub. Co. Dordrecht, Holland, 1974, Chapter 1.
22. H. A. Benesi, R. M. Curtis and H. P. Studer, J. Catal., 10 (1968) 328.
23. G. B. McVicker, R. L. Garten and R. T. K. Baker, Fifth Canadian Symposium on Catalysis, October 26-27, 1977, Calgary, Alberta, p. 346.
24. Y. Ti Yu Yao and J. T. Kummer, J. Catal., 46 (1977) 388.
25. L. L. Murrell and D. J. C. Yates, in press.

DISCUSSION

H.T. RIJNTEN : You indicated in your paper that after impregnation of $RuCl_3$ on Al_2O_3 from an aqueous solution, low dispersion and maldistribution occur. A low dispersion might be understood from t polymeric Ru species in aqueous solution. Can you indicate why maldistribution occurs ?

L.L. MURRELL : Maldistribution occurs due to the depletion of
ruthenium in solution, due to the pH changes from the hydrolysis
of the Al_2O_3 support before uniform contacting of the monolith
support with the solution can be accomplished.

F.S. STONE : The results showing very high dispersion of Ru on MgO
powder are certainly very interesting and clearly one would like to
be able to define better the properties which give rise to it. It
would help in understanding the origin of this behaviour if one could
discriminate between the effects of surface OH groups on the one
hand and basic centres on the other. For MgO outgassed at 500°C,
as in the present work, there is a marked overlap of these two effects.
MgO powder outgassed below 500°C remains very hydroxylated. However,
if outgassed above 500°C the oxide reveals its intrinsic surface
basicity increasingly as the outgassing temperature is raised. Thus
a series of preparations of Ru/MgO catalysts using MgO powders
which have been pre-outgassed at, say, 350°C, 500°C, 650°C, 800°C
and 950°C respectively before the non-aqueous impregnation of the
ruthenium would show whether the high dispersion in the final Ru/MgO
catalyst depends on both the above effects or on only one of them. Provided the MgO powder is slowly heated in a good dynamic vacuum
during outgassing, the MgO surface area could be maintained at a
similar high value for all the preparations, thereby minimizing
any changes in texture.

L.L. MURRELL : I would agree that these experiments would be
interesting to do. Studies showed that the ruthenium dispersion was
not greatly influenced for a fully-hydroxylated MgO sample dried at
110°C compared to the 500°C calcination reported in our paper.

A. OZAKI : I am in favor of your interpretation to form a surface
ruthenate of Mg. In this case some part of MgO is dissolved into
solution to form the precipitate.

D.E. WEBSTER : Ruthenium containing catalysts which are stable under
prolonged oxidation and reduction conditions are known, but much
more difficulty is experienced with producing stable ruthenium
catalysts to operate under continuous short oxidation and reduction
cycles. Is there any reason to expect the high dispersion catalysts
prepared by the non-aqueous route will perform better in such

conditions ?

L.L. MURRELL : This comment is certainly inkeeping with observations in our laboratories. Short oxidation-reduction cycles do lead to significant sintering of the ruthenium phase for acetone prepared catalysts. The mechanism of sintering of ruthenium for reduction-oxidation cycles is consistent with the observations of G.B. Mc Vicker, R.L. Garten, and R.T.K. Baker, for IrO_2 migration on BaO-doped alumina for cycled reducing-oxidizing conditions (To be Published in J. Catalysis).

H. CHARCOSSET : 1. Have you considered the possibility of an incomplete reduction of ruthenium to the metallic state in your catalysts, in particular when supported by Al_2O_3 ? 2) Have you compared the chemisorption data with electron microscopy data, in typical examples at least ?

L.L. MURRELL : Incomplete reduction of ruthenium on alumina and on MgO is not greatly different than found for ruthenium on silica based on the similarity of the infrared spectra of CO chemisorbed on Ru on these three supports. Transmission electron spectroscopy has been attempted but is experimentally quite difficult due to the opaqueness of the support.

C.J. WRIGHT : You proposed a mechanism for the stabilization of ruthenium upon the MgO surface under oxidizing conditions. Under reducing conditions does the ruthenium have the same resistance to sintering and are you able to propose a physical mechanism for any stabilization that occurs ?

L.L. MURRELL : Ruthenium on magnesium oxide show evidence for a strong metal-support interaction when ruthenium is reduced to the metal in that the CO chemisorption shows little evidence of multiple chemisorption, i.e., $M(CO)_x$, x = 2-4, even for a highly dispersed ruthenium phase. Ruthenium on Al_2O_3 in comparison, has $M(CO)_x$ complexes with values of x = 2-4. The model that we propose is interaction of Ruthenium rafts or small clusters with the basic oxide ions of the magnesium oxide surface.

M.V. TWIGG : You have discussed results obtained using acetone and acetonitrile. Why were these solvents selected, and would you care to comment on the advantages and disadvantages associated with other non-aqueous solvents.

L.L. MURRELL : Acetone and acetonitrile were selected because of the excellent solubility of hydrated ruthenium chloride in these two solvents. Alcohol solvents although effective for dissolving the metal salt lead to poor adsorption characteristics. An exhaustive study of high diaelectric solvents was not undertaken as the two solvents discussed in the paper proved effective in solving the limitations of aqueous preparations.

CATALYST ACTIVATION BY REDUCTION *

N.PERNICONE and F.TRAINA

Montedison S.p.A., DIPI/Attività Catalizzatori, Novara, Italy

ABSTRACT

After having strictly defined the concept of activation by reduction, the industrial procedures for the reduction of the catalysts for high and low temperature CO conversion and for ammonia synthesis are reported and the basic principles on which they are based are discussed.

Emphasis is given to the difficulty of obtaining laboratory-scale data that can be easily connected with such industrial procedures.

The experimental techniques used for basic and applied research on catalyst reduction are briefly discussed, in particular thermogravimetry and temperature programmed reduction.

Finally the now available knowledge about the reduction mechanisms of the above-mentioned industrial catalysts is summarized.

1. THE CONCEPT OF ACTIVATION

The manufacture of an industrial catalyst (refs.1-2) involves some unit operations, which can be grouped into three main classes (Chemical Preparation, Thermal Treatments and Forming). The class "Thermal Treatments" includes three main unit operations: Drying, Calcination and Activation. There is no universally accepted definition for these unit operations. For instance, there is some confusion between drying and calcination: some people distinguish them only on the basis of the temperature level, while others pay attention to the purpose of the operation. Other misunderstandings may occur about calcination and activation. In an attempt to reach a more uniform terminology among catalysis specialists, we propose the following defi-

---

* Published in Pure Appl. Chem., 50, 9-10 (1978), p. 1169.

nitions.

By drying we strictly mean a thermal treatment performed with the aim of eliminating the absorbed water, though we admit that also the hydration water may be partially or completely eliminated. Drying can be performed in air or inert gas atmosphere or under vacuum at temperatures usually not much higher than 100°C.

By calcination a thermal treatment carried out in an oxidizing atmosphere should be strictly understood. However we prefer to mean by calcination any thermal treatment carried out with the following purposes: to decompose some compounds present in the catalyst, generally with evolution of gaseous products, and/or to allow solid state reactions among different catalyst components to occur and/or to make the catalyst sinter. The importance of calcination is straightforward, as it can influence all the catalyst properties, especially porous structure and mechanical strength.

Having so defined drying and calcination, it is possible to restrict the definition of activation to thermal treatments carried out in the reactor itself, either industrial, pilot or laboratory scale, usually in special gaseous atmospheres. Such treatments are required only for some catalysts. As a further restriction we do not consider a kind of activation the regeneration of poisoned catalysts, even if it is carried out "in situ". Therefore the term "activation" is due to fresh catalysts only.

2. ACTIVATION BY REDUCTION

It is a matter of fact that in most cases the activation consists in a reduction by suitable gaseous reducing agents. This is needed because it is very troublesome for the manufacturer to prepare catalysts where the oxidation states of the metallic elements are lower than those existing in air atmosphere. In the few cases where the reduction is carried out by the manufacturer, a passivation procedure is then needed to handle the catalyst in air and finally an activation by reduction, though shorter, must be performed.

The most important catalytic processes, from an industrial point of view, involving catalyst activation by reduction, are summarized in Table 1.

TABLE 1

Some catalytic processes involving catalyst activation by reduction

Catalytic process	Catalyst reduction *
Steam reforming of hydrocarbons	NiO → Ni
High temperature CO conversion	Fe_2O_3 → Fe_3O_4
Low temperature CO conversion	CuO → Cu
Methanation	NiO → Ni
Ammonia synthesis	Fe_3O_4 → Fe

* Only the main compound to be reduced is reported here.

It should be stressed that such activation, though not performed by the manufacturer, is in fact the final stage of catalyst preparation and that its incorrect performance may have dramatic effects on catalyst efficacy. To consider only one example, if exothermic reductions are carried out at too high rates, the catalyst temperature may exceed the maximum allowable, with consequent decrease of catalyst surface area and/or undesirable solid state reactions. To avoid such dangers, usually the manufacturer specifies the reduction procedure in detail, having taken into account the characteristics of the plant. We will give here some examples of these procedures, after having discussed the principles on which they are based.

3. PROCEDURES FOR THE REDUCTION OF SOME INDUSTRIAL CATALYSTS

3.1 High temperature CO conversion catalyst

Taking into account the chemical composition of various commercial catalysts, the reactions that may occur during reduction in the converter are reported in Table 2, together with their standard enthalpy values.

As it is necessary to avoid catalyst overheating during reduction (otherwise a decrease of catalytic activity, due to sintering, and/or of mechanical strength will result), the following general conclusions can be drawn from Table 2 about the influence of catalyst composition

TABLE 2

Reactions occurring during reduction of high temperature CO conversion catalysts.

Reaction	ΔH at 25°C, Kcal/mole
$2\alpha\text{-FeOOH} = \alpha\text{-Fe}_2O_3 + H_2O$	+ 9.3
$2\gamma\text{-FeOOH} = \gamma\text{-Fe}_2O_3 + H_2O$	+ 5.0
$3\alpha\text{-Fe}_2O_3 + H_2 = 2Fe_3O_4 + H_2O$	− 2.3
$3\alpha\text{-Fe}_2O_3 + CO = 2Fe_3O_4 + CO_2$	− 12.1
$3\gamma\text{-Fe}_2O_3 + H_2 = 2Fe_3O_4 + H_2O$	− 20.3
$3\gamma\text{-Fe}_2O_3 + CO = 2Fe_3O_4 + H_2O$	− 30.1
$2CrO_3 + 3H_2 = Cr_2O_3 + 3H_2O$	−163.6
$2CrO_3 + 3CO = Cr_2O_3 + 3CO_2$	−193.1

on the reduction process:

a) the content of CrO_3 should be as low as possible (in practice, a maximum content of 0.5% can be easily tolerated);

b) the presence of $\gamma\text{-Fe}_2O_3$, hence of $\gamma\text{-FeOOH}$, in high amounts should be avoided:

c) the presence of $\alpha\text{-FeOOH}$ (goethite) is very useful, due to its endothermic decomposition in a range of temperature where Fe_2O_3 reduction has already started.

From the same Table it is also possible to conclude that the reduction of this catalyst should be carried out more by H_2 than by CO.

However, about the composition of the reducing gas, there is another important argument to be taken into account, namely the possibility of Fe_3O_4 reduction to Fe (this should be carefully avoided, as metallic Fe catalyzes the highly exothermic reaction of methanation). To eliminate such risk, a sufficient amount of steam must be added to the reducing gas to have a H_2O/H_2 ratio higher than the equilibrium one at the working temperature (for instance $(H_2O/H_2)_{eq} = 0.09$ at 400°C). Steam has also the function of balancing the higher reducing power of CO with respect to H_2 for Fe_3O_4 (this is a second reason for having a low CO content in the reducing gas).

In addition to the above considerations, based on thermodynamic data, the reduction procedures recommended by catalyst manufactures re also on other factors, like operating restraints typical of the plan and ease of operation. In practice, they strongly depend on the reduction conditions of the reforming catalysts. We will report here the main stages of the reduction procedure of the Montedison MHTC catalyst in a shift converter following a modern steam reformer.

After having purged the converter with nitrogen, the catalyst is heated firstly in nitrogen stream up to 150°C, then in superheated steam up to 300°C (both these feeds come from the reformer). The reduction starts at this temperature when a H_2-steam mixture comes from the reformer and continues with process gas (H_2-CO-CO_2-N_2), always containing much steam, up to 350°C (steam has also the function of increasing heat transfer). If during reduction sudden increases of temperature occur, it is necessary to lower the process gas content, or, exceptionally, to stop it completely until the temperature decreases below the recommended limit (500°C can be tolerated for short periods). However, with proper operation the maximum temperature is unlikely to exceed 400°C, with benefit for catalyst performances.

Other reduction procedures involve the use of wet process gas starting at lower temperatures. This may cause steam condensation in the catalyst bed, with possible damage for mechanical strength.

3.2 Low-temperature CO conversion catalyst

The reactions that may occur during reduction in the converter are reported in Table 3, together with their standard enthalpy values (the main components of commercial catalysts are CuO and ZnO, together with some amount of amorphous alumina).

TABLE 3

Possible reactions during reduction of low-temperature CO conversion catalyst

Reaction	ΔH at 25°C, Kcal/mole
$CuO + H_2 = Cu + H_2O$	− 20.7
$CuO + CO = Cu + CO_2$	− 30.6
$ZnO + H_2 = Zn + H_2O$	+ 25.4
$ZnO + CO = Zn + CO_2$	+ 15.5

The reduction of ZnO should not occur (ΔG = + 19.2 kcal/mole at 200°C); however, attention has been drawn to the possible formation of brass (refs. 3-4). In fact, this solid state reaction has a negative ΔG value that can compensate, especially at low Zn concentrations, the positive ΔG of ZnO reduction.

The low-temperature CO conversion catalyst requires very careful reduction to give its best performances. Therefore the reduction procedure does not appreciably depend on the characteristics of the plant, contrary to that of the high temperature conversion catalyst. In fact, it is performed independently of the start-up of both reforming and high temperature shift converters. We will report here the main features of the reduction procedures recommended by some manufacturers.

After having purged the catalyst free from air with an inert gas (nitrogen or natural gas), the temperature is raised, always in inert gas stream (space velocity 500-1000 h^{-1}) to 70-120°C. Then hydrogen is gradually added up to 0.3-0.5 mole % and later the temperature is slowly increased up to 180°C. At this temperature the hydrogen content is raised to 1.5-2 mole % in 3-5 hours. The temperature is slowly increased up to 200-220°C and finally the hydrogen content is raised to about 10 mole%.

All manufacturers stress that it is necessary to decrease the hydrogen feed or even to stop it completely at any time, should any sudden increase of catalyst temperature occur. It may be useful to remark that the use of steam, to obtain a better temperature control, is not recommended, because it seems to have a negative influence on catalytic activity.

3.3 Ammonia synthesis catalyst

The main reaction occurring during the reduction of the ammonia synthesis catalyst is

$$Fe_3O_4 + 4H_2 = 3Fe + 4H_2O \qquad \Delta H = + 35.8 \text{ Kcal}$$

Small enthalpy differences, due to the presence of foreign cations (Al, Mg, Ca) in the magnetite lattice and of small amounts of other iron oxides in the catalysts (α-Fe_2O_3, FeO, ferrites), can be neglected, as a first approximation.

Owing to its endothermicity, this reduction is performed in the plants following rather different guidelines with respect to the two CO conversion catalysts. A powerful electric heater is needed to startup the reduction and to sustain it almost till its end. To decrease power consumption, the reduction is usually performed with synthesis gas (H_2-N_2 3:1), not with pure H_2, to take advantage of the exothermicity of ammonia synthesis. After the synthesis reaction has started, a careful temperature control is obviously needed.

The most important factor that influences the final catalytic properties is the water concentration in the gas in contact, during the reduction process, with the already reduced catalyst. In fact, owing to the high reactivity of iron surface towards oxygenated compounds, repeated oxidation-reduction reactions occur in water-containing hydrogen atmosphere with consequent recrystallization and growing of iron crystallites. To minimize such phenomena, the following operating criteria are adopted, when compatible with plant's characteristics:

- low reduction rate
- high space velocity
- low pressure
- effective drying of feed gas
- small catalyst granules

In any case, the measurement of the water content of the exit gases has to be done, as catalyst manufacturers usually specify the maximum value allowed. As an example, we will report here the main directions for the reduction of the Montedison MSFN catalyst in a multibed converter.

After having purged the converter free from air, the catalyst temperature is slowly increased by means of the heater in a stream of synthesis gas at 70-100 Atm. The temperature of the first bed is increased over 300°C, while those of the other beds are kept below this value. When the catalyst in the first bed has started its reduction, the operating conditions (mainly synthesis gas space velocity and heat supply) are controlled in such a way as to have a water content between 2000 and 3000 ppm in the exit gas. As an additional restraint, the bed temperature should not exceed 480°C. When the first bed catalyst is almost completely reduced (80-90%), the other beds are reduced in turn in the same way. At the end of the reduction stage the heater is taken out of service and the pressure is slowly increased until the normal working value is reached. During this operation the reduction of the residual unreduced magnetite slowly occurs.

3.4 Relation with applied laboratory-scale research

From the above-reported description of the reduction procedures of some industrial catalysts, it is clear that it is very difficult, if not impossible, to strictly define the operating conditions at the catalyst level. To justify this statement, let us summarize the main features of the reduction of the above-mentioned catalysts.

In the more complex case (high temperature CO conversion) three different chemical processes (without considering the physical ones) are simultaneously involved:

- hydroxide decomposition
- oxide reduction
- CO conversion

In the intermediate case (ammonia synthesis) we have simultaneously:

- oxide reduction
- ammonia synthesis

Even in the simpler case (low temperature CO conversion), where only reduction takes place, the reduction is performed with both temperature and hydrogen partial pressure variable over a wide range.

In fact, the reductions of the industrial catalysts are often performed under conditions varying very irregularly, with the aim ensuring that temperature and/or water concentration in the bed are under control. An additional complication is given by the use of high pressures

The difficulties of reproducing on a laboratory scale such complex processes and of obtaining meaningful data have caused a marked lack of applied research on this subject. What is usually done is the comparison of the catalysts in the same experimental conditions, obviously chosen in such a way as to obtain results that can be easily interpreted. Now we will briefly discuss the experimental techniques used for these investigations.

4. EXPERIMENTAL TECHNIQUES FOR RESEARCH ON CATALYST REDUCTION

4.1 Thermogravimetry (TG)

As catalyst reduction involves usually oxygen abstraction, apart from other simultaneous decomposition processes, like dehydration, thermogravimetry is the most suitable technique, and the most widely used indeed, to get information useful both in applied and basic research.

The modern commercial thermogravimetric balances have eliminated most of the experimental difficulties commonly encountered by researchers until about 20 years ago. Ease of use and high sensitivity (up to 10^{-7} g) make now feasible investigations previously very difficult. However one must be completely aware of the many experimental variables that may strongly influence the results obtainable with a thermogravimetric balance. Such experimental variables can be summarized as follows:

- geometry of the balance
- shape of the pan
- mass and shape of the sample
- flow rate of the reducing gas (pure or mixed with inert gas)
- rate of temperature increase
- way of reaching constant temperature (for isothermal runs).

It may be seen that only variables connected with the actual experiment have been considered, while other important factors can be detected in the previous history of the sample. Obviously, we do not intend to deeply discuss here such thermogravimetric problems; however, some typical results will be reported later.

We would recall here some difficulties connected with the possibility of using thermogravimetry in conditions simulating the actual reduction of some industrial catalysts, namely that commercial thermogravimetric balances are usually not qualified for work under pressure and/or with high water concentrations in the reducing gas.

Very often it is necessary to obtain additional information about the processes occurring during catalyst reduction; this goal can be reached by physical methods of analysis of the gases leaving the thermobalance. Connections of thermobalances with quadrupole mass spectrometers, gas chromatographs, infrared spectrometers, etc. have been already effected (ref. 5). They are strongly recommended when processes different from reduction and from dehydration may simultaneously occur.

We will limit our discussion to the collection of thermogravimetric data, which can be obtained by either dynamic heating or isothermal method. In the case of catalyst reduction studies, the former is usually employed for fast characterization of the reduction process, mainly on a comparative basis. Mathematical methods of data treatment have been developped for the dynamic heating method (ref. 6); all these are based on the equation

$$\frac{d\alpha}{dT} = \frac{Z}{q} e^{-E/RT} (1-\alpha)^n$$

where α = degree of transformation, T = temperature (°K), Z = frequency factor, q = rate of temperature increase, E = activation energy, n = reaction order.

However, the physical meaning of the kinetic parameters obtained by such treatments of the experimental data is very uncertain, so that it is probably preferable to rely on the data immediately available from the thermogram, namely starting reduction temperature (T_s), maximum reduction rate (R_m) and its corresponding temperature (T_m). The best data can be obtained if the differential thermogravimetric (DTG) curve is also available, as in most modern instruments. We stress that, if a correct comparison among different catalyst samples is to be done, the previously mentioned experimental factors must be kept strictly constant.

The isothermal method is more frequently used for basic studies on reduction mechanisms. The experimental result consists in a curve of degree of transformation versus time at constant temperature. The experimental curve must be fitted to a theoretical one obtained on the basis of reasonable hypotheses concerning the reduction mechanism.

Various models are presently available for the mathematical treatment of isothermal thermogravimetric data (ref. 7). Though they are valid for any gas-solid reaction occurring with loss or increase of weight, we will briefly discuss them only in the aspects concerning oxide reduction. All these models assume that the nucleation process, namely the initial formation of germs of the metallic or reduced oxide phase, is very fast with respect to the progress of the reduction, so that the presence of an induction period cannot be explained.

The simplest model, therefore more frequently used, is the unreacted shrinking core model. It assumes that the sample particle is completely non-porous, that the reduction occurs at the sharp interface between already reduced and unreduced solid and that in the particle, of simple geometric shape, there are no temperature gradients. In this very favourable case one can distinguish between reductions controlled by reaction at the phase boundary or by ash diffusion (however a shift from one controlling step to another usually occurs when temperature is changed, with possible existence of a wide range of intermediate states).

If the catalyst particles are porous and internal diffusion is very fast with respect to reduction, the homogeneous model should be applied, where reduction is visualized as occurring throughout the solid phase. However the mathematical treatment is very complex, so that some simplifications have been introduced. In the best known simplified model the particle is considered like an aggregate of small spheres, each of which is reduced according to the unreacted shrinking core model.

In the cases where the particle is porous and the reduction is fast with respect to ash diffusion, a modification of the unreacted shrinking core model has been developped, assuming that the reaction zone has a finite thickness.

For all these models more complex cases have been considered, assuming, for instance, irregular shape of the particle, temperature gradients inside and outside the particle, various reaction orders, particle contraction or expansion during reduction, various adsorption mechanisms of the reducing gas, more than one reducing agent, influence of particle size distribution, formation of cracks during reduction, etc.

The increasing complexity of the data treatment typical of these more sophisticated models may sometimes discourage the catalysis researcher; furthermore, he is aware that the conditions for the validity of the various theoretical models are often very far from the actual reduction procedures used for the industrial catalysts (for instance, the frequently occurring retarding effect of water is not taken into account). However, this seems to be the main road along which one can hope to reach, soon or later, a comprehensive picture of the complex phenomena occurring during the reduction of industrial catalysts, with consequent benefit for the rational optimization of this process.

4.2 Temperature programmed reduction (TPR)

This technique has been recently developped by Shell researchers (ref. 8). The principle, on which TPR is based, can be summarized as follows. If a hydrogen-inert gas (except helium) mixture (H_2 usually about 5%) is allowed to pass, successively, over the first detector of a thermal conductivity cell, through the catalyst sample placed in a furnace connected to a linear temperature programmer, through an absorption trap able to collect any gaseous product of catalyst reduction, finally over the second detector of the T.C. cell, and the sample temperature is linearly increased, when hydrogen will be consumed owing to catalyst reduction, the T.C. cell will be unbalanced and a signal, proportional to the difference of hydrogen concentration between inlet and outlet of the sample, will be received by the recorder. If the gas flow is constant, such difference is proportional to the rate of hydrogen consumption, namely to the rate of catalyst reduction.

In practice, the TPR profile is quite similar to that arising from DTG, but with the great advantages that the effect of any other reaction simultaneous with reduction is eliminated and that, as the reducing gas passes through the sample, not over, like in thermogravimetry, the elimination of external diffusion effects is much easier. Moreover the sensitivity of TPR measurements is often higher than that of thermogravimetric measurements and, last but not least, the required equipment is relatively inexpensive.

On the other hand, some difficulties may arise if accurate quantitative measurements are needed, as the calibration procedures are still open to objections (according to Shell researchers an accuracy of $\pm$ 10% can be reached). Moreover, like thermogravimetry, runs at high pressures and high water concentrations are hardly feasible.

As TPR is a relatively recent technique, the influence of the various operating variables on the TPR data is still to be investigated. When a deeper knowledge of the capabilities of this technique will be available, it will surely represent a powerful tool to obtain information on catalyst reduction mechanisms.

4.3 Other techniques

While at present thermogravimetry and temperature programmed reduction are undoubtedly the most widely used techniques in research on catalyst reduction, in special cases the techniques hereafter mentioned may find application.

<u>Differential thermal analysis</u> (DTA). This well-known technique may give qualitative information in addition to the thermogravimetric data, if one can take advantage of the modern instruments simultaneously monitoring both thermogravimetric and DTA curves (however, when precise DTA data are required, the use of an indipendent DTA equipment is recommended). In fact, when complex simultaneous phenomena occur during catalyst reduction, the comparison of thermogravimetric and DTA curves may help to give an insight into the whole process.

DTA could be useful also to detect the total thermal effect of the reduction process in industrial conditions, however, as in the case of thermogravimetry, only appropriately built equipment can withstand high pressures.

<u>Magnetic measurements</u>. When the non reduced and/or the reduced phases include ferromagnetic or ferrimagnetic materials (Fe, Co, Ni, Fe_3O_4, etc.), one can take advantage of the possibility of measuring the saturation magnetization as a function of time to follow the reduction process with high sensitivity. However this technique has been scarcely used.

<u>Mössbauer spectroscopy</u>. The use of this powerful technique in catalyst reduction is practically confined to iron-containing catalysts. The main advantage of Mössbauer spectroscopy with respect to thermogravimetry and TPR is that it allows one to distinguish simul-

taneous reduction processes due to different oxides. On the other hand, as the recording of each Mössbauer spectrum takes some hours, it is necessary to cool the sample for the measurement, so that the reduction process is discontinuous.

5. REDUCTION MECHANISMS OF SOME INDUSTRIAL CATALYSTS

The title of this chapter is probably too ambitious, as, owing to the previously mentioned inadequate knowledge on catalyst reduction carried out in industrial conditions, only the information coming from basic research, carried out very often on single catalyst components and in easy experimental conditions, is available at present. Yet some interesting conclusions can be drawn from such basic knowledge. In addition, we will report hereafter the results of reduction characterization of some industrial catalysts.

5.1 High temperature CO conversion catalyst

As we have previously seen in Chapter 3.1, the activation of this catalyst involves mainly the reductions of γ-Fe_2O_3 and α-Fe_2O_3 to Fe_3O_4. The mechanisms of these reductions have been studied in Montedison some years ago (ref. 9). Isothermal thermogravimetric runs were performed in the temperature range 250-350°C in pure H_2 and H_2-He mixtures. Samples were in the form of fine powders, so that the influence of ash diffusion was minimized (according to ref. 10 it begins above 8000 Å particle size). Qualitative and quantitative analyses of the crystal phases in samples taken at various degrees of reduction were carried out by X-ray diffraction.

As concerns the reduction of γ-Fe_2O_3, it occurs through the solid solution $Fe_{3-x}O_4$ ($0 \leqslant x \leqslant \frac{1}{3}$) without phase change. The kinetic equation is of the type $\ln(1-\alpha) = -Kt$, with an activation energy of 29 Kcal/mole. Two simultaneous reduction mechanisms are involved:

a) Fe^{2+} ions formed on the surface migrate through the lattice to occupy vacant octahedral sites;

b) electron trasfer from Fe^{2+} ions on the surface to Fe^{3+} ions in the bulk, therefore regenerating Fe^{3+} ions on the surface and allowing the reduction to continue.

More important for our purposes is the reduction of α-Fe_2O_3, which is usually the main component of the high temperature CO conversion catalysts. It occurs with different mechanisms, depending on the amount of lattice disorder present in the sample. In highly-disordered samples α-Fe_2O_3 is firstly completely reduced to Fe_3O_4, which is later reduced to Fe. For ordered samples (natural or treated at high temperature) the reductions of α-Fe_2O_3 to Fe_3O_4 and of Fe_3O_4 to Fe simultaneously occur. In both cases at the beginning of the reduction an outer shell of Fe_3O_4 is formed, with a degree of disorder similar to that of the parent α-Fe_2O_3 phase. As the reduction of α-Fe_2O_3 proceeds through the migration of Fe^{2+} ions and electrons from the outer surface to the Fe_2O_3-Fe_3O_4 interface and as this migration is easier in the case of disordered lattice, in well-ordered samples the reduction of the outer spinel phase begins when a large amount of α-Fe_2O_3 is still to be reduced. In the case of the high temperature CO conversion catalysts the lattice disorder is obviously high (the surface area is of the order of 100 m^2/g), so that the risk of Fe formation is minimized.

Information about the influence of water vapour on the reduction of α-Fe_2O_3 to Fe_3O_4, though very important for industrial reductions, is scanty. According to a recent paper (ref. 11), the reduction rate is lowered by water vapour for Mg- and Li-doped α-Fe_2O_3, while for pure α-Fe_2O_3 it is firstly decreased, then increased. On this subject research work concerning the high temperature CO conversion catalyst is clearly needed.

Now we will discuss some results obtained during our investigations on the reduction behavior of the Montedison MHTC catalyst. In Table 4 we report some TPR data concerning laboratory preparations of the high temperature CO conversion catalyst, as function of both Cr_2O_3 concentration and method of preparation.

The method of preparation of all the samples reported in Table 4 is based on precipitation from $FeSO_4$ and Na_2CO_3 aqueous solutions, while chromium is added as Cr^{VI}. The following main conclusions can be drawn from Table 4 (samples A-B-C):

TABLE 4

TPR data of high temperature CO conversion catalysts (laboratory preparations).

Sample	Cr_2O_3 (%)	Shoulder temp.(°C)	α-Fe_2O_3 red.		Fe_3O_4 red.
			T_s (°C)	T_m (°C)	T_s (°C)
A	-	-	195	335	342
B	7.5	264	196	338	374
C	7.5	270	197	325	373
D	7.5	-	221	350	433

The definitions of T_s and T_m have been given in chapter 4.1

a) the presence of Cr in the catalyst has practically no influence on the reduction of α-Fe_2O_3 to Fe_3O_4, while the reduction of Fe_3O_4 to Fe is slightly hindered (samples A and B have been prepared by the same method);

b) as the shoulder on the α-Fe_2O_3 reduction peak is present only in Cr-containing samples, it has been attributed to the reduction of Cr^{VI} to Cr^{III} (this has been confirmed by chemical analysis, that gives a Cr^{VI} concentration of 0.5-1% in samples B and C).

Moreover, the data obtained for sample D show that it is possible, through special methods of catalyst preparation, not only to eliminate the harmful presence of Cr^{VI}, but also to obtain such a good distribution of Cr^{III} in the Fe oxides as to delay both α-Fe_2O_3 and Fe_3O_4 reductions, the latter by about 90°C. It is expected that a better distribution of Cr^{III} in the catalyst should appreciably improve its life.

Some data concerning the reduction behavior of the Montedison industrial catalyst MHTC are reported in Table 5.

The good agreement of TPR and TG data is firstly to be noticed (compare T_m α-Fe_2O_3 red. and T_s Fe_3O_4 red. of runs 1 and 2). The attribution of the α-FeOOH decomposition peak comes immediately from comparison of run 1 with run 2 and of runs 2-3 with run 4. The determination of T_s α-Fe_2O_3 red. in TG runs was not possible owing to overlapping with the α-FeOOH decomposition peak (see the TPR value of T_s α-Fe_2O_3 red. and the TG values of T_m α-FeOOH dec.). The in-

TABLE 5

Reduction data of the Montedison catalyst MHTC (size 0.2-0.3 mm)

Run	Technique	Gas	α-FeOOH dec.		α-Fe$_2$O$_3$ red.		Fe$_3$O$_4$ red.	
			T_s (°C)	T_m (°C)	T_s (°C)	T_m (°C)	T_s (°C)	T_m (°C)
1	TPR	H$_2$ 5% N$_2$ 95%			195	325	370	
2	TG	H$_2$ 5% N$_2$ 95%	125	235		330	380	640
3	TG	H$_2$	125	225		290	315	385
4	TG	H$_2$	130	225				

The rate of temperature increase was 2°C/min for all the runs. The definitions of T_s and T_m have been given in chapter 4.1.

fluence of H$_2$ concentration on the reduction of α-Fe$_2$O$_3$ to Fe$_3$O$_4$ and especially on the reduction of Fe$_3$O$_4$ to Fe is quite evident (the same phenomenon was reported in ref. 9). From the relatively low values of T_s Fe$_3$O$_4$ red. it is clear the need of excess water vapour during industrial reductions, where catalyst temperature may well exceed 400°C.

Finally, we will notice the strictly similar reduction behavior of laboratory and industrial high temperature CO conversion catalysts, when prepared by same method (compare sample B of Table 4 with run 1 of Table 5). The only difference is that no shoulder due to the presence of Cr^{VI} was detected in the industrial catalyst.

5.2 Low temperature CO conversion catalyst

As it was shown in chapter 3.2, the activation of this catalyst involves the reduction of CuO to Cu. CuO and ZnO are practically the only crystalline phases detected in the commercial Cu-Zn-Al oxide catalysts. However, in some of them the presence of small amounts of the following crystalline compounds can be detected by X-ray diffraction: γ-alumina, zinc or copper aluminate, basic zinc carbonate, basic copper carbonate (the presence of the two last compounds is clearly due to low calcination temperature).

The reduction degree of ZnO at temperatures below 230°C should be negligible; in fact, the formation of brass reported in ref. 3 occurs in appreciable amount at temperatures higher than

250°C, while at lower temperatures it requires a very long time. Therefore during catalyst activation, unless the catalyst is incautiously overheated, the reduction of ZnO can be neglected.

The reduction of pure CuO has been studied by many Authors; Bond (ref. 29) and Schoepp (ref. 30) seem to have performed the most comprehensive thermogravimetric investigations. The method of preparation was found to have a strong influence on the reduction behavior. The reduction kinetics were satisfactorily interpreted in terms of both instantaneous and constant-rate nucleations (ref. 31).

Though the basic knowledge concerning the reduction of pure CuO may be an useful starting-point, we clearly need information about the reduction of CuO dispersed over a $ZnO-Al_2O_3$ mixed carrier. This kind of information is almost completely lacking. We are aware only of a Bulgarian DTA study (ref. 32) on the reduction of two commercial catalysts not containing alumina ($CuO-ZnO$ and $CuO-ZnO-Cr_2O_3$). The starting reduction temperatures were, respectively, 130°C and 150°C with hydrogen concentration 2.6% and heating rate 6°C/min. The CuO reduction was found to occur in two stages: firstly to Cu_2O, then to Cu. However, from our TG and TPR experiments we never obtained any evidence of a two-stages reduction process.

We are investigating the influence of various factors concerning catalyst composition and preparation procedure on the reduction behavior. Some preliminary results are reported in Table 8.

TABLE 8

TPR data of pure and supported CuO catalysts.

Sample	Chemical composition	T_s (°C)	T_m (°C)
S	CuO	180	250
LTC 4	CuO	110	198
LTC 6	CuO-ZnO	140	200
LTC 1	$CuO-ZnO-Al_2O_3$	157	202

The definitions of T_s and T_m have been given in chapter 4.1

While sample S has been prepared by thermal decomposition of $Cu_2(OH)_2CO_3$, the preparation of sample LTC 4 has been performed with the procedure used for the Montedison catalyst. It may be noticed that the strong influence of the preparation method on the reduction behavior of CuO, already reported by Bond (ref. 29), has been confirmed. By comparison of the TPR data of the samples of the LTC series, all prepared with the same procedure, it is possible to conclude that the retarding effect of ZnO addition concerns only the nucleation of metallic Cu. A further retarding effect on Cu nucleation is given by the addition of Al_2O_3.

The reduction behavior of industrial catalysts is also under investigation in our laboratories. Some preliminary results are reported in Table 9.

TABLE 9

TPR data of some low temperature CO conversion catalysts.

Catalyst	T_s (°C)	T_m (°C)
Montedison	158	218
A	127	177
B	105	158
C	129	195

The definitions of T_s and T_m have been given in chapter 4.1. Appreciable differences exist among the values of T_s and T_m of the examined catalysts. Obviously, as the preparation method of catalysts A, B and C is not known, we cannot draw any conclusion from the above-reported data, which only suggest the need of further studies. However, we were successful in preparing a catalyst of lower reducibility; we think that this feature should make easier the temperature control during industrial reduction.

5.3 Ammonia synthesis catalyst

As we have seen in chapter 3.3, the activation of this catalyst occurs by the endothermic reduction of Fe_3O_4 to Fe, which proceeds without intermediate phases (FeO is not stable below 570°C).

As the catalyst (promoted fused magnetite) is practically non-porous, the unreacted shrinking core model could be used to handle the isothermal thermogravimetric data. Unfortunately the catalyst particle shape is very irregular, while the model is strictly valid only for spherical particles of uniform size; a mathematical treatment for various spherical particle size distributions is however available (ref. 12).

The classical unreacted shrinking core model has been applied to the reduction of a commercial ammonia synthesis catalyst by some Polish researchers (ref. 13). Cube-shaped catalysts particles were used. It was found that the reduction process proceeds in the intermediate region between chemical reaction and ash diffusion controls. However, the temperature range investigated (450-550°C) is at the upper limit of the usual reduction temperatures of the ammonia synthesis catalysts (some of them begin to reduce below 350°C). The same Polish group (ref. 14) developped a mathematical model for the constant-rate reduction performed at temperatures slowly increasing. Though the starting temperature chosen is rather high (400°C) and the total reduction time too short (15 hours), and though the interactions among neighbouring particles were not considered, this work represents the first attempt towards the mathematical modeling of catalyst reduction in industrial conditions.

However, it is our opinion that the basic knowledge of the reduction mechanisms of promoted magnetites is still too incomplete to allow a sound development of the applied research. With this concept in mind, we have carried out some investigations with the aim of clarifying the basic phenomena occurring during magnetite reduction. The problem was attacked through an isothermal thermogravimetric study (ref. 15) of pure magnetite reduction in the low-temperature range (220-330°C) that should allow both to minimize the influence of ash diffusion and to work with the low water concentrations required in industrial reductions. The experiments were carried out by means of a sensitive electrobalance; powdered magnetite with particle size in the range 0.5-3 μm was used (according to ref. 16 the influence of ash diffusion should be negligible with particle size below 200 μm).

It was found that the reduction rate decreases with the increasing of sample mass in the range 0-30 mg; it also decreases when water vapour (1500 vpm) is added to the reducing gas (pure hydrogen). This behavior can be explained in terms of interparticle diffusion phenomena. In fact, as in TG experimental conditions there is no hydrogen flow through the powder, a water concentration gradient may easily arise between the interior of the sample and the hydrogen stream. In other words, the magnetite heap tends to react as a whole grain from the outer layers towards the interior. This is confirmed by the presence of many unreacted particles in samples reduced up to 65%. However, this interparticle diffusion should be considered more properly, as suggested in ref. 17, a surface diffusion consisting, owing to the high reactivity of finely divided iron towards oxygenated compounds, of repeated oxidation by water-reduction by hydrogen steps. This is also in accordance with the relatively high value of the activation energy (14.5 kcal/mole). It may be observed that the situation occurring in these experiments (interaction of reduced particles with water vapour coming from the neighbours) reminds what occurs, on a larger scale, during reduction in industrial conditions, where, however, also ash diffusion is to be taken into account. Unfortunately no mathematical model is available, as yet, to describe such a complex behavior.

It was also found that an induction period is always present if dry hydrogen is used and that it disappears if water vapour is added. As mixing of magnetite with Pt black has no influence on the induction period, the rate-determining step in the initial stage of the reduction should not be the formation of metallic germs, but rather the hydroxylation of the oxide surface. Surface hydroxyls should indeed make easier the migration of dissociated hydrogen (ref. 18).

The substitution of Fe^{3+} with Al^{3+} ions in magnetite decreases the reduction rate and makes the induction period to disappear (ref. 19). The latter fact could be related to a higher amount of hydroxyls connected to Al, the former to the formation, during reduction, of $FeAl_2O_4$ groups, unreducible in our experimental conditions, that could decrease the advancing of the reacting interface. In fact, it was demonstrated that $FeAl_2O_4$ groups included in the lattice of α-Fe occur in the reduced samples (ref. 20).

We have also carried out isothermal thermogravimetric runs to measure the reduction rate of some commercial catalysts on samples of different particle size (1-5 μm and 0.3-0.4 mm). The influence of sample mass on the reduction rate was much lower in the case of 0.3-0.4 mm samples. It means that for these samples interparticle diffusion is negligible with respect to the intraparticle one. The induction period, like in the Al-containing magnetites, is not present. Values of the reduction rate $K_{0.5}$ (slope of the reduction curve at 50% conversion) of some commercial catalysts (particle size 0.3-0.4 mm) are reported in Table 6.

TABLE 6

Reduction rate of some commercial catalysts for ammonia synthesis.

Catalyst	$K_{0.5} \cdot 10^3$ (min^{-1})	Experimental conditions	
MSFN	1.63	Temperature	345°C
A	0.68	Sample mass	20 mg
B	1.40	Reducing gas	pure H_2
C	3.15		

It may be seen that appreciable differences exist among the examined catalysts, though the chemical composition is practically the same. This fact shows the importance of the preparation method, whose details obviously are not disclosed by manufacturers, with respect to the reduction process. From a practical point of view, a very low reducibility makes easier the control of water concentration in the bed, but may prolong too much the reduction period, while the opposite holds if the reducibility is very high.

For a better characterization of the reduction behavior of commercial catalysts for ammonia synthesis, we have also carried out dynamic heating thermogravimetric reduction runs. The results obtained are reported in Table 7. There are some catalysts (MSFN, F, H) that begin to be reduced, in our experimental conditions, at about 350°C, while for others (D, E, G) the starting reduction temperature is about 380°C. In the DTG profiles of the former group a shoulder centered at about 420°C is present, which is completely absent in those of the latter. As in catalysts MSFN, F and H a small amount of FeO

TABLE 7

TG reduction data of some commercial catalysts for ammonia synthesis.

Catalyst	T_s(°C)	T_m(°C)	$R_m \cdot 10^3$ (min^{-1})
MSFN	345	590	9.9
D	385	575	12.3
E	375	570	13.8
F	345	580	9.5
G	380	595	9.9
H	355	600	9.8

Experimental conditions: Rate of temperature increase 5°C/min
Reducing gas pure H_2
The definitions of T_s, T_m and R_m have been given in chapter 4.1.

is present (estimated 1-5% from X-ray diffraction powder pattern) and as FeO is usually reduced at temperatures lower than Fe_3O_4, the lower T_s values of catalysts MSFN, F and H can be explained. This is confirmed by the results obtained by other Authors (ref. 21), who studied the reduction of FeO-containing ammonia synthesis catalysts by Mössbauer spectroscopy. These Authors found also different reduction mechanisms depending on FeO content (4% and 27%).

If the R_m values reported in Table 7 are considered, it may be noticed that the catalysts D and E are more easily reducible than the others (the same holds for catalyst C in Table 6). We have found that in the less easily reducible catalysts FeO and $CaFe_3O_5$ are present (these two compounds are always present together, as shown also in ref. 22), while in catalysts C, D and E two unidentified crystal phases (probably complex ferrites) have been detected. The same two phases have been found in a catalyst prepared in our laboratory by very fast cooling of fused promoted magnetite. This is a clear indication of the influence that the method of preparation may have on the reduction behavior of ammonia synthesis catalysts.

The final degree of reduction has been reported (ref. 23) to be a crucial point for the activity of ammonia synthesis catalysts. For promoted catalysts the iron specific activity always increases with the reduction degree, but, due to the possible occurring of sintering phenomena, the catalytic activity may reach a maximum value, then decrease. For a better understanding of this problem, a discussion of the structure of the reduced catalyst is certainly useful. According to our (refs. 20 and 24) and Hosemann and coworkers (refs. 25-26) results, in alumina-containing reduced catalysts $FeAl_2O_4$ groups are endotactically built into the α-Fe lattice (one $FeAl_2O_4$ group takes the place of seven Fe atoms), causing paracrystalline distortions. Other Authors (ref. 27) suggest that Al_2O_3, not $FeAl_2O_4$, is present in samples reduced over 400°C. However, the X-ray diffraction lines of $FeAl_2O_4$ were clearly detected by us after high temperature annealing in inert atmosphere of samples reduced at 500°C. Moreover, the presence of Al_2O_3 groups was excluded on the basis of SIMS experiments (ref. 28).

Another debated argument is the constitution of the $FeAl_2O_4$ groups (are single groups or clusters, and in the latter case what is their size ?). Hosemann has recently demonstrated (ref. 26) that the presence of paracrystallinity can be explained only if no more than two $FeAl_2O_4$ groups are joined together. On the contrary, other Authors (ref. 27), on the basis of Mössbauer spectroscopy data, stated that the size of the inclusions in the α-Fe lattice, either $FeAl_2O_4$ or Al_2O_3, should be at least 30 Å. We too suggested the presence of small clusters of $FeAl_2O_4$ (ref. 20). However, if according to Hosemann's data (ref. 26), we assume an average size of 15 Å for each $(FeAl_2O_4)_2$ cluster, in the case of our alumina-promoted samples about 250 clusters are included in each α-Fe crystallite. In this circumstance it may be calculated that about 3% of the Fe atoms are contacting the $(FeAl_2O_4)_2$ clusters, while the remaining 97% is in the same structural configuration as in pure α-Fe. Such a small percentage could not have a detectable influence on both Curie point and Mössbauer parameters of α-Fe. Moreover, Hosemann questions (ref. 26) the parallelism between Fe-Al alloys and reduced alumina-promoted Fe catalysts, on which other Authors (ref. 27) have based their conclusions. If we take into account the structure of Al-containing magnetites, where the Al^{3+} ions are randomly distributed in the spinel lattice, and also the reduction mechanism (shell by shell from the external surface towards the interior), for reduction temperatures below 500°C it does not seem reasonable to assume that a group like $FeAl_2O_4$ can easily diffuse in the α-Fe lattice to reach other groups and form large clusters. This may be another argument in support of Hosemann's results.

The disorder introduced in the alumina-cointaining catalysts through their reduction has also some relevancy with their catalytic activity. In fact, we have found (ref. 24) that the specific activity of iron increases in the presence of alumina, which is not therefore a structural promoter only.

While the alumina-containing catalysts have been studied rather deeply, this is not the case of the industrial catalysts, which contain also K, Ca, Mg and Si oxides. Clausen et al. (ref. 21) found evidence of some unreduced Fe compound in their Mössbauer spectra of catalysts reduced at 550°C and suggested that it could involve Ca. In fact, we have detected $CaFe_3O_5$ in some reduced commercial catalysts. It is to be expected that interesting information could be obtained on the mechanisms of the reduction process and its correlations with catalytic activity if thorough studies were accomplished on magnetites promoted with the various oxides used in the industrial practice.

CONCLUSIONS

After having discussed the industrial procedures for the reduction of some commercial catalysts (high and low temperature CO conversion and ammonia synthesis), the experimental techniques for laboratory-scale investigations on the reduction process and the currently available basic knowledge on the reduction mechanisms, we will summarize here the present status of the problem.

The degree of knowledge of the reduction mechanisms of the above-mentioned industrial catalysts should be defined insufficient. However, it is slightly better for the ammonia synthesis catalyst. As concerns the reduction mechanisms of the pure main compounds (α-Fe_2O_3, CuO, Fe_3O_4), the degree of knowledge is fairly good, however there is wide space for improvement. For instance, more information on the influence of the composition of the various reducing gas mixtures used in the industrial practice on the reduction rate of these pure compounds should be obtained.

For the complete knowledge of the reduction behavior of many commercial catalysts much is still to be done. Information is lacking about the influence of the other catalyst components on the reduction of the main compound (only the influence of Al_2O_3 on the reduction of Fe_3O_4 is fairly known). Also the influence of the gaseous phase composition is not sufficiently known, except for the effect of water vapour on the reduction of Fe_3O_4.

We have remarked the experimental difficulties that hamper the collection of meaningful data about catalyst reduction. Most of the published thermogravimetric data, though representing the main source of information, should be considered with much caution, as not all the experimental factors that may have a strong influence on the obtained results were previously clearly recognized. Temperature-programmed reduction seems to be a very promising technique, but it requires deeper studies before a reliable evaluation of its capabilities can be done. Many mathematical models are available to handle the kinetic data concerning oxide reduction in gas-solid systems, but their limitations, due to the need of avoiding mathematical difficulties, are still too many. However, some progress in this field is to be expected in the near future.

In the general remarks on the previous Symposium on catalyst preparation the editors of the proceedings noticed the almost complete lack of reports about catalyst activation. The present paper probably explains some reasons of such a disappointing fact. It is to be hoped that, starting with this Symposium, and still more in the forthcoming ones, catalyst activation will receive increasing attention in catalysis research.

REFERENCES

1 F. Traina and N. Pernicone, Chim. Ind. (Milan), 52(1970)1.
2 G. Berrebi and P. Bernusset, in B. Delmon, P.A. Jacobs and
   G.Poncelet (Eds.), Preparation of Catalysts, Elsevier, Amsterdam,
   1976, p. 13.
3 T. Van Herwijnen and W.A. de Jong, J. Catal., 34(1974)209.
4 G.Leherte, R.Derie and P.H. Duvigneaud, in B.Delmon, P.A. Jacobs
   and G.Poncelet (Eds.), Preparation of Catalysts, Elsevier,
   Amsterdam, 1976, p. 303.
5 I. Buzas (Ed.), Thermal Analysis, vol. 3, sect. 6, Heyden Son
   Ltd., London, 1975.
6 A. Blazek, Thermal Analysis, Van Nostrand Reinhold Co., London,
   1973, p. 61.
7 B.S. Sampath and R. Hughes, Chemical Engineer, (1973)485.
8 J.W. Jenkins, B.D. McNicol and S.D. Robertson, Chem. Tech., (1977)
   316.
9 U.Colombo, F. Gazzarrini and G. Lanzavecchia, Mater. Sci. Eng.,
   2(1967)125.
10 J.J. Heizmann, P. Becker and R. Baro, Mem. Sci. Rev. Metallurg.,
   70(1973)625.
11 R.P. Viswanath, B. Viswanathan and M.V.C. Sastri, React. Kinet.
   Catal. Letters, 2(1975)51.
12 H.G. Mc Ilvried and F.E. Massoth, Ind. Eng. Chem. Fundam.,
   12(1973)225.
13 A. Baranski, A. Bielanski and A. Pattek, J. Catal., 26(1972)286.
14 A. Baranski, A. Fulinski, A. Pattek and A. Reizer, Bull. Acad.
   Pol. Sci. Ser. Sci. Chim., 24(1976)729.
15 G. Liberti, G. Servi, A. Guaglio, C. Ruffino and N. Pernicone,
   J. Thermal Anal., 6(1974)183.
16 S.M. Loktev, I.P. Mukhlenov, I.F. Darovskikh, L.I. Zvezdkina,
   G.L. Yakovleva and E.I. Dobkina, Khim. Prom., (1970)108.
17 Z.P. Popovich and M.V. Tovbin, Ukr. Khim. Zhur., 36(1970)726.
18 J.E. Benson, H.W. Kohn and M. Boudart, J. Catal., 5(1966)307.
19 N. Pernicone, G. Liberti and G. Servi, in C. Eyraud and M. Escoubes
   (Eds.), Progress in Vacuum Microbalance Techniques, vol. 3,
   Heyden, London, 1975, p. 304.
20 G. Fagherazzi, F. Galante, F. Garbassi and N. Pernicone, J. Catal.,
   26(1972)344.
21 B.S. Clausen, S. Mørup, H. Topsøe, R. Candia, E.J. Jensen,
   A. Baranski and A. Pattek, J. Phys., 37(1976)C6-245.
22 P.D. Rabina, T.Y. Malysheva, L.D. Kuznetsov and V.A. Batyrev,
   Kinetics Catalysis (English transl.), 11(1970)1030.
23 I.V. Uvarova, M.T. Rusov and N.P. Samchenko, Kinetics Catalysis
   (English transl.), 10(1969)456.
24 N. Pernicone, G. Fagherazzi, F. Galante, F. Garbassi, F. Lazzerin
   and A. Mattera, in J.W. Hightower (Ed.), Catalysis, North-Holland
   Publ. Co., Amsterdam, 1973, p. 1241.
25 R. Hosemann, A. Preisinger and W. Vogel, Ber. Bunsenges. Physik.
   Chem., 70(1966)796.
26 H. Ludwiczek, A. Preisinger, A. Fischer, R. Hosemann, A. Schoenfeld
   and W. Vogel, J. Catal., 51(1978)326.

27 H. Topsøe, J.A. Dumesic and M. Boudart, J. Catal., 28(1973)477.
28 R. Buhl and A. Preisinger, Surface Sci., 47(1975)344.
29 W.D. Bond, J. Phys. Chem., 66(1962)1573.
30 R. Schoepp and I. Hajal, Bull. soc. chim. France, (1975)1965.
31 B. Delmon, Introduction a la cinetique heterogene, Editions Technip, Paris, 1969, p. 393.
32 D. Shishkov, D. Ivanov and G. Radoeva, J. Appl. Chem. URSS (English transl.), 44(1971)263.

DISCUSSION

Z.G. SZABO : You have projected the activity of ammonia synthesis catalysts as a function of the degree of reduction. Several years ago, we have found that the maximum activity of nickel reduced from NiO was also not at 100% reduction. Ni was more active if it was supported on its own oxide. Even this maximum changed if the NiO was doped with foreign ions, pointing to an electronic interaction between the metallic Ni and its supporting oxide. (cf. II International Congress on Catalysis, Paris, 1960). Can the unreduced Fe_3O_4 be considered as a kind of support ?

N. PERNICONE : I think that small amounts (max $\sim$ 5%) of unreduced oxide can hardly be regarded as a support for the metal. Though your remark about the Ni-NiO system is very interesting, in the case of Fe catalysts the activity maximum found at incomplete reduction is to be attributed to subsequent sintering phenomena.

S.P.S. ANDREW : You suggested that it is important that a low partial pressure of water is employed during reduction of NH_3 synthesis catalyst in order to obtain a high activity. There are important theoretical consequences depending on whether what is important is the absolute partial pressure of water or its relative partial pressure with respect to that of hydrogen. I favour the latter.

N. PERNICONE : I agree completely with you. In fact, I was speaking in terms of vpm of water in the gaseous phase (synthesis gas).

A.V. KRYLOVA : What do you think about the practical use of adding quantities of transition metal oxides to accelerate the reduction of the ammonia synthesis catalyst ? We added a same small quantity of NiO (0.8%) to the precipitated promoted iron catalyst prepared by two methods : 1) impregnation (on the surface) and 2) coprecipitatio (in the bulk). We found (figure below) that in the first case, the rate of the reduction (curve 2) and the activity of the catalyst are higher, while in the second case they are lower (curve 3) than for a sample without NiO (curve 1). In conclusion we think that the addition of NiO and some other transition metal oxides (for example CoO) may be useful but it is necessary to use a well-defined method for the addition of these oxides.

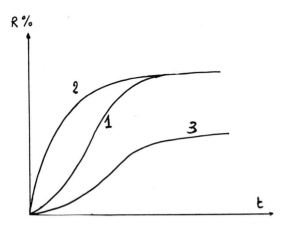

N. PERNICONE : If industrial catalysts are considered, there is no need for an acceleration of the reduction, because difficulties could arise to maintain a low water concentration. If desired, reduction can be initiated at lower temperatures in the presence of small amounts of FeO, as shown in Table 7. The results you have reported for Ni-promoted catalysts are very interesting, but are for a method of preparation (precipitation) which is not practiced industrially. As Ni^{++} ions can substitute Fe^{++} ions in the spinel lattice, both the extent and the order of this substitution may be influenced by the way of Ni addition and influence both reduction rate and activity of the promoted catalyst.

N. BAERNS : 1) As you mentioned, temperature programmed reduction is not feasible at high pressures. Could you comment on the main experimental obstacles to high pressure application of TPR ?
2) Can you comment on the rate law on which the rate constant is based, that you gave for the reduction of the ammonia catalyst ?

N. PERNICONE : 1) The difficulty of performing high pressure TPR is mainly due to the need of both strictly constant gas flow rate and uniform conditions at the two thermal conductivity detectors.

These requirements are not easy, though not impossible, to be fulfilled at pressure of, for instance, 100 atm.

2) The definition of $K_{0.5}$ is given in ref. 15. It is not a rate constant, but the experimental reduction rate at 50% conversion.

B. DELMON : You quite rightly pin-pointed the difficulties in studying the reduction of catalysts during activation. I completely agree with your wish that catalyst activation should receive increasing attention in catalysis research.

You suggested that investigators might hesitate to study the corresponding phenomena, because several reactions occur simultaneously.

However, it becomes increasingly clear that there is much interaction between simultaneously occurring reactions, namely a "coupling" of the processes (1).

In a field quite similar to catalyst activation, namely the dehydration-reduction of goethite to magnetite, we conclusively showed how the occurrence of one reaction has an effect on the other, the ultimate products having quite different structures and textures (2). There are coupling phenomena in the reaction of $CaCO_3$ with SO_2 and O_2 (3). We more than suspect that several coupling effects occur during the activation of cobalt-molybdenum hydrodesulfurization catalysts (4).

Because of the powerful action that we can exert on the texture and structure of the final catalyst by creation or suppression of coupling, I think that research in this area should be strongly encouraged.

(1) V.V. Boldyrev, M. Bulens, B. Delmon, The Control of the Reactivity of Solids, Elsevier, Amsterdam, in press.

(2) M.L. Garcia-Gonzalez, P. Grange, B. Delmon, in "Reactivity of Solids" (J.Wood, O. Lindqvist, C. Helgessan, N.G. Venneberg, Editors), Plenum Press, New York, 1977, 755-759.

(3) G. Van Houte, B. Delmon, Several papers submitted for publication.

(4) J.M. Zabala, P. Grange, B. Delmon, C.R. Acad. Sci. Paris, Série C, 279, 725-728 (1974).

N. PERNICONE : I am very glad for such forthcoming increase of research work on simultaneous chemical processes.

M. HOUALLA : You have reported that the influence of H_2O and Al_2O_3 is to decrease the induction period for the reduction of Fe_2O_3, which you relate to "spillover" phenomenon by the OH groups. Other studies on the reduction of NiO^x attribute an opposite role to H_2O. Did you try to assess your hypothesis by reducing Fe_2O_3 with carbon monoxide, for example ?

x See paper of B. Delmon and M. Houalla, in these Proceedings.

N. PERNICONE : From a comparison of the reduction behavior of Fe_3O_4 (refs. 15 and 19 of our paper) and NiO (R. Frety, L. Tournayan and H. Charcosset, Ann. Chim., 9 (1974), 341), the following conclusions can be drawn :
a) in both cases, water has an inhibiting effect on the progression of the reaction interface;
b) the initial reduction rate is increased by H_2O in the case of Fe_3O_4, but decreased in the case of NiO.
It is probable that the nucleation stage depends very much on the experimental conditions. We have not yet studied the reduction of Fe_3O_4 by CO.

H. CHARCOSSET : When studying the reduction of unsupported nickel oxide by hydrogen we found it very useful to carry out electron microscopic examination of partially reduced samples, having submitted them to selective extraction of the metal phase by means of a solution of Br_2 in CH_3OH. Have you applied such a method to iron catalysts and could you give some detail about the results, for instance, where was the iron located in a partially reduced iron oxide particle ?

N. PERNICONE : The metal phase in partially reduced magnetites was dissolved in 0.01 N HCl. This method gives as good results as the one you have mentioned. More details are reported in ref. 15. We have shown that the reduction of magnetite particles proceeds in an irregular way from a macroscopic point of view (see Fig. 7b of ref. 15).

THE ACTIVATION OF IRON CATALYST FOR AMMONIA SYNTHESIS

A. BARAŃSKI, M. ŁAGAN, A. PATTEK and A. REIZER
Jagellonian University, Cracow, Poland
and
L.J. CHRISTIANSEN and H. TOPSØE
Haldor Topsøe Research Laboratories, Lyngby, Denmark

SUMMARY

The activation of ammonia catalysts is essentially a reduction of non-porous magnetite particles with hydrogen to yield the internal porous structure of the iron catalyst. The reduction is studied by means of a "core-and-shell" model where the gas-solid reaction proceeds on the interface between the core and shell. Effective diffusion coefficients are calculated from experimental pore size distribution measurements. Different reaction rate expressions were tested on experimental reduction curves for cases with and without water addition, and it is shown that it is necessary to include the adsorption of water in the rate expression. Finally, some limitations in the model are described.

INTRODUCTION

The preparation and activation of an iron ammonia synthesis catalyst involve essentially two steps [1]. Iron oxides are first fused together with promoters such as Al_2O_3, CaO, and K_2O, to yield a non-porous "magnetite" with negligible surface area. This structure is subsequently activated by reduction to yield the active, highly porous structure with a surface area in the range 5-20 m^2/g.

In an earlier study [2] a model for this reduction process was proposed. According to this model a partly reduced catalyst is assumed to consist of an unreduced core of oxide surrounded by a porous shell of reduced material, and the reaction proceeds on the interface between the core and shell. The reactant H_2 and the product H_2O diffuse in and out through the shell.

Recent structure investigations [3,4] show that ammonia synthesis catalysts are not single phase systems and that the core-and-shell picture for the reduction process is not representative for all catalyst compositions. However, for catalysts with low wüstite contents the core-and-shell model was found applicable.

In the present study, the necessity of including the adsorption of water in the kinetic expression is shown. The core-and-shell model is further extended to include film diffusion besides diffusion of water and hydrogen in the shell. In order to test the model, kinetic measurements are performed over a range of temperatures and water concentrations, and pore size distributions are measured direct on reduced catalysts. These latter measurements allow calculations of effective diffusion coefficients.

Another aim of this type of study is to calculate the ratio of the partial pressures of water and hydrogen at the interface where the reduction takes place. It is likely that this ratio determines the particle size of the resulting metallic iron phase and, therefore, will be determining also for the activity of the final catalyst.

EXPERIMENTAL

Regular grains (1.00-1.19 mm) of the multiple promoted Topsøe KMI catalyst were used in the experiments. The H_2-N_2 gas mixture was obtained by catalytic decomposition of ammonia followed by drying at liquid nitrogen temperature. In some experiments water was added to this gas mixture. The kinetic experiments were carried out using a spring microbalance as described in [2]. Pore size distribution curves were obtained by means of a Carlo Erba porosimeter, model 1500. The catalysts used for these experiments had first been reduced in a specially built reactor. This reactor allowed the reduced catalysts to be placed in a container which could be sealed off inside the reactor. In this way the reduced catalysts could be weighed and transferred to the porosimeter without being exposed to air.

RESULTS

The kinetic curves for reduction in dry N_2:H_2 gas at temperatures between 450°C and 550°C are shown in Fig.1. Adding water to the reduction gas decreases the rate of reduction as shown in Fig.2 for reduction at 550°C.

Pore size distribution curves for catalysts reduced at different temperatures in dry H_2-N_2 are shown in Fig.3. The pore volumes are referred to the mass of reduced catalyst. The catalyst had in all cases been reduced to R≈0.8. Two distinct maxima are observed in

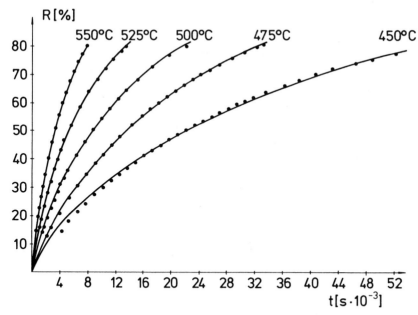

Fig. 1. Reduction in dry H2-N2 at different temperatures. The points represent the experimental data, and the curves are calculated by the model.

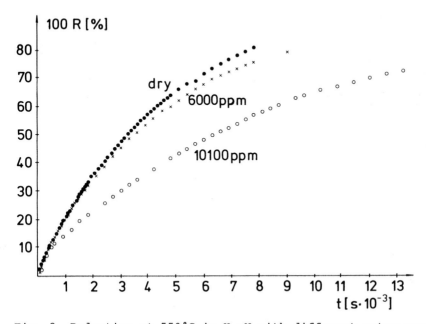

Fig. 2. Reduction at 550°C in H_2-N_2 with different water contents.

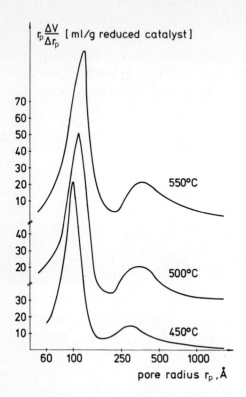

Fig.3. Pore size distribution curves for catalysts reduced at different temperatures in dry H_2-N_2.

the curves. The maximum around 100-120 Å is in agreement with the results of Nielsen [5]; it can be seen that the position of the maximum increases somewhat with increasing temperature. The second maximum around 260-430 Å has not been reported previously. The intensity and the position increase somewhat with rise in temperature.

On the basis of the measured pore size distribution the effective diffusion coefficients, D, of hydrogen and water in the porous shell are calculated by use of the formula:

$$D = L \cdot \sum_{j=1}^{n} \varepsilon_j \left(\frac{1}{D_B} + \frac{1}{D_{K,j}} \right)^{-1} \qquad (1)$$

The symbols are explained in the list included at the end. The bulk diffusion coefficients, D_B, were calculated according to

Satterfield and Sherwood [6] and the labyrinth factor, L, equal to 0.45 was taken from Nielsen [1].

The effective diffusion coefficients for hydrogen and water are shown as a function of temperature in Fig. 4 for reduction in dry H_2-N_2. The data for reduction at 550°C in H_2-N_2 containing different amounts of water are also shown. The apparent activation energy is 4.0 ± 0.7 kcal/mol for hydrogen and 2.6 ± 1.1 kcal/mol for water.

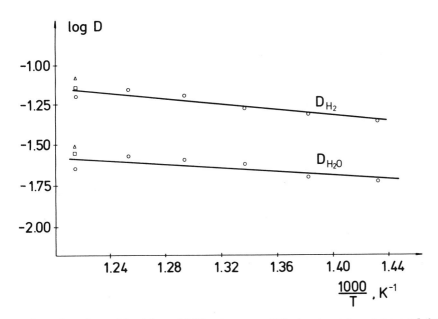

Fig. 4. The effective diffusion coefficients of water and hydrogen at different temperatures and water concentrations.
o: dry gas,   □: 6000 ppm H_2O,   △: 10100 ppm H_2O.

MATHEMATICAL MODEL FOR THE REDUCTION PROCESS

The mathematical model for the reduction process takes its starting point in the work of Spitzer et al. [7].

It is assumed that the reduction of a single particle can be described as a core-and-shell model, that the reaction proceeds on the interface between the core and the shell, and that the effective diffusion coefficients are independent of the reduction degree. Two gas phase components take part in the reaction: hydrogen as reactant, and water as product.

The transport of hydrogen through the shell is governed by

the basic diffusion equation. The amount of hydrogen that diffuses through the shell in unit time can be found by integration of this equation from the particle surface to the interface. By using the ideal gas law to convert from concentration to pressure, the following equation is found :

$$N_{H_2} = -D_{H_2} \cdot \frac{1}{R_g T} \cdot 4\pi \cdot x_i \cdot x_o \cdot \frac{p_{H_2}^o - p_{H_2}^i}{x_o - x_i} \tag{2}$$

The gas film resistance will also be included here, since the flow rates encountered often are small.

The flow of hydrogen from the bulk gas phase to the surface of the particle is given by the film transport equation:

$$N_{H_2} = \frac{K_{G,H_2}}{R_g T} \cdot 4\pi \cdot x_o^2 \cdot \left(p_{H_2}^o - p_{H_2}^b\right) \tag{3}$$

Hydrogen is consumed at a reduction rate, V, at the interface. This gives the equation

$$-N_{H_2} = 4\pi \cdot x_i^2 \cdot V \tag{4}$$

Combining equations (2) and (4), and equations (3) and (4), eliminates the surface concentration and the mole flow, if the two resulting equations are combined. The resulting equation is:

$$p_{H_2}^b - p_{H_2}^i = V \left\{ \frac{R_g T}{K_{G,H_2}} \left(\frac{x_i}{x_o}\right)^2 + \frac{R_g T}{D_{H_2}} \frac{x_i}{x_o} (x_o - x_i) \right\} \tag{5}$$

The reduction degree is now introduced. It is defined through the equation :

$$R = 1 - \left(\frac{x_i}{x_o}\right)^3 \tag{6}$$

and insertion of this equation in (5) gives :

$$p_{H_2}^b - p_{H_2}^i = V R_g T \left\{ \frac{1}{K_{G,H_2}} (1-R)^{2/3} + \frac{x_o}{D_{H_2}} \left((1-R)^{1/3} - (1-R)^{2/3}\right) \right\} \tag{7}$$

The equation for water can be derived in a similar way :

$$p_{H_2O}^i - p_{H_2O}^b = V R_g T \left\{ \frac{1}{K_{G,H_2O}} (1-R)^{2/3} + \frac{x_o}{D_{H_2O}} \left((1-R)^{1/3} - (1-R)^{2/3}\right) \right\} \tag{8}$$

These two equations form the basic equations, since no assumptions about the actual form of the rate expression have been made, and since they express the partial pressure at the interface as a function of the bulk pressures and the reduction degree.

The rate of reduction, V, can be expressed by the consumption of oxygen, i.e. by the equation:

$$V = c_o \frac{dx_i}{dt} \qquad (9)$$

Insertion of (6) in this equation gives:

$$V = \frac{c_o x_o}{3(1-R)^{2/3}} \cdot \frac{dR}{dt} \qquad (10)$$

The variables in this equation can be separated and the equation integrated as:

$$t = \frac{c_o x_o}{3} \int_o^R \frac{dR}{(1-R)^{2/3} \cdot V} \qquad (11)$$

The expression for V is a function of the partial pressures of hydrogen and water at the interface. Insertion of (7) and (8) in the rate expression gives after some rearrangements an expression which can be used direct in the integral in eq.(11).

Langmuir-Hinshelwood rate expression

The following rate expression takes into account the adsorption of water, but not that of hydrogen:

$$V = \frac{(K_r/R_g T) \cdot P_{H_2O}^i}{1 + (K_{A,H_2O}/R_g T) \cdot P_{H_2O}^i} \cdot (1 - P_{H_2O}^i / P_{H_2}^i K_e) \qquad (12)$$

Insertion of equations (7) and (8) in this equation gives after some rearrangements the following quadratic equation:

$$AV^2 + BV + C = 0 \qquad (13)$$

where

$$A = -\frac{1}{K_r} \left\{ \frac{K_{A,H_2O}}{K_{G,H_2O}} \cdot (1-R)^{2/3} + \frac{K_{A,H_2O}}{D_{H_2O}} \cdot x_o \left((1-R)^{1/3} - (1-R)^{2/3} \right) \right\} \qquad (14)$$

$$B = -\frac{1}{K_r} \left\{ \left(1 + \frac{K_{A,H_2O}}{R_g T} \cdot P_{H_2O}^b \right) + \frac{1}{\alpha}(1-R)^{2/3} + \frac{x_o}{\beta}\left((1-R)^{1/3} - (1-R)^{2/3} \right) \right\} \qquad (15)$$

$$C = \frac{1}{R_g T} \cdot \left(P_{H_2}^b - (P_{H_2O}^b / K_e) \right) \qquad (16)$$

$$\frac{1}{\alpha} = \frac{1}{K_{G,H_2}} + \frac{1}{K_e \, K_{G,H_2O}} \qquad (17)$$

$$\frac{1}{\beta} = \frac{1}{D_{H_2}} + \frac{1}{K_e \, D_{H_2O}} \qquad (18)$$

Given the partial pressures in the bulk phase and the reduction degree this equation can be solved as a simple quadratic equation, where only the positive root is used.

If no adsorption of H_2O is considered ($K_{A,H_2O} = 0$), a simple first order reaction in hydrogen results and the integral in equation (11) can be solved analytically. The resulting expression is in that case identical to that derived by Seth and Ross [8] with the exception that the parameters have a different physical meaning. However, if the kinetic data are analyzed using the reversible first order kinetics, the parameters have no physically meaningful values. The activation energy (21 kcal/mol) for the effective diffusion coefficient for water is unacceptably high. Therefore, it seems that the adsorption of water must be considered and, in the following, the kinetics given by equation (12) will be tested.

The effective diffusion coefficients were calculated from the porosimeter data, and the mass transfer coefficients were calculated by using the correlations given by Szekely et al. [9]. The parameters K_r and K_{A,H_2O} are then determined by a non-linear least squares fit. The integral in equation (11) cannot be solved analytically and an ordinary Gauss method has therefore been used to solve the integral numerically.

DISCUSSION

The values of the parameters K_r and K_{A,H_2O} obtained by fitting the data in Fig.1 are shown in Fig. 5. The kinetic curves calculated from these parameters are shown in Fig. 1. The agreement between the measured and the calculated values is quite good, especially for temperatures above 450°C. The mean deviation expressed as percentage of the reduction times, amounts to 2.2 % for the curve at 450°C and about 0.60 % for the remaining reduction curves. It should be noted that the deviations between the measured and the calculated values seem to be systematic rather than random.

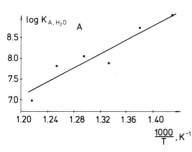
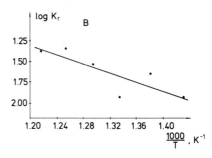

Temperature dependence of $\log K_{A,H_2O}$ (Fig.5A) and $\log K_r$ (Fig.5B).

In order to test the physical significance of the two parameters, K_r and K_{A,H_2O}, the calculated values have been drawn in Arrhenius type plots in Fig.s 5 A and 5 B. Although there is some scatter in the points, they seem to form straight lines with reasonable values for the slopes. The activation energy for the rate constant, K_r, is equal to 12.6 ± 6.3 kcal/mol. This value is quite close to the one found by Quets et al. [10]. The heat of adsorption of water is estimated to be 39.0 ± 10.8 kcal/mol, which seems quite reasonable. Consequently, it can be concluded that for reduction in dry H_2-N_2 the model seems both to represent the data well and to give reasonable values of the estimated parameters.

In the following, the values of the water concentration at the interface and the importance of the water adsorption term will be discussed briefly. The partial pressure of water, $p_{H_2O}^i$, at the interface increases with increasing temperature and reduction degree from 5×10^{-4} atm at 450°C for R = 0.1 to 10^{-2} atm at 550°C for R = 0.8. The adsorption term in the denominator of equation (12) increases with increasing reduction degree, but decreases with increasing temperature. At 450°C the term varies from 1.3 for R = 0.1 to 4.5 for R = 0.8. At 550°C the variation is from 0.2 for R = 0.1 to 1.0 for R = 0.8. So it seems that the inhibiting influence of the water is significant, even for the very high reduction temperatures.

The application of the model to explain the data for reduction in H_2-N_2 with added water shows again that the agreement is better for the higher reduction temperatures. However, the results obtained until now are rather uncertain, and further experiments will be needed.

This might also indicate that some of the assumptions in the evaluation of the model are not fulfilled. It was mentioned in the Introduction that the catalyst encountered might be multi-phase, and the assumption of a single core-and-shell model should therefore be further investigated. This investigation should include a multiple core-and-shell model [11] and, preferably, also take into account the irregular shapes of the grains and the variation of the pore size distribution with reduction degree [12].

LIST OF SYMBOLS

A, B, C	Parameters in equation (13)
c_o	Oxygen concentration in core, mol/cm^3
D	Effective diffusion coefficient, cm^2/sec
D_B	Bulk diffusion coefficient, cm^2/sec
$D_{K,j}$	Knudsen diffusion coefficient for radius x_j
K_G	Film mass transfer coefficient, cm/s
K_A	Adsorption constant, cm^3/mol
K_e	Chemical equilibrium constant
K_r	Rate constant, cm^3/mol
L	Labyrinth factor
N	Flow, mol/sec
n	Number of experimental points
p^b	Partial pressure in bulk phase
p^i	Partial pressure at interface
p^o	Partial pressure at particle surface
R	Reduction degree, defined in equation (6)
R_g	Gas constant, atm cm^3/mol/°K
T	Temperature, °K
V	Reaction rate, mol/cm^2/sec
x_i	Core radius, cm
x_o	Particle radius, cm
α	Parameter defined in equation (17)
β	Parameter defined in equation (18)
ε_j	Porosity in j-th pore volume fraction

REFERENCES

1. A.Nielsen, An Investigation on Promoted Iron Catalysts for the Synthesis of Ammonia, 3rd edn., Jul. Gjellerups Forlag, Copenhagen, 1968.
2. A.Barański,A.Bielański and A.Pattek, J.Catal.,26(1972)286.
3. B.S. Clausen, S. Mørup, H. Topsøe, R.Candia, E.J. Jensen, A. Barański and A. Pattek, J.Physique,37(1976)C6-245.
4. E.J. Jensen, H. Topsøe, O. Sørensen, R. Candia, F. Krag, B.S. Clausen, S. Mørup, Scand.J.Metallurgy,6(1977)6.
5. Reference 1, p.205.
6. C.N. Satterfield and T.K. Sherwood, The Role of Diffusion in Catalysis, Addison-Wesley Publishing Comp., Reading-Paolo Alto - London, 1969, p.21.
7. R.H. Spitzer, F.S. Manning and W.O. Philbrook, Trans.Met.Soc. AIME,236(1966)726.
8. B.B.L. Seth and H.M. Ross, Trans.Met.Soc.AIME, 233(1965)180.
9. J.Szekely, J.W. Ewans and H.Y. Sohn, Gas-Solid Reactions,Academic Press, New York - San Fransisco - London, 1976, p.13.
10. J.M. Quets, M.E. Wadsworth and J.R. Lewis, Trans.Met.Soc.AIME, 221(1961)1186.
11. J.Y.Park and O. Levenspiel, Chem.Eng.Science,30(1975)1207.
12. A. Barański, R. Candia and H. Topsøe, Unpublished results.

DISCUSSION

N. PERNICONE : The temperature range you have chosen for your investigations is 450-550°C. These are rather high temperatures with respect to those used for the industrial reduction of ammonia synthesis catalysts. Did you prefer a higher temperature range to have higher reduction rates, which are therefore more easily measurable, or for other reasons ?

A. BARANSKI : Experiments have been carried out at temperatures lower than 450°C, but they are not included here. However, the rate of reduction of industrial ammonia catalysts is quite low, below around 400°C, and it is therefore mainly the temperature region above this value which is of industrial importance. We have included data over a rather wide temperature range, since in this way the temperature dependence of the parameters in the kinetic equation can be obtained and checked for physical significance. Futhermore, during the industrial reduction of a catalyst, the temperature also changes over quite a wide range.

S.P.S. ANDREW : I have had the experience of using the authors' form of mathematical analysis to the reaction of a number of impure solid substances. (NH_3 synthesis catalyst - because of the promotors - may be called impure magnetite). I have always found that the experimental curve followed the theoretical up to about

80% conversion but the experimental rate was markedly less than the theoretical above this conversion. Did you find anything similar ?

A. BARANSKI : Yes, we have the same experience and only the data up to 80% reduction have been reported here. The discrepancy above this reduction degree was reported in an earlier paper (ref.2 in the paper). The lower reduction rate at high reduction degrees is most likely due to a difficultly reducible promoter-rich phase which forms as the reduction of the mixed magnetites progresses (see ref. 3 in the paper).

J.W.E. COENEN : In the discussion of your paper you proceed to test the physical significance of two parameters : the adsorption equilibrium constant for water adsorption and the rate constant. From their temperature dependence you derive the adsorption heat and the activation energy, and you conclude their values to be reasonable. But you could go one step further and check also the absolute value of these constants for reasonableness. In other words, what about entropy effects ?
For instance, the adsorption constant may be written as :
$$K = \exp(\Delta S_a^o/R) \exp(-\Delta H_a^o/RT).$$

Since the value of K and the energy factor is known, you can calculate ΔS^o and consider whether its value is acceptable. Since you find a high heat of adsorption, the adsorption should be immobile which means that the entropy loss would involve complete loss of translational entropy : ΔS_a^o should be at least -30 cal/K mol, for a standard state of 1 atm (that is partial pressure expressed in atmospheres).

A. BARANSKI : The calculated value of the adsorption entropy amounts to -14.6 cal/K mol. Adsorption of molecules from the gas phase onto a surface results in a decrease in entropy. If the adsorption is mobile like for xenon on mercury, only one degree of freedom of translational motion is lost and between -2.0 and -5.4 e.u. is found for the adsorption entropy (1). The calculated value of -14 e.u. for the water adsorption entropy does not seem unreasonable.
(1) J.H. de BOER, The Dynamical Character of Adsorption, Oxford and Clarendon Press, 1968, pp. 112-117.

PREPARATION AND CHARACTERIZATION OF SMALL IRON PARTICLES SUPPORTED ON MgO

H. TOPSØE[a], J.A. DUMESIC[b], E.G. DEROUANE[c], B.S. CLAUSEN[d], S. MØRUP[d], J. VILLADSEN[a] and N. TOPSØE[a]

[a] Haldor Topsøe Research Laboratories, DK-2800 Lyngby (Denmark).
[b] Department of Chemical Engineering, University of Wisconsin, Madison, Wisc. 53706 (U.S.A.).
[c] Facultés Universitaires N.D. de la Paix, Laboratoire de Catalyse, B-5000 Namur (Belgium).
[d] Laboratory of Applied Physics II, Technical University of Denmark, DK-2800 Lyngby (Denmark).

ABSTRACT

The preparation of small metallic iron particles from ion exchanged magnesium hydroxycarbonate and coprecipitated starting materials is reported. The effect of different activation procedures prior to reduction is studied, and the nature of the starting material is found to play a more important role than the activation procedure. The smallest iron particles were prepared starting from coprecipitated structures. These results are discussed in terms of a model for the structure and genesis of the various catalysts. The structural information is obtained mainly from Mössbauer spectroscopy and X-ray diffraction. The study leads to a possible explanation of the nature of the metal-supported interaction in these catalyst systems.

1 INTRODUCTION

The search for means of preparing small, thermally stable iron particles has for many years been motivated both by practical and theoretical interest. From a practical viewpoint, one has searched for catalysts with a high specific iron surface area. Such catalysts would naturally be of interest, for example, in the ammonia or Fischer-Tropsch synthesis. Theoretically, small catalyst particles are also of great current interest owing to their special surface structures and electronic properties. Small metallic iron particles are, however, difficult to stabilize, since iron sinters easily at hydrogen reduction temperatures (~ 700 K). The stabilization can be achieved by textural promotion as is done, for example, in the classical ammonia synthesis catalyst [1] or by dispersion on a thermostable high surface area support, as will be discussed in the present paper.

In order to stabilize metallic iron particles, the support interaction should be optimum, i.e. neither so strong that the reduction to the metallic state fails to occur, nor so weak that high mobility results in rapid sintering.

Boudart et al. [2] found that magnesium oxide is an ideal support for the preparation of small iron particles with sizes below 10 nm. Reduction to metallic iron was possible, and the resulting small iron particles showed little tendency toward sintering. A large fraction of the metallic iron surface area was accessible to gases. The catalysts were used in an investigation of the particle size dependence of the ammonia synthesis reaction [3]. They were prepared by exchange of magnesium cations with iron ions at the surface of magnesium hydroxy-carbonate (MHC) crystals, followed by drying and direct reduction in hydrogen [2]. Bussière et al. [4] and Dutartre et al. [5] also supported iron on magnesium oxide. However, their catalysts were prepared by coprecipitation of iron and magnesium nitrates, followed by calcination and reduction at high temperatures. The metallic iron particle size was reported to be larger than those in the MHC-based catalysts. Also, the observed properties of these samples were rather different from those of the samples prepared from the MHC precursor [2,6]. Ferromagnetic resonance (FMR) was used by Derouane and Monseur [7] to show that different thermal treatments of ion exchanged, MHC-based structures resulted in different magnetic phases.

In view of the above studies, the present paper deals with a further examination of the influence of both the nature of the starting material and its activation on the ultimate structure of iron supported on magnesium oxide.

We have used different physical and chemical techniques to provide information about the chemical state and dispersion of the various iron- and magnesium-containing phases present during the catalyst genesis. In particular, the results of Mössbauer spectroscopy, X-ray diffraction, CO chemisorption, and N_2 physical adsorption (BET), reveal a complex interaction among these different phases which appears essential for the preparation of small, metallic iron particles.

2 EXPERIMENTAL

2.1 Sample preparation and nomenclature

The magnesium hydroxycarbonate (MHC) catalysts were prepared according to the method described in [2]. The dried ion exchanged MHC will be denoted x%Fe-MHC, where x denotes the weight percent of iron in the reduced catalyst. These samples will be referred to as pre-precursors. x%Fe-MHC-y labels the same catalyst after the pretreatment or activation procedure described by "y". Some catalysts were prepared as described in [2] by direct reduction (y = H_2) of the pre-precursors according to the program: 2-5 h at 370 K; 12-16 h at 520 K; 2-5 h at 600 K; 2-5 h at 650 K, followed by at least 20 h at 700 K. Other catalysts were made by first decomposing the pre-precursors to give different oxidic precursors before the final reduction. These decompositions consist of slow activation in either air (y = air) or vacuum (y = vac) according to the above time-temperature program, or a direct calcination at 770 K in air for 24 h (y = calc). The catalysts resulting from reduction of these precursors will be denoted: y = air,H_2; y = vac,H_2; etc.

For the supported iron catalysts prepared by coprecipitation (CP) of the magnesium and iron(III) nitrates with ammonium hydroxide, an analogous nomenclature will be used having the general form x%Fe-CP-y.

2.2 Mössbauer spectroscopy

The Mössbauer spectra were obtained with a constant acceleration spectrometer using a source of ^{57}Co in Rh. The experiments on the reduced samples were performed *in situ* in an ultrahigh vacuum compatible cell allowing temperatures between 78 K and 100 K and application of magnetic fields.

2.3 X-ray diffraction

X-ray results were obtained with a Philips vertical diffractometer equipped with a proportional counter and graphite monochromator. CuKα radiation was employed at a scan speed of $\frac{1}{2}°$(2θ) per minute.

2.4 CO chemisorption and total surface area measurements

CO chemisorption as described in [2], and BET surface area measurements were carried out using a constant volume system.

3 RESULTS

3.1 Pre-precursors

The X-ray diagrams of the MHC pre-precursors were all typical of the MHC structure. Chemical analysis and TGA results showed that for the different iron loadings the formulas can be approximately described as $3(Mg,Fe)CO_3(Mg,Fe)(OH)_2 \cdot 3\frac{1}{2}H_2O$, although the ideal formula of the pure Mg compound has been definitively shown to be $4MgCO_3 \cdot Mg(OH)_2 \cdot 4H_2O$ [8]. The structure of the CP pre-precursors did, however, seem to vary with the iron concentration. The 0%Fe-CP sample was quite pure $Mg(OH)_2$. However, as the iron content increased, the structures contained NO_3^- and crystal water in increasing amounts.

The X-ray diagram of the 16%Fe-CP showed that the system is mainly single-phased with a structure (*sjögrenite* type) not unlike the various MHC samples. In addition, minor amounts of $Mg(OH)_2$ were detected.

3.2 Precursors

In Fig.1 we show the Mössbauer spectra of the 8%Fe-MHC-vac, the 16%Fe-MHC-air, and the 16%Fe-CP-calc precursors at 293 K and 78 K. The 293 K spectra (Figs. 1a, 1c, and 1e) all contain a paramagnetic Fe^{3+} (2-line) component. Such a component could originate from both superparamagnetic and paramagnetic iron species.

The 8%Fe-MHC-vac sample shows a significant, magnetically ordered component at room temperature (spectrum 1a); this component does not significantly increase in relative intensity as the temperature is lowered (spectrum 1b). These spectra can be interpreted in terms of a paramagnetic component and a magnetic (6-line) component: the former component is typical of Fe(III) in MgO [9-11], and the latter component has quite broad lines and shows "structure" at 78 K indicative of Fe on several different sites. This may be interpreted as iron present in $Mg_xFe_{3-x}O_4$ ferrite phases of various compositions. The X-ray results are in agreement with these findings, since both a spinel phase with lattice constant (a = 8.398 Å) typical of $Fe_3O_4/MgFe_2O_4$ and an iron containing MgO phase (a = 4.236 Å) are observed together with a quite pure MgO phase (a = 4.218 Å). This lattice parameter is typical of small particles of MgO. The presence of several magnetically ordered phases was also confirmed by FMR [7]. The 16%Fe-MHC-air sample (spectra 1c and 1d) shows superparamagnetic behavior. This is seen from the appearance of a magnetically split component as the temperature is lowered to 78 K. The 16%Fe-CP-calc sample prepared

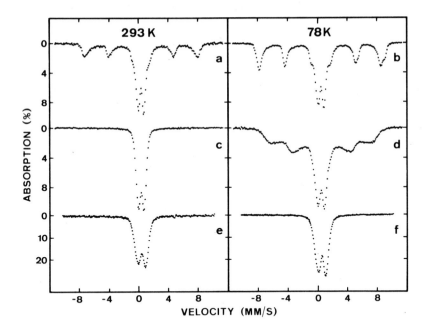

Fig. 1. Mössbauer spectra of precursors obtained at 293 K and 78 K. (a) and (b) 8%Fe-MHC-vac; (c) and (d) 16%Fe-MHC-air; (e) and (f) 16%Fe-CP-calc.

by coprecipitation (spectra 1e and 1f) is, however, the only sample that does not show any magnetic ordering even at 78 K. The X-ray patterns show a single MgO type phase with lattice constant a = 4.219 Å. A 12%Fe-MHC-calc sample showed a Mössbauer spectrum quite similar to that of the 16%Fe-MHC-air sample. However, some magnetic splitting was present already at 293 K.

3.3 Reduced catalysts

The Mössbauer spectra obtained *in situ* at 670 K of the progressively reduced pre-precursors and of the reduced precursors, are shown in Fig.2. These spectra all show the presence of iron in two forms: metallic iron (superparamagnetic and/or ferromagnetic) and Fe^{2+} in magnesium oxide. The relative amount of iron which is reduced to the metallic state is quite similar in the different samples and appears to be independent of both the nature of the pre-precursor and their prior activation pretreatments. The fraction of reduced iron has also been observed to depend only slightly on metal loading [2].

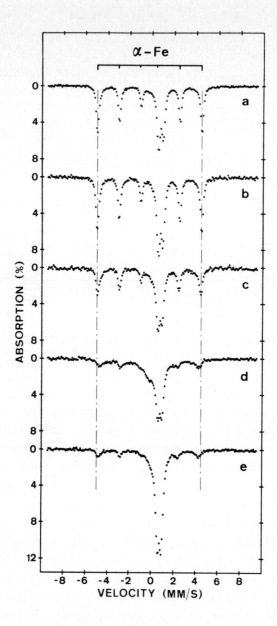

Fig. 2. *In situ* Mössbauer spectra of reduced MHC- and CP-based catalysts in H_2 at 670 K. (a) 16%Fe-MHC-air,H_2; (b) 16%Fe-MHC-H_2; (c) 8%Fe-MHC-vac,H_2; (d) 16%Fe-CP-calc,H_2; (e) 16%Fe-CP-H_2.

It is, however, interesting that the metallic iron particles in the CP-based catalysts are much smaller than those in the MHC-based ones. This is seen from the Mössbauer spectra, where only the 16%Fe-CP-calc,H_2 (spectrum 2d) and the 16%Fe-CP-H_2 (spectrum 2e) samples show small particle behavior, i.e. a large superparamagnetic metallic iron peak (at approximately -0.2 mm s^{-1}) is present, and a magnetic hyperfine field smaller than that of bulk α-Fe is observed due to the presence of collective magnetic excitations [12,13]. These two effects are not dominant in the MHC catalysts (spectra 2a, 2b, and 2c), where the metallic iron spectral parameters are typical of bulk α-Fe.

In order to obtain information concerning the Fe^{2+} ions, spectra (not shown here) were also recorded at 78 K. At this temperature two quadrupole split Fe^{2+} doublets were observed. The relative intensities of these two doublets were about equal for the 16%Fe-MHC-H_2 sample. However, for CP-based catalysts, the doublet with the small quadrupole splitting was the dominant one, independent of the activation procedure. The spectra can be interpreted as Fe^{2+} being present in clusters in the MgO rather than in solid solution [14]. The doublets with the large and small quadrupole splittings originate from the "interior" and "exterior" atoms of the cluster, respectively. Thus the cluster size appears to be smaller for the CP-based catalysts and is only slightly affected by various pretreatments.

The metallic iron particle size could not be measured directly from the X-ray diffraction patterns, since these could not be recorded *in situ*. The number of exposed surface iron atoms can, however, be determined from CO chemisorption using a 2:1 Fe:CO surface stoichiometry [15]. This stoichiometry may not hold for very small iron particles as shown recently [16], but will suffice for the present estimates of particle size. In order to calculate the particle size from the chemisorption measurements, the fraction of iron reduced to the metallic state must be known. These fractions can be directly determined from the *in situ* Mössbauer spectra obtained in hydrogen, and the resulting particle sizes are given in Fig.3. The results show that for both the CP- and the MHC-based catalysts the metallic iron particle size increases with increasing metal loading. It is seen that the iron particle size in the CP-based catalysts is much smaller than in the MHC-based catalysts with similar iron loading, and is also much smaller than that reported by other investigators [4]. It is interesting that upon reduction of the different pre-

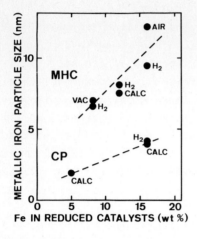
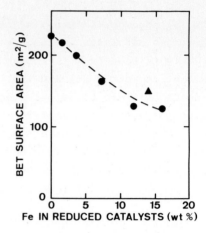

Fig. 3. Metallic iron particle size as function of iron loading and activation procedure for MHC- and CP-based catalysts. The activation procedures ("y") are specified in the figure.

Fig. 4. Total surface areas of x%Fe-MHC-H_2 (●) and x%Fe-CP-H_2 (▲) catalysts as function of iron loading.

cursor structures (y = calc,H_2, y = vac,H_2, or y = air,H_2) or by direct reduction (y = H_2) of the pre-precursor, the resulting metallic iron particle sizes are rather similar.

The dependence of the total surface areas on iron loading is depicted in Fig.4. It is seen that the particle size of the (Mg,Fe)O particles (the dominant phase) increases greatly in the presence of iron.

Figure 5 shows a slow scan of the (220)-reflections of the (Mg,Fe)O phase in 16%Fe-MHC-H_2 (Fig.5a) and 16%Fe-CP-H_2 (Fig.5b). The bars represent the line width, Γ_{BET}, which should be expected from the average particle size calculated from the measured surface areas. The figure shows that the observed line width is much larger for the MHC-based catalyst than that calculated from the surface area, whereas a good agreement is obtained in the CP-based catalyst indicating a single (Mg,Fe)O phase. The source of excessive line broadening in the MHC-based catalyst is undoubtedly the existence of a distribution of (Mg,Fe)O phases with different composition and crystallite sizes.

The above type of behavior is quite general for all the different catalysts investigated.

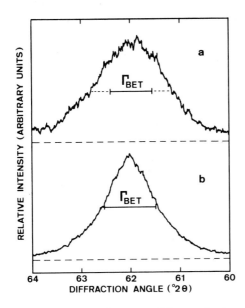

Fig. 5. X-ray diffraction profiles of the (220)-reflections of the (Mg,Fe)O phase in reduced catalysts. (a) 16%Fe-MHC-H_2; (b) 16%Fe-CP-H_2. The expected line widths, Γ_{BET}, calculated on the basis of measured total surface areas are also shown.

4 GENESIS OF CATALYSTS

On the basis of the results, schematic models are proposed for the genesis of the MHC- and CP-type catalysts. Figure 6 illustrates the situation where the catalyst is produced via a calcined precursor structure. A more general discussion will be given in section 5.

4.1 MHC catalysts

The preparation of MHC catalysts can be described according to the sequence presented in Fig.6, as follows :
- During the ion exchange of the initially large (approx.300×20 nm [2]) crystals of MHC, only the surface region of these crystals becomes "exchanged" with iron (Fig.6a).
- The structure which forms upon decomposition (Fig.6b) of the "surface exchanged" MHC crystals retains some of the characteristics of the pre-precursor structure. Rather large iron rich MgO particles form at the exterior regions of the pre-precursor particles. At the regions corresponding to the interior of the pre-precursor, very small MgO particles free of iron are formed.

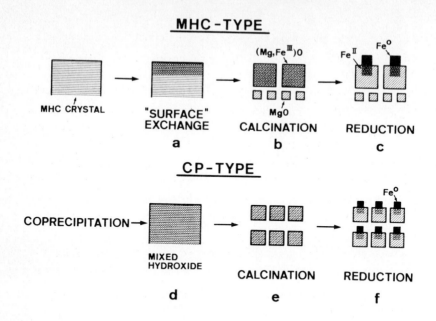

Fig. 6. Schematic models for the genesis of the MHC- and CP-type catalysts.

- During the reduction of this precursor structure (Fig.6c), or the direct reduction of the pre-precursor, metallic iron particles are produced. These particles are, however, quite large due to the very high local iron concentration, which is much higher than the average concentration. As a result of the reduction process, the catalyst produced from the MHC pre-precursor consists of several types of phases, the most important of which are :
    - a quite pure MgO phase with small particle size
    - MgO phase(s) with a high content of Fe^{2+} present as large clusters
    - metallic iron particles with broad particle size distribution and large average size.

4.2 CP catalysts

The preparation of the coprecipitated catalysts is thought to proceed as follows :
- The coprecipitation procedure ensures that the mixed hydroxynitrate pre-precursor will have a homogeneous iron distribution (Fig.6d)

with a much lower local iron concentration than in the MHC pre-
precursor of the same overall metal loading.
- Similarly, the precursor structure (Fig.6e) will form a single
iron-magnesium oxide phase with iron concentration and particle
size lying between the iron rich and iron poor (Mg,Fe)O regions
formed in the MHC-based precursors.
- As this homogeneous precursor is reduced (Fig.6f), quite small
particles should result as compared to that of MHC-based catalyst
with similar metal loading. Also, it is expected, due to the homogeneous structure of the oxidic compounds before reduction, that
the resulting metallic iron particle size distribution of the reduced catalyst should be quite narrow. The reduced catalysts will
therefore contain the following phases :
- a MgO phase of medium particle size with low content of
  Fe^{2+} present in small clusters
- metallic iron particles with narrow particle size distribution and small average size.

5 DISCUSSION

The production of small metallic iron particles using MgO as a
support may be understood in terms of the various iron- and magnesium-containing compounds encountered in the genesis of the reduced
catalysts. The fact that the precursors starting from CP pre-precursors are much more homogeneous and have lower local iron concentration is seen from both the Mössbauer and the X-ray results. The
X-ray patterns of the CP precursors, in contrast to those of the
MHC precursors, showed single phase behavior and line widths equal
to Γ_{BET}. The low local iron concentration in the CP precursors was
seen from the Mössbauer spectra, where magnetic ordering was absent
even at 78 K, in contrast again to the situation encountered for
the MHC precursors.

These facts imply that the decomposition of the pre-precursor
structure occurs before a distribution of iron can be attained. As
a result, the MHC-based catalysts formed are quite inhomogeneous,
whereas the CP-based catalysts will be rather homogeneous. In this
way the systems will retain a "memory" of the nature of the pre-
precursor which cannot be "erased" by subsequent treatments, since
inter-particle equilibrium will not be established. The similarity
in structure and particle size of the reduced catalysts activated

in different ways suggests that under reducing conditions, *intra-particle* equilibrium seems to be established, hence erasing differences in the previous precursor structures.

Thus the two main factors affecting structure and metallic iron particle size in the reduced catalysts are the metal loading and the type of pre-precursor.

Both chemisorption and Mössbauer results show that very small metallic iron particles can be produced from the coprecipitated pre-precursors. For example, the metallic iron particle size in the 16%Fe-CP catalysts is much smaller than those of the 16%Fe-MHC catalysts. Indeed, from the results of Boudart *et al.* [2,6] it appears that the metal loading must be reduced to ca. 1% in the MHC-based catalyst system in order to produce the small average metallic iron particle size reported here for the 16%Fe-CP catalysts.

From a comparison between the Mössbauer spectra of the 1%Fe-MHC-H_2 catalyst reported in [2], and of the 16%Fe-CP-calc,H_2 shown in Fig.2d, we can get information regarding the particle size distribution. Both samples have about the same average particle size and fraction of superparamagnetic metallic iron at 670 K. From a study of the magnitude of the hyperfine field exhibited by the fraction of the metallic iron that is magnetically split, the particles at the upper end of the particle size distribution can be studied. As mentioned in section 3.3, the Mössbauer lines of the CP catalyst have shifted due to the presence of collective magnetic excitations. Thus, the large metallic iron particles in this sample are still quite small. This is in contrast to the behavior of the 1%Fe-MHC-H_2 catalyst where the results do not indicate the presence of collective magnetic excitations. The particles at the upper end of the particle size distribution must, therefore, be quite large, and the particle size distribution must be rather broad.

In the previous study [2] of directly reduced MHC-based catalysts (x%Fe-MHC-H_2), the fraction of the iron reduced to the metallic state was found to be almost independent of iron loading. The present results confirm these findings, but show also that the fraction of reduced metal is independent of both the type of pre-precursor and precursor. This behavior is quite in contrast to the behavior of, for example, silica- and alumina-supported catalysts (for a review see [17]), where the fraction of reduced metal varies with metal loading. The presence of stable Fe^{2+} clusters in MgO may be the reason for the rather constant degree of reduction of the metal in the Fe/MgO system and may furthermore be the cause of the similarity in structure and metallic iron particle size of catalysts reduced from different precursors.

The strong interaction of part of the iron with the support, forming clusters, may also explain why small, thermostable iron particles are easily formed when MgO is used as a support. In this way MgO may provide a unique and highly desirable support-interaction for producing small metallic iron particles.

REFERENCES

1. A.Nielsen, An Investigation on Promoted Iron Catalysts for the Synthesis of Ammonia, 3rd edn., Jul. Gjellerups Forlag, Copenhagen, 1968.
2. M.Boudart, A.Delbouille, J.A.Dumesic, S.Khammouma and H.Topsøe, J.Catal., 37(1975)486.
3. J.A.Dumesic, H.Topsøe, S.Khammouma and M.Boudart, J.Catal. 37(1975)503.
4. P.Bussière, R.Dutartre, G.A.Martin and J-P.Mathieu, C.R.Acad. Sci.(Paris), sec.6, 280C,(1975)1133.
5. R.Dutartre, M.Primet and G.A.Martin, Reac.Kinet.and Catal.Letters, 3(1975)249.
6. M.Boudart, J.A.Dumesic and H.Topsøe, Proc.Natl.Acad.Sci.(USA), 74(1977)806.
7. E.G.Derouane and P.Monseur, Chem.Phys.Letters, 44(1976)114.
8. M.Akao, F.Marumo and S.Iwai, Acta Cryst. B30(1974)2670.
9. H.Wiedersich, in "Proceedings of the Second Symposium on Low Energy X- and Gamma Sources", (P.S.Baker and M.Gerrard, Eds.) Austin, TX, 1967.
10. V.G.Bhide and B.R.Tambe, J.Mater.Sci., 4(1969)955.
11. N.R.Large, R.W.Wilkinson and R.J.Bullock, U.K.Atomic Energy Authority Research Group Report, AERE-R 5580, Chem.Division, Atomic Energy Research Establishment, Harwell, Berkshire,1968.
12. S.Mørup and H.Topsøe, Appl.Phys., 11(1976)63.
13. S.Mørup, H.Topsøe and J.Lipka, J.Phys.(Paris), 37(1976)C6-287.
14. V.V.Kurash, V.I.Gol'danskii, T.V.Malysheva, V.S.Urusov, L.M. Kuznetsov and L.A.Moskovkina, Inorg.Materials(USSR), 7(1971)1395.
15. P.H.Emmett and S.Brunauer, J.Amer.Chem.Soc. 56(1934)35.
16. N.Topsøe, H.Topsøe and J.A.Dumesic, Unpublished results.
17. J.A.Dumesic and H.Topsøe, Advan.Catal. 26(1977)121.

DISCUSSION

L. GUCZI : You said that Fe^{2+} clusters are responsible to prevent the migration of Fe^{o}. Can the interaction between MgO and Fe^{2+} clusters be the reason for the hindrance of this migration ?

H. TOPSØE : In order to stabilize small metal particles on a support, one needs a strong interaction between the metal and the support. This interaction may be so strong that part of the atoms

are not reduced to the metallic state. In the present case, the
fraction of the iron atoms which are not reduced to the metallic
state, is present as Fe^{2+}-rich clusters in the MgO. We believe,
as illustrated in Fig. 6, that the metallic iron particles are
interacting with these clusters and in this way are prevented from
migration.

L.L. VAN REYEN : In view of the similarity in the crystal structures
of the magnesium-iron-hydroxy nitrates and carbonates used as pre-
precursors for the preparation of iron/magnesia catalysts and the
corresponding nickel-aluminium layer compounds from which nickel/
alumina catalysts are made, do you have experience about the relative
merits of carbonates or nitrates as starting materials with respect
to the activity and stability of the iron/magnesia catalysts prepared therefrom ?

H. TOPSØE : In an earlier investigation (ref. 3 in the paper) the
ammonia synthesis activity of magnesium hydroxycarbonate-based
catalysts was studied. It was found that the turnover number decreased with decreasing particle size. A similar behavior has been
found for the coprecipitated catalysts. However, some differences
were observed between the two types of catalyst.
These differences can be understood taking into account the differences in particle size distribution found, as discussed in the present paper.
With respect to the stability of these catalysts, they appear to
be quite thermostable, since no significant changes in chemisorption
or catalytic properties occurred after prolonged exposure to the
reaction mixture.

M. CARBUCICCHIO : Have you made a computer fit to the spectra for
16% Fe-CP-calc.-H_2 and 16% Fe-CP-H_2 (Fig. 2 d and e) in order to
individualise, among other things, the position at -0.2 mm/sec of
the superparamagnetic iron peak ? From the figures, this is not
apparent.

H. TOPSØE : Yes. The spectra in Fig. 2 have been computer-fitted,
but the fits are not shown. The positions of the superparamagnetic
as well as those of the hyperfine split metallic iron peaks have
shifted from zero velocity due to the second order Doppler shift
(the spectra were obtained at 670°K).

E.J. NEWSON : 1) Were the co-precipitated catalysts (CP) prepared by heterogeneous or homogeneous co-precipitation ? Can you give actual pH ranges used in the preparation ?
2) What were the corresponding activity results for the MHC catalysts and the CP catalysts ? Which reaction was used ? Presumably activity tests were carried out at identically similar reaction conditions ?

H. TOPSØE : The precipitation behaviour of iron and magnesium is quite different, which makes homogeneous precipitation methods difficult. The precipitation was carried out with excess of NH_4OH without following the pH. However, the X-ray and Mössbauer data showed that the resulting catalysts were very homogeneous in structure and not just a mixture of components.
With regard to the second part of the question, we refer to the answer to L.L. VAN REYEN.

J.W. HIGHTOWER : Do you observe the formation of the very stable ferrite $MgFe_2O_4$ in any of your work ?

H. TOPSØE : In some of the oxidic precursors, $MgFe_2O_4$ may be present. As mentioned in the paper, the analyses of the spectra shown in Fig. 1 are quite difficult. However, they point to one or more spinel phases being present. The magnetic components could be due to a mixture of $MgFe_2O_4$ and Fe_3O_4 or to the presence of mixed phases of the composition $Mg_xFe_{3-x}O_4$. In the reduced catalysts, $MgFe_2O_4$ is clearly not present, since trivalent iron is no longer observed.

A. KUBASOV : It is known that the surface of MgO is usually covered with carbonate-like species. Such species change the surface properties rather drastically, but it is difficult to remove them from the surface. Therefore, I suppose that such species exist on MgO in your case too. What do you think about the possible influence of them on the preparation of your catalysts ?

H. TOPSØE : We have done TGA studies of the decomposition of both the MHC- and CP-based catalysts. These results show that the carbonate decomposes at a quite high temperature and, as you suggest, you may have carbonate species left on the surface. This has also been shown to be the case by IR studies. However, the difference in the nature of the decomposition behavior of the MHC- and CP-precursors does not seem to influence greatly the structure of the final catalysts.

A. OZAKI : Is there any difference in particle size of Fe when you reduce by a N_2-H_2 mixture ?

H. TOPSØE : We have tried to reduce the catalysts in both very pure hydrogen and in N_2-H_2 mixture and found no significant differences in the metallic iron particle size. However, the initial activity toward NH_3 synthesis varied greatly, depending on whether or not nitrogen was present in the reduction gas (see ref. 3 in the paper). The results were interpreted in terms of a nitrogen induced reconstruction of the surface of the metallic iron particles (see also : J.A. Dumesic, H. Topsøe, and M. Boudart, J. Catal. 37, 513-622 (1975), and M. Boudart, J.A. Dumesic, and H. Topsøe, Proc. Natl. Acad. Sci. U.S.A. 74, N°3 (1977), Chemistry p. 806-810.

RUTHENIUM CATALYSTS FOR AMMONIA SYNTHESIS PREPARED BY DIFFERENT METHODS

A. OZAKI, K. URABE, K. SHIMAZAKI and S. SUMIYA
Research Laboratory of Resources Utilization, Tokyo Institute of Technology, Nagatsuta, Midori-ku, Yokohama (Japan)

ABSTRACT

Ru/Al_2O_3 catalysts promoted by alkali were prepared by impregnation of Al_2O_3 with 1) $RuCl_3$-KNO_3 (or $CsNO_3$) mixture, or 2) $K_4Ru(CN)_6$ followed by reduction with hydrogen. The nitrate anion was mostly coverted to NH_3 giving highly efficient and water-resistant ammonia catalysts. In particular, the cesium oxide-promoted catalyst was as active as that promoted by metallic potassium. In the hydrogen treatment of $K_4Ru(CN)_6/Al_2O_3$, three or more CN ligands were converted to CH_4 and NH_3, suggesting that K(I) in addition to Ru(II) is reduced to metal by hydrogen, which is supported by hydrogen evolution upon water treatment as well as drastic poisoning by water vapor.

INTRODUCTION

It has been shown that the transition metals are remarkably promoted by addition of metallic potassium for the ammonia synthesis (ref. 1 and 2) as well as for the isotopic equilibration of nitrogen (ref. 2 and 3) and that ruthenium is the most active element for this catalyst system (ref. 1 and 3). The role of potassium has been shown to be an electron donor, as evidenced by remarkable increase in the activity with decrease in the ionization potential of alkali metal as the promoter, i.e. Cs > K > Na (ref. 4). In agreement with this idea, Raney ruthenium containing the electropositive aluminum metal has been found to be an efficient ammonia catalyst that is active even at 100°C after addition of potassium (ref. 5).

It seems obvious, however, that such catalysts promoted by metallic alkali are subject to irreversible poisoning by water, suggesting a possible drawback of the catalyst from a practical point of view. It would be desirable to get a water-resistant and efficient catalyst. Thus alternative methods have been investigated for preparation of the ruthenium catalyst. The present paper deals with two methods of potassium addition to Ru/Al_2O_3, in which potassium compounds were used as the starting material.

CATALYST PREPARATION

1. Oxide-promoted catalysts.

Potash(or cesium oxide)-promoted ruthenium catalysts were prepared by impregnating alumina with a mixture of $RuCl_3$-KNO_3(or $CsNO_3$)(1/10,mol /mol). The alumina sample obtained from Tokai Konetsu Co. was in a form of cylindrical pellet(3.5mm both in diameter and in length) with a coaxial open hole and had a specific surface area of 180 m^2/g. The alumina, in the form of pellets or crushed grains of 10-20 mesh, was immersed in the mixed aqueous solution of 0.02 mol and evaporated to dryness on a water bath. The Ru/Al_2O_3 ratio was adjusted to give 2.0%(w/w)Ru. The dried material was treated with circulating hydrogen at increasing temperatures, first at 100°C until no hydrogen consumption was observed and finally at 450°C for 40hr, during which the gaseous products of reduction were removed by a liquid nitrogen trap.

Four catalysts were used as follows:

Catalyst Number	Impregnation with	Support
N-1	$RuCl_3$-KNO_3	grains
N-2	$RuCl_3$-$CsNO_3$	grains
N-3	$RuCl_3$-NH_4NO_3	grains
X	$RuCl_3$	pellet

2. Cyanide-promoted catalysts.

Two different samples,A and B, of $K_4Ru(CN)_6$ were prepared according to the known methods. The sample A was prepared by adding KCN to K_2RuO_4 which was prepared by alkali fusion of Ru metal and KNO_3 (ref. 6 and 7). The sample B was prepared by adding KCN to $RuCl_3$ (ref. 8). Both samples were purified by recrystallization from water to give a colorless tetragonal platy crystal. The former method gave a better yield (@ 50% in Ru).

Both samples were examined by elemental analysis for their identity as shown in Table 1. The sample B was dehydrated in advance.

TABLE 1

Elemental analysis of samples A and B

Sample		H	C	N	K
A	found	1.21	15.42	18.01	34.00
	calc. as trihydrate	1.30	15.41	17.96	33.45
B	found	0.05	17.41	19.73	-
	calc. as anhydride	0.00	17.43	20.31	37.82

There is a good agreement with the calculated value for $K_4Ru(CN)_6$.

The sample A was further identified by X-ray diffraction and IR absorption. Since $K_4Ru(CN)_6 \cdot 3H_2O$ and $K_4Fe(CN)_6 \cdot 3H_2O$ are isomorphous (ref. 9), the identity of the sample A was confirmed by a comparison of X-ray diffraction patterns with a known sample of $K_4Fe(CN)_6 \cdot 3H_2O$. The IR absorption spectrum of the sample A was in accord with the literature for $K_4Ru(CN)_6 \cdot 3H_2O$ (ref. 10).

The samples A and B were supported on the 10-20 mesh grains of alumina described above to give a 2.0%(w/w) Ru/Al_2O_3, i.e., the alumina was impregnated with A or B from its aqueous solution of 0.04 mol/l by evaporation to dryness. The supported samples were subjected to decomposition in circulating H_2 or $3H_2/N_2$ mixture at increasing temperatures up to 450°C.

Four catalyst samples were used as follows;

Number	$K_4Ru(CN)_6$	Reducing gas
C1	B	$3H_2-N_2$
C2	B	H_2
C3	A	H_2
C4	A	$3H_2-N_2$

PROCEDURES

The hydrogen reduction and the activity measurements were carried out in a closed circulating system comprising a liquid nitrogen trap. The ammonia synthesis runs were carried out at 600torr using the $3H_2/N_2$ gas. The numbers of surface ruthenium atoms were estimated by chemisorption of hydrogen which was determined by extrapolation of linear part of the isotherm at 0°C.

RESULTS AND DISCUSSIONS

1. Stoichiometry of reduction by hydrogen

1-1 Hydrogen treatment of nitrate-containing catalysts. A rapid consumption of hydrogen started at around 100°C and was nearly completed at 200°C. The total consumption up to 450°C for Cat.N-2 was 3.9 mol H_2/mol Ru which corresponded to 94% of the calculated value for a reaction;

$$RuCl_3 + 10CsNO_3 + 41.5H_2 \rightarrow Ru + 3CsCl + 7CsOH + 10NH_3 + 23H_2O \qquad (1)$$

It is obvious that most of the nitrate is reduced to NH_3 and H_2O. On the other hand, the reduction of Cat.X ($RuCl_3/Al_2O_3$) was much slower and the total consumption up to 450°C was 1.38 mol H_2/mol Ru or 92% of the calculated value for the reaction:

$$RuCl_3 + 1.5H_2 \rightarrow Ru + 3HCl \qquad (2)$$

The deviation from 100% would be due partly to the loss during preparation. At any rate the presence of nitrate clearly results in the faster and larger consumption of hydrogen.

1-2 **Hydrogen treatment of cyanide-containing catalysts.** The reaction of unsupported $K_4Ru(CN)_6$ with hydrogen was found to be very slow even at 500°C (2 mmol /mol Ru hr), while the hydrogen consumption was readily observed at 380°C with the alumina-supported catalysts as illustrated in Fig. 1 for Cat. C-2.

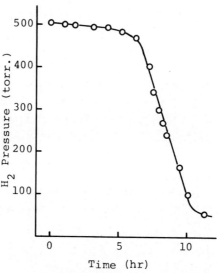

Fig.1. Time course of hydrogen consumption at 380°C for catalyst C-2.

The initial slow reaction (66 mmol H_2/mol Ru hr) is accelerated after about 6 hr by a factor of about 25. It is noticeable that the rate of reaction is independent of hydrogen pressure, suggesting that a solid phase process is rate-limiting. The temperature was raised stepwise (400, 420 and 450°C) after the rate at a temperature was slowed down.

After the decomposition of Cat. C-1 at 450°C the products collected in the trap were identified by massspectroscopy to be methane and a much larger amount of ammonia, indicating that the ammonia synthesis reaction took place because the reducing gas was $3H_2/N_2$. Thus the decomposition was made in H_2 for Cat. C-2 and the total hydrogen consumption up to 450°C was determined to be 10.7 mol/mol Ru, a value corresponding to 3CN/Ru converted to CH_4 and NH_3. In the case of Cat. C-3 decomposed in H_2, the amounts of CH_4 and NH_3 formed were determined in addition to the hydrogen consumption as follows;

CH_4/Ru : 4.8, NH_3/Ru : 4.6, H_2/Ru : 16.5

The hydrogen consumption corresponds to 4.7 CN/Ru, in a beautiful agreement with the values of CH_4/Ru and NH_3/Ru. In the case of Cat. C-4 decomposed in $3H_2/N_2$, both the CH_4/Ru and CN/Ru values are found to be 4.6.

It is clear from the above that more than two CN ligands are hydrogenated in the decomposition, which implies that some cation other than Ru(II) is reduced by hydrogen, as represented for K(I) by

$$K_4Ru(CN)_6 + 10.5H_2 \rightarrow Ru + K + 3KCN + 3CH_4 + 3NH_3 \qquad (3)$$

or

$$K_4Ru(CN)_6 + 14 H_2 \rightarrow Ru + 2K + 2KCN + 4CH_4 + 4NH_3. \qquad (4)$$

In fact the reduction of KCN by H_2 is not impossible thermodynamically when the products are removed continuously. The alumina support might be partly transformed to cyanide during the decomposition, resulting in a reduction to form metallic aluminum, on which no information is available. No information is also available about the reason of the difference in extent of reduction observed for Cat. C-2 and C-3.

2. Ammonia synthesis activities

The rates of ammonia synthesis over the catalysts are summarized in Fig. 2 as Arrhenius plots. Those catalysts are commonly 2%(w/w)Ru on the same alumina support, while differences may be found in the extent of dispersion as well as in the nature of promoter. The amounts of hydrogen chemisorption at 0°C are given in Table 2 together with the rates of ammonia synthesis per g cat. at 300°C and the apparent activation energies (Ea).

TABLE 2

Catalytic properties of catalysts

Catalyst	N-1	N-2	N-3	X	X	C-1	C-2	C-3	C-4
Promoter	K_2O	Cs_2O	none	none	K	cyanide			
H_2 mlSTP/g.cat	0.68	1.08	0.31	0.13	—	0.98	1.08	1.22	1.35
NH_3 mlSTP/g.cat.hr	1.9	8.3	0.076*	0.039*	4.3	1.9	0.67	0.70	1.1
Ea Kcal/mol	28	28	17	17	27	27	28	26	26

* estimation by extrapolation

Since little difference is observed in Ea on those alkali-promoted catalysts, the observed difference in the catalytic activity seems to come from the extent of dispersion. However the variation in hydrogen chemisorption is too small to explain the difference. The hydrogen chemisorptions on the cyanide series of catalysts are generally larger, whereas the catalytic activities are lower than those on Cat. N-1 and N-2. If metallic potassium is formed during the decomposition

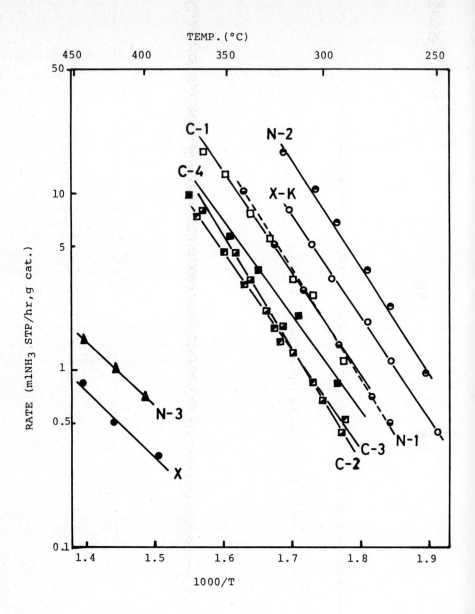

Fig. 2. Arrhenius plots of rate of ammonia synthesis.

of Cat. C-2 or C-3 as suggested by the stoichiometry, hydrogen can be chemisorbed on K in addition to on Ru, giving rise to larger chemisorptions. In fact, as described later, the chemisorption value on Cat. C-2 decreased by about 30% after a treatment with water vapor. The larger chemisorption on Cat. C-3 or C-4 than on Cat. C-2 can be due to larger amount of potassium as suggested by the stoichiometry. Thus the real chemisorption on Ru would be less than that observed on the cyanide series of catalysts.

The high activity on Cat. N-1 or N-2 is remarkable. In comparison with Cat. N-3 which has no promoter, Cat. N-1 and N-2 are respectively, 11 and 31 times more active in terms of the rate per chemisorption. When Cat. X was activated by addition of metallic potassium (K/Ru=2.9mol /mol), the extent of promotion at 300°C was 110 times as shown in Table 1. Since the potassium addition may give rise to an increase in dispersion of ruthenium on alumina as was the case with unsupported ruthenium (ref. 2), the difference in promoting efficiency between metallic potassium and cesium oxide would be smaller. Thus cesium oxide is a promising promoter which is stable in the presence of water vapor. The efficiency sequence of promoters, $K > Cs_2O > K_2O$, is in accord with the order of electro-positivity, in conformity with the electron donation to ruthenium as has been suggested (ref. 4). The high activity of Cat. N-2 arises firstly from the high electropositivity of Cs_2O and secondly from the increased dispersion of ruthenium. Presumably the low temperature reduction which is realized in the presence of nitrate is responsible for the higher dispersion.

Although metallic potassium is efficient as the promoter and is likely formed in Cat. C-2 or C-3, the specific activity per chemisorption for Cat. C-2 or C-3 is much less than that on Cat. N-2. On the other hand it is to be noted that the decomposition of $K_4Ru(CN)_6/Al_2O_3$ made in H_2-N_2 mixture results in a higher activity than that in H_2, as is clear form the comparison of Cat. C-1 or C-4 with C-2 or C-3. This result seems to be caused by a cooperative incorporation of K and N_2 into Ru metal (ref. 11), which gives rise to a higher potassium content. Hence the lower activity of Cat. C-2 or C-3 seems to be caused by a less extensive interaction of K with Ru or a possible loss of K during the decomposition of $K_4Ru(CN)_6$.

3. Effect of water vapor treatment.

The catalyst N-1 was treated with a circulating CO/N_2 (=3/1) mixture containing saturated water vapor at 450°C for 24 hr, during which CH_4 and CO_2 were formed. After evacuation at 400°C for 2 hr, the ammonia synthesis activity was found to be 64% of the initial value, while it was recovered to 86% after hydrogen treatment at 450°C for 40 hr. The amount of hydrogen chemisorption changed little in the above treatments. In contrast, the Cat. X with K was found to suffer a drastic poisoning by water vapor. After a treatment with circulating water vapor (15 torr) at 450°C for 14 hr (until the finish of hydrogen evolution) followed by evacuation at 400°C for 2 hr, the ammonia synthesis activity was found to be 6% of the initial value and recovered to 15% after hydrogen treatment at 450°C for 40 hr, with the final activity being comparable to that of Cat.N-1. Interestingly the amount of hydrogen chemisorption on Cat. X increased by a factor of 2.8 in the above treatments. The addition of potassium followed by the ammonia synthesis runs presumably gave rise to a corrosive chemisorption of nitrogen to form a compound (ref. 11), which was decomposed by water resulting in the increased dispersion. In fact, a large nitrogen uptake has been observed on Ru-K/Al_2O_3 catalysts (ref. 12), whereas little uptake on Cat. N-1.

Since metallic potassium is likely formed in the decomposition of cyanide series of catalysts, the effect of water vapor would be revealing. Cat. C-2 was treated with circulating water vapor at 440°C until the finish of pressure increase. The evolved gas was confirmed to be hydrogen (0.67 mol H_2/mol Ru), suggesting that 1.3 mol K/mol Ru is formed in Cat. C-2, which is near to 1 moleK/moleRu as estimated from the hydrogen consumption during the decomposition. The amount of hydrogen chemisorption decreased to 72% of the initial value, indicating that the initial value involved the chemisorption on K.

It may be concluded from the above results that metallic potassium (or aluminum) is formed by hydrogen reduction of $K_4R(CN)_6$/Al_2O_3 resulting in an extensive promotion of Ru. It is recalled that an iron catalyst derived from ferrocyanide was used for ammonia synthesis (ref. 13). Although the formation of potassium was not detected at that time (ref. 14), it might be the reason of high activity.

REFERENCES

1 A. Ozaki, K. Aika and H. Hori, Bull. Chem. Soc. Jap., 44, 3216 (1971)
2 K. Urabe, K. Aika and A. Ozaki, J. Cat., 42, 197 (1976)
3 K. Urabe, A. Ohya and A. Ozaki, to be published
4 K. Aika, H. Hori and A. Ozaki, J. Cat., 27, 424 (1972)
5 K. Urabe, T. Yoshioka and A. Ozaki, J. Cat., in press
6 F. Krauss, Z. Anorg. Allg. Chem., 165, 59 (1927)
7 K. Masuno, S. Waku, Nippon Kagaku Zasshi 83, 161 (1962)
8 J.L. Howe, J. Am. Chem. Soc., 18 981 (1896)
9 V.A. Pospelov and G.S. Zhdanov, Zh. Fiz. Khim, SSSR, 21, 405 (1947)
10 I. Nakagawa and T. Shimanouchi, Spectrochim. Acta, 18, 101 (1962)
11 A. Ohya, K. Urabe and A. Ozaki, Chem. Lett., 1978, 233
12 K. Urabe, K. Shiratori and A. Ozaki, J. Cat., submitted
13 A. Mittasch and E. Kuss, Z. Elektrochem., 34, 159 (1928)
14 A. Mittasch, E. Kuss and O. Emert. Z. Anorg. Allg. Chem., 170, 193 (1928)

DISCUSSION

J.G. van OMMEN : It is not strange that Cs_2O and K metal have a comparable promoting effect on Ru ? Couldn't it be, that you actually compare Cs_2O or CsOH (when Cs_2O reacts with water it forms CsOH) with K_2O or KOH. In that case, in my opinion it is easier to understand the comparable promoter effect of CsOH and KOH because they are both strong bases.

You also showed that poisoning by water vapor was irreversible because potassium is converted to KOH. But couldn't it be possible that water vapor also destroyes the ruthenium surface by reconstructing it, and so that the lower activity is caused by this effect ?

In your paper you also mention an activity sequence of promotors, that differs from the sequence presented in Fig. 2. From this figure, the sequence is $Cs_2O>K>K_2O$, while in the text the sequence is $K>Cs_2O>K_2O$. Can you explain this ?

Can you also agree that if one compares the overall activity of the Cs_2O promoted catalyst with the K metal promoted one, the first one gives the better catalysts.

A. OZAKI : K exhibits definitely a higher promoting effect than K_2O and, after poisoning with H_2O, the promoting effect approaches that of K_2O. The sequence $K>Cs_2O>K_2O$ is based on the activity per surface atom, the latter being determined by H_2 chemisorption. An activity increase due to desintegration of particle was found with K-promoted Ru. (J. Cat. $\underline{38}$, 430(1975)). If it takes place, an increase in activity is observed in the second run.

P.E.H. NIELSEN : Have you observed any support effect in going from carbon supported Ru-K catalyst to alumina supported Ru-K catalyst ?

A. OZAKI : These two catalysts should not be compared in terms of the activity per surface Ru. But evaluation of Ru dispersion on carbon remains difficult so that no comparison has been made.

K. JOHANSEN : 1) Which is the space velocity used in your activity measurements in Table 2 ? 2) Have you any high pressure activity measurements ?

A. OZAKI : 1) 2000-3000 hr^{-1}. 2) The rate of NH_3 formation on Ru-K/Ac tends to approach a plateau value at high pressures.

J.W. HIGHTOWER : From Table 2 and Fig. 2, I notice that the least active unpromoted catalysts (X and N-3) have the lowest apparent activation energy. Because the rates in these cases are quite low, I do not think that diffusion limitations can be responsible for these results. The pre-exponential factor must be somehow greatly altered by the promotor. Can you comment on these rather surprising observations ?

A. OZAKI : The promotor effect of K on Ru apparently results in a remarkable increase in the pre-exponential factor. This must be caused by a structural change in Ru surface, but we cannot give any further comment on this at present.

Z.G. SZABO : You have put emphasis on the nitrate effect, and not without reason. During the reduction, NH_3 is first formed and afterwards NH_4NO_3. Under your experimental conditions it produces N_2O. We also used this decomposition for making highly active catalysts. It is blowing up the precipitate, resulting in very high surface development. As the activation energies are nearly the same, it points to an increased pre-exponential factor. But this effect must operate at the most appropriate stage, during the preparation. This is perhaps the reason why previously added NH_4NO_3 is not advantageous.

A. OZAKI : Thank you for your comment.

LIMITING FACTORS ON THE STRUCTURAL CHARACTERISTICS OF Ru/SiO$_2$ AND RuFe/SiO$_2$ CATALYSTS

L. GUCZI, K. MATUSEK, I. MANNINGER, J. KIRÁLY and M. ESZTERLE
Institute of Isotopes of the Hungarian Academy of Sciences, H-1525 Budapest, P.O.Box 77, Hungary

ABSTRACT

The effect of different treatments on the surface characteristics of supported ruthenium catalysts has been investigated. It was established that not only at calcination followed by hydrogen treatment, but also at the reduction at 773 K, hydrogen is strongly bonded to the surface and incorporated into sub-surface layers. Repeated oxygen-hydrogen treatment removes this hydrogen and along with the disintegration of ruthenium particles the activity of catalysts is enhanced. On iron addition metallic iron is not formed. Nevertheless, the agglomeration of ruthenium particles is hindered at the preparation and the catalyst is stabilized by the presence of Fe_2O_3.

1. INTRODUCTION

In recent years much effort has been spent on supported ruthenium catalysts for their being active in direct conversion of NO_x into nitrogen [1]. Investigations on the preparation of these catalysts were mainly focused on the metallic dispersion and the catalyst stabilization under high temperature oxidizing condition.

When the catalyst prepared from $RuCl_3 \cdot 3H_2O$ supported on alumina or on silica was calcined at high temperature the formation of small RuO_2 particles has been observed by Mössbauer spectroscopy [2]. However, the following reduction by hydrogen yielded metals with large crystallite size. Similar effect was also found by Dalla-Betta [3]. Since at high temperature calcination the formation of highly volatile RuO_4 may not be excluded stabilization and fixation of ruthenium in form of ruthenate (in combination with Ba, La [4] and with Mg [5]) was proposed. However, Mössbauer data obtained on this stabilized catalyst after repeated net reducing and oxidizing cycle show a loss in ruthenium stabilization which is likely due to the separation of ruthenium metal from the stabilizing BaO phase [6].

The metallic dispersion and the catalytic activity of supported ruthenium catalysts, however, can be altered not only by calcination. At high temperature reduction in hydrogen stream the sintering of metallic phase was observed [7] but this was not the sole reason for decreased activity. Montarnal et al. [8] have shown that Ru/Al_2O_3 reduced in hydrogen at 623 K revealed much lower initial activity in ammonia decomposition that the one calcined at 773 K. However, on reduced catalyst the after the first cycle stationary activity became higher than on the oxidized form. This phenomenon predicts an effect different from that due to solely particle size variation, otherwise initial activity for reduced catalyst should have been higher. Oxidation itself at medium temperature range affects the catalytic activity either by perturbing the surface Ru atoms [9,10] or simply by removing the carbonaceous material formed at exposure to carbon monoxide [11].

The present paper is concerned with two problems. First, we wish to study the hysteresis observed in the initial activity of reduced ruthenium by using different treatments and applying a combination of different methods for determination of metal load, dispersion etc. Second, similarly to an earlier study [12] at the addition of iron as a second metal to Ru/SiO_2 the influence of iron on the stabilization of ruthenium will be investigated. As test reaction H_2-D_2 exchange, ethane-D_2 exchange and the hydrogenolysis of ethane will be used.

2. EXPERIMENTAL

2.1. Catalyst

$RuCl_3$ solution at pH = 1 is used for impregnation of silicagel (SAS, specific surface is 560 m^2 g^{-1}). After drying it at 393 K overnight the catalyst is calcined at 573 K for 1 hour followed by reduction in hydrogen at 773 K for 3 hours. In the following sections, if otherwise not stated, this standard treatment is used. For the preparation of bimetallic catalyst $RuCl_3/SiO_2$ sample before the calcination step is impregnated with $Fe(NO_3)_3$ solution in different concentrations at pH = 1. Metal concentration is determined by X-ray fluorescence method using RuKα line. (1 Ci ^{241}Am source and Sn-target is used.) Average particle size is measured by X-ray diffraction

2.2. Mössbauer spectroscopy

^{57}Fe in 88% enrichment has been incorporated into the catalyst and the Mössbauer spectra are recorded at constant acceleration mode using multichannel analyzer. ^{57}Co in Cr matrix as source and stainless steel as reference sample are applied. Deconvolution of spectra also carried out using least square fitting program [13].

2.3. Adsorption and catalytic reactions

All adsorption measurements are carried out in dynamic apparatus using nitrogen and helium carrier gases. TPD method appears to be suitable to monitor the nature of hydrogen adsorbed on the surface.

Isotope exchange between deuterium and ethane as well as between hydrogen and deuterium is analysed by AEI MS 10 C2 mass spectrometer connected directly to the catalytic apparatus. The deuterated peaks are corrected for naturally occuring carbon-13 and for fragmentation.

Catalytic hydrogenolysis of ethane is investigated in a glass circulating apparatus connected to gas chromatograph.

3. RESULTS AND DISCUSSION

In table 1 the treatment, metal concentration and X-ray diffraction data are presented. It is obvious from the data that particle size increases under oxygen treatment and at the same time some loss in Ru metal concentration occurs.

TABLE 1.

Change of metal concentration and crystallite size on different treatments

Cat. No	Treatment	Ru, content wt%	Fe, content wt%	X-ray (nm)
Ru[O]	impregnation pH = 1	0.52	-	-
Ru[H300]	H_2 at 573 K	0.52	-	amorphous
Ru[H500]	H_2 at 773 K	0.52	-	crystalline in trace
Ru[S]	standard	0.45	-	18.5 ± 2.5
Ru[O400]	O_2 at 673 K and H_2 at 773 K	0.45	-	21.5 ± 3.0
Ru[S]	spent	0.41	-	amorphous
RuFe[S] (I)	standard	0.49	0.1	amorphous
RuFe[S] (II)	standard	0.49	0.2	amorphous
RuFe[S] (II)	spent	0.35	0.2	amorphous
RuFe[S] (III)	standard	0.49	0.7	amorphous
RuFe[S] (III)	spent	0.35	0.7	amorphous

At high temperature hydrogen treatment trace of X-ray diffraction line can be observed, which means that sintering of metal particles occurs under this condition. When Ru/SiO_2 and $RuFe/SiO_2$ catalysts are compared the main characteristic is that ruthenium particle size drastically decreases on iron addition (see line 4 and e.g. line 7). On spent catalysts regardless of whether it contains iron or not the ruthenium concentration further decreases (compare lines 4 and 6) and simultaneously the large crystallites of ruthenium are disintegrated and the sample becomes amorphous. The change in metal concentration are illustrated by X-ray fluorescence spectra taken at different reatments as shown in figure 1.

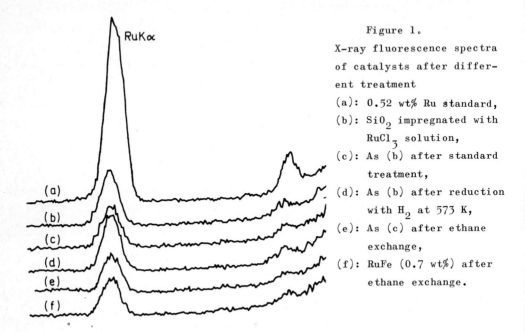

Figure 1.
X-ray fluorescence spectra of catalysts after different treatment
(a): 0.52 wt% Ru standard,
(b): SiO_2 impregnated with $RuCl_3$ solution,
(c): As (b) after standard treatment,
(d): As (b) after reduction with H_2 at 573 K,
(e): As (c) after ethane exchange,
(f): RuFe (0.7 wt%) after ethane exchange.

When catalysts prepared either by standard treatment or by reduction with hydrogen at 773 K are tested in ethane hydrogenolysis it is found that in both cases the catalytic activity measured at 523 K using 10:1 hydrogen--ethane mixture is very low (in the range of 10^{-11} mol s^{-1} g_{cat}^{-1}). Since the crystallite size is different for the two catalysts we may assume that the low catalytic activity is devoted to some extent to the high temperature hydrogen treatment but the main effect is not the sintering. Furthermore, activity enhancement of about two orders of magnitude is experienced on the repeated oxygen (20 Torr) and hydrogen (40 Torr) cycles between hydrogenolysis runs and the activity levels off after about 10 cycles.

For the elucidation of this effect it is assumed as working hypothesis that beside the difference in crystallite size between Ru/SiO_2 and $RuFe/SiO_2$ the presence of hydrogen strongly bonded to the catalyst is the reason for the low initial activity. TPD data shown in table 2 indicate that if the treatment in hydrogen is carried out at 773 K, the intensity of a second peak desorbed at 773 K increases and this sort of hydrogen cannot be removed at lower temperature, i.e. at high temperature hydrogen treatment part of metal sites remains covered.

TABLE 2

Oxygen adsorption and hydrogen TPD data

Cat. No	Treatment	O_2 adsorbed /umol g_{cat}^{-1}	TPD (Hydrogen)	
			Peak a 423 K	Peak a 773 K
Ru	Standard	trace	-	-
Ru	H_2 at 573 K	8.2	-	-
Ru	Repeated O_2-H_2 cycl.	4.17	-	-
Ru	H_2 at 573 K[a]	-	2.3	0
Ru	H_2 at 773 K[b]	-	2.3	4.6
RuFe(I)	H_2 at 573 K	7.8	-	-
RuFe(I)	Repeated O_2-H_2 cycl.	7.2	-	-
RuFe(I)	H_2 at 573 K[a]	-	2.4	1.6
RuFe(I)	H_2 at 773 K[b]	-	3.5	7.0
RuFe(II)	Standard	2.6	-	-
RuFe(II)	H_2 at 573 K	9.9	-	-
RuFe(II)	H_2 at 573 K[a]	-	4.1	2.4
RuFe(II)	H_2 at 773 K[b]	-	5.2	10.9
RuFe(III)	H_2 at 573 K	4.4	-	-
RuFe(III)	Spent	4.1	-	-

a) Catalyst treated in H_2 at 573 K and cooled down in H_2.
b) Catalyst treated in H_2 at 773 K and cooled down in H_2.

Oxygen adsorption gives only approximation to the number of surface sites and it is not as sensitive as other methods. Nevertheless, in some cases it is used bearing in mind that the exact stoichiometry is still not clear. Kubicka [14] has found a stoichiometry O/Ru is 2, while Gonzales [15] was using O/Ru equal to 1. In table 2 the data are summarized from which it appears that oxygen adsorption generally follows the particle size change determined by other methods. Ru/SiO_2 treated by standard procedure reveals large crystallite size and our method is not sensitive enough to trace oxygen adsorption (line 1). When RuFe/SiO_2 is treated in the similar way oxygen adsorption can be measured but this amount is about four times less than the one determined after hydrogen treatment only at 573 K (see line 11). That is, iron exerts stabilization effect on Ru particles during the formation of metal from $RuCl_3$. This effect is significant only at calcination because when reduction takes place under mild conditions (573 K in hydrogen) there is only a little difference in oxygen adsorption (compare lines 2 and 6). It is, however, significant that at 0.2 wt% Fe addition not only the oxygen adsorbed but the amount of hydrogen recovered by TPD measurements pass through maximum.

In order to further clarify the effect of hydrogen, reduction of Ru/SiO_2 catalyst was carried out in deuterium at 773 K. After evacuation 50 Torr hydrogen gas was admitted to the catalyst and at 473 K the deuterium removal from the surface was measured by the increase of HD peak intensity. The rate of D atoms entering the hydrogen is 6.6×10^{-8} mol s^{-1} g_{cat}^{-1}. The rate of ethane hydrogenolysis measured after evacuation is 6.6×10^{-12} mol s^{-1} g_{cat}^{-1}. At the following oxygen treatment the rate of the enter of deuterium remains the same but the rate of hydrogenolysis increase up to 1.43×10^{-9} mol s^{-1} g_{cat}^{-1}. The integral amount of deuterium removed was about twice as high as can be expected on the basis of metal surface area. Considering the simultaneous increase in catalytic activity it seems reasonable to assume that deuterium is incorporated into the sub-surface layers and at the surface "corrosion" under the effect of cyclic oxygen treatment it becomes accessible for hydrogen and for the reaction mixture. In the light of these experiments the result obtained by Montarnal [8] can be understand. During the reduction procedure hydrogen is incorporated into a few sub-surface layers and under the effect of reaction mixture particles are disintegrated with the simultaneous removal of hydrogen atoms strongly bonded to ruthenium atoms. This idea is further supported by two other experimental facts. When oxygen-hydrogen treatment is carried out in X-ray diffractometer and the Ru line at $44.0°$ is measured, the line practically disappears after about 10 cycles of this treatment. This is in agreement with that we found when fresh and spent Ru/SiO_2 are compared (see table 1). It means that the increase in particle size is rather due to the oxygen-hydrogen treatment than to the catalytic reaction.

The increase in the number of surface Ru atoms is proved if catalytic H_2-D_2 and ethane-D_2 exchange are studied during this initial stabilization period. In table 3 the results are presented. Although H-D exchange proceeds much faster than ethane-D_2 reaction, the two reactions change parallel after each oxygen-hydrogen treatment. Since H-D exchange presumably does not have structure sensitivity and directly proportional to the number of surface site the parallelism is surely ascertained to the increase of working metallic surface.

We have shown that on iron addition not only the initial particle size of the calcined and reduced catalyst is smaller but the oxygen adsorption on spent catalyst is practically not altered, either. To study the effect of iron in RuFe/SiO$_2$ catalyst Mössbauer spectroscopy on catalyst containing iron ^{57}Fe form has been applied. In figure 2 the results are presented.

TABLE 3

The rates (in mol s^{-1} g$_{cat}^{-1}$) of ethane-D$_2$ and H$_2$-D$_2$ exchange on Ru/SiO$_2$ and RuFe/SiO$_2$ catalysts as a function of the number of runs

No. runs	0 wt% Fe		0.1 wt% Fe		0.2 wt% Fe	
	C$_2$H$_6$/D$_2$	H/D exch.	C$_2$H$_6$/D$_2$	H/D exch.	C$_2$H$_6$/D$_2$	H/D exch.
1	4.2x10^{-11}	7.2x10^{-4}	5.6x10^{-8}	1.2x10^{-5}	5.8x10^{-8}	6.6x10^{-7}
2	8.1x10^{-11}	1.1x10^{-3}	4.9x10^{-7}	1.4x10^{-5}	8.7x10^{-8}	3.4x10^{-6}
3	1.1x10^{-10}	2.1x10^{-3}	2.4x10^{-6}	1.4x10^{-5}	2.1x10^{-7}	6.6x10^{-6}
4	1.6x10^{-10}	2.2x10^{-3}	3.5x10^{-6}	2.0x10^{-5}	3.2x10^{-7}	1.5x10^{-5}
5	1.8x10^{-10}	-	3.6x10^{-6}	2.0x10^{-5}	7.4x10^{-7}	2.1x10^{-5}
6	7.2x10^{-10}	-	-	-	8.9x10^{-7}	-
7	8.1x10^{-10}	-	-	-	-	-
8	1.4x10^{-9}	-	-	-	-	-
9	2.1x10^{-9}	-	-	-	-	-
10	4.2x10^{-9}	-	-	-	-	-

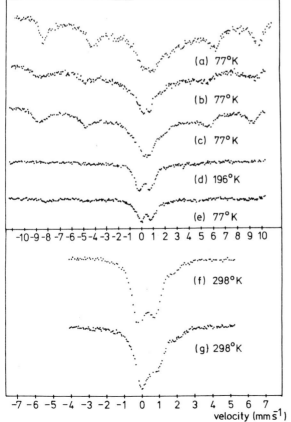

Figure 2.

Mössbauer spectra

(a) 0.5 wt% FeCl$_3$/SiO$_2$;
(b) 0.5 wt% Fe(NO$_3$)$_3$/SiO$_2$;
(c) 0.5 wt% Fe(NO$_3$)$_3$ + 0.52 wt% RuCl$_3$/SiO$_2$;
(d) 0.7 wt% Fe(NO$_3$)$_3$ + 0.52 wt% RuCl$_3$/SiO$_2$; reduced at 773 K in H$_2$;
(e) 0.7 wt% Fe(NO$_3$)$_3$ + 0.52 wt% RuCl$_3$/SiO$_2$; reduced at 773 K in H$_2$;
(f) 0.7 wt% Fe(NO$_3$)$_3$ + 0.52 wt% RuCl$_3$/SiO$_2$; reduced at 773 K in H$_2$;
(g) 0.2 wt% Fe(NO$_3$)$_3$ + 0.52 wt% RuCl$_3$/SiO$_2$; reduced at 773 K in H$_2$.
(d)-(e) measured in air,
(f)-(g) in hydrogen.

Comparing the upper 3 spectra it can be established that the six peaks in the spectrum at 77 K is a result of weak spin-spin interaction due to the Fe^{3+} ions sitting in large distance from each other; this is not effected by the anions and by the presence of ruthenium. Since at 298 K no peak was found after two million counts in each channel we may assume that the Debye-Waller factor is small, consequently Fe^{3+} ions are weakly bonded to the surface. After reduction in hydrogen the most characteristic is the decrease of intensity of peaks which appears at high velocity, i.e. which results from the paramagnetic relaxation. It shows a stronger spin-spin interaction, therefore metallic particles are agglomerated to some extent. When measurements are carried out in hydrogen (sample has not been exposed to air after reduction) metallic iron is not formed, only some amount of Fe^{2+} (in about 6% concentration). Formation of RuFe bimetallic phase has not been detected at all, although the Mössbauer parameters for this alloy are well known [16]. We may say, therefore, that unlike the PtFe system [17] it is not the super-structure phase which increases the metallic dispersion. Since on calcination in oxygen RuO_2 and Fe_2O_3 are formed and this latter cannot be reduced into metallic phase as indicated by Mössbauer spectra, the strong interaction between Fe_2O_3 oxide phase and ruthenium formed after reduction is the reason why the migration of metallic ruthenium particles to form large crystallites is hindered. The valence state of iron at oxygen-hydrogen treatment is not changed, either. In table 4 it is shown that at oxygen treatment at 573 K the Fe^{2+} content is reduced and it remains the same at the following hydrogen treatment. The reduction at elevated temperature restores the original Fe^{2+} content.

TABLE 4

Mössbauer parameters for $RuFe/SiO_2$ catalyst at different treatments

Sample	$C_{Fe^{2+}}$	Fe^{2+}			Fe^{3+}			$d\chi^2/df$
		δ, mm s^{-1}	Δ, mm s^{-1}	Γ, mm s^{-1}	δ, mm s^{-1}	Δ, mm s^{-1}	Γ, mm s^{-1}	
1	0.067 (0.015)	0.88 (0.04)	1.93 (0.08)	0.69 (0.09)	0.30 (0.06)	1.04 (0.01)	0.98 (0.01)	1.22
2	0.026 (0.01)	1.09 (0.03)	2.22 (0.05)	0.44 (0.08)	0.30 (0.01)	1.04 (0.01)	1.02 (0.01)	1.03
3	0.068 (0.01)	1.00 (0.02)	1.97 (0.05)	0.58 (0.06)	0.31 (0.01)	0.91 (0.01)	1.10 (0.02)	1.69

Sample 1: Standard treatment; Sample 2: O_2-H_2 treatment at 523 K; Sample 3: H_2 treatment at 873 K; Calculation: constrained doublet for Fe^{2+} and Fe^{3+} with equal linewidth and amplitude; $C_{Fe^{2+}} = [Fe^{2+}]/([Fe^{2+}]+[Fe^{3+}])$; figures in paranthesis are the errors of calculated.

Since Mössbauer data give clear evidence that iron in metallic phase is not formed direct metal-to-metal interaction cannot be operative. This is shown by the distribution of deuterated products of ethane which are the same regardless of the iron content. The sole effect is the increase in the number of active sites on the effect of iron. Further important evidence [18] for this statement is that in ethane and n-butane hydrogenolysis no change in energy of activation and selectivity has been observed, the only effect is the increase in activity. Unlike the $PtFe/SiO_2$, where the formation of bimetallic system was also proved by the change in activation energy and selectivity, here only the morphology and structure are affected indirectly by iron addition by hindering the migration of metallic particles during the preparation and reaction. Consequently ruthenium is stabilized by the presence of iron without any electron interaction.

4. CONCLUSIONS

Two main conclusions can be obtained from the results:

i) Hydrogen after the reduction period is incorporated also into the sub--surface layers of Ru particles and it is strongly bonded. Repeated oxygen--hydrogen cycle can be used not only to disintegrate the particles but to remove the hydrogen bonded strongly and this procedure results in the enhancement of catalytic activity regardless of the nature of reaction.

ii) Iron addition hinders the agglomeration of particles, although iron itself is not reduced into metallic phase. This effect can be indicated by the constancy of kinetic parameters in all reactions studied and the increase in activity is exclusively due to the larger number of surface ruthenium atoms.

REFERENCES

1. M. Shelef and H.S. Gandhi, Ind. Eng. Chem. Prod. Res. Develop. 11(1972)2
2. C.A. Clausen III and M.L. Good, J. Catal. 38(1975)92
3. R.A. Dalla-Betta, J. Catal. 34(1974)57
4. M. Shelef and H.S. Gandhi, Plat. Metals. Rev. 18(1974)2
5. S.J. Tauster, L.L. Murrell and J.P. DeLuca, J. Catal. 48(1977)258
6. C.A. Clause III and M.L. Good, J. Catal. 46(1977)58
7. A.A. Davydov and A.T. Bell, J. Catal. 49(1977)332
8. A.G. Friedlander, Ph.R. Courty and R.E. Montarnal, J. Catal. 48(1977)312
9. K.C. Taylor, J. Catal. 30(1973)478
10. M.F. Brown, and R.D. Gonzales, J. Phys. Chem. 80(1976)1731
11. R.A. Dalla-Betta and M. Shelef, J. Catal. 48(1977)111
12. L. Guczi, K. Matusek and M. Eszterle, paper submitted, J. Catal.
13. I. Dézsi, D.L. Nagy, L. Guczi and M. Eszterle, Reaction Kinetics and Catalysis Letters, Vol.8(1978)301

14. H. Kubicka, Reaction Kinetics and Catalysis Letters, 5(1976)223
15. R.D. Gonzales, personal communication
16. H. Ohno, J. Phys. Soc. Japan, 31(1971)92
17. L. Guczi, K. Matusek, M. Eszterle, Paper presented at EUCHEM Conference, Namur, 1977
18. L. Guczi, K. Matusek, M. Eszterle, F. Till and D.L. Nagy, to be published in Acta Chim. Acad. Sci. Hung.

DISCUSSION

H. CHARCOSSET : 1) With respect to your H_2-TPD experiments (Table 2), we have observed over Pt/SiO_2 the same phenomena as described by you for Ru/SiO_2 : 2 peaks are obtained after cooling down from 773K instead of one after cooling from about 500K. The same experiments performed over a Pt wire in UHV conditions of desorption did not show this effect. We arrived to the conclusion that the support or(and) its impurities were involved in the experiments on Pt/SiO_2. Have you considered a possible effect of the support in your case, instead of penetration of H in the sub-surface layers ?
2) You suppose that O_2-H_2 cycles disintegrate the Ru particles. This could be proved by microscopic examinations. Have you experimental data with regard to that possibility ?

L. GUCZI : We have measured in the TPD experiments on Pt/SiO_2 only one peak at 423K. The blank SiO_2 showed a small peak at 923K but the one obtained in case of Ru/SiO_2 is higher than on the support itself. Moreover, the temperature is lower for Ru/SiO_2 than for SiO_2. Nevertheless, the peak coming at 773K for Ru/SiO_2 is characteristic for the strength of the hydrogen sorption rather than for the absolute amount. However, you are right that we need additional experiments concerning the deuterium exchange with the support itself.
The disintegration of Ru particles has been proved by X-ray diffraction. The spent Ru/SiO_2 after having undergone several O_2-H_2 cycles, does not show X-ray lines and it becomes amorphous. At the same time the rate of H_2/D_2 exchange increases indicating that the number of active sites responsible for the catalytic activity also increases.

B. DELMON : At the end of your conclusion, you write "Iron addition hinders the agglomeration of particles, <u>although</u> iron itself is not reduced into metallic phase". I wonder whether this "although"

should not be "because". What is your opinion concerning the
interpretation I suggest, namely that oxides quite often efficiently
play the role of "spacers" or of grain growth inhibitors ?

L. GUCZI : Since there is no metal-metal interaction as a result of
the fact that iron cannot be reduced beyond the Fe^{2+} state, we
believe that Fe_2O_3 particles which are very finely dispersed interact
with ruthenium atoms and prevent their agglomeration to form larger
crystals. Thus the substitution of the word "although" by "because" is
correct.
At the present stage we still do not know if this effect starts at
the nucleation to form particles. However, the migration of small
ruthenium clusters are prevented by iron oxide.

M. CARBUCICCHIO : Have you interpreted the paramagnetic central
contribution of the Mössbauer spectra obtained for unreduced
catalysts (Fig. 2a, b, c) ?

L. GUCZI : These spectra are typical for isolated paramagnetic Fe^{3+}
ions (probably in the form of some ferric aquo-complex) with long
spin-relaxation times. The hyperfine field corresponding to the
outmost lines is 572 $\pm$ 2 kG.

P.A. SERMON : TPD results (Gombeis et al., Naturwisenschaffen, (1977))
have shown that H_2 spillover onto Al_2O_3 from Pt may be desorbed
at high temperature. Were your results for Pt/SiO_2 (mentioned to
H. Charcosset) obtained under identical reduction and desorption
conditions as those for Ru/SiO_2, or might the H_2 taken up by
Ru/SiO_2 at 600-700K and desorbed at 773K be spillover rather than
subsurface ? If this is not true, are there sufficient "internal"
Ru atoms to accommodate the additional 2H atoms per "surface" Ru
atom (suggested by TPD and deuterium work) in a subsurface state,
particularly in view of the high Ru dispersion indicated by O_2
chemisorption and XRD ?

L. GUCZI : We may exclude the effect of spillover in the case of
Pt/SiO_2 and Ru/SiO_2. The experimental facts are as follows : 1) on
Pt/SiO_2, H/Pt is equal to 1.3 if the O/Pt value is equal to 1.
It means that considerable amount of hydrogen is not taken up by the
support. 2) If we reduce a 5% $FePt/SiO_2$ having a Fe/Pt ratio

∿ 2.2, the amount of iron above the stoichiometric ratio cannot be reduced. The reason is probably that hydrogen activated at Pt can reduce the iron which is in contact with platinum and hydrogen cannot migrate far away.

We believe in our idea based upon the facts that O_2-H_2 titration increases the number of surface sites (parallelism between H/D axchange C_2H_6-D_2 exchange and the disappearance of the Ru line at 44°) that disintegration of the particles occurs.

R. MONTARNAL : You mentioned our results (your ref.8) on NH_3 decomposition (300°C) with respect to the need for activation by NH_3 at relatively high temperatures (700-800°C), of a Ru catalyst reduced in H_2.

On the contrary, simple calcination of precursors of the Ru catalysts immediately gives catalytic activity for NH_3 decomposition at 300°C.

If I have well understood, you attribute the need of activation for the reduced catalyst to the presence of strongly inhibiting adsorbed atoms which are eliminated by activation by NH_3 at 700°C, giving simultaneously disintegrated small Ru particles. It seems to us that we can agree with your interpretation which is a more precise description of what we called reconstruction phenomenon under the influence of reactants.

L. GUCZI : The term reconstruction used in your paper can be influenced by two factors : one is the removal of hydrogen strongly bound to Ru particles and the other is the disintegration of Ru particles into smaller crystallites.

J. SCHEVE : Can you explain the loss of ruthenium during calcination and reaction ? If due to sublimation of ruthenium oxide it should also occur during the O_2-H_2 cycle.

L. GUCZI : We did not look deeply for the reason of the loss of Ru from the surface. Probably the main loss occurs during the calcination step because of the high temperature treatment in oxygen atmosphere. Additional loss seems to be of minor importance.

P.G. MENON : Your results on Ru catalysts are very similar to our observations on Pt-Al_2O_3 catalysts, which I mentioned earlier in connection with the paper of Scholten. Exposure of the catalyst to high temperature (500°C) in H_2 results in a stronger chemisorption of H_2 on the catalyst. But a few H_2-O_2 cycles at 20°C seem to anneal or smooth out the heterogeneity of the catalyst. It is remarkable that this effect is seen for both supported Pt and Ru catalysts. Perhaps this phenomenon is of a more general occurrence in supported metal catalyst than hitherto suspected.

L. GUCZI : Just one remark: the O_2-H_2 cycle was carried out at the reaction temperature.

A. OZAKI : Deuterium displacement with SiOH must be taken into account when you work on metal/SiO_2 catalyst.

L. GUCZI : It might be possible that the excess amount of deuterium entering the gas phase is partly due to the SiOH groups. However according to TG-MS data after treatment at high temperature for a long time, the major part of the structural OH groups is destroyed. So their contribution seems to be negligible.

PREPARATION ASPECTS OF Ru-SUPPORTED CATALYSTS AND THEIR INFLUENCE ON THE FINAL PRODUCTS

A. Bossi, F. Garbassi, A. Orlandi, G. Petrini and L. Zanderighi[+]
"G. Donegani" Research Institute, Montedison SpA, Novara (Italy)
[+]Università di Milano, Istituto di Chimica Fisica, Milano (Italy)

ABSTRACT

A wide investigation of the influence of some preparation aspects, such as the nature of the support, the acidity of the precursor solution and the activation temperature, was carried out on Ru catalysts supported on SiO_2, Al_2O_3, SiO_2-Al_2O_3 and MgO.

For this purpose several bulk and surface techniques, such as thermal analysis, chemisorption, transmission electron microscopy, X-ray photoelectron spectroscopy and Auger electron spectroscopy, were employed.

The dependence of the metal dispersion on the homogeneity of the metal distribution is discussed, together with the connection between the latter and the precursor solution acidity. A bimodal distribution of Ru-supported particles, massive metal crystallites and bidimensional sheets, was found. Binding energy measurements demonstrated that only on silica is Ru in a metallic state, while on the other supports an oxidized ruthenium species is prevalent, suggesting an interaction of the bidimensional sheets with the support.

INTRODUCTION

The increased yield in a metal-catalyzed reaction is in principle related to an increase of the specific surface of the metal. When a high surface area is obtained by dispersing the metal on a suitable support, changes in activity and selectivity ascribed to the so called "crystallite-effect" and/or to the "support effect" can be observed (ref. 1).

Due to the dispersion, defined as the ratio between surface and total amount of the metal atoms, particles of different size and shape can be present on the support surface. According to their physical behaviour such particles might be roughly classified in four different species, with different chemical and catalytic properties : metal crystals, metal crystallites (refs. 2 and 3), metal clusters (refs. 4 and 5), sheets or rafts (refs. 6-8). Some of them give rise to charge transfer or chemical bonding with the support (refs. 9 and 10). The procedure followed during the preparation of the dispersed metal can influence the presence of one or more of such species on the support surface. Thus it is important to characterize the physico-chemical state of the metal in such catalysts and to connect these properties with the preparation procedure.

In this work, we have investigated the influence of some aspects of preparation on the physico-chemical characteristics of supported ruthenium, which is known to be a catalyst for many chemical reactions (refs. 11-16).

EXPERIMENTAL

Sample preparation

Commercial microspheroidal supports like SiO_2 (S samples, 680 m^2/g), Al_2O_3 (A samples, 320 m^2/g) and $SiO_2-Al_2O_3$ (SA sample, 620 m^2/g) were used, while MgO (M sample, 230 m^2/g) was prepared by decomposition of $Mg(OH)_2$ in vacuo. The precursor salt was a commercial ruthenium trichloride (Rudi-Pont), containing an excess of Cl^- (ref. 16)[+].

Aqueous solutions (sol. 1), 1/10 diluted (sol. 2) and concentrated (sol. 3) acetic solutions and 1/10 diluted (sol. 4) nitric solutions were used for the impregnation. MgO was impregnated with an aqueous solution of $RuNO(NO_3)_x$, where $x \simeq 3$. Impregnations were made by the incipient wetness method, followed by 12 hrs drying at 383 K after overnight equilibration. The catalysts were then activated with H_2 in a fluidized bed (10 1/h) according to the following procedure : 4 hrs heating from room temperature to the activation temperature (673, 773 or 873 K), 2 hrs rest at this temperature and then discharging in air, after cooling. Only S1 and A1 samples were discharged and kept in an inert atmosphere.

Chemical analysis

The Ru content in the supported samples was determined by atomic absorption spectroscopy after dissolution of the metal as Ru^{3+} (ref. 17). The chlorine content was determined by potentiometric titration with $AgNO_3$ (ref. 18).

Surface chemical compositions and binding energies

The surface chemical analysis was performed on all samples both with Auger electron spectroscopy (AES) and X-ray photoelectron spectroscopy (XPS, MgKα radiation) in a commercial PHI instrument. Targets were prepared from powder samples by insertion in pure In foil (ref. 19). Quantitative analysis was based on XPS peak areas, corrected according to the relative photoelectric cross-sections

---

[+]The precursor analysis has given a Cl/Ru ratio of 3.77, with 2.85 H_2O.

(ref. 20), rather than on AES peak-to-peak heights, where empirical sensitivity factors must be used (ref. 21). Auger analysis however proved useful in revealing contaminant traces.

The following peak areas, O_{1s}, Al_{2p}, Si_{2p}, Mg_{2p}, Cl_{2p} and $Ru_{3p3/2}$ were measured. The last was choosen instead of the stronger $Ru_{3d5/2}$ transition because of its overlap with the C_{1s} contamination peak. Electrostatic charging occurred during measurements on insulator samples, affecting the experimental binding energy values. A flood gun was used to remove this effect and the C_{1s} peak at 285.0 eV was assumed as reference. In such determinations, the $Ru_{3d5/2}$ transition was used, after subtraction of the carbon contribution.

Total and metal surface area

Total surface area and pore size distribution measurements were carried out by N_2 adsorption at 77 K in a static volumetric apparatus.

Metal surface areas were determined by means of O_2 chemisorption at room temperature, both by the static and pulse (ref. 22) technique, with good agreement between the two methods. Heating for 4 hrs at 673 K in H_2 flow, followed by a degassing in vacuum for 45 min and a cooling to room temperature preceded all static measurements. In the case of pulse experiments the degassing was carried out in He.

Thermal analysis

The reduction of pure $RuCl_3$ and of ruthenium chloride supported on SiO_2 and Al_2O_3 was carried out in a horizontal Linseis L-81 thermobalance. 100-150 mg of the previously dried (at 383 K) samples were put in a dry H_2 flow (10 l/hr) at room temperature until a constant weight was reached. Then the sample was heated at a rate of 10K/min up to 800 K in H_2 while flow recording the weight loss as a function of the temperature.

Under the same experimental conditions, DTA runs carried out on the supported samples did not reveal any thermal effect corresponding to the halide reduction, due to the very low Ru content.

Transmission electron microscopy (TEM)

TE micrographs were obtained using a Philips EM-300 microscope, supplied with the high resolution stage. Direct and extractive replica observations were performed (ref. 23). Electron diffraction experiments proved the presence of ruthenium on the target.

RESULTS AND DISCUSSION

Table 1 lists some qualitative observations of phenomena occuring during impregnation and drying. The dark-red aqueous solution of ruthenium chloride does not change its colour on contact with S and SA supports, but becomes black with A support. By increasing the acidity of the precursor solution no colour change is observed. This seems to indicate that only alumina gives a strong interaction with the ruthenium salt, with the precipitation of hydrolysis products. By washing the samples after drying the extractability of Ru from the catalyst increases with the acidity of the precursor solution and appears greater on silica than on alumina samples.

TABLE 1

Behaviour of the Ru-supported catalysts during impregnation and drying

Sample	Colour after impregnation	Colour after drying at 383 K	Ru extractability
S1	dark-red	black	low
S2	red-violet	brown	high
S3	red-violet	brown	high
S4	dark-red	black	low
A1	black	black	absent
A2	dark-violet	black-violet	high
A3	black	black	low
A4	black	black	very low
SA1	dark-red	black	low
M1	black	gray	absent

The influence of the support and of the reduction temperature on the metal distribution into the catalyst particle is shown in Tables 2 and 3, where bulk and surface chemical analysis data are collected. In all S samples the amount of chlorine is very low, not detectable on the surface by XPS and decreases as a function of the reduction temperature. On the contrary on A samples it seems that all the initial chlorine is present on dried samples, with a slight lowering due to subsequent reduction. Moreover the surface Cl/Ru ratio decreases remarkably suggesting the formation of Cl-poor hydrolysis products.

Assuming the ratio between the concentrations of external and total ruthenium (Ru_{ex} and Ru_t respectively) as a measure of the homogeneity of the metal distribution in a single catalyst pellet and calling it DIRS (Distribution Index Relative to the Surface), we can note that DIRS values greater than 1, as for A and SA samples, indicate an interaction between the support and the precursor.

TABLE 2

Influence of the support on the metal distribution

Sample	Bulk chemical analysis[+]			Surface chemical analysis[+]			DIRS
	Ru	Cl	Cl/Ru	Ru	Cl	Cl/Ru	
S1-673	3.0	0.9	0.3	2.5	–	–	0.8
A1-673	2.9	9.9	3.4	27.5	10.1	0.4	9.5
SA1-673	3.1	0.9	0.3	13.2	–	–	4.2
M1-673	3.2	–	–	3.0	–	–	0.9

TABLE 3

Influence of the reduction temperature on the metal distribution

Sample	Bulk chemical analysis[+]			Surface chemical analysis[+]			DIRS
	Ru	Cl	Cl/Ru	Ru	Cl	Cl/Ru	
A1-383	2.8	11.1	4.0	25.7	16.1	0.6	9.2
A1-673	2.9	9.9	3.4	27.5	10.1	0.4	9.5
A1-773	3.2	7.1	2.2	26.6	9.6	0.4	8.3
A1-873	3.2	6.7	2.1	22.2	10.7	0.5	6.9
S1-383	3.1	8.2	2.7	3.0	–	–	1.0
S1-673	3.0	0.9	0.3	2.5	–	–	0.8
S1-773	3.3	0.7	0.2	3.9	–	–	1.2
S1-873	3.4	0.5	0.1	3.0	–	–	0.9

[+]Data expressed as atoms $\times 10^4$/g. cat.

The interaction of the ruthenium solution with the basic sites of alumina is likely to determine the precipitation of hydrolysis products, as insoluble oxychlorides or hydroxides, on the external surface of the particle. These compounds inhibit the further diffusion of Ru within the particle pores with consequent increase of the metal concentration at the external particle shell. By increasing the acidity of the solution the basic sites are neutralized and the DIRS value decreases to about 1 in the sample A2 (Table 4).

TABLE 4

Influence of the precursor acidity on the metal distribution

Sample	Bulk chemical analysis[+]			Surface chemical analysis[+]			DIRS
	Ru	Cl	Cl/Ru	Ru	Cl	Cl/Ru	
A1-673	2.9	9.9	3.4	27.5	10.1	0.4	9.5
A2-673	3.2	8.5	2.6	4.4	8.9	2.0	1.4
A3-673	2.9	7.6	2.7	11.2	8.9	0.8	3.9
A4-673	2.4	6.3	2.7	6.8	–	–	2.9
S1-673	3.0	0.9	0.3	2.5	–	–	0.8
S2-673	3.1	0.9	0.3	2.5	–	–	0.8
S3-673	3.3	0.9	0.3	3.3	–	–	1.0
S4-673	3.2	0.8	0.3	2.5	–	–	0.8

The interaction between unreduced ruthenium and support is proved by the thermogravimetric runs reported in fig. 1. The experimental results for ruthenium chloride, ruthenium dioxide and samples S1 and A1 are expressed as mg/100 mg Ru. A correction has been made for the catalysts due to the weight loss of the support according to the equation :

$$\Delta W_{Ru} = \frac{\Delta W_{cat} - \Delta W_s X_s}{X_{Ru}}$$

where ΔW_{cat} and ΔW_s are the weight losses referred to 100 mg of catalyst and support respectively; X_s and X_{Ru} are the weight fractions of support and metal in the catalyst. The reduction temperature of Ru in the samples S1 and A1 is lower than in the pure trichloride, but similar to RuO_2. The weight loss of the sample S1 at 520 K indicates the total elmination of Cl and complete reduction of the precursor to metal. On the sample A1 the chlorine loss is still incomplete at 773 K. All these results are in good agreement with the chemical analysis data.

Fig. 2 shows the binding energy values corresponding to the $Ru_{3d5/2}$ transition as a function of the reduction temperature of the samples. As reference, values corresponding to some pure compounds are reported (ref. 24). The analysis of these results produces some interesting findings. S samples, when reduced at high temperature, become similar and show binding energy values between Ru and RuO_2, indicating that the most abundant species of ruthenium on the surface is not in a metallic state. Conversely the sample kept in an inert atmosphere after reduction shows only metallic Ru on the surface. It seems that the highly dispersed ruthenium is particularly active with oxygen at room temperature.

On A samples binding energy values near to those of RuO_2 are always observed in the sample kept in an inert atmosphere. As a reaction between ruthenium and residual water cannot be postulated for thermodynamic reasons, an interaction between the metal and the support must take place (ref. 25). This suggests that a measurable fraction of Ru may be present in a physical state able to produce charge transfer effects between metal and support, like clusters, sheets or dispersed atoms.

The sample SA1 has a binding energy very similar to those of the A series whereas that of sample M is higher. In both cases Ru is therefore present in a high oxidation state.

N_2 adsorption isotherms have shown that the preparation parameters have only a little influence on the total surface area, pore volume, pore shape and pore size distribution. The metal dispersion data obtained by O_2 chemisorption and TEM are collected in Tables 5, 6 and 7.

TABLE 5

Influence of the support on the metal dispersion

Sample	Oxygen chemisorption		TEM
	Dispersion	d(Å)	D(Å)
S1-673	0.43	22	86
A1-673	0.26	34	207
SA1-673	0.33	28	210
M1-673	0.37	24	–

The apparent particle size, d, was calculated by oxygen chemisorption at room temperature assuming a spherical model and an adsorption stoichiometry O/Ru = 2

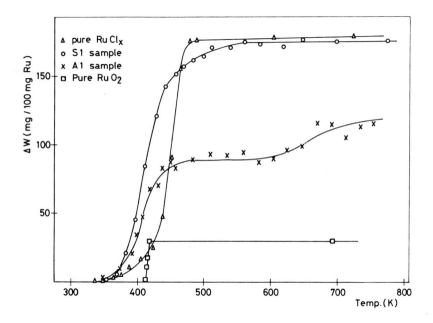

Fig. 1. Weight loss curves of ruthenium dioxide, ruthenium chloride and samples S1 and A1 after drying at 383 K.

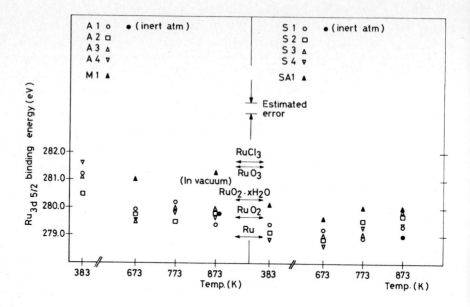

Fig. 2. $Ru_{3d5/2}$ binding energy in function of the thermal treatment. Binding energy values for Ru metal and compounds are reported.

(ref. 26). The ruthenium site area (9.03 $Å^2$) was calculated as the average value for the (001), (110) and (100) planes (ref. 27).

As a consequence of the influence of acidity on the metal distribution of the samples there is a dispersion increase with the precursor acidity. An influence the anion is apparent from the results of the samples A2 and A4. Nevertheless th influence disappears after high temperature treatment.

TABLE 6

Influence of the precursor acidity and of the reduction temperature on the dispersion of Al_2O_3-supported ruthenium

Sample	Oxygen chemisorption Dispersion	d(Å)	TEM D(Å)
A1-673	0.26	34	207
A1-773	0.26	34	221
A1-873	0.26	34	-
A2-673	0.15	62	300
A2-773	0.26	34	87
A2-873	0.35	26	-
A3-673	0.19	48	280
A3-773	0.21	44	82
A3-873	0.33	28	-
A4-673	0.42	22	253
A4-773	0.42	22	76
A4-873	0.45	20	-

The significant variation of the particle size against reduction temperature, particularly evidenced by TEM (Tables 6 and 7), must be attributed to the different heating rates during the activation process. A competition between the reduction and agglomeration rates occurs, so that with high heating rates small particles are obtained.

Finally, the particle size data from TEM and oxygen chemisorption differ greatly in all samples. This behaviour is in agreement with other observations (ref. 8). Bearing in mind that X-ray diffraction spectra did not give rise to definite metal patterns, we suggest that ruthenium is present in the examined samples in a bimodal distribution, as massive crystallites and bidimensional sheets of atoms causing a high dispersion degree.

CONCLUSIONS

From the analysis of the experimental results, some conclusions concerning the preparation of Ru-supported catalysts can be drawn.

When the support surface exchanges the hydroxyl groups with the precursor solution anions, the formation of a film of Ru hydroxide or oxychloride occurs on the external shell of the support particles, inhibiting the successive diffusion of

TABLE 7

Influence of the precursor acidity and of the reduction temperature on the dispersion of SiO_2-supported ruthenium

Sample	Oxygen chemisorption Dispersion	$d(Å)$	TEM $D(Å)$
S1-673	0.43	22	86
S1-773	0.44	20	168
S1-873	0.40	22	80
S2-673	0.45	20	–
S2-773	0.42	22	–
S2-873	0.48	18	–
S3-673	0.37	24	210
S3-773	0.32	28	110
S3-873	0.35	26	75
S4-673	0.35	26	152
S4-773	0.37	24	–
S4-873	0.37	24	–

ruthenium into the pores. This phenomenon results in a large inhomogeneity of the metal distribution, with DIRS values greater than 1. By increasing the acidity of the precursor solution, it is possible to neutralize the basic sites, producing a more homogeneous catalyst (DIRS $\simeq$ 1).

The final metal particle size depends on the kind of support and on the acidity of the precursor solution. When an interaction between solution and support occurs, a low dispersion degree is obtained, as a consequence of a high metal concentration on the external shell of the catalyst particle. With acidification, a higher degree of dispersion is reached, together with a more homogeneous metal distribution. Thus, control of the above parameters allows the desired dispersion degree to be achieved.

On the supports considered, a bimodal distribution of Ru particles is present : metal crystallites and bidimensional sheets. On alumina, silica-alumina and magnesia the highly dispersed ruthenium appears in an oxidized state, while on silica the most abundant species is metallic Ru. The presence of an oxidized form of Ru suggests an interaction between the metal and the semiconductor like a charge transfer or a weak chemical bond. The nature of such oxidized Ru species and their dependence on the preparation parameters has not yet be ascertained.

ACKNOWLEDGEMENTS

We are indebted to Prof. S. Pizzini for his interest in this work and to R. Bertè, G. Mittino, G. Morelli and L. Pozzi for the technical assistance in the experimental work.

REFERENCES

1 R.L. Moss, in R.B. Anderson and P.T. Dawson (Ed.), Experimental Methods in Catalytic Research, Vol. III, Academic Press, New York, 1976,pp.43-94.
2 F.L. Williams and M. Boudart, J. Catalysis, 30(1973)438-443.
3 R. Bouwman, G.J.M. Lippits and W.M.H. Sachtler, J. Catalysis, 25(1972)350-361.
4 R.P. Messmer, S.K. Knudson, K.H. Johnson, J.B. Diamond and C.Y. Yang, Phys. Rev. B, 13(1976)1396-1415.
5 P. Fantucci and P. Balzarini, EUCHEM Conference, Namur, 1977.
6 L. Spenadel and M. Boudart, J. Am. Chem. Soc., 64(1960)204-207.
7 P. Debye and B. Chu, J. Am. Chem. Soc., 66(1962)1021-1027.
8 E.B. Prestridge, G.H. Via and J.H. Sinfelt, J. Catalysis, 50(1977)115-123.
9 F. Solymosi, Catal. Rev., 1(1967)233-255.
10 J. Escard, C. Leclère and J.P. Contour, J. Catalysis, 29(1973)31-39.
11 G.C.A. Schuit and L.L. van Reijen, Adv. Catalysis, 10(1958)242-317.
12 J.H. Sinfelt and D.J.C. Yates, Nature, 229(1971)27.
13 J.L. Carter, J.A. Cusumano and J.H. Sinfelt, J. Catalysis, 20(1971)223-229.
14 P.H. Emmett, in E. Drauglis and R.I. Jaffee (Ed.), The Physical Basis for Heterogeneous Catalysis, Plenum Press, New York, 1974, p.3.
15 H.H. Storch, N. Golumbic and R.B. Anderson, The Fischer-Tropsch and Related Syntheses, Wiley, New York, 1951, p.309.
16 W.P. Griffith, The Chemistry of the Rarer Platinum Metals, Interscience, London, 1967, p.136.
17 N. Bottazzini, L. Fenoggio and A. Gozzi, Montedison Internal Rep. 11/77 (1977).
18 N. Bottazzini and A. Gozzi, Montedison Internal Rep. 13/77 (1977).
19 G.E. Theriault, T.L. Barry and M.J.B. Thomas, Anal. Chem., 47(1975)1492-1493.
20 J.H. Scofield, Lawrence Livermore Laboratory Rep. UCRL-51326 (1973).
21 P.W. Palmberg, Anal. Chem.,45(1973)549A-556A.
22 N.E. Buyanova, A.P. Karnaukhov, N.G. Koroleva, I.D. Rotner and O.N. Chernyavskay Kinet. Catal., 13(1972)1364-1369.
23 G. Dalmai-Imelik, C. Leclercq and I. Mutin, J. Microscopie, 20(1974)123-132.
24 K.S. Kim and N. Winograd, J. Catalysis, 35(1974)66-72.
25 C.A. Clausen and M.L. Good, J. Catalysis, 46(1977)53-64.

26 H. Kubicka, React. Kinet. Catal. Letters, 5(1976)223-228.
27 H. Kubicka, J. Catalysis, 12(1968)223-237

DISCUSSION

L.L. MURRELL : The observation of the apparent oxidization of ruthenium on Al_2O_3 compared to ruthenium on SiO_2 is apparently not a drastic difference because of the very similar infrared spectra of CO chemisorbed on Ru on Al_2O_3, SiO_2, and MgO. Could you comment on the extent of the degree of "oxidation" of the reduced Ru rafts on Al_2O_3 observed in your work ?

A. BOSSI : We have not yet made infrared spectra on Ru on SiO_2 and MgO. From the preliminary results on Al_2O_3 it appears that the amount of non metallic ruthenium depends on the history of the sample. About the degree of "oxidation" of non metallic ruthenium, XPS data give obviously an average value for both types of Ru particles. Binding energy values on Al_2O_3 are significantly different from those on SiO_2, with a chemical shift similar to that of "non-defective" RuO_2 (see K.S. Kim & N. Winograd, J. Catalysis 35 (1974), 66).

H. CHARCOSSET : In a general way your results are in agreement with ours (paper B 8): e.g. with regard to the difference in reducibility of Ru deposited either on Al_2O_3 or on SiO_2. Could you comment on the chemisorption stoichiometries of different gases on Ru ? Have you carried out H_2 or CO chemisorption measurements on your catalysts and compared the results with those obtained with O_2 as adsorbate ?

A. BOSSI : According to our experimental results, on pure ruthenium the adsorption stoichiometries appear to be H/Ru = 1 and O/Ru = 1 (see also R.A. Dalla Betta, J. Catalysis, 34, (1974), 57, and K.C. Taylor, J. Catalysis 30 (1973), 478). On the other hand, on well dispersed supported ruthenium,the O/Ru ratio appears to be related to the binding energy data presented in the paper (Fig. 2). We carried out CO adsorption experiments on Ru/Al_2O_3 at different temperatures in a static system and we found that the amount of CO adsorbed significantly increases in a time range of 16 hours. It is our opinion that an interaction between CO and bulk ruthenium takes places. Work is in progress in order to obtain detailed information on this phenomenon.

A. OSAKI : If you have any data on H_2 consumption during reduction, it would be helpful.

A. BOSSI : We have not made such measurements.

J.W. GEUS : Can you explain in some detail how you distinguish by means of bright and dark field electron micrographs globular and sheet-like (denoted rafts by the ESCA workers) particles ?
This cannot be done by comparing the contrast of the particles in bright-factor micrographs as the contrast is mainly due to the particles being oriented in a Bragg reflection position or not.

S.R. TENNISON : We have found "high transparency" crystallites similar to those found by the authors when studying platinum and ruthenium supported on graphite, a system where there is no possibility for support interaction. The transparency was such that the support structure could be observed through up to 5 or 6 superimposed 40 Å Pt crystallites in the high resolution transmission micrographs (x 1.600.000). However, when the sample was tilted by ± 30° the results showed the crystallites to be approximately spherical. This demonstrated the risks involved in assuming raft formation simply on the basis of transparency in the HRTEM results. A comparison of crystallite sizes determined by X-ray line broadening and electron microscopy also demonstrated the problems involved in a reliable detection of supported metal particles smaller than approximately 15 Å. For highly dispersed catalysts, XRD consistently gives significantly lower average crystallite sizes than HRTEM.

HIGH SURFACE AREA OXIDE SOLID SOLUTION CATALYSTS

A.P.HAGAN,[+] M.G.LOFTHOUSE,[++] F.S.STONE and M.A.TREVETHAN[+++]
School of Chemistry, University of Bath, Bath BA2 7AY, England

SUMMARY

NiO-MgO and CoO-MgO catalysts (1 to 10 mole % NiO or CoO) have been prepared in high surface area form (40 to 300 m^2g^{-1}) by impregnation of $Mg(OH)_2$ with Ni or Co nitrate, followed by thermal decomposition and annealing in vacuo at $1000°C$. The resulting solids have been characterized by a combination of X-ray diffraction, electron diffraction and magnetic susceptibility studies for their bulk properties and by UV-visible reflectance spectroscopy and adsorption for surface properties.

Lattice parameters, magnetic moments and Weiss constant values, all determined after high temperature outgassing and without exposure to the atmosphere, are fully consistent with the vacuum-annealed catalysts being true substitutional solid solutions $Ni_xMg_{1-x}O$ and $Co_xMg_{1-x}O$, with Ni^{2+} and Co^{2+} in octahedral coordination. Reflectance studies, however, reveal the presence of Ni and Co ions in other coordinations, notably Co^{2+} in tetrahedral coordination in $Co_xMg_{1-x}O$, and these ions are considered to be located in the surface regions of the crystallites. The effects of adsorbing O_2, CO_2 and H_2O vapour on $Ni_xMg_{1-x}O$ and $Co_xMg_{1-x}O$ are described and discussed. A calorimetric study of the adsorption of NO and CO and their chemical interaction on $Ni_xMg_{1-x}O$ is also reported.

INTRODUCTION

Oxide solid solution catalysts are important for fundamental studies of the relationship between electronic properties and catalysis. This is especially the case when the solute is a transition metal oxide and the solvent an insulating, diamagnetic oxide. The value of such solid solutions in catalytic studies was first recognized [1,2] in regard to identifying how electronic interactions between

---

[+] Present address: Shell U.K. Oil Ltd., Shell Haven Refinery, Stanford-le-Hope, Essex, England
[++] Present address: DuPont (U.K.) Ltd., Instruments Products Division, 64 Wilbury Way, Hitchin, Herts., England
[+++] Present address: Unilever Research Ltd., Port Sunlight Laboratories, Wirral, England

cations affect catalytic action. The scope is now much wider [3-6], and includes studies of the effects of solvent ionicity and the use of more complex solvent oxides (spinels, scheelites, perovskites) in which site symmetry and point defect concentrations can be varied by appropriate substitutions.

The majority of solid solution catalysts so far prepared and characterized have been oxides of low surface area (LSA), typically 1 to 20 $m^2 g^{-1}$.* Catalytic studies can readily be carried out on such solids, but quantitative chemisorption work and surface spectroscopy are difficult. Thus there is a need to prepare and characterise oxide solid solutions of higher surface area (20 to 300 $m^2 g^{-1}$). Secondly, with such areas solid solution catalysts would in principle become competitive with technically-used <u>supported</u> mixed oxide catalysts. The relevance of this is that supported mixed oxides are often difficult to stabilize on account of chemical and physical changes in the active component (e.g. irreversible oxidation or reduction, crystallization, compound formation with the support) during use at conventional catalytic temperatures, typically 250-750°C. With solid solution catalysts these undesirable characteristics are less likely to occur. Moreover, high temperature applications of transition metal oxide catalysis (T > 750°C) become feasible.

The research to be described relates to $Ni_x Mg_{1-x} O$ and $Co_x Mg_{1-x} O$ ($0 < x < 0.1$) and covers work carried out over several years [7]. The choice was influenced by the fact that both NiO-MgO and CoO-MgO had already been studied as catalysts for N_2O decomposition by Cimino et al. [2,3,8,9] using low surface area solid solutions (SA < 20 $m^2 g^{-1}$). Those studies, which had been accompanied by thorough work on characterization by X-ray, magnetic and spectroscopic methods [10,11], had shown that the catalytic activities of both Ni^{2+} and Co^{2+} were enhanced if they were present as dilute solid solutions of NiO or CoO in MgO. Related catalytic work on other reactions has since followed [12,13]. Our choice of NiO-MgO and CoO-MgO was also influenced by the fact that Anderson and co-workers [14,15] had shown that $Mg(OH)_2$ could be decomposed to give MgO in a high surface area (200-300 $m^2 g^{-1}$) form, stable at 1000°C, <u>provided</u> water vapour was rigorously excluded. Since Ni^{2+} and Co^{2+} ions are known to have high diffusion rates in bulk MgO at 1000°C, these studies afforded a very good background for an attempt to prepare high surface area solid solutions. We adopted as a major premise the desirability of high temperature preparation to ensure that cation diffusion rates should not be a limiting factor in dispersing the solute ions in the MgO structure.

Although there has been other work on high surface area NiO-MgO and CoO-MgO

---

*In this paper, we divide the surface area range arbitrarily into three groups: (1) low surface area, LSA, < 20 $m^2 g^{-1}$; (2) medium surface area, MSA, 20-100 $m^2 g^{-1}$; (3) high surface area, HSA, 100-300 $m^2 g^{-1}$.

[16-19], the catalyst characterization has not been reported in sufficient detail to judge whether the oxides were in fact solid solutions or merely mixtures containing incompletely dispersed transition metal ions. Recently, however, a comprehensive characterization study on HSA, MSA and LSA CoO-MgO has been undertaken by Dyrek, Bielanski and co-workers [20-22]. This work has overlapped our own studies [23], and the important conclusion from both groups is that tetrahedral Co^{2+} ions are present in both MSA and HSA $Co_xMg_{1-x}O$. Bielanski et al. [21,22] believe that these ions are distributed throughout the bulk of the crystallites, whereas we have linked them with the surface region. Also within the last few years there has been an esr study by Kuznetsov et al. [24] identifying the presence of tetrahedral Co^{2+} in CoO-MgO. No comparable work has yet appeared on MSA and HSA NiO-MgO.

EXPERIMENTAL

Preparation of high surface area (HSA) solid solutions

Solid solutions $Ni_xMg_{1-x}O$ (designated MN) and $Co_xMg_{1-x}O$ (designated MCo) were prepared by decomposition of $Mg(OH)_2$ which had been impregnated with an aqueous solution of $Ni(NO_3)_2$ or $Co(NO_3)_2$. $Mg(OH)_2$ was precipitated from $Mg(NO_3)_2$ solution with NH_4OH, dried at $120°C$ and then ground and sieved to 60 mesh. The hydroxide was impregnated with an approximately equal volume of Ni or Co nitrate solution, whose concentration depended on the nickel or cobalt concentration ultimately required in the oxide solid solution. The impregnated $Mg(OH)_2$ was dried at $120°C$ (Ni impregnation) or at $50°C$ (Co impregnation), the lower temperature for cobalt being necessary to avoid premature decomposition of nitrate. The mixture was then ground and reslurried in water. This procedure was repeated twice more in order to achieve complete mixing through the hydroxide mass. The dried, nitrate-impregnated $Mg(OH)_2$ was placed in a platinum boat in the assembly shown in Fig.1 and evacuated. With Trap 1 at 77 K, the furnace temperature was raised slowly to $350°C$ with continuous evacuation and these conditions were maintained for 12 h. The sample was isolated by closing the stopcock and the furnace was cooled; Trap 1 was then warmed and condensed water was pumped away. Trap 1 was cooled again to 77 K, the whole system was returned to high vacuum (10^{-6} Torr, as monitored by the ionization gauge) and the furnace temperature was increased in stages to $700°C$. Trap 2 was cooled to 77 K and the temperature raised to $1000°C$; Trap 3 was then cooled to 77 K. These conditions ($1000°C$, continuous pumping, all 3 traps at 77 K) were maintained for 12 h in order to allow time for ionic diffusion to homogenize the oxide solid solution. With the temperature still at $1000°C$, the sample was magnetically withdrawn into the copper block in vacuo and rapidly cooled. The essential requirement for producing a high surface area oxide is adequate removal of water vapour when the precursor material is undergoing decomposition. Development of high surface area in a NiO-MgO or CoO-MgO mixture can be achieved merely by heating to 300°C. The purpose of the 1000°C vacuum anneal is to complete internal dehydroxylation and to form true oxide solid solution (whilst still maintaining the surface area). This is a much more demanding requirement than mere production of high surface area.

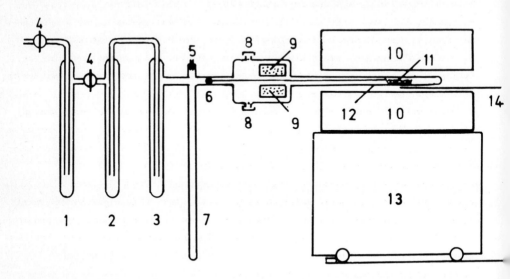

Fig.1. Apparatus for preparation of high surface area solid solutions.
1,2,3 - Traps; 4 - Stopcocks; 5 - Ionization gauge; 6 - Magnetic follower; 7 - Receiver tube for magnetic follower; 8 - O-ring seal; 9 - Copper block; 10 - Furnace; 11 - Sample in platinum boat; 12 - Mullite tube; 13 - Trolley for furnace; 14 - Thermocouple.

High purity is not a pre-requisite for preparing the HSA solid solutions, but because of the requirements of the magnetic and spectroscopic studies, the present research has been conducted with very pure starting materials, including the preparation of $Mg(NO_3)_2$ by dissolving specpure Mg in Aristar HNO_3 and use of specpure Ni and Co nitrate. Chemical analysis of the solid solutions was made by dissolving them in HNO_3 followed by EDTA titration and either atomic absorption (Ni) or spectrophotometric analysis (Co).

In this paper we report on NiO-MgO and CoO-MgO solutions with nominal nickel or cobalt concentrations of 1, 3, 5, 7.5 and 10 mole per cent. They are designated for convenience as MN 1, MN 3, MN 5, MN 7.5 and MN 10 for NiO-MgO and MCo 1, MCo 3, MCo 5, MCo 7.5 and MCo 10 for CoO-MgO. The actual molar percentages (always known from chemical analysis) differ slightly from the nominal percentages, but they are only listed when the precise values are relevant.

Techniques and methods of study

1. Surface area determination.

Surface areas were determined by the BET method (N_2, 77 K) after outgassing at $500^\circ C$.

2. X-ray diffraction and electron diffraction.

XRD was carried out with Ni-filtered Cu Kα radiation. For powder photographs a Philips 114.6 mm diameter Debye-Scherrer camera was used. Air-exposed specimens were studied conventionally. To study $800^\circ C$ or $1000^\circ C$ outgassed specimens, a special device was used [25] which permitted grinding and transfer to capillaries in vacuo, prior to sealing.

X-ray diffractometry of air-exposed solid solutions was carried out using a Philips goniometer spectrometer with a flat sample. For X-ray diffractometry in vacuo a Rigaku-Denki high-temperature attachment was used with a Pt sample holder: the specimens were outgassed at $800^\circ C$ and cooled in vacuo.

Electron diffraction was carried out using an A.E.I. EM802 electron microscope fitted with a high temperature attachment. After installation of the HSA oxide specimen and evaporation of a gold film (as internal standard) the oxide was heated briefly at $700^\circ C$ in the vacuum of the microscope before recording the diffraction pattern.

3. Gravimetry and magnetic susceptibility.

A vacuum-mounted Cahn RG microbalance was used for TG studies and for magnetic susceptibility measurements by the Gouy method [26]. The specimen was contained in a cylindrical Pt gauze bucket which facilitated removal of water vapour during high temperature outgassing, thereby avoiding loss of surface area.

4. Diffuse reflectance spectroscopy.

UV-visible reflectance spectra were measured with a Pye Unicam SP 700C spectrometer using a sample cell attached to a vacuum manifold [27]. Gases could be dosed in situ and the sample could be outgassed in the cell up to $1000^\circ C$.

5. Adsorption calorimetry.

Heats of adsorption were determined using an all-glass resistance thermometer calorimeter of a design previously described [28]. The oxide could be outgassed at $500^\circ C$ in situ.

RESULTS

Surface areas

Surface areas of solids prepared as described in the preceding section ranged from 275-300 $m^2 g^{-1}$ for MN 1 and MCo 1 to 100-125 $m^2 g^{-1}$ for MN 10 and MCo 10. The high surface area develops already at $250^\circ C$, as exemplified by the data for the formation of MCo 3 in Table 1. These results refer to Co-impregnated $Mg(OH)_2$ which had been heated in a silica tube under conditions closely matching those in the assembly of Fig.1, but with interruptions at various temperatures and cooling to 77 K to determine the surface area. The surface area rises abruptly as Mg(OH)$_2$ decomposes (250°C); some decrease in area occurs at 400-600°C, but thereafter it is constant to 1000°C.

The high surface area is essentially stable in vacuo at 1000°C. If cooled and exposed to air at 20°C, the oxides readily take up water vapour, oxygen and CO_2, but the high surface area oxide can be regained by a subsequent outgassing. However, if <u>heated</u> in open air, permanent loss of area occurs. After 15 h at 1000°C, areas which were in the HSA range 300-125 m^{2}g^{-1} (depending on Ni or Co content) become typically 100-20 m^{2}g^{-1} (MSA range). The present paper deals mostly with HSA solid solutions, but occasional reference will be made to MSA solid solutions.

TABLE 1

Development and maintenance of surface area in the formation of HSA MCo 3

Temp. of heating (°C)	20	200	250	300	350	400	500	600	800	1000
Surface area (m^{2}g^{-1})	34	40	237	243	245	234	207	179	174	180

Lattice parameters

X-ray analysis of the 1000°C-prepared HSA materials revealed a set of lines all of which could be assigned to a single cubic MgO-type phase. Short contact with the atmosphere produced no extra lines, but gravimetric measurements showed that water vapour was readily adsorbed. As a skin of hydroxide can affect the lattice parameter of a small oxide crystal [29], it was considered important for correct determination of a_o to take the X-ray pattern in vacuo after a high-temperature outgassing in situ. For MN this was done at 20°C in the high-temperature diffractometer after cooling from an 800°C-outgassing. For MCo it was done in two ways: (a) by a Debye-Scherrer photograph after a 1000°C-outgas, grinding in vacuo and transfer to a capillary which was then sealed, (b) by electron diffraction after a short 700°C outgas in situ. Nelson-Riley extrapolations were used for the X-ray determinations of a_o: the accuracy of the determination is necessarily limited by the breadth of the high-angle reflections. Results are shown in Table 2 and compared graphically with those for highly-sintered LSA solid solutions in Fig.2.

Magnetic measurements

The temperature dependence of the magnetic susceptibility χ is given by $\chi_M = \chi_o + C/(T + \theta)$, where χ_o is the Van Vleck temperature independent paramagnetism (TIP), C the Curie constant (related to the effective magnetic moment μ_{eff} by the equation $C = \mu^2_{eff}/8$), and θ the Weiss constant. An assumption that $\chi_o = 0$ will not greatly affect the determination of μ_{eff}, but may significantly affect θ. We therefore used an iterative procedure to evaluate μ_{eff}, θ and χ_o

TABLE 2

Lattice parameters (a_o) of HSA solid solutions determined in situ after high-temperature outgassing

Sample	[Ni] by chemical analysis/mol%	Lattice parameter a_o/Å	
MN 1	1.19	4.2114	
MN 3	3.34	4.2097	
MN 5	5.55	4.2089	
MN 7.5	7.66	4.208$_1$	
MN 10	10.06	4.207$_4$	
Sample	[Co] by chemical analysis/mole %	Lattice parameter a_o/Å *	
		Debye-Scherrer	Electron diffraction
MCo 1	0.94	-	-
MCo 3	3.14	4.212	4.213
MCo 5	4.50	4.213	4.214
MCo 7.5	7.63	4.215	4.216
MCo 10	9.54	4.217	4.217

* We are indebted to Dr.C.Otero Arean for these measurements

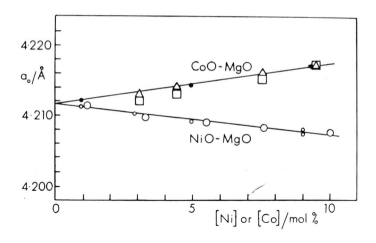

Fig.2. Variation of lattice parameter of CoO-MgO and NiO-MgO with Co and Ni content. Triangles - HSA MCo, by X-ray diffraction; Squares - HSA MCo, by electron diffraction; Solid small circles - LSA MCo, from ref.11; Large open circles - HSA MN, by X-ray diffraction; Small open circles - LSA MN, from ref.10.

from determinations of χ at about 20 temperatures between 77 and 373 K. Samples of HSA MN and MCo were outgassed in situ on the magnetic balance at 800°C and cooled in vacuo. Helium was then admitted for the susceptibility measurements. Results are given in Table 3. μ_{eff} has values typical of Ni^{2+} and Co^{2+} in octahedral coordination, and θ is quite low.

As already mentioned, the HSA oxides readily take up water vapour from the atmosphere. If left in air, bulk rehydration will occur. Specific study of this was made by allowing MN and MCo to stand in air over saturated NH_4Cl solution for 30 days. XRD of the product showed the presence of both the cubic oxide and the hexagonal hydroxide, and TG showed a weight loss of 25-27% on heating

TABLE 3

Magnetic properties of HSA MN and MCo outgassed at 800°C

Sample	μ_{eff}/B.M.	$-\theta$/K	$10^5 \chi_o$/erg gauss^{-2} mol^{-1}
MN 1	3.42	40	50
MN 3	3.59	35	30
MN 5	3.41	45	35
MN 7.5	3.30	50	40
MN 10	3.25	60	40
MCo 1	5.40	70	30
MCo 3	5.14	50	120
MCo 5	5.42	50	60
MCo 7.5	5.18	55	30
MCo 10	5.39	70	20

from 200° to 800° in vacuo, suggesting that the solid was about 80% hydroxide. This bulk hydration of HSA oxide on contact with moist air affects the magnetic parameters, as shown in Table 4. These hydrated oxides were then outgassed in situ on the magnetic balance in a series of stages, with cooling for a sequence of magnetic susceptibility measurements (77 to 373 K) at each stage. Table 4 shows how the true oxide values are gradually recovered. This effect with MCo shows the importance of measuring the magnetic properties of HSA oxide in the outgassed state.

Diffuse reflectance spectra

HSA MN outgassed at 1000°C (Fig.3, Curve 1) shows four principal regions of d-d absorption with band maxima (reflectance minima) at 23,000, 14,400, 8,300 and 5,000 cm^{-1}. Absorption in the first three of these regions is expected for Ni^{2+} in octahedral symmetry (transitions from $^3A_{2g}(F)$ to $^3T_{1g}(P)$, $^3T_{1g}(F)$ and $^3T_{2g}(F)$ respectively), but the band at 5,000 cm^{-1} (at the spectrometer limit) is unexpected.

Dosing H_2O vapour at 20°C (Fig.3, Curve 2) did not significantly affect the octahedral bands at 14,400 and 8,300 cm^{-1} but greatly decreased the absorption at 5,000 cm^{-1}. Simultaneously the broad absorption at 23,000 cm^{-1} markedly decreased, leaving a band at higher $\bar{\nu}$. This shows clearly that the broad absorption at 23,000 cm^{-1} is a composite absorption, one part (at higher $\bar{\nu}$) being the octahedral band, the other (at lower $\bar{\nu}$) being a band which, like that at 5,000 cm^{-1}, is destroyed by H_2O vapour. An outgassing recovered the initial spectrum (Fig.3, Curve 3). CO_2 also decreased preferentially the 5,000 cm^{-1} band, but less effectively than H_2O. In this case there was very little effect on the other bands. The change

TABLE 4

Magnetic properties of HSA MN and MCo on hydration and dehydration

Sample	Condition*	μ_{eff}/B.M.	$-\theta$/K	$10^5 \chi_o$/erg gauss^{-2} mol^{-1}
MN 3	OG 800°	3.59	35	30
MN 3	H	3.21	5	5
MN 5	OG 800°	3.41	45	35
MN 5	H	3.38	5	35
MN 5	H & OG 400°	3.30	50	45
MN 5	H & OG 800°	3.38	55	50
MN 10	OG 800°	3.25	60	40
MN 10	H	3.32	-5	15
MN 10	H & OG 400°	3.20	80	55
MN 10	H & OG 800°	3.24	90	50
MCo 3	OG 800°	5.14	50	120
MCo 3	H	3.73	35	10
MCo 5	OG 800°	5.42	50	60
MCo 5	H	4.22	35	15
MCo 5	H & OG 250°	4.45	35	20
MCo 5	H & OG 400°	4.78	55	10
MCo 5	H & OG 800°	5.05	60	5
MCo 10	OG 800°	5.39	70	20
MCo 10	H	4.50	40	10
MCo 10	H & OG 400°	4.92	60	0
MCo 10	H & OG 800°	5.10	55	25

*OG 800° - outgassed at 800°C; H - hydrated;
H & OG 250° (400°, 800°) - hydrated and then outgassed at 250°C (400°C, 800°C)

was ultimately reversible on outgassing, but a feature of the CO_2 spectra was a small fluctuation of the background absorption. On desorbing CO_2 the background at 15,000 - 35,000 cm^{-1} initially increased, but by a 600°C-outgas this trend had been reversed.

Oxygen adsorption (Fig.3, Curves 3-5) developed a broad background absorption, most intense between 15,000 and 35,000 cm^{-1}. This made difficult any conclusion about the influence of O_2 on the d-d bands, but if there is any effect it must be small. Outgassing in stages (Fig.3, Curves 5',6,7) showed that the oxygen effect was still significant after a 500°C desorption, but could be completely reversed at high temperature. Exposure to air gave the expected effects of a combination of O_2, H_2O and CO_2, namely a decrease in the 5,000 cm^{-1} band (the effects of H_2O and CO_2), a change in the 24,000 cm^{-1} band so as to shift the maximum and produce a shoulder (the effect of H_2O), along with a general increase in absorption between 17,000 and 32,000 cm^{-1} (the effect of O_2).

Turning to MCo, Fig.4 shows a series of spectra taken at stages during the formation of the solid solution. Little change occurs below 500°C, even though,

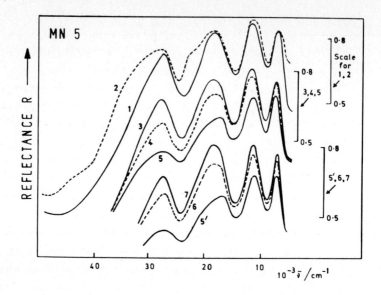

Fig.3. Reflectance spectra of HSA NiO-MgO solid solution (MN 5).
1 - Outgassed 17 h at 1000°C; 2 - After admission of H_2O vapour (20 Torr);
3 - Outgassed 17 h at 1000°C; 4 and 5 - After admission of O_2 at 0.25 Torr (4) and 180 Torr (5); 5' - As 5, but with scale shifted vertically; 6 - Outgassed 3 h at 500°C; 7 - Outgassed 17 h at 800°C. Reflectance scales at right-hand side.

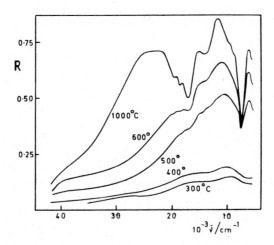

Fig.4. Reflectance spectra at stages during the preparation of HSA CoO-MgO (MCo 5). Sequence of spectra from bottom to top: Co-impregnated $Mg(OH)_2$ outgassed 17 h at 300°, further 17 h at 400°C, further 5 h at 500°C, further 6 h at 600°C, respectively. Prior to recording top spectrum, sample was given further 17 h at 700°C, 17 h at 800°C and 17 h at 1000°C, always in vacuo.

recalling Table 1, the surface area is fully developed. At 500°C and above, however, a large general decrease in absorption (increase in reflectance) occurs and a spectrum develops which exhibits four d-d bands, the one at highest $\bar{\nu}$ being a triplet. For the 1000°C-treated oxide, which was pale blue in colour, the bands (reflectance minima) occur at 7,000 cm^{-1}, 9,000 cm^{-1} (shoulder), 13,500 cm^{-1} and 16,800, 18,300, 19,500 cm^{-1} (triplet). This spectrum is certainly not that expected for octahedral Co^{2+}, which exhibits bands at 8,500 and 19,800 cm^{-1}, but contains instead features typical for tetrahedrally-coordinated Co^{2+} [30].

Exposure to H_2O vapour at 20°C (Fig.5, Curves 1 and 2) produced a dramatic change towards a typically octahedral Co^{2+} spectrum, and the solid became pink. Outgassing and heating very readily regenerated the blue colour, and the prominent features of the original spectrum were almost completely recovered on outgassing at 400°C. CO_2 produced the changes shown in Fig.5, Curves 3 and 4. At 1 Torr, there was decrease in absorption at 7,000 cm^{-1}, at 9,000 cm^{-1} and in the triplet region, and the 13,500 cm^{-1} band was replaced by a shoulder at 15,000 cm^{-1}. Further exposure (not illustrated in Fig.5) continued to decrease the original bands and a shoulder developed at 8100 cm^{-1}. Even at high exposures (p > 200 Torr for several hours), however, the effect was less pronounced than with H_2O vapour at

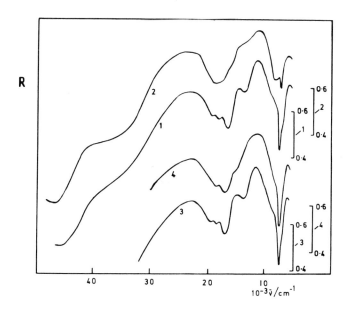

Fig.5. Effect of H_2O vapour and CO_2 on reflectance spectrum of MCo 5.
1 - Outgassed 24 h at 1000°C; 2 - After admission of H_2O vapour at 20°C (2 Torr);
3 - Outgassed 17 h at 1000°C; 4 - After admission of CO_2 at 20°C (1 Torr).

low exposures (p ~ 2 Torr for a few minutes). Progressive outgassing retraced the spectral changes exactly, but a high temperature (800°C) was necessary to recover completely the original features, notably the transformation of the 15,000 cm^{-1} shoulder into the 13,500 cm^{-1} band.

The effect of oxygen at 20°C on the spectrum of 1000°C-outgassed MCo is shown in Fig.6. As with MN, the principal effect was the development of a broad background absorption. Stepwise outgassing led to a complete recovery of the original spectrum; the colour, which had become dark brown in oxygen, reverted to blue.

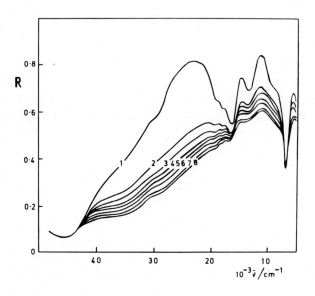

Fig.6. Effect of oxygen at 20°C on reflectance spectrum of MCo 5.
1 - Outgassed 17 h at 1000°C; 2 - 0.08 Torr O_2; 3 - 0.040 Torr O_2; 4 - 1.20 Torr O_2; 5 - 4.0 Torr O_2; 6 - 12 Torr O_2; 7 - 50 Torr O_2; 8 - 160 Torr O_2, after 17 h.

The effect of O_2 on the 7,000 cm^{-1} band was examined in detail using a very slow scanning speed; this was undertaken because of an apparent discrepancy with our previously reported result [23] which indicated a decrease in absorption at 7,000 cm^{-1} on exposure to oxygen. The slow-scan results, transposed into a Kubelka-Munk function, now lead us to the conclusion that oxygen does not decrease the 7,000 cm^{-1} band, any apparent effect in experiments using normal scanning being an artefact caused by the complicating influence of the change in background absorption.

The effect of air on 1000°C-outgassed MCo was to give spectral changes which were a combination of those reported for H_2O, CO_2 and O_2, as expected, and the blue colour changed to reddish-brown. Outgassing did not produce marked spectral changes until the temperature exceeded 400°C, but thereafter the broad background absorption decreased progressively.

Adsorption calorimetry

Air-exposed samples of HSA NiO-MgO were outgassed at 500°C and the adsorptions of NO and CO were studied volumetrically and calorimetrically at 0°C. Nitric oxide coverage* on MN 10 at 0°C tended to a limiting value of ~ 4%, as compared with ~ 1% for pure MgO (Fig.7). MN 5 showed an intermediate value. The heat of adsorption of NO, however, was not greatly affected by the presence of nickel. MgO, MN 5 and MN 10 all show heats between 20 and 12 kcal mol^{-1} (Fig.8), although there is a trend whereby the heat falls less rapidly with coverage as nickel content increases. A fraction of the adsorbed gas could be removed by evacuation at 0°C, and readsorption occurred with heats of adsorption equal to those of the previous limiting values (12-15 kcal mol^{-1}).

Carbon monoxide coverages were much less than those of NO. At p = 0.5 Torr, where NO coverages on MN 5 and MN 10 were ~ 1%, CO coverages were ~ 0.2%. Heats of adsorption were also somewhat lower, with initial values of 16 kcal mol^{-1} in both cases.

Heats of adsorption of NO and CO were also studied on MSA MN 5 and MN 10 (surface areas of 46 and 42 $m^2 g^{-1}$ respectively), with generally similar results to those on HSA MN. Interaction of NO and CO was studied calorimetrically on MSA MN 10. After outgassing for 15 h at 500°C, NO was adsorbed to saturation at 0°C and then the reversible fraction was removed by pumping for 15 h at 0°C. CO was then dosed to the surface carrying presorbed NO and the uptake and the heat of adsorption were measured. The total uptake amounted to a coverage increase of 0.1% and the heat of adsorption was 95 kcal mol^{-1}. This is to be compared with only 16 kcal mol^{-1} for the surface without presorbed NO. After outgassing again at 500°C, the experiment was repeated in the reverse sequence, NO being dosed to the surface carrying presorbed CO. The initial dose of NO registered a heat of adsorption of 36 kcal mol^{-1}, as compared with 15-20 kcal mol^{-1} for the surface without presorbed CO. Subsequent doses registered normal heats (15 and 12 kcal mol^{-1}).

This clear evidence of interaction of CO and NO at 0°C manifests itself as a true catalysis of the CO + NO reaction at higher temperatures. The threshold temperature for sustained catalysis of the CO + NO reaction over HSA MN 10 in a

---

* The definition of 100% coverage is taken as 1 molecule adsorbed per anion-cation pair in the surface.

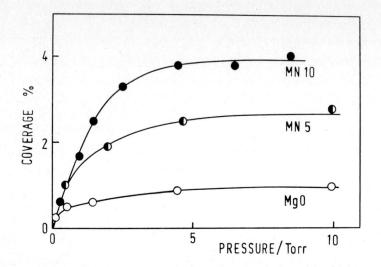

Fig.7. Adsorption isotherms of NO at $0°C$ on HSA MgO, MN 5 and MN 10.

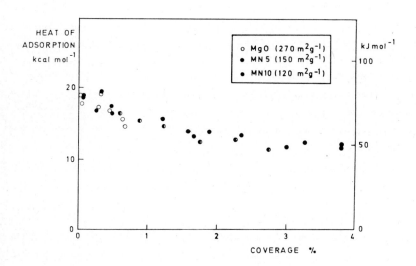

Fig.8. Heats of adsorption of NO at $0°C$ on HSA MgO, MN 5 and MN 10.

flow system* was found to be $\sim 300°C$, the products being CO_2 and N_2O. N_2O production was maximum at $400°C$, and above $500°C$ N_2 was the principal nitrogenous product.

---
*We are indebted to Mr.J.Greenwood for carrying out these experiments.

DISCUSSION

The X-ray diffraction results, supplemented by the electron diffraction results in the case of CoO-MgO, afford good evidence that the HSA oxides are true solid solutions. The lattice parameter variations (Fig.2) are the same as those reported for LSA oxides prepared in air at $1000^{\circ}C$. Moreover, the two straight lines converge at zero solute concentration to a_o = 4.2116 Å, the lattice parameter of well-crystallized MgO [29]. Although Co and Ni have the same metallic radii, their divalent ionic radii in 6-coordination are significantly different. Substitution of Ni^{2+} for Mg^{2+} in MgO leads to a decrease of a_o because of the d^8 electronic configuration of Ni^{2+} and consequent high crystal field stabilization when octahedrally coordinated. By the same token, the increase of a_o on substituting Co^{2+} shows that the d^7 ion must be in the high spin state $[(t_{2g})^5(e_g)^2]$.

A more exacting test for true solid solution is the magnetic behaviour. Here again it was important to study material which had been outgassed at high temperature and maintained in vacuo or an inert atmosphere. When this was done (Table 3), the values of μ_{eff} (3.25 to 3.59 B.M. for $Ni_xMg_{1-x}O$ and 5.14 to 5.42 B.M. for $Co_xMg_{1-x}O$) were entirely consistent with expectations for octahedrally-coordinated Ni^{2+} and octahedrally-coordinated spin-free Co^{2+} respectively. This agrees with the X-ray results, where the a_o values implied simple substitutional solid solution with good distribution of solute in the bulk. Magnetic data give additional information on solute distribution through the magnitude of the Weiss constant. The low values of $-\theta$ confirm that the magnetic ions are well dispersed and that clustering (if any) must be on a very minor scale.

Table 4 shows the effects on the magnetic properties of hydration in air and subsequent dehydration. The MN results are those one would expect: μ_{eff} remains essentially unaltered since Ni^{2+} is still octahedrally coordinated in the hydroxide but the exchange interaction (as reflected in the Weiss constant) is weakened because of the expanded and lower symmetry structure. The normal value is regained on outgassing. The MCo results, however, reveal a surprising effect. μ_{eff} falls to very low values on hydration, especially with the very dilute MCo 3. It is not possible to account for μ_{eff} values of 3.73 and 4.22 B.M. on the basis of a change in coordination, e.g. tetrahedral Co^{2+}, which would in any case be unexpected on hydration. Moreover, 3.73 B.M. is below the spin-only value for spin-free Co^{2+}, and the OH^- anion does not have a strong enough field to give spin-pairing of Co^{2+}. We interpret this result as showing that hydration is accompanied by oxidation of some of the cobalt to Co^{3+}, with increased Dq leading to spin-pairing in the d^6 ion (μ_{eff} = 0). The overall magnetic moment is therefore drastically reduced. It is clear that the oxidation is more complete the more dilute the solid solution. The changes are reversed on outgassing, but it is noteworthy that an outgassing at $400^{\circ}C$, although more than adequate to regenerate the surface area (cf.Table 1), does not regenerate the original conditions of wholly divalent Co^{2+}. This result highlights the fact that even minor contamination of $Co_xMg_{1-x}O$ with atmospheric gases will lead to values of μ_{eff} rather below the octahedral value for high spin Co^{2+} due to superficial oxidation.

The lattice parameter and magnetic results reflect primarily the properties of the bulk. The data so far discussed are in complete accord with the HSA oxides being true solid solutions with the solute ions (Ni^{2+} and Co^{2+}) well distributed in the bulk and correctly substituted there for Mg^{2+} ions.

To obtain information more specifically on the surface region of the crystallites, we turn to the reflectance spectroscopy and adsorption results. The spectrum of $1000°C$-outgassed HSA MN (Fig.3) shows d-d absorption at $5,000$ cm^{-1} which is additional to that expected for Ni^{2+} in O_h symmetry. This absorption is selectively reduced by contact with H_2O vapour, as also is the broad absorption at ca. $24,000$ cm^{-1}. We believe that the absorption at $5,000$ cm^{-1} and at ca. $24,000$ cm^{-1} must be due to Ni^{2+} species specifically associated with the surface. A tentative assignment is that the species contributing are recessed Ni^{2+} ions in 5-fold square pyramidal (C_{4v}) symmetry on the {100} faces. By analogy with the spectra of 5-fold Ni^{2+} complexes [31,32], a C_{4v} fit is consistent with high spin Ni^{2+} and $Dq = 1000$ cm^{-1}, the apical angle caused by recession of the cation from the cube face [33] being taken as $100°$ [34]; the $5,000$ and $24,000$ cm^{-1} absorptions can then be assigned as the transitions from $^3B_1$ to the lower-lying $^3E(F)$ and to the $^3E(P)$ levels respectively. Additional bands expected between these absorptions [32] are overlapped by the octahedral spectrum. Other coordinations (4-fold T_d, 3-fold D_{3h}) have been considered, but they give less satisfactory fits than 5-fold C_{4v}. Oxygen contact does not significantly change either the $5,000$ or the $24,000$ cm^{-1} bands, but produces general absorption. This we ascribe to limited chemisorption producing O_2^- or O^- adsorbed anions and Ni^{3+} cations, and thereby generating intense intervalence absorption (Ni^{2+}/Ni^{3+}). Some oxygen is certainly bound quite strongly, since the original spectrum is not fully regained on outgassin at $500°C$ (Fig.3).

The $500°C$-outgassed surface exemplified by Fig.3, Curve 6 is the type of surface on which the NO and CO adsorptions were carried out. Fig.7 shows that the presence of Ni^{2+} ions gives rise to a specific NO adsorption, since replacement of only 5% of the Mg^{2+} ions in MgO leads to a threefold increase in coverage, and the further increase on going to MN 10 indicates that the surface concentration of Ni^{2+} is indeed changing broadly in parallel with total Ni^{2+} concentration. The calorimetry experiments (Fig.8), however, show that the heat of adsorption of NO on MN 5 and MN 10 is not significantly greater than on MgO. Whilst this does not exclude the interpretation that the additional adsorption is restricted to the nickel ions (or Ni plus adjacent oxide ions), it suggests that the effect of Ni ions may be to activate NO for surface diffusion on to the MgO matrix. Cimino et al. [3,35]

have postulated a rather similar role for the transition metal ion in N_2O decomposition over LSA solid solutions, with oxygen migrating over the surface of the matrix after break-up of N_2O. The results with CO were broadly similar to those with NO, except that coverages and heats were smaller, so the same conclusions follow.

A further important result to emerge from the calorimetry was the clear evidence for strong chemical interaction between CO and NO in the adsorbed state, even at $0°C$. The relatively lower heat in the case of NO dosed to CO_{ads} (as compared to CO dosed to NO_{ads}) is due to the chemical reaction heat being diluted by that of the simple adsorption of NO on the free surface, NO limiting coverages being much higher than those of CO. The nickel ions are able to lower the threshold temperature for the CO + NO reaction, as compared to MgO, and also they lead to sustained decomposition of the N_2O intermediate once the temperature for clearing adsorbed oxygen is exceeded.

In sharp contrast to HSA MN, the reflectance spectrum of $1000°C$-outgassed HSA MCo (Fig.5, Curve 1) is not recognisable as the spectrum of octahedral Co^{2+} ions known to be present in the solid from the magnetic study. Two main d-d absorptions are expected for spin-free Co^{2+} in O_h symmetry viz. $^4T_{1g}(F) \rightarrow {}^4T_{2g}(F)$ and $^4T_{1g}(F) \rightarrow {}^4T_{1g}(P)$, and in single crystal $Co_xMg_{1-x}O$ these occur at 8,500 and 19,800 cm^{-1} respectively [36]. Absorption at these values is present, but is evidently dominated by more intensely absorbing chromophores. The bands at 7,000 cm^{-1} and at 16,000-20,000 cm^{-1} (triplet) are characteristic of Co^{2+} in T_d symmetry, being the $^4A_2(F) \rightarrow {}^4T_1(F)$ and $^4A_2(F) \rightarrow {}^4T_1(P)$ transitions respectively [30,37], with the latter the source of the blue colour. Tetrahedral coordination produces intense absorption. The band at 13,500 cm^{-1} is not expected for either octahedral or tetrahedral Co^{2+}. Let us note that all the 'non-octahedral' absorption is greatly reduced on exposure to H_2O vapour and a recognizably octahedral spectrum results. CO_2 also attenuates the non-octahedral bands. Oxygen develops a very intense general absorption, obviously producing Co^{3+} ions and intervalence (Co^{2+}/Co^{3+}) absorption, but does not affect the non-octahedral bands.

Bearing in mind the magnetic evidence of octahedral Co^{2+} in the bulk, and the entirely normal values for a_o at all the Co concentrations studied, we conclude that the non-octahedral Co^{2+} is principally present in the surface layers of the HSA crystallites. It is able to dominate the octahedral spectrum because the centres responsible are non-centrosymmetric and hence have high intrinsic absorption. The tetrahedral spectrum could arise in various ways, either from the near-tetrahedral sites expected on {110} faces [33], or from 'disproportionation' of 5-coordinated surface cations on {100} faces [23], or from occupancy of the actual tetrahedral holes of the cubic close-packed anion structure.

Bielanski et al. [21,22] propose that Schottky defect formation leads to tetrahedral Co^{2+} throughout the bulk of the crystallites of their CoO-MgO. Our vacuum-prepared HSA MCo appears to have a conventional octahedral distribution for the Co^{2+} ions in the bulk, but to have tetrahedral Co^{2+} in the surface region. A mechanism for developing tetrahedral Co^{2+} in the ccp anion structure of the surface region of $Co_xMg_{1-x}O$ is shown schematically in Fig.9. Schottky defects generated in the surface region during the treatment at $1000°C$ lead to an extension of the ccp structure (Fig.9b). The cations then redistribute in the anion array (Fig.9c) so as to give cation occupation partly in octahedral holes and partly in tetrahedral holes, always with the proviso that adjacent face-shared oxygen polyhedra would not be occupied. This process requires no valency change. Such an expanded anion

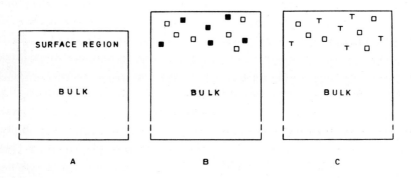

Fig.9. Schematic representation of formation of tetrahedral Co^{2+} in surface region of HSA CoO-MgO solid solution.
A - Cubic close packed (ccp) anions; cations in octahedral holes, no cations in tetrahedral holes (NaCl structure).
B - Expanded ccp structure with Schottky defects (anion and cation vacancies) in surface region. Cations still in octahedral holes but some octahedral holes now unoccupied (cation vacancies, shown shaded). No tetrahedral holes occupied.
C - Expanded ccp structure but with cations reorganized in surface region. Anion vacancies (open squares) as in B, but cations redistributed among a mixture of octahedral and tetrahedral (T) holes, cf.spinel structure.

array would convert rapidly to hydroxide on exposure to H_2O vapour, anion vacancies ($\square$) being destroyed by the process $O^{2-} + \square + H_2O \rightarrow OH^- + OH^-$: all cations would remain divalent but would become octahedrally coordinated [cf.$Mg(OH)_2$ or $Co(OH)_2$]. If water vapour and oxygen were admitted, the array would again be completed but in this case by both OH^- and O^{2-}, as in an oxyhydroxide [cf.$CoO(OH)$]: all cations would become octahedral and oxidizable cations (Co^{2+}) would be oxidized (Co^{3+}). If oxygen were admitted alone, penetration to fill the anion vacancies would probably be limited at room temperature to one or two layers only. A limited number of Co^{2+} ions would be oxidized, commensurate with the uptake of oxygen to form O^{2-}:

$$4Co^{2+} + O_2 + 2\square \rightarrow 4Co^{3+} + 2O^{2-}$$

The oxidized cations in this case would be preferentially those Co^{2+} ions in **octahedral sites**, because of the greater gain in crystal field stabilization energy (by analogy with the fact that Co_3O_4 is a normal and not an inverse spinel). The tetrahedral Co^{2+} ions would be unaffected.

This model is sufficient to explain our observations on HSA MCo. H_2O exposure leads to an octahedral Co^{2+} spectrum with no evidence for Co^{3+}. Atmospheric exposure ($H_2O + O_2$) generates Co^{3+} and destroys tetrahedral Co^{2+}, as evidenced by the magnetic results and the intervalence absorption. Exposure to oxygen does not affect tetrahedral Co^{2+}, but creates Co^{3+} (intervalence absorption). The motion of a Co^{2+} ion from an octahedral site to a tetrahedral site which is inherent in the model requires the passage through a 'squeeze' position of D_{3h} symmetry and then an approach in C_{3v} symmetry. The band at 13,500 cm^{-1}, not previously assigned, seems likely to be associated with the Co^{2+} ion in one of these symmetries, other bands from the same chromophore being overlapped by the tetrahedral spectrum.

The model is consistent with the esr results reported by Kuznetsova et al. [24] and Boreskov [19]. Boreskov refers to Co ions of Types I and II. Type I ions are regarded as surface tetrahedral Co^{2+} and are not affected by oxygen; Type II ions are octahedral and are readily oxidized and reduced. Following Boreskov [38], these are now components of the above model. Our picture is also close to that advanced by Bielanski et al.[21,22], except that we do not consider the defect structure in our HSA MCo to involve the bulk at all significantly. Their reflectance spectra and those of Dyrek et al.[20] appear to have been recorded only in air (or after exposure to air) and thus Co^{3+} will always be present. Also a band at 5,260 cm^{-1} in these circumstances [21] could be due to a H_2O combination band from adsorbed atmospheric water rather than tetrahedral Co^{2+}. Similarly in their magnetic work inference about tetrahedral Co^{2+} is difficult if Co^{3+} ions are present. In our work we hope to have shown that the intriguing complexity of the **HSA CoO-MgO system can be rationalized further** by conducting spectroscopic and magnetic studies in a controlled environment and one which includes the ability to measure reflectance spectra in high vacuum.

REFERENCES

1 E.G.Vrieland and P.W.Selwood, J.Catalysis, 3 (1964) 539.
2 A.Cimino, M.Schiavello and F.S.Stone, Disc.Faraday Soc., 41 (1966) 350.
3 A.Cimino, Chim.Ind.(Milan), 56 (1974) 27.
4 F.S.Stone, J.Solid State Chem., 12 (1975) 271.
5 A.W.Sleight and W.J.Linn, Ann.N.Y.Acad.Sci., 272 (1976) 22.
6 R.J.H.Voorhoeve, J.P.Remeika and L.E.Trimble, Ann.N.Y.Acad.Sci., 272 (1976) 3; R.J.H.Voorhoeve, D.W.Johnson, J.P.Remeika and P.K.Gallagher, Science, 195 (1977) 827.
7 M.G.Lofthouse, Thesis, Univ.Bristol (1972); A.P.Hagan, Thesis, Univ.Bristol (1974); C.Otero Arean, Thesis, Univ.Bath (1976); M.A.Trevethan, Thesis, Univ.Bath (1977).

8   A.Cimino, R.Bosco, V.Indovina and M.Schiavello, J.Catalysis, 5 (1966) 271;
    A.Cimino, V.Indovina, F.Pepe and M.Schiavello, J.Catalysis, 14 (1969) 49;
    A.Cimino, V.Indovina, F.Pepe and F.S.Stone, Gazz.Chim.Ital., 103 (1973) 935.
9   A.Cimino and F.Pepe, J.Catalysis, 25 (1972) 362.
10 A.Cimino, M.Lo Jacono, P.Porta and M.Valigi, Z.Phys.Chem.(Frankfurt), 55 (1967) 14.
11 A.Cimino, M.Lo Jacono, P.Porta and M.Valigi, Z.Phys.Chem.(Frankfurt), 70 (1970) 166.
12 V.Indovina and F.Pepe, personal communication.
13 F.Pepe and F.S.Stone, J.Catalysis, in press.
14 P.J.Anderson and R.F.Horlock, Trans.Faraday Soc., 58 (1963) 1993.
15 P.J.Anderson and P.L.Morgan, Trans.Faraday Soc., 60 (1964) 930.
16 S.Z.Roginskii and V.A.Seleznev, Kinetika i Kataliz, 8 (1967) 1342.
17 N.P.Keier, I.S.Sazonova and R.V.Bunina, Kinetika i Kataliz, 10 (1969) 1036.
18 V.V.Popovskii, G.K.Boreskov, V.S.Muzykantov, V.A.Sazonov, G.I.Panov, V.A.Rotchin, L.M.Plyasova and V.V.Malakhov, Kinetika i Kataliz, 13 (1972) 727.
19 G.K.Boreskov, Proc.6th Int.Congr.Catalysis, London, July 12-16, 1976, Chem.Soc., London, 1977, p.204.
20 K.Dyrek, Bull.Acad.Polon.Sci., Ser.Sci.Chim., 21 (1973) 675; K.Dyrek and V.A.Shvets, Bull.Acad.Polon.Sci., Ser.Sci.Chim., 22 (1974) 315.
21 A.Bielanski, Z.Kluz, M.Jagiello and L.Waclawska, Z.Phys.Chem.(Frankfurt), 97 (1975) 207.
22 A.Bielanski, Z.Kluz and A.Wojtaszczyk, Bull.Acad.Polon.Sci., Ser.Sci.Chim., 25 (1977) 721.
23 A.P.Hagan, C.Otero Arean and F.S.Stone, Proc.8th Int.Symp.Reactivity of Solids, Gothenburg, June 14-19, 1976, Plenum Press, New York, 1977, p.69.
24 L.I.Kuznetsova, G.K.Boreskov, T.M.Yurieva, V.F.Anufrienko and N.G.Maksimov, Doklady Akad.Nauk, S.S.S.R., 216 (1974) 1323; N.G.Maksimov, L.I.Kuznetsova, V.F.Anufrienko and T.M.Yurieva, Izv.Akad.Nauk Neorg.Materialy, 12 (1976) 1219.
25 C.Otero Arean and F.S.Stone, to be published.
26 T.A.Egerton, A.Hagan, F.S.Stone and J.C.Vickerman, J.C.S.Faraday I, 68 (1972) 723
27 A.Zecchina, M.G.Lofthouse and F.S.Stone, J.C.S.Faraday I, 71 (1975) 1476.
28 R.L.Gale, J.Haber and F.S.Stone, J.Catalysis, 1 (1962) 32.
29 A.Cimino, P.Porta and M.Valigi, J.Amer.Ceram.Soc., 49 (1966) 152.
30 R.Pappalardo, D.Wood and R.C.Linares, J.Chem.Phys., 35 (1961) 2041.
31 M.Ciampolini, Inorg.Chem., 5 (1966) 35; M.Ciampolini, N.Nardi and G.P.Speroni, Coord.Chem.Rev., 1 (1966) 222.
32 M.Gerloch, J.Kohl, J.Lewis and W.Urland, J.Chem.Soc.(A), (1970) 3269.
33 J.Haber and F.S.Stone, Trans.Faraday Soc., 59 (1963) 152.
34 L.Sacconi, Coord.Chem.Rev., 8 (1972) 351.
35 A.Cimino, F.Pepe and M.Schiavello, Proc.5th Int.Congr.Catalysis, Florida, Aug.20-26, 1972, North Holland /American Elsevier, 1973, p.125.
36 W.Low, Phys.Rev., 109 (1958) 256.
37 F.Pepe, M.Schiavello and G.Ferraris, J.Solid State Chem., 12 (1975) 63.
38 G.K.Boreskov, Proc.6th Int.Congr.Catalysis, London, July 12-16, 1976, Chem.Soc., London, 1977, p.214.

DISCUSSION

K. JOHANSEN : In Figure 2 you have a linear relationship for lattice parameters as a function of the concentration of Ni and Co. Does this continue for higher concentrations ? Or, in other words, is it an ideal solution ?

F.S. STONE : The linear relationship does continue for higher concentrations. This has been shown (1-3) for low surface area

(LSA) solid solutions and the inference is that the solutions are ideal. In our work with high surface area (HSA) solid solutions we have explored as far as 15-20 mole per cent Ni or Co, and we find no discrepancy with this. By 15-20 mole per cent the areas have decreased to values outside the arbitrarily-defined HSA range, i.e. they are below 100 m^2g^{-1}.

1. J. Robin, Ann. Chim. Acta, 10(1955), 389.
2. W.C. Hahn and A. Muan, J. Phys. Chem Solids, 19(1961), 339.
3. A. Cimino, M.Lo Jacono, P. Porta and M. Valigi, Z. Phys. Chem. (Frankfurt), 55(1967) 14; 70(1970) 166.

L. GUCZI : Your finding of tetrahedral cobalt at the surface by reflectance spectroscopy strengthens our previous hypothesis explaining the behaviour of α-Cr(III) oxide under reducing conditions (J. Kalman and L. Guczi, J. Catal., 47(1977) 371). We have found a decrease of catalytic activity after long treatment with hydrogen. The activity, however, could be restored instantaneously by contacting the catalyst with air at room temperature. We explained it by assuming the formation of Cr(II) in tetrahedral positions, which required large energy. The reverse transition, from tetrahedral Cr(II) to octahedral Cr(III), takes place very easily. Your experimental data seem to me to verify our assumption.

F.S. STONE : There is indeed some analogy between Cr(II) and Co(II), since both are known, for instance, to prefer tetrahedral coordination in the spinel structure. However, as regards ease of conversion to octahedral coordination, it is worth noting that for our tetrahedral Co(II) ions in MgO, mere exposure to oxygen at room temperature does not destroy the tetrahedral spectrum. Exposure to air destroys it, and we show clearly that the key component in air responsible for this is water vapour. The result is then octahedral coordination for the cobalt, and the oxygen produces some oxidation to Co(III). The corresponding process with Cr(II) may be easier. If there is tetrahedral Cr(II) in the surface of reduced $\alpha-Cr_2O_3$, oxygen may be able on its own to produce octahedral Cr(III) : certainly there is a large gain in crystal field stabilization energy to assist it. The existence of chromium in different coordinations on the surface of the corundum structure ($\alpha-Al_2O_3$ and $\alpha-Cr_2O_3$) is a topic which has interested us also (1,2).

1. F.S. Stone, Chimia, 23(1969),490.
2. F.S. Stone and J.C. Vickerman, Proc. Roy. Soc. (London), A 354 (1977), 331.

J.W. HIGHTOWER : During the NO adsorption experiments, was there any evidence of NO decomposition such as is frequently observed when NO reacts stoichiometrically with a reduced transition metal atom to liberate N_2/N_2O ?

F.S. STONE : We have not observed NO decomposition to give N_2 or N_2O at 0°C, the temperature of the adsorption experiments.

L. RIEKERT : If I understood correctly, it is assumed that the different adsorption capacities of the different solids (MgO, MN 5 and MN 10) are due to kinetics, since the heat of adsorption is essentially the same for all the solids.
(a) Would this mean that the isotherms observed are not equilibrium isotherms, even though they exhibit saturation ?
(b) Would not the heat of adsorption determine only the slope of the isotherm near the origin, whereas the amount adsorbed at saturation is independent of the heat of adsorption and should only depend on the stoichiometry or on the structure of the adsorption layer ?

F.S. STONE : The NO adsorption isotherms are not equilibrium isotherms, since they are not reversible at 0°C. We assume that they are activation-energy limited, as you say. However, it is not uncommon to find in non-equilibrium chemisorption that "saturation" coverages do relate to heats of adsorption, higher coverages being associated with higher molar heats. The fact that this does not happen in the present case therefore seemed to justify the speculation about nickel ions facilitating chemisorption on the solvent matrix, but it is certainly no more than speculation.

J.A. RABO : In the case of adsrobing NO on Ni^{2+}-containing MgO surface did you observe by IR or ESR the nature of NO; namely, does it form the NO^+ cation ?

F.S. STONE : We have observed a specific ESR spectrum to develop when NO is contacted with the nickel-containing MgO specimens, but we have not identified the adsorbed species responsible. The IR spectra of NO adsorbed on these MgO-based oxides are complex (1), and several species are certainly present. However, in view of the basic nature of these oxides, I would not think it likely that the NO^+ cation is one of them.

1. L. Cerruti, E. Modone, E. Guglielminotti and E. Borello, J.C.S. Faraday I, 10(1974), 729.

TENTATIVE CLASSIFICATION OF THE FACTORS INFLUENCING THE REDUCTION STEP IN THE ACTIVATION OF SUPPORTED CATALYSTS

B. DELMON, M. HOUALLA
Groupe de Physico-Chimie Minérale et de Catalyse
Université Catholique de Louvain, Louvain-la-Neuve (Belgium)

ABSTRACT

The communication is an attempt to classify the various parameters which affect the reduction step in the activation of supported oxides. Discussion will mainly be conducted with reference to hydrodesulfurization catalysts. The following groups of parameters will be discussed : dispersion, interaction between active oxides and support, reduction conditions, additives. Extensive reference will be made to mechanisms known in the field of the reaction of solids.

1. INTRODUCTION

Before it begins to work in a catalytic reactor, a catalyst has undergone a succession of transformations, first, during its preparation, starting from raw materials, and ending with a product which can be shipped to the user, and second, while converting the transportable material into the form which is active in the reactor. Activation, which usually designates the final steps in the conversion of the catalyst into the working form, may correspond to different transformations, according to cases, for example dehydration, decomposition, reduction or sulfidation, either isolated or combined. All correspond to chemical changes occuring during reactions which involve solids. The science of the chemical transformations of solids has now gained the status of a new branch of chemistry. It is therefore logical to approach the chemical reactions involved in the activation of catalysts by using, as a background, the large body of the existing knowledge in the field of the chemistry of solids in general.

The present communication examines, from the standpoint of solid state chemistry, the factors which influence the reduction step in activation, namely the reduction of supported oxides into metals or lower oxides. The factors controlling reactions of solids, in general, begin to be discerned [1-3]. Reductions of oxides are one of the groups of reactions of solids which have been the object of the most intensive study. It is felt that time comes where the knowledge gained on the reduction of oxides may really help understand the reduction of the special class of oxides which supported catalysts constitute.

The main objective of the present communication is to propose a *classification* of the factors which control the reduction of supported oxides. Science needs such attempts to ordering, especially when basic knowledge has still to be fortified, because such rational groupings help distinguish between alternative hypotheses. While presenting this classification, some *typical informations* will be examined. Thus, in addition to reaching the main objective, we hope to delineate cases where a systematic study has contributed important pieces of information with respect to the reactivity of supported oxides in reduction. Too few examples of in-depth studies have already been published for a successful methodology to begin to shape out. However, we believe that the present communication can give some hints in this last respect.

The Organizing Committee of this Symposium very rightly indicated that the preparation and activation of catalysts constitute a whole interacting <u>sequence of steps</u>. The behaviour of the catalyst at one step is conditioned by all previous steps, and, in turn, alters all subsequent steps and, ultimately, the active form. The present authors have kept this remark in their minds. Because this Symposium was centered on impregnation and activation, impregnated samples have been considered.

Many situations can exist as a consequence of the reactions taking place during impregnation, drying and calcination. One major point of discussion in this communication will be the parameters which depend on the *preparation steps* which precede reduction.

The communication will discuss results obtained by ourselves and other investigators in the Groupe de Physico-Chimie Minérale et de Catalyse (GPCMC), as well as data reported by other investigators. Because of our special interest in *hydrodesulfurization and hydrotreating catalysts*, illustrations will be taken preferentially from that field, although other examples may be occasionnaly discussed, when arguments and observations can be readily extrapolated to HDS catalysts. One feature which distinguishes hydrotreating catalysts from others, e.g., hydrogenation catalysts, with respect to reduction, is the simultaneous presence of <u>two</u> oxides on the surface of the carrier. The situation is more complicated, and hence, more difficult to analyze but, on the other hand, this complexity unveils new phenomena, possibly having some general bearing, which deserve discussion.

In the case of hydrotreating catalysts, the sequence of preparation steps is the following :

(a) Adsorption or deposition of the precursor or precursors on the carrier.
(b) Drying, which, in addition to the removal of water, may lead to crystallization of the precursors in the pores and on the outer surface of the carrier.

(c) Decomposition of the precursor salts (dehydration, decomposition to oxides), namely "calcination".
(d) Possibly, if double impregnation is used, again a sequence of steps similar, respectively, to (a), (b) and (c).
(e) Activation, namely reduction and sulfidation.

Our discussion will rest on the assumption that the catalysts are prepared according to this general sequence. We will examine, successively, the influence of the following groups of parameters on the reduction of the supported oxides :

(i) the parameters which depend on the starting supported oxides, namely the solid resulting from steps (a) through (d);

(ii) the parameters which depend on the reduction conditions, with some reference to the role of H_2S in the reduction-sulfidation process (step e);

(iii) the role of additives with reference to their influence on the parameters mentioned under (i) and (ii).

2. INFLUENCE OF THE VARIOUS PREPARATION STEPS ON THE CHARACTERISTICS OF THE STARTING SUPPORTED OXIDE

This first section intends to summarize the specific bearing of each preparation step (i.e. impregnation, drying, calcination) on the characteristics of the supported oxide which will be reduced. Two subsequent sections will examine in detail the corresponding parameters with respect to *dispersion* and to *interaction between active oxide and support*.

2.1. Influence of the impregnation step

To a large extent, impregnation determines the *repartition* of the deposited substance in the body of the carrier (in pores as on the outer surface, uniformly all through the pellet or bead, or only in the porosity of the outer parts of the pellet). At a small scale, it determines the *dispersion* (uniform deposition of isolated ions, uniform layer of deposited substance on the walls of the pores, or isolated clusters; small or large clusters). Impregnation also influences the nature of *interaction* existing between the support and the impregnating precursor which, in turn, may determine the kind of interaction ultimately taking place between the oxide and the support. Moreover, the *surface of the support*, and even, in principle, its *texture*, may be *altered*, as a consequence of the contact with the impregnating solution. Impregnation can constitute a considerable trauma for the support. Impregnating solutions, because of their content in various ions, may have a strong corrosive power and can bring about ion exchange as well as partial dissolution and redeposition.

The influence of impregnation on repartition and dispersion of the precursor, and the nature of the interaction between precursor and support are discussed in various contributions to this Symposium. One will mention, here with reference to HDS catalysts, that Mo from ammonium paramolybdate can adsorb on the support, either through an interaction with surface hydroxyl groups (as MoO_4^{2-}) (4-7), or, very likely, through mere crystallization on the surface. Similarly salts of group VIII metals can either exchange on the surface or crystallize (4, 8, 9, 10). Modification of the surface of the support and of its texture during impregnation has been reported (11, 12). In other instances, it can be inferred from data obtained in other fields, e.g. the destruction of structures of crystalline alumino-silicates in solution of moderately low pH (13), or the fact that cation exchange capacity measurements (C.E.C.) lead to values well exceeding those obtained by less destructive methods (14,15).

2.2. Influence of the drying step

The effects of drying on repartition and dispersion of the active phase begin to be well documented (16, 17). The contributions to the present Symposium dealing with impregnation often describe effects which are due to impregnation *plus drying*.

A factor which, in our opinion, would deserve more attention is the ability of the partially dried or dehydrated salt to wet the support. We believe that the choice of the impregnating salt and, possibly, of the drying procedure may have a strong effect on the characteristics of the catalyst.

2.3. Influence of the calcination step

Calcination of the deposited precursor brings about several transformations and solid state reactions which can be summarized as follows :
- decomposition of the dried precursor;
- solid state reactions of the supported oxide (e.g. crystallographic change) or oxides (e.g. reactions between oxides);
- solid state transformations of the support;
- reactions between deposited oxides and the support;
- sintering of the various phases (supported phases, support).

The product resulting from this last step in catalyst preparation is the reactant of the reduction step.

We must now examine in detail the characteristics of the solids and their bearing on the reduction. A rather satisfactory classification is to consider separately the parameters related to *dispersion* and those describing the type of *interaction* between the supported oxides and the solid carrier.

443

3. INFLUENCE OF DISPERSION ON REDUCTION
3.1. Dispersion characteristics of supported oxides

Generally speaking, one should use a broad definition of *dispersion*, namely the distribution, at a large and small scale, of the active phase over the carrier, and also the shape and orientation of the crystallites or aggregates of the deposited oxide.

Fig. 1 suggests the various situations, either with respect to repartition in the depth of each pellet or bead (Fig. 1a) or to the dispersion on the walls (Fig. 1b). In the present section, we shall not be concerned with the nature of the interaction between active oxide and support (Fig. 1c).

3.2. Active phase loading

The previous sections have suggested how the various stages of catalyst preparation may influence dispersion when catalysts of identical composition are considered. At this point, it is necessary to examine the effects of the loading in active phase on the dispersion. Indeed for a given carrier and depending on active phase-support interaction, increasing the active phase loading will modify the degree of dispersion of the oxide. The final repartition of

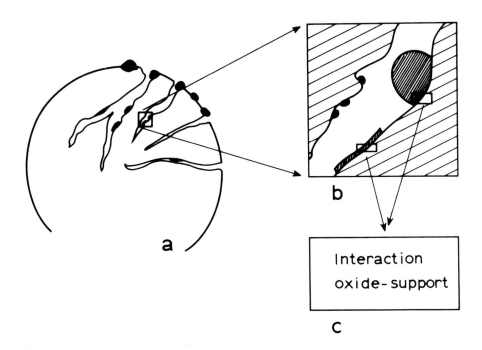

Fig. 1 : Repartition of the active phase over the carrier.

the active phase might vary from a "monolayer" type arrangement to aggregate formation.

Direct observation of the evolution of the active phase disperseness can be made by differents techniques (electron microscopy, X-Ray-diffraction lines broadening, E.S.C.A.).

3.3. Kinetic consequences of dispersion

In principle, the reactivity of a solid is strongly dependent upon its dispersion. The reactivity of supported oxides is known to be profoundly altered, in comparison to that of the bulk oxide. In order to understand the reduction step in the activation of supported catalysts, it is necessary to distinguish what may be attributed to dispersion and what is due to the interaction between the active phase and the support. We shall examine, in the present section, some consequences of the mere *dispersion*.

3.3.1. Theoretical consideration. The decrease of reactivity often observed, when supported oxides are compared to unsupported ones, seems contradictory to common sense, because one speculates that dispersed solids should be more reactive. We shall intend to clarify this problem in the present paragraph, supposing that no special interaction between active phase and support takes place.

One can imagine that the effect of dispersion is mainly to lessen the possibility of nucleus formation either because of the small size of the aggregates of the active phase, or because of a lack of nucleus forming sites on many of these aggregates(19). Thus, when the nucleation stage in reduction is a rate-limiting step, it is to be expected that modifications of nucleation brought about by the disperseness of the active phase will have considerable kinetic consequences.

The effect of particle size on reduction is actually complex. We have shown in Fig. 2 the results of a mathematical analysis, which describes the evolution of the kinetics of the transformation, as a function of particle size a_o, in otherwise identical experimental conditions. The values of parameter $A_{S(O)}$ reflects the relative rates of nucleus formation to nucleus growth(20). It turns out to be proportional to the cube of the radius of the particles, a_o. An increase of the particle size by a factor of 10 will, then, increase $A_{S(O)}$ by a factor of 1,000. Fig. 2 shows that decreasing the particle size (starting from $A_{S(O)} = 1,000$, for example) will first increase the reduction rate until a certain particle size threshold is attained, after which the reaction is slowed down.

We do not believe that the slow reduction of finely dispersed oxides can <u>always</u> be explained according to the preceding picture. However, this should be taken into account before looking for other interpretations, or unless the assumption of an inhibition of reaction by carrier-active phase interaction is confirmed by suitable physico-chemical investigations.

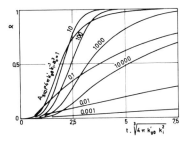

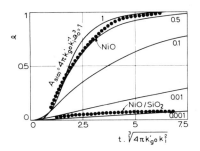

Fig. 2 : Theoretical evolution of the kinetics of reactions where nucleus formation is rate limiting, k'_{gO} : rate of nucleus formation; k_i : rate of nucleus growth (ref. 20).

Fig. 3 : Comparison of reduction of a bulk and silica supported nickel oxide at 265°C (α:extent of the reduction to metal. Dotted lines : experimental results) (ref. 21). Solid lines : theoretical curves for different value of $A_{S(O)}$.

3.3.2. Experimental results in relation with HDS catalysts. As indicated in Fig. 3, the previous picture could very satisfactorily account for the difference of reactivity in reduction between bulk and silica supported nickel oxide (21). A reduction of the size of the NiO particles on SiO_2, in comparison to unsupported NiO, by a factor of 7 (from $\sim$ 0.25 μm for bulk NiO (21)), would be quite sufficient for explaining the difference.

The situation of NiO/SiO_2 is formally very similar to that of unpromoted HDS type catalysts in their oxidic form. The case of MoO_3/SiO_2 can be cited as an example.

Fig. 4 shows clearly a decrease of the reducibility of MoO_3 supported on SiO_2 when MoO_3 content is lowered (e.g. smaller particles of MoO_3) (22). Taking into account the picture proposed above, it seems likely that the phenomenon can be related to the modifications of the reactivity of the oxide with dispersion.

If this interpretation is true, one should not attribute the variations of the degree of reduction of MoO_3 supported on Al_2O_3 only to the interaction with the support (23-26).

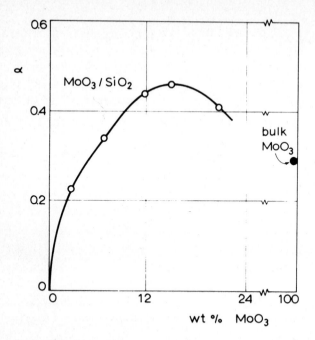

Fig. 4 : Effect of MoO_3 content on the extent of reduction reached in 10 hours by a series of MoO_3/SiO_2 catalysts (T = 400°C; p_{H_2} = 1 atm.) (ref. 22).

4. INFLUENCE OF THE INTERACTION ACTIVE OXIDE-SUPPORT

4.1. Overall picture

The interaction between the supported oxide and the carrier, although determined by all steps in the preparation, depends more strongly on the phenomena which have taken place <u>during calcination</u>. We shall therefore refer to the phenomena mentioned in the above corresponding paragraph for attempting a classification of the possible situations; these situations may *coexist* on a given catalyst, because different chemical reactions may occur simultaneously.

The following cases will be considered (Fig. 5) :

- weak forces between supported oxide and carrier;
- aggregates of deposited oxide, electronically interacting with the carrier;
- monolayer of deposited oxide;
- aggregates covalently or ionically bound with the support;
- solid solution of active phase in the support present either on <u>all</u> the surface of the carrier or in patches;
- new compound present on the whole surface of the carrier or in domains.

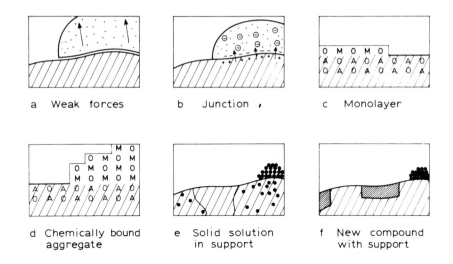

Fig. 5 : Various types of active oxide-support interaction.

4.1.1. Very Weak forces. Ideally, when the interaction is due to weak forces (VAN DER WAALS forces), the role of the support is only to disperse the catalyst. In principle, weak forces could also be sufficient to bring about some preferential orientation of the lattice of the supported oxide (Fig. 5a). Graphite (27) and, perhaps, sometimes silica (28) could be examples of carriers which exert only a weak action on the supported substance.

4.1.2. Electronic interaction. In principle, semi-conducting or conducting substances deposited on a semi-conducting support may form a *junction* with the latter, even if no chemical bond (in the conventional sense) is formed, by tunnelling or similar phenomena (Fig. 5b). Electron transfer across the boundary would change the electron density of deposited aggregates, with the effect being more pronounced for small aggregates; this could change the reactivity.

4.1.3. "Adhesion" type interaction. There is certainly a whole spectrum of possible situations between the weak interaction discussed above and strong chemical interaction.

For example, certain oxides, like Mn_2O_3 and MoO_3, are reported to wet spontaneously each other, while others (MoO_3 and Co_3O_4) tend to form aggregates (29). A possible explanation of the wetting phenomenon may well be found in relatively weak forces, similar to those mentioned for interpreting adhesion phenomena. Obviously, stronger interaction can also bring about wetting.

4.1.4. Chemical interaction : one monolayer or less. This category of situations comprises those where all the atoms or molecules of the supported substance are exposed on the surface. One case is when the deposited oxide acts as a dope in very small concentration on the surface. The upper limit is encountered when the active phase combines with the carrier as a continuous monolayer (Fig. 5c). A typical example of the latter is the formation of a monolayer of MoO_3 on the surface of γ-Al_2O_3 as follows :

$$Al_2O_3 \begin{bmatrix} -OH \\ -OH \end{bmatrix} + \begin{matrix} HO \\ HO \end{matrix} Mo \begin{matrix} O \\ O \end{matrix} \rightarrow Al_2O_3 \begin{bmatrix} -O \\ -O \end{bmatrix} Mo \begin{matrix} O \\ O \end{matrix} + 2H_2O$$

Many evidences of this phenomenon have been presented (5, 30-34). We shall take here another example which, although not completely elucidated, may well illustrate this section.

Silica is generally considered as an inert carrier and is expected to act merely as a dispersing agent. However, recent evidences (12, 35), including results obtained in GPCMC, seem to indicate that its inertness was probably overestimated. E.S.C.A studies of a series of MoO_3/SiO_2 catalysts containing up to 20 wt % MoO_3, prepared by pore volume impregnation, followed by drying at 110° and calcination at 500°C, show a shift towards higher values of the binding energy (B.E.) of the Si_{2p} line, with increasing active phase loading (Fig. 6).

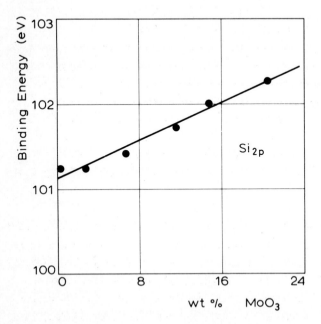

Fig. 6 : Evolution of the Si_{2p} binding energy as a function of MoO_3 content in a series of MoO_3/SiO_2 catalysts (ref. 22)

These results, which cannot be deduced from previously reported data (28) on a MoO_3/SiO_2 catalyst containing 12 wt % MoO_3, suggest an electron transfer from silica to molybdena. Unfortunately a direct verification from the Mo lines (e.g. Mo $3d^{3/2}$ and $3d^{5/2}$) is impossible, because of their broadening, in comparison with those of MoO_3. This broadening, however, also indicates that MoO_3 supported on silica suffers an interaction from the latter.

Such an electronic effect would be similar to what has been observed between MoO_3 and $\gamma-Al_2O_3$ (36-41), where a slight shift (<0.5 eV), towards higher binding energies, of the Mo $3d^{3/2}$, $3d^{5/2}$ levels has been reported.

4.1.5. Chemical interaction : aggregates. When the active phase loading exceeds the amount which corresponds to a monolayer, it is likely that the excess oxides will present features more similar to those of bulk oxides, since the possible interaction with the support will exert relatively less and less importance when the ratio (active phase volume)/(contact area between active phase and support) decreases (Fig. 5d).

4.1.6. Chemical interaction : solid solutions. Depending on the temperature of calcination, the oxide and the carrier, when their structure are compatible, can dissolve in each other to different extents, ranging from an ionic exchange confined to the interface to a multiple layer thick solid solution and, ultimately a homogeneous one (Fig.5e). Occurence of the different situations and their bearing on reactivity have been discussed in a case not related to catalysts, but quite illustrative, namely that of the NiO-CuO system (42). Examples of formation of homogeneous solid solution involving substances known as supports are abundant in literature $Cr_2O_3-Al_2O_3$ (43) CoO-MgO (44). The changes in reactivity which may occur when solid solutions are formed are discussed in detail in a general frame (3). While the possibility that solid solutions formation in the case of hydrotreating catalysts has not been studied, such a phenomenon cannot be disregarded. For example Co_3O_4 and $CoAl_2O_4$, having both a spinel structure, may form a solid solution in the usual conditions of preparation of catalyst.

4.1.7. Chemical interaction : new compounds. The formation of a new compound can be either restricted to the near surface domains or extended to the bulk of the solid (Fig. 5f). Examples illustrating the first situation are not lacking in literature. Many authors have reported that Co^{2+} and other transition metal ions (Ni^{2+}, Mn^{2+}, Cu^{2+}) can combine with transition alumina to form cobalt pseudo aluminate or a surface spinel (45-48). In other instances, one may observe the formation of a well defined solid. It has, in effect, been reported that, at high loading of MoO_3 on Al_2O_3 (20 wt % MoO_3) $Al_2(MoO_4)_3$ crystals can be detected (6, 49).

4.1.8. <u>Mixed situations</u>. A major difficulty, in describing the supported oxide which undergoes activation, is the fact that many of the previously described situations *coexist* in a given catalyst.

This can be illustrated by taking again the example of MoO_3 deposited on SiO_2, which is considered to be a relatively simple one (12). Depending on the composition, at least 4 different species have been detected (Fig. 7).

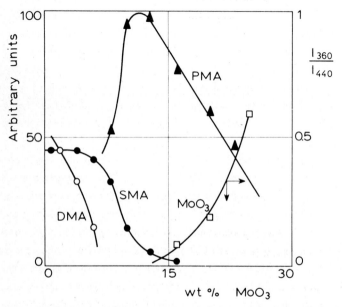

Fig. 7 : Product distribution in MoO_3/SiO_2 catalysts as determined from acidimetric titrations : SMA (●) silicomolybdic acid, DMA (o) dimolybdates; PMA (▲) polymolybdates. The estimation of MoO_3 (□) is derived from the intensity ratio of the reflectance bands at 360 and 440 cm^{-1} (ref. 12).

The following products, ranging from free MoO_3 to silica bound molybdena, have been detected, namely :

. MoO_3,
. dimolybdates (DMA),
. polymolybdates (PMA),
. silicomolybdic acid (SMA),
. molybdenum bound to the surface of the support, in a manner similar to the monolayer model proposed for MoO_3/Al_2O_3 system (31).

The situation for the oxidic form of HDS catalysts supported on alumina is still more complicated (6, 45, 50, 51).

Another additional complication arises when two or several different oxides are deposited on a support, namely the possibility that they *combine together*. Although not perfectly fitting into the classification which we have adopted in this communication, this phenomenon has to be mentioned here. In some HDS catalyst precursors, it has been indicated that Co and Mo can combine in the form of cobalt molybdate (22, 50, 51, 52).

4.2. Influence of the interactions between deposited oxide and support on reduction : general comment

When only weak interactions are at stake, the major action of the carrier is simply to disperse the supported oxide. In addition, the presence of the support may, as a consequence of a narrow porosity, impede the access of the reducing gas to the active oxide, or modify the thermal stability and resistance to sintering of the supported phase.

The situation is obviously more complicated if a chemical interaction takes place. However, the behaviour, in reduction, of solids which exhibit characteristics similar to those of supported catalysts, begins to be known (3). This is true for the behaviour of solids, the surface of which has been modified by ion exchange or by formation of an adhering layer, or which have partially reacted to form a solid solution near the surface (2, 3, 53). The relative reactivities of solid solution, compound oxides (oxides associating in a given stoichiometry two or several metals) and pure oxides have also been discussed (2,3), and the various aspects of the modifications that the presence of a foreign phase (i.e. the carrier) can bring about on the reactivity of oxides have been examined and classified (1, 3). It may thus be hoped that the reduction of supported oxides is amenable to a full interpretation.

4.3. Influence of the interactions between deposited oxide and support on the reduction of HDS catalysts precursors

In the present paragraph, we shall attempt to analyze the phenomena which may occur during the reduction of the oxidic form of HDS catalyst. The complexity of the $CoMo/Al_2O_3$ systems practically reflects all possibilities. The discussion will therefore be centered on this case, but simpler situations, namely those of oxides supported on SiO_2, will also be commented.

Evidence has been presented (6, 45, 51, 52) for the existence of at least six different species in the $CoMo/Al_2O_3$ system in its oxidic form (Fig. 8) :
- free cobalt oxide,
- $CoAl_2O_4$ or "cobalt aluminate",

- MoO_3 monolayer,
- free MoO_3,
- $Al_2(MoO_4)_3$,
- $CoMoO_4$.

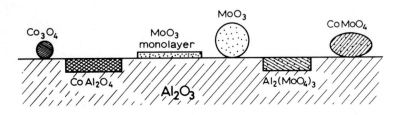

Fig. 8 : Species eventually present in the $CoMo/Al_2O_3$ system.

It is suspected that, among all the preparation parameters which determine the presence and amount of these species, the *order* of impregnation (if double impregnation is used) and the active phase *loading* are of prime importance. We shall conduct the discussion in steps, considering first the Co/support and the MoO_3/support systems, and then the complete CoMo/support catalyst.

4.3.1. Reduction of cobalt oxide/support. The effects of disperseness of group VIII metal oxides on supports have been fully discussed above, with NiO as an example.

The only additional feature is the formation of a compound with the carrier, cobalt aluminate in the present case (not reducible at moderate temperature (54)). The fact that all the deposited oxide cannot be reduced, even when nucleation is induced in some way, reflects the presence of a compound with the carrier. Let us take again the simple example of NiO/SiO_2 (19) for illustrating this point. Part of nickel oxide seems to be withdrawn from reduction (Figs. 9 and 10). The amount of unreducible nickel seems to be almost independant of nickel loading (Fig. 10). It seems that the effect cannot be attributed to dispersion, as copper, which has been said to very efficiently promote nucleus formation (19), cannot make the reaction complete (Fig.9). This result suggests that even quite inert carriers may combine in some way with supported oxides and prevent them from reacting. Although apparently not forming during impregnation (19), a nickel silicate might be the unreactive compound (55). However an alternative explanation, such as the presence of a fraction of the active oxide encapsulated in the pores of the carrier, cannot be dismissed in principle.

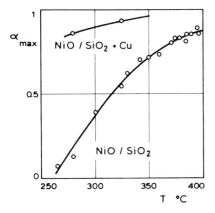
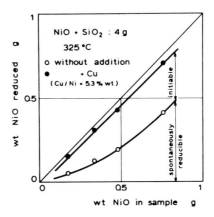

Fig. 9 : Apparent maximum α_{max} of the degree of reduction vs. temperature, for pure and promoted NiO/SiO$_2$ (ref. 19).

Fig. 10 : Influence of NiO content on the amount of spontaneously reducible, initiable and unreducible NiO at 325°C (ref. 19).

4.3.3. Reduction of molybdenum oxide/support. Three different situations may occur in the MoO$_3$/Al$_2$O$_3$ system (Fig. 8).

Concerning the reduction of "free" MoO$_3$, an explanation has been proposed, which takes into account the effect of dispersion (23, 24). One salient feature of the reduction of MoO$_3$ is that it proceeds according to the sequence

$$Mo_{6+} \rightarrow Mo_{4+} \rightarrow Mo$$

The reduction of a monolayer will obviously have completely different kinetics, in comparison with that of the bulk oxide. This aspect has been investigated when MoO$_3$ is deposited on various supports (23, 56), with the case of γ-Al$_2$O$_3$ having been the object of the most extensive research (23, 25, 26, 34, 57, 58). Results of reduction experiments have been interpreted by assuming an epitaxial monolayer of aluminium molybdate (34, 56), or a surface interaction compound having a distribution of bonding strengths (26). It must however, be noted that a comparison of the reduction results of MoO$_3$/SiO$_2$ (22) and MoO$_3$/Al$_2$O$_3$ (26) suggests that there are no essentially different features, when account is taken of the *incomplete* presence of a monolayer in the former case (Fig. 11).

Actually, the most conspicuous feature in the reduction of the MoO$_3$ monolayer is the stabilization of Mo^{+5} (23-26, 59). It should, however, be remarked that in HDS catalysts, this stabilization is not strong enough, and the reduction can go further, either because of the presence of sulfur containing molecules (H$_2$/H$_2$S) or, in "complete" catalysts (containing Co or Ni, in addition to Mo), because of the action of these elements in promoting reduction (1, 2, 3, 23, 25).

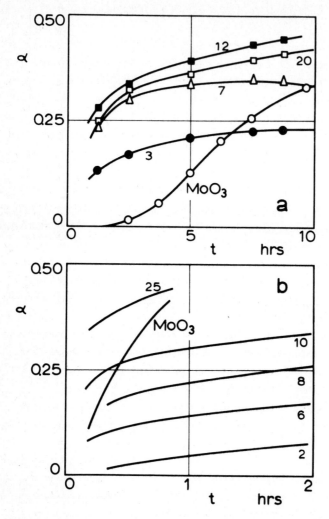

Fig. 11 : Effect of MoO_3 content on the reduction of supported MoO_3 (α: fraction of MoO_3 reduced to Mo; figures on the curves refer to wt % MoO_3; MoO_3 curve correspond to reduction of bulk MoO_3 at the same temperature.
a. MoO_3/SiO_2, T = 400°C, p_{H_2} = 1 atm. (ref. 22)
b. MoO_3/Al_2O_3, T = 500°C, p_{H_2} = 1 atm. (ref. 26)

Few results are available concerning the reduction of $Al_2(MoO_4)_3$. Massoth (26) has reported that the ease of reduction of bulk $Al_2(MoO_4)_3$ is somewhat intermediate between that of MoO_3 and of 10 % Mo/Al_2O_3.

Consequently, taking into consideration the possible influence of dispersion on the reduction of $Al_2(MoO_4)_3$, we could infer that the presence of this compound might, in principle, account for the difficulty to reduce the Mo/Al_2O_3 system.

4.3.3. Reduction of the Co-Mo oxide/support system.

Two new problems have to be discussed, with reference to the complete system, where cobalt and molybdenum oxides are present.

The first concerns the presence of *cobalt molybdate*. The formation of a well defined $CoMoO_4$ on the surface of Al_2O_3 is likely to modify the reducibility of cobalt promoted catalyst $Co-Mo/Al_2O_3$ since X.P.S. studies [28] indicate a deeper reduction of the former.

The other complication which arises when cobalt and molybdenum oxides are simultaneously present on the surface of the carrier is the possible *mutual effect* in reduction. We have already indicated the promoting effect of cobalt (or nickel) on the reduction of MoO_3 monolayer. This effect is also exerted on "free" MoO_3 [23, 25], as it is on bulk MoO_3 [60, 61]. On the other hand, molybdenum oxides seem not to affect the reduction of groupe VIII metal oxides [62].

5. INFLUENCE OF REDUCTION CONDITIONS

Generally speaking, the assessment of the influence of the parameters related to the reduction conditions requires a consideration of the different steps of the reduction process, namely :

(i) coming into contact of the reducing agent with the solid;

(ii) adsorption;

(iii) nucleation;

(iv) reaction (interface progress).

On the other hand, many parameters are necessary for describing the reduction conditions :

- composition of the reducing gas (nature of reducing agent, impurities, presence of the gaseous product of reaction, etc...);
- temperature;
- time sequence of various conditions (progressive variation of composition and/or temperature).

Not all aspects can be discussed here. We shall mainly restrict our review to the case of the activation of HDS catalysts and to parameters which have been shown to influence this activation.

5.1. Role of H_2O

5.1.1. Inhibition of nucleation.

Charcosset et al (63), from their studies of the influence of water on copper catalyzed reduction of NiO, have inferred that H_2O inhibits the reduction by blocking nucleus forming sites.

Similar effects probably take place when Co/Al_2O_3 and Mo/Al_2O_3 are reduced with a mixture of H_2-H_2O (4.5 torr H_2O) (64). The degree of reduction with H_2-H_2O is always lower than with pure H_2 (Fig. 12). When, in the course of the reaction, H_2-H_2O is changed by hydrogen, the reduction exhibits an induction period, which is absent when the reduction stage is initiated with pure H_2. The appearance of an induction stage would possibly be related to a progressive restoration of the reactivity of nucleus forming sites previously inhibited by the water.

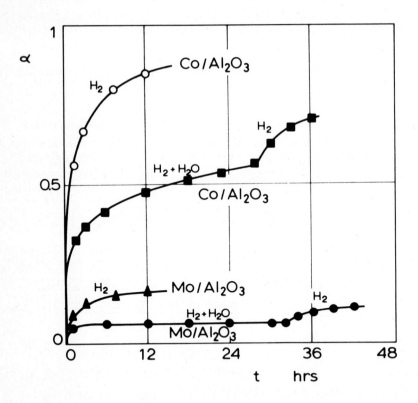

Fig. 12 : Reduction of Mo/Al_2O_3 and Co/Al_2O_3 in dry and wet H_2
(T = 400°C, $p_{H_2} \simeq 1$ atm.; x : fraction of MoO_3 or Co_3O_4 reduced to metal)
(ref. 64).

5.1.2. Transfer of active species. The water may be indirectly responsible for the acceleration of reduction, by providing a "transport vehicle" for the reducing species activated over a promoter (65, 66). Water, indeed, plays an important role in the spill-over of hydrogen activated on a group VIII metal onto WO_3 in the partial reduction of the latter (65, 66) and similar effects probably occur with MoO_3.

5.1.3. Change of the disperseness of the product. The water may also act in a quite different way. One can envisage that, depending on temperature, nature of the metal, and H_2/H_2O ratio, water may influence the disperseness of the reduced phase, by modifying the interaction metal-support. The following situations might be considered (67) :

(i) the water is adsorbed on the carrier. In this case, H_2O will eventually modify the roughness of the support surface. The water will also increase the mobility of the metal, as the polarizability of the water is usually less than that of the ions of the carrier, thus leading to an easier sintering;

(ii) the water can oxidize the metal at the interface, increasing, generally, the adherence of the metal to an oxidic support.

The discussion can be developed along similar lines, when oxygen, rather than water, is present at the surface or in the gas phase.

5.2. Role of H_2S in reduction-sulfidation

The present paragraph is a little outside the scope of the communication, as it deals with more than merely reduction, namely reduction-sulfidation. However, we believe that the corresponding results should be mentioned here, because of their direct relevance to the activation of HDS catalysts.

The results of a series of temperature programmed reduction or reduction-sulfidation on MoO_3 shown in Fig. 13 (curves 1, 2) indicate clearly an easier reduction of MoO_3 in the presence of H_2S (68). In effect, the reduction starts at a much lower temperature than when it is carried out with pure H_2; it then proceeds until the formation of MoO_2 with only traces of MoS_2. Total sulfidation to MoS_2 occurs at higher temperatures.

5.3. Influence of the succession in time of different reduction conditions

It has been emphasized that the reaction of solids has a different course if the succession of reduction conditions (the activation program) is different. This can be demonstrated theoretically when nucleation is partially rate-determining (69), and experiments in the reduction of NiO fully substantiate the conclusion (70, 71).

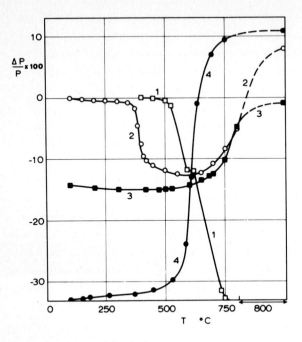

Fig. 13 : Reduction and reduction-sulfidation of MoO_3 and MoO_2 (ref. 68) :
curve 1 : $MoO_3 + H_2$; curve 3 : $MoO_2 + H_2-H_2S$
curve 2 : $MoO_3 + H_2-H_2S$; curve 4 : $Mo + H_2S$.

Results obtained in the preparation of *unsupported* HDS catalysts suggest how critical the activation cycle may be. If reduction-sulfidation is realized by first reducing MoO_3 to MoO_2 with a H_2-H_2O mixture, and only then introducing H_2S, the sulfidation begins at temperatures, similar to those observed with MoO_3, but the sulfidation does not go to completion and stops when an approximately equimolecular mixture of MoO_2 and MoS_2 (both identified in X-ray diffraction) is observed (Fig. 13 curve 3) (68). There is no undisputable explanation for that phenomenon. It might simply be attributed to the fact that reduction of MoO_3 in the presence of H_2S takes place at lower temperature and, hence, gives a more reactive MoO_2. Alternatively, one might speculate that nucleation of MoS_2 takes place more easily from MoO_3 than from MoO_2. A third explanation is that we have an example of a *coupling* of reactions, namely the fact that two different reactions (here : reduction and sulfidation) promote themselves mutually, when occuring simultaneously. Some instances of such coupling exist in other fields of solid state chemistry (3, 72, 73).

6. INFLUENCE OF ADDITIVES

Additives (so called "minor components") are widely used in catalysts. The fact that they so often play an important role has fascinated investigators since catalysis has begun to develop as a science. When trying to assess their possible effects, one is led to suspect that some explanations have been overemphasized and others, equally reasonable ones, almost forgotten. Modification of the electronic properties of metals or semiconductors by foreign elements dispersed in the bulk or on the surface has been the object of innumerable discussions in academic laboratories. On the other hand, catalyst manufacturers are faced with less fashionable effects of accidental impurities, the action of which has rarely been investigated, and the effect of which, if deleterious, has to be corrected more or less empirically, by the use of additives.

A first problem, which is far from being solved, is to distinguish between the effects which correspond to a *promotion* of the active sites (by changing the local atomic arrangement, the electronic density, etc...) by additives, and those which merely reflect the fact that the additives have influenced in some way the *preparation steps*. In the first cases, one has to take, as a background, the theories of catalysis, while in the second, the science of the reactions of solids will serve as a reference.

The present paragraph will try to consider the second category of effects. It is probably not possible to enumerate exhaustively the various effects that additives may exert on the preparation steps. The classification made for the reaction of solids (3), broadly speaking, certainly constitutes an excellent frame. We shall only cite here some typical effects, with relation to HDS catalysts and the additives used or mentioned in relation to these catalysts.

The best classification is the one which considers the possible effects of additives on the three categories of parameters examined above, namely those related to dispersion, to active phase-support interaction, and to the development of the activation reaction.

6.1. Dispersion

The role of soluble additives in modifying dispersion during impregnation is discussed in other contributions. Let us only speculate on the role that insoluble acidic or basic sites on the support of the catalyst might have in helping dispersion or, conversely, in promoting nucleation of a deposited cluster. We do not believe that this effect has received sufficient attention.

Additives may also play an important role during the decomposition of the precursor salt, by changing the kinetics of the reactions and, also its *spatial* course (because of more or less numerous nuclei are being formed) and consequently, the dispersion and texture of the product (3, 74). In a similar respect, additives are known in ceramic science and technology to prevent (role of "spacers" (75)) or promote sintering according to cases.

6.2. Active phase-support interaction

It is believed that the role of certain additives in HDS catalysts is to change the reactivity of the surface of the support with the active oxide. This belief is supported, for example, by studies indicating a change in the distribution of sites of different symmetries in alumina, when it is doped with Zn^{2+} or Ga^{2+} (76).

Additives may inhibit interaction between active phase and support. An example can be found in the reported influence of fluorine and boron on the interaction between NiO and Al_2O_3 (77). Conversely, the additives may promote active phase-carrier interaction. The influence of the incorporation of a small amount of Al_2O_3 in a SiO_2 matrix, on the interaction between NiO and support can be considered as an example of the promoting effect of the additive (Fig. 14) (78). Another possibility is that additives may simply act as a "glue" between the active phase and the carrier. The corresponding phenomena belong to the field of adhesion, and are far from clear presently.

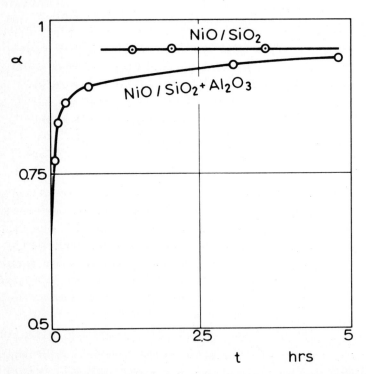

Fig. 14 : Reduction of NiO supported on silica and silica alumina; Si/Si+Al=95wt.%); 10 wt % NiO; T=400°C; p_{H_2} = 1 atm; α: fraction of NiO reduced to metal (ref.78).

6.3. Development of the reactions taking place during activation

The role of additives in the reduction of oxides has been extensively studied (1, 2, 3, 20, 66), and special attention has been focussed on the oxides which are present in the oxidic form of HDS catalysts (23, 25, 79, 80, 81). Many effects have been discussed in the previous sections. Let us simply indicate that Cu, in NiO/SiO_2, (Figs. 9, 10) may constitute a typical example of the effect of additives, otherwise inert, on the reduction step.

6.4. Mixed situations

Although the preceding paragraphs have attempted to classify the various effects of additives, it should be remarked that many of them may modify several features of the catalyst to be activated and the activation reaction. For example, addition of an alkali metal ion on an Al_2O_3 carrier can vary the porosity distribution of the carrier, modify the acidity, change the dispersion of the active phase during impregnation, affect the interaction oxide-support, influence the surface reactivity with respect to hydrogen and, possibly, provoque an alteration of the texture of the reduced solid.

7. CONCLUSION

The factors which influence the reduction step of supported catalysts are multiple. Elaborating a simple and exhaustive classification which accounts for all the different situations is a difficult task. However, we have suggested a possible approach. We have restricted the discussion to the major factors likely to have a pronounced effect on reduction and which are relatively well documented. This means that some special factors have been either disregarded or only briefly mentioned. Others, like spill-over processes, have been thoroughly reviewed several times and no further discussion was necessary.

Concluding this communication, it seems that one remark may emerge, namely that much information is available in the field of the science of the reactivity of solids which can be used almost directly for explaining the reaction of *supported* solids. The opinion has been often expressed that the making of catalysts is an art, much like cooking. This is certainly not the opinion of the scientists interested in the bases of preparation of catalysts. Transdisciplinary co-operation can help making the activation of catalysts more scientifically footed.

ACKNOWLEDGEMENTS

Parts of the experimental work carried out in GPCMC, and the postdoctoral fellowship of one of us (M.H.) were supported by the "Services de la Programmation de la Politique Scientifique" in the frame of the "Actions Concertées Interuniversitaires Catalyse".

REFERENCES

1. B. Delmon, in J.S. Anderson, M.W. Roberts and F.S. Stone (Eds.), Reactivity of Solids, Chapman and Hall, London, 1972, pp.567-576.
2. B. Delmon, in P. Barret (Ed.), Reaction Kinetics in Heterogeneous Chemical Systems, Elsevier, Amsterdam, 1975, pp.640-675.
3. V.V. Boldyrev, M. Bulens, B. Delmon, The Control of the Reactivity of Solids, Elsevier, Amsterdam, in press.
4. J.H. Ashley and P.C.H. Mitchell, J. Chem. Soc. (A) (1969) 2730-2735.
5. J.Sonnemans and P. Mars, J. Catal. 31 (1973) 209-219.
6. N. Giordano, J.C.J. Bart, A. Vaghi, A. Castellan and G. Matinotti, J. Catal. 36 (1975) 81-92.
7. A. Iannibello and F. Trifiro, Z. Anorg. Allg. Chem. 413 (1975) 293-304.
8. F. Bozon-Verduraz, M. Tardy, G. Bugli and G. Pannetier, in B. Delmon, P.A. Jacobs, G. Poncelet (Eds.), Preparation of Catalysts, Elsevier, Amsterdam, 1976, pp.265-274.
9. J.P. Brunelle, A. Sugier and J.F Lepage, J. Catal. 43(1976) 273-291.
10. T.A. Dorling, B.W.J. Lynch and R.L. Moss, J. Catal. 20(1971) 190-201.
11. J.W.E. Coenen in B. Delmon, P. Grange, P.A. Jacobs and G. Poncelet ('Eds.), 2nd Symposium on Preparation of Catalysts, Louvain-la-Neuve, 1978.
12. A. Castellan, J.C.J. Bart, A. Vaghi and N. Giordano, J. Catal. 42(1976) 162-172.
13. M.A. Vicente, M. Razzaghe, M. Robert, Clay Minerals 12(1977) 101-112.
14. P. Rouxhet, P.O. Scokart, P. Canesson, C. Defossé, L. Rodrique, F.D. Declerck, A.J. Léonard, B. Delmon, J.P. Damon, in M. Dekker (Ed.), Colloid and Interface Science, Academic Press, New York, 1976, pp.81-94.
15. J.P. Damon, B. Delmon, J.M. Bonnier, J. Chem. Soc., Faraday Trans.I, 73(1977) 372-380.
16. J. Cervello, E. Hermana, J.F. Jimenez and F. Melo, in B. Delmon, P.A. Jacobs, G. Poncelet (Eds.), Preparation of Catalysts, Elsevier, Amsterdam, 1976, pp. 251-263
17. G.K. Boreskov, in B. Delmon, P.A. Jacobs, G. Poncelet (Eds.), Preparation of Catalysts, Elsevier, Amsterdam, 1976, pp.223-245.
18. T. Fransen, P.C. Van Berge and P. Mars, in B. Delmon, P.A. Jacobs, G. Poncelet (Eds.), Preparation of Catalysts, Elsevier, Amsterdam, 1976, pp.405-416.
19. A. Roman and B. Delmon, J. Catal. 30(1973) 333-342.
20. B. Delmon, Introduction à la cinétique hétérogène, Technip Publi., Paris, 1969, pp. 443-444.
21. A. Roman, Thesis, Grenoble, 1971.
22. D. Pirotte, P. Gajardo, P. Grange and B. Delmon, in preparation.
23. J. Masson, B. Delmon and J. Nechtschein, C.R. Acad. Sci. Ser. C 266(1968) 428-430.
24. J. Masson, B. Delmon and J. Nechtschein, C.R. Acad. Sci. Ser. C 266(1968) 1257-1259.
25. J. Masson, J. Nechtschein, Bull. Soc. Chim. Fr.(1968) 3933- 3938.
26. F.E. Massoth, J. Catal. 30(1973) 204-217.
27. G.C. Stevens and T. Edmonds, in B.Delmon, P. Grange, P.A. Jacobs, G. Poncelet (Eds.), 2nd International Symposium on Preparation of Catalysts, Louvain-la-Neuve, 1978.
28. A. Cimino and B.A. Angelis, J. Catal. 36(1975) 11-22.
29. J. Haber and J. Ziolkowski, in J.S. Anderson, M.W. Roberts and F.S. Stone, (Eds.), Reactivity of Solids, Chapman and Hall, London, 1972, p.782.
30. J.M.J.G. Lipsh and G.C.A. Schuit, J. Catal. 15(1969) 174-178.
31. M. Dufaux, M. Che, C. Naccache, J. Chim. Phys. 67(1970) 527-534.
32. V.H.J. De Beer, T.H.M. Van Sint Fiet, J.F. Engelen, A.C. van Haandel, M.V.J. Wolfs, C.H. Amberg and G.C.A. Schuit, J. Catal. 27(1972) 357-368.
33. G.C.A. Schuit, B.C. Gates, A.I. Che J. 19(1973) 417-438.
34. W.K. Hall and M. Lojacono, in G.C. Bonds, P.B. Wells, E.C. Tompkins (Eds.), 6th International Congress on Catalysis, The Chemical Society, London, 1977, pp. 246-254.

35 P. Biloen and G.T. Pott, J. Catal. 30(1973) 169-174.
36 P. Gajardo, R.I. Declerck-Grimée, G. Delvaux, P. Olodo, J.M.Zabala, P. Canesson, P. Grange and B. Delmon, J. Less Common. Met. 54(1977) 311-320.
37 R.I. Declerck-Grimée, P. Canesson, R.M. Friedman and J.J. Fripiat, J. Phys. Chem., in press.
38 P. Canesson, C. Defossé, R.I. Declerck-Grimée and B. Delmon, 5th Ibero-American Symp. on Catalysis, Lisbon, 1976, paper A3-8.
39 G.C. Stevens and T. Edmonds, in P.C.H. Mitchell' and A. Seamen (Eds.), Chemistry and Uses of Molybdenum Climax, 1976, pp.155-159.
40 A. Miller, W. Atkinson, M. Barber and P. Swift, J. Catal. 22(1971) 140-142.
41 A.W. Armour, P.C.H. Mitchell, B. Folkesson and R. Larsson, J. Less Common Met. 36(1974) 361.
42 P. Grange, H. Charcosset and Y. Trambouze, J. Thermal Anal. 1(1969) 311-317.
43 Ch. Marcilly, B. Delmon, J. Catal. 24(1972) 336-346.
44 A.P. Hagan, C.O. Arean and F.S. Stone in J. Wood, O. Lindqvist, C. Helgesson and N.G. Vannerberg (Eds.), Reactivity of Solids, Plenum Press, New York, 1977, pp.69-74.
45 M. Lojacono, A. Cimino and G.C.A. Schuit, Gazz. Chim. Ital 103(1973) 1281-1295.
46 M. Lojacono, M. Schiavello and A. Cimino, J. Phys. Chem. 75(1971) 1044-1050.
47 M. Lojacono, M. Schiavello, D. Cordischi and G. Mercati, Gazz. Chim. Ital. 105(1975) 1165-1176.
48 M. Lojacono and M. Schiavello, in B. Delmon, P.A. Jacobs, G. Poncelet (Eds.), Preparation of Catalysts, Elsevier, Amsterdam, 1976, pp. 473-484.
49 G.N. Asmolov and O.V. Krylov, Kinet. Katal. 11(1970) 1028-1033.
50 J. Medema, C. Van Stam, V.H.J. De Beer, A.J.A. Konings and D.C. Koningsberger, submitted to publication.
51 V.H.J. De Beer, M.J.M. Van der Aalst, C.J. Machiels and G.C.A. Schuit, J. Catal. 43(1976) 78-89.
52 M. Lojacono, M. Schiavello, V.H.J. De Beer and G. Minelli, J. Phys. Chem. 81(1977) 1583-1588.
53 M. Bulens, Ann. Chim. 1(1976) 13-25.
54 P. Ratnasamy, A.V. Ramaswamy, K. Bannerjee, D.K. Sharma and N. Ray, J. Catal. 38(1975) 19-32.
55 J.W.E. Coenen, B.G. Linsen, in B.G. Linsen (Ed.), Physical and Chemical Aspects of Absorbants and Catalysts, Academic Press, New York, 1970, p.479.
56 T. Fransen, P.C. Van Berge and P. Mars, React. Kinet. Catal. Lett. 5(1976) 445-452.
57 P. Sontag, D.Q. Kim and F. Marion, C.R. Acad. Sci. Ser. C 259(1964) 4704-4707.
58 T. Kabe, S. Yamadaya, M. Oba and Y. Miki, Int. Chem. Eng. 12(1972) 366.
59 N. Giordano, J.C.J. Bart, A. Castellan and A. Vaghi, in P.C.H. Mitchell (Ed.), Chemistry and Uses of Molybdenum, Climax, Reading, 1973, pp.194-200.
60 J.M. Zabala, P. Grange and B. Delmon, C.R. Acad. Sci. Ser. C 279(1974) 561-563.
61 D.K. Lambiev, T.T. Tomova and G.C. Samsonov, Powder Metallurgy Int. 4(1972) 17.
62 J.M. Zabala, M. Mainil, P. Grange and B. Delmon, C.R. Acad. Sci. Ser. C 280(1975) 1129-1132.
63 H. Charcosset, R. Frety, P. Grange and Y. Trambouze, C.R. Acad. Sci. Ser. C (1968) 1746-1748.
64 P. Gajardo, P. Grange and B. Delmon, to be published.
65 J.E. Benson, H.W. Kohn and M. Boudart, J. Catal. 5(1966) 307-313.
66 H. Charcosset, B. Delmon, Ind. Chim. Belg. 38(1973) 481-495.
67 J. Geus, Communication EUCHEM Meeting, Namur,(1976).
68 J.M. Zabala, P. Grange and B. Delmon, C.R. Acad. Sci. Ser. C 279(1974) 725-728.
69 B. Delmon, ref.[20], pp.414-436.
70 B. Delmon, ref.[20], pp. 485-498 and 258-269.

71 B. Delmon and A. Roman, J. Chem. Soc. 69(1973) 941-948.
72 M.L. Garcia-Gonzalez, P. Grange and B. Delmon, C.R. Acad. Sci. Ser. C 280 (1975) 1439-1441.
73 M.L. Garcia-Gonzalez, P. Grange and B. Delmon, in J. Wood, O. Linqvist, C. Helgesson and N.G. Vannerberg(Eds.), Reactivity of Solids, Plenum Press, New York, 1977, pp.69-74.
74 V.V. Boldyrev, Wissenshaft u. Fortshritt 28(1978) 17-21.
75 D.L. Trimm, in B. Delmon, P. Grange, P.A. Jacobs and G. Poncelet (Eds.), 2nd International Symposium on Preparation of Catalysts, Louvain-la-Neuve, 1978.
76 A. Cimino, M. Lojacono and M. Schiavello, J. Phys. Chem. 79(1975) 243-249.
77 H. Lafitau, E. Neel and J.C. Clement, in B. Delmon, P.A. Jacobs and G. Poncelet (Eds.), Preparation of Catalysts, Elsevier, Amsterdam(1976), pp.393-401.
78 M. Houalla, B. Delmon, unpublished results.
79 W. Verhoeven and B. Delmon, C.R. Acad. Sci. Ser. C 262(1966), 33-35.
80 W. Verhoeven and B. Delmon, Bull. Soc. Chim. Fr. (1966) 3065-3073 and 3073-3079.
81 H. Charcosset, R. Frety, A. Soldat, Y. Trambouze, J. Catal. 22(1971) 204-212.

DISCUSSION

J.W. HIGHTOWER : Your bi-layer model for the optimally active 0.25 Co/Mo catalyst suggests that the Mo is covered by a Co layer and may not be readily available to the gas phase. This picture may be naive in view of recent work we have done with some Mo-containing Haldor Topsøe catalysts both with and without Co for the metathesis of propylene near room temperature. Mo is the active ingredient in this system, and the activity at constant Mo loading is independent of the presence of Co. Furthermore, NO poisons this reaction and can be used as a titrating agent to determine the active site concentration. The infrared spectra of NO adsorbed on these materials show bands that can be attributed to NO on both Mo (2 bands) and Co (1 band). Hence, both Co and Mo must be readily available to the gas phase.

B. DELMON : We are not yet in a position to propose a precise model of the bilayer, and our answer, presently, is only speculative. Measurements of the intensity of the XPS lines suggest that Co associated with this bilayer lies closer to the surface as opposed to Co in $CoAl_2O_4$. On the other hand a rough estimate indicates that this bilayer associates approximately 1 Co to several Mo atoms (3 to 4). If this picture is correct, one may conceive an arrangement in which Co of the bilayer does not shield Mo and we arrive at the conclusion that both atoms should be readily accessible.

J.T. RICHARDSON : I have a question and a comment. The question is : what is the evidence that bulk Co_3O_4 exist on the surface in the presence of MoO_3 ? It has always been my interpretation that with 10-15 wt % MoO_3, cobalt that is not present as $CoAl_2O_4$ or $Co_xMo_{1-x}O_3$ is CoO.

The comment : magnetic measurements indicate that commercial catalysts have different cobalt states after calcining and after sulfiding plus regeneration. Perhaps we should turn our attention to regeneration systems since these still retain their activity.

B. DELMON : As concerns your question, there are many evidences that Co_3O_4 is present on the catalyst containing a high proportion of Co (r = Co/(Co+Mo) = 0.75 - 1.00). We can cite first the blackening of the sample which is characteristic of the presence of Co_3O_4. Another evidence stems from the sensitivity of the U.V. reflectance spectrum of the samples to the presence of Co_3O_4 (ref. 45 of our paper) illustrated by the presence of a shoulder near 750 nm. Moreover electron diffraction measurements performed on these samples give a further proof on the existence of Co_3O_4. Another reason coming from XPS measurements lies in the high intensity ratio of the principal to the satellite peaks of Co 2p 3/2 ($p=I_p/(I_p+I_s)$) : it is $p \simeq 0.7$ for r = 0.75-1.00, whereas one should obtain, and actually obtains, $p \simeq 0.58$ for Co(II) (e.g. in $CoAl_2O_4$ and $b.CoMoO_4$). However, the values observed do not correspond to pure Co_3O_4 (p=0.82). One might estimate that, in the catalyst with r=1.0, only 50 % of Co is present as Co_3O_4. We cannot indicate in which form exactly is the remaining Co. But we partly agree with your observation when we consider catalysts with smaller proportion of $r \leq 0.55$. Co is essentially in the form of Co(II), but probably not as CoO. Concerning your comment, we completely agree with you. We think that measurements made on fresh catalysts, or on catalysts having worked only a few hours are blurred by the presence of species which have nothing to do with the catalyst activity. Long time working, and, still better, regeneration, would reveal the really important phases, because they presumably develop at the expense of the other ones.

F.P.J.M. KERKHOF : 1) We have studied the binding energy of the Si(2p) electron for a series of WO_3/SiO_2 catalysts and found a constant value (1). Moreover, the Si(2p)/O(1s) intensity ratio was constant and the same as in pure SiO_2, which fact supports the "bulk" character of the XPS technique for high surface area supports. So, in our view, it is very difficult to detect the presence of a surface layer of Mo-Si-O compound by XPS. Could you comment on this ?

2) Is it possible that the maximal reduction for the $\pm$ 15 % MoO_3/SiO_2 catalyst must be attributed to the water evolved during reduction, poisoning the reduction ?

B. DELMON : It is true that the variation of the binding energy of the Si(2p) level is quite important, especially if one takes into account that several silica layers contribute to the Si(2p)XPS signal. Having this in mind, we have stressed in our communication the need of a direct verification of the hypothesis of an electron transfer from silica to molybdena by checking the shift in the Mo(3d) lines. Unfortunately such attempts were frustrated, in the case of the MoO_3/SiO_2 samples examined, by the extensive broadening of the Mo(3d) spectrum.

Concerning the difficulty to detect the presence of a Mo-Si-O surface compound by XPS, we think that in principle it is possible to observe the formation of this kind of surface layer when its occurrence leads to a shift in the Mo(3d) binding energy and obviously it represents a considerable fraction of total Mo content.

2) Your remark refers to our Fig. 4. We must remark that the maximum degree of reduction we report is quite an empirical value. The reduction after 10 hrs depends on the kinetics of the process, and, among other parameters, on the nature of the species (bulk MoO_3, MoO_3 in interaction with SiO_2), on the dispersion of the species, and on diffusion phenomena (especially with respect to H_2O). Hence, this value only indirectly and imperfectly characterizes the deposited species on the degree of dispersion. In our above mentioned investigations on MoO_3/SiO_2 with analytical electron microscopy, the upper limit of MoO_3 in interaction with SiO_2 was estimated to be about 7 %. The shape of the curve between 7 and 20% or, at least, the ascending branch between 7 and 15%, could thus be attributed to dispersion. The decrease between 15 and 20% may be attributed, either to the inversion of the effect of particle size growth indicated in Figs. 2 and 3, or to diffusion phenomena arising as a consequence of the constriction of the pores or of the complete blocking of pores.

J.F. LEPAGE : On the basis of your models, can you explain the changes in relative activity observed in hydrogenation of olefins or aromatics, in HDS, when the carriers of the Co-Mo catalysts are changed, the Co/Mo ratio remaining "the same" ? Is it possible as with metallic catalysts to invoke "facile" or "difficile" reactions with respect to the form, the size or the composition of the active "sulfided deposists" ?

B. DELMON : We do not know whether it will turn out that certain reactions in hydrodesulfurization are "facile" and other "difficile". But, whichever the model used for explaining the activity of HDS catalysts, we can expect the selectivity to change with the form and size of the sulfide particles, with the composition of the catalysts, and with the nature of the support (1). If we forget the acidic properties of the support, and consider only the reactions supposed to be catalyzed by the sulfides, the problem which is still not solved is whether the support is directly involved in the architecture of the active phase, or if it mainly plays a role in achieving a given dispersion, repartition in size, contact between sulfides, etc.. during preparation. It is clear that Al_2O_3 brings about a good dispersion because it favors the formation of a CoMo "bilayer" and that SiO_2 is much less efficient in this respect. The difference between the two carriers may lie in the fact that the CoMo "bilayer" is interacting with Al_2O_3, while, on SiO_2 it is only weakly interacting and consequently Co and Mo tend to form $CoMoO_4$. The "bilayer" would be a good precursor of a finely (inter)dispersed active sulfided phase, whereas $CoMoO_4$ is a very poor precursor (2,3). So, in our opinion at least part of the difference in catalytic behaviour must be attributed to differences in the genesis of the active systems, rather than to specific interaction between the final sulfided species and supports.

1. B. Delmon : A.C.S. meeting Petroleum Div. Preprints New Orleans (1977), p.503.
2. G. Hagenbach, Ph. Courty and B. Delmon, J. Catal. 23 (1971), 295 and J. Catal. 31 (1973), 264.
3. P. Grange and B. Delmon, J. Less Comm. Metals. 36 (1974), 353.

V. KOSTROV : What do you think about the theoretical evolution of the inhibition of the kinetics of nucleation ? For example, what is the influence of steam on kinetics of reduction of Co-Mo/alumina catalysts ?

B. DELMON : We were referring to the kinetic consequences of the intervention of two quite different processes in many reactions of solids, namely nucleation and interface progress (as growth of nuclei). Taking a series of samples with different grain sizes, but characterized by identical rates of nucleus formation k_g on the surface and identical rates of interface progress, the rate-limiting step is k_i for large particles, k_g for small ones. This explains the variation of the overall rates with particle size (proportional to $A_{S(0)}^{1/3}$ indicated in Fig. 2 (1). Our suggestion is that the decrease in reduction rates observed with supported oxides, in comparison to unsupported ones, can, in certain cases, be attributed to the fact that nucleation becomes more rate-limiting.

But you are quite right in indicating that some factors could intervene, such as inhibition by H_2O, etc... Both activation effects (e.g. by group VIII metals) and inhibition can be observed in the reduction of oxides (2, 3, 4).

1. B. Delmon, Introduction à la cinétique hétérogène, Edition Technip, Paris, 1969; Kinetika geterogennykh reaktsii, Editions Min., Moscou, 1972.
2. J. Masson, B. Delmon, J. Nechtschein, C.R. Acad. Sci., Paris, Série C, 266, (1968), 428-430.
3. A. Roman, B. Delmon, J. Catal., 30, (1973), 333-342.
4. H. Charcosset, B. Delmon, Ind. Chim. Belg., 38, (1973), 481-495.

PREPARATIVE CHEMISTRY OF COBALT-MOLYBDENUM/ALUMINA CATALYSTS

A. IANNIBELLO and P.C.H. MITCHELL

Stazione Sperimentale per i Combustibili, S. Donato Milanese, Italy and Department of Chemistry, The University, Whiteknights, Reading, U.K.

ABSTRACT

The catalytic properties of $CoO-MoO_3/Al_2O_3$ catalysts are affected by their method of preparation. As a basis for understanding these effects we have investigated the chemical changes in the preparation of the catalysts and the structures of the cobalt and molybdenum oxo-species formed. Our main results are:

(a) Adsorption of molybdate by alumina involved exchange of surface -OH groups for $[MoO_4]^{2-}$ ions.

(b) Some aluminium molybdate was formed in the preparation of MoO_3/Al_2O_3 catalysts ($>$10g MoO_3/100g Al_2O_3) and remained in unwashed catalysts.

(c) The predominant oxomolybdenum(VI) species in the MoO_3/Al_2O_3 catalysts depended on the molybdenum concentration and were: strongly bound tetrahedral $[MoO_4]$ ($<$5g MoO_3/100g Al_2O_3), less strongly bound tetrahedral $[MoO_4]$ ($>$5g MoO_3/100g Al_2O_3), and octahedral $[MoO_6]$ ($>$12g MoO_3/100g Al_2O_3). Washing the catalysts with water before calcining removed octahedral and less strongly bound tetrahedral oxomolybdenum species.

(d) Catalysts prepared by the pore-filling (PF) technique contained a smaller proportion of strongly bound molybdenum than those prepared by the equilibrium adsorption (EA) technique. The local pH in the pores is greater in the PF technique and so the extent of reaction of molybdate with alumina is less than in the EA technique.

(e) Addition of a cobalt(II) nitrate solution to a MoO_3/Al_2O_3 catalyst caused part of the cobalt to be adsorbed and part to be precipitated as a surface hydroxide.

(f) The $CoO-MoO_3/Al_2O_3$ catalysts contained tetrahedral and octahedral oxocobalt species. The proportion of the octahedral species was greater in catalysts prepared by cobalt-impregnation of calcined MoO_3/Al_2O_3 catalysts.

(g) There was interaction of the oxocobalt and oxomolybdenum species in the catalysts.

INTRODUCTION

Catalysts comprising cobalt (or nickel) and molybdenum oxo-species on γ-alumina are used in hydrodesulphurisation (abbreviated hds) of petroleum and other fossil fuels[1]. Their activities are affected by the method of preparation (impregnation, washing, drying, calcining, and sulphiding)[2]. For example, the catalytic activity in hds depends on the method of impregnation. The most active catalysts are prepared by double impregnation of γ-alumina, first with ammonium molybdate solution and then with cobalt(II) [or nickel (II)] nitrate solution [2a]. Catalysts prepared by kneading boehmite with solutions of molybdenum(VI) and cobalt(II) salts are less active apparently because migration of promoter ions into the alumina lattice is favoured during the phase transformation of boehmite to γ-alumina which occurs when the catalyst is calcined [2c,d,e].

The presence of cobalt in $CoO-MoO_3/Al_2O_3$ catalysts causes a synergic increase of hds activity (promoter effect) [1,2]. To understand the effect of cobalt we must distinguish between active and inactive cobalt[3]. Active cobalt is at the catalyst surface and in the oxide form of the catalyst is in octahedral co-ordination by oxide ions[3,4]. Active cobalt is sulphided under hds conditions and intercalates in MoS_2 in the active catalyst[5]. The inactive cobalt is locked in the alumina lattice and is tetrahedrally co-ordinated by oxide ions in both the oxide and sulphided forms of the catalysts. We expect therefore that the most active catalysts are those which are prepared in such a way that the amount of octahedral cobalt is optimised.

As a basis for understanding how preparative variables affect properties of the catalysts we have investigated chemical changes in the preparation of the catalysts and the structures of the cobalt and molybdenum oxo-species formed. (We follow the convention of expressing the concentrations of cobalt and molybdenum in the catalysts in terms of oxides (CoO and MoO_3) but we do not thereby imply that these oxides are present as well-defined phases). We show in this paper that impregnation of calcined MoO_3/Al_2O_3 with cobalt gives the greatest proportion of octahedral cobalt.

EXPERIMENTAL

Gamma-alumina (Ketjen 000-1.5E; surface area, 300 $m^2 g^{-1}$; water pore volume, 0.72 $cm^3 g^{-1}$; particle diameter, 0.15 - 0.30 mm) was impregnated with aqueous solutions of ammonium heptamolybdate at room temperature (ca. 20°C) by: (a) equilibrium adsorption (EA series) [ammonium heptamolybdate solution (100 cm^3) stirred with γ-alumina (1 g) until no further uptake of molybdenum or pH change and then filtered], (b) pore-filling (PF series) [ammonium heptamolybdate solution (0.7 cm^3) added to γ-alumina (1 g) and left for 48 h]. Catalysts of both series were washed with distilled water (3 x 75 cm^3, 45 min) (some samples only), dried at 110°C, and calcined in air at 550°C. The CoO-MoO$_3$/Al$_2$O$_3$ catalysts were prepared by the pore-filling method from solutions of cobalt(II) nitrate and the various MoO$_3$/Al$_2$O$_3$ catalysts dried at 110°C and calcined at 550°C.

Molybdenum was determined by atomic absorption (changes of concentration of the impregnating solutions, EA series catalysts) and by X-ray fluorescence (solids).

U.v.-visible spectra of the solids were measured by the reflectance technique against a magnesium oxide standard, and i.r. spectra in KBr discs. Magnetic susceptibilities were determined by the Gouy method at room temperature.

RESULTS AND DISCUSSION

We have prepared catalysts containing various amounts of cobalt and molybdenum supported on γ-alumina in order to determine the effect of variables in the preparation (method of impregnation, washing, drying, and calcining) on the composition of the catalysts and the structures of the cobalt and molybdenum species.

(a) MoO$_3$/Al$_2$O$_3$ Catalysts - (i) Preparation

Two methods were used to impregnate the γ-alumina support with ammonium heptamolybdate: equilibrium adsorption (EA series) and pore-filling (PF series). In the EA method γ-alumina was left in contact with a solution of the molybdate until there was no further change of the molybdate concentration and the liquid phase was removed by filtration. In the PF method, catalysts of desired compositions were prepared by adding to γ-alumina solutions containing the requisite amount of molybdate in a volume just sufficient to fill the pores; any excess of moisture was removed by evaporation, not filtration. We describe first our results with the EA method.

When γ-alumina was added to a solution of ammonium heptamolybdate the pH (ca. 7.0) rapidly decreased (pH 4-5) and then increased slowly to a steady value (pH 5-6). The pH changes were a convenient means of following the reaction to equilibrium (constant final pH). We attribute the fast initial decrease to exchange of ammonium ions with protons of the alumina (the presence of ammonium ions in the dried catalysts was shown by their i.r. spectra) and the slow subsequent increase to exchange of molybdate ions with surface hydroxyl groups[6].

$$Al-OH + [MoO_4]^{2-} = Al-OMoO_3^- + OH^- \text{ or} \qquad (1a)$$
$$AlO-H + [MoO_4]^{2-} = AlO-MoO_3^- + OH^- \qquad (1b)$$

The increase of pH was slower (ca. 0.5 h) than we would expect for bulk diffusion in the alumina pores (see below) or for the establishment of acid-base and molybdate-polymerisation equilibria in solution (fast reactions). The final constant pH was lower the greater the initial (and equilibrium) concentration of molybdate becuase of release of protons in the equilibrium

$$[Mo_7O_{24}]^{6-} + 4H_2O = 7[MoO_4]^{2-} + 8H^+ \qquad (2)$$

The species which adsorbs on the alumina is the $[MoO_4]^{2-}$ ion (see below). As this species adsorbs, the above equilibrium shifts to the right so releasing protons the concentration of which increases with the amount of molybdate adsorbed, i.e. with the total molybdate concentration.

The isotherm of molybdenum concentration change ('adsorption isotherm') (Fig. 1) was similar to that determined previously with more points[3]. Uptake of molybdenum tailed off at ca. 12 g MoO_3/100 g Al_2O_3. There was an inflexion at ca. 5 g MoO_3 corresponding to the plateau in the more detailed isotherm[3].

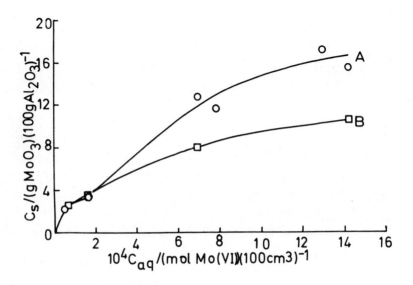

Fig. 1. Reaction of ammonium heptamolybdate with γ-alumina. Amount of Mo(VI) reacted (adsorbed) (C_s) from change of concentration of the impregnating solution determined by atomic absorption (curve A) and from analysis of calcined MoO_3/Al_2O_3 catalysts by X-ray fluorescence (curve B) plotted versus equilibrium concentration (C_{aq}) of Mo(VI) in the impregnating solution.

(ii) Composition of the samples

The concentration of molybdenum in the MoO_3/Al_2O_3 samples was assessed in two ways: from the change of concentration of molybdenum in the solution during the adsorption experiments (determined by atomic absorption, abbreviated AA, on filtered and centrifuged solutions) and by analysis of solid samples (by X-ray fluorescence, abbreviated XRF). The results by the two methods agreed at low molybdenum concentrations, but at higher concentrations XRF gave lower analyses than AA (Fig. 1). The difference is due to the loss of molybdenum as a finely divided solid which passed through the filter. Before the AA determination of molybdenum in the filtrate, suspended solid was removed by centrifuging. This solid (equivalent in weight to <0.05 g MoO_3), which was formed by reaction of molybdate with alumina at the higher molybdate concentrations, probably contained aluminium molybdate (i.r. spectrum, shoulder at ca. 920 cm^{-1}, cf. aluminium molybdate peak at 915 cm^{-1}; amorphous to X-ray diffraction). A similar reaction is the formation of insoluble aluminium phosphates during adsorption of phosphate on alumina from solutions of Na_2HPO_4 [7]. Although the preceeding observations were incidental to our main work they are of some significance in catalyst preparation. Formation of the finely divided solid implies some breakdown of the alumina surface in contact with the aqueous molybdate solution. If this solid is retained in the catalysts (as in the PF procedure, see below) it could cause inhomogeneity in their compositions and properties. The analytical data give the total amount of molybdenum converted to an insoluble form, i.e. the total amount reacted, and not necessarily the amount _adsorbed_ by the alumina.

(iii) Effect of washing

The effect of washing with distilled water on the molybdenum content of catalysts prepared by the EA and PF methods was studied in order to discover how much molybdenum was irreversibly bound to the alumina (Fig. 2). For the EA catalysts, there was no loss of molybdenum from the first three samples (which contained up to ca. 7% MoO_3) on washing; but part of the molybdenum was washed from the other samples and so these contained some less strongly bound molybdenum.

A portion of each PF catalyst was washed with water after equilibration but before drying (and so the pores still contained the impregnating solution). The pH of the wash liquid was determined as a function of time; for all samples the pH increased (by ca. 0.7 units) to a constant value during ca. 1 min presumably because of diffusion of OH^- ions from the pores into the bulk of the solution. The final value of the pH which is determined by the equilibria in solution was lower at higher molybdenum and ammonium ion concentrations as in the equilibrium adsorption experiments.

The concentrations of MoO_3 in the catalyst before and after washing are shown in Fig. 2. The amount of molybdenum retained in the EA series catalysts after washing is greater than in the PF series catalysts. Thus the amount of strongly bound molybdenum is less for the PF series catalysts. We can understand this result on the basis of exchange of molybdate ions with surface hydroxyl groups of alumina [equations (1a) and (1b)]. In the EA preparative procedure hydroxide ions diffuse from the pores into the bulk of the impregnating solution but this is not possible in the PF procedure since _all_ of the impregnating solution is in the pores. Thus the local pH in the pores will be greater in the PF procedure and so the uptake of molybdate by reaction with the alumina will be less. (In general the uptake of molybdate decreases as the pH of the impregnating solution increases [6,8]).

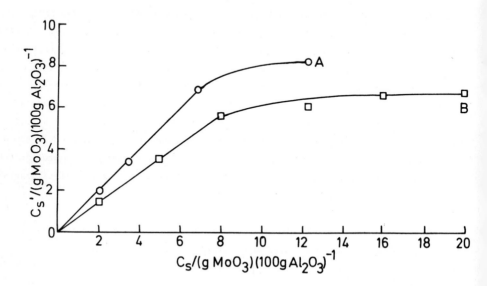

Fig. 2. Effect of washing on MoO_3/Al_2O_3 catalysts. Amount of molybdenum in the catalysts after washing (C_s') plotted versus amount before washing (C_s) for catalysts prepared by the equilibrium adsorption (EA) method (curve A) and the pore-filling (PF) method (curve B) determined by X-ray fluorescence.

(iv) <u>Drying and calcination and structures of the molybdenum species</u>

The MoO_3/Al_2O_3 catalysts were dried in air at 110°C for 24 h and calcined in air at 550°C for 24 h. Their u.v. diffuse reflectance spectra (Fig. 3) showed the following features.

<u>EA and PF catalysts: washed; calcined at 550°C</u> The spectra were characteristic of tetrahedral oxomolybdenum species. The u.v. absorbances were the same for those EA and PF catalysts which contained the <u>same amount</u> of molybdenum. Thus both series contained the same strongly adsorbed [MoO_4] species with the same degree of dispersion.

<u>EA and PF catalysts: unwashed; calcined at 550°C</u> The PF series catalysts with MoO_3 > 10 g showed an additional u.v. peak at ca. 32,000 cm^{-1} and so

contained some octahedral oxomolybdenum species formed presumably by thermal decomposition of bulk ammonium molybdate deposited in the pores when the catalysts were dried. The lower-molybdenum PF catalysts and the EA catalysts had spectra characteristic of tetrahedral $[MoO_4]$ species.

EA and PF catalysts: unwashed; dried 110°C The spectra showed one peak at ca. 37,000 cm^{-1}; thus the main species is tetrahedral molybdenum(VI), $[MoO_4]$, [9]. The u.v. absorbance was greater for EA than for PF catalysts with the same amount of molybdenum which suggests that the oxomolybdenum species is more dispersed in the EA catalysts[10]. This agrees with our observation that the amount of strongly bound molybdenum is greater in the EA catalysts. The PF catalysts with the highest molybdenum concentrations (16, 20 g MoO_3) showed tetrahedral oxomolybdenum spectra; evidently the tetrahedral $[MoO_4]^{2-}$ ion is deposited (as ammonium molybdate) from the solution in the pores when the catalysts are dried. This agrees with our suggestion (see above) of a high (alkaline) local pH in the pores the effect of which would be to shift the molybdate solution equilibria over to the $[MoO_4]^{2-}$ ion.

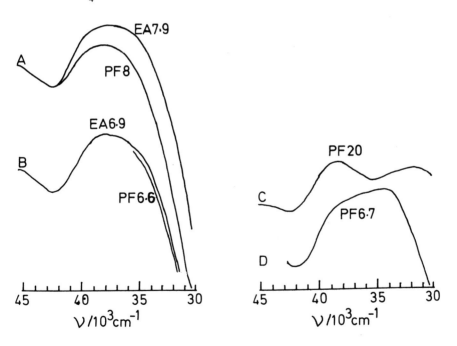

Fig. 3. Representative u.v. reflectance spectra (vs. MgO blank) of MoO_3/Al_2O_3 catalysts: A, unwashed, dried 110°C; B, washed, dried 110°C; C, unwashed, calcined 550°C; D, washed, calcined 550°C. Catalysts prepared by equilibrium adsorption (EA) and pore-filling (PF) methods. Numbers on the curves are the Mo(VI) concentrations [(g MoO_3)/(100 g Al_2O_3)].

EA and PF catalysts: washed; calcined at 550°C The spectra were characteristic of tetrahedral [MoO_4] species. Thus washing removes the species which otherwise would form octahedral oxomolybdenum(VI) in the high-molybdenum PF catalysts.

(b) CoO-MoO_3/Al_2O_3 Catalysts

The catlaysts contained 8 g MoO_3/100 g Al_2O_3 and up to ca. 4 g CoO. They were prepared by the pore-filling (PF) method in three different ways: (i) simultaneous impregnation with a solution containing both molybdate(VI) and cobalt(II) ions, (ii) adding cobalt(II) solution to MoO_3/Al_2O_3 catalysts dried at 110°C, and (iii) to catalysts calcined at 550°C. The u.v.-visible reflectance spectra and magnetic moments of the cobalt-containing catalysts dried at 110°C and calcined at 550°C were measured.

(i) Effect of cobalt on the molybdenum spectra Selected spectra are shown in Fig. 4. For the 110°C dried catalysts prepared by simultaneous impregnation, and by adding cobalt to 110°C dried MoO_3/Al_2O_3 samples, the effect of cobalt is to shift the Mo-O peak to lower wavenumbers. The reason for this is not clear but is apparently not due to a cobalt contribution to the spectrum since the effect is not observed for catalysts prepared by adding cobalt to 550°C calcined MoO_3/Al_2O_3 catalysts. For the 550°C calcined catalysts the effect of cobalt is to reduce the intensity of the Mo-O peak and shift it to lower wavenumbers. The changes in the spectra increase with increasing concentration of cobalt in the catalysts. Our interpretation is that cobalt is interacting with the oxomolybdenum species in the catalyst presumably by forming Co-O-Mo bonds. The interaction as judged from peak broadening is less in the catalysts prepared by double impregnation and by adding cobalt(II) to 110°C dried MoO_3/Al_2O_3 catalysts than in those prepared by adding cobalt(II) to 550°C calcined MoO_3/Al_2O_3 catalysts. The broadening of the molybdenum spectra to lower wavenumbers could indicate an increased proportion of octahedral oxomolybdenum species in the cobalt-containing catalysts but the change is small and the major molybdenum species remains tetrahedral.

(ii) Spectra and magnetic properties of cobalt Selected cobalt spectra are shown in Fig. 4. The spectra of all preparations of 110°C dried CoO-MoO_3/Al_2O_3 samples were identical (peaks at]7.7, 7.2, 6.2 x 10^3 cm^{-1}). In the spectra of the high-cobalt samples there was an additional shoulder at ca. 20,000 cm^{-1} indicating the presence of some octahedral cobalt(II). The spectra of the blue 110°C dried catalyst corresponded to the spectrum of blue cobalt(II) hydroxide in which cobalt(II) is octahedrally co-ordinated by oxide rather than to tetrahedrally co-ordinated cobalt(II). Thus it appears that at least part of the cobalt in our catalysts is deposited as a surface precipitated hydroxide[11].

Variations in the room temperature magnetic moments were only just outside the experimental precision (± 0.1 BM). For the catalysts prepared by simultaneous impregnation of alumina, and by the addition of cobalt to 110°C dried MoO_3/Al_2O_3 catalysts, the magnetic moments decreased slightly when the dried catalysts were calcined (4.6 to 4.4 and 4.9 to 4.8 BM respectively). The magnetic moments of the catalyst prepared by adding cobalt to calcined MoO_3/Al_2O_3 catalysts increased significantly after calcining (4.7 to 5.1 BM) in agreement with spectroscopic indications that these catalysts contained more octahedral cobalt.

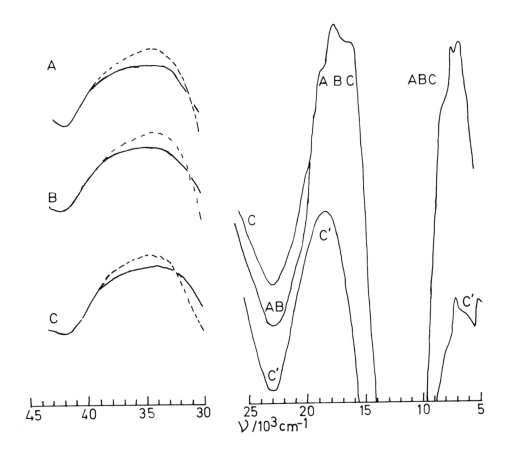

Fig. 4. U.v. and visible reflectance spectra (vs. MgO blank) of MoO_3/Al_2O_3 (broken lines) and $CoO-MoO_3/Al_2O_3$ (full lines) catalysts (3.4 g CoO, 8g MoO_3, 100 g Al_2O_3). U.v. spectra assigned to oxoMo(VI) species and visible spectra to crystal field transitions of Co(II). Catalysts prepared by double impregnation of γ-Al_2O_3 (A), adding Co(II) to 110°C dried MoO_3/Al_2O_3 (B) and 550°C calcined MoO_3/Al_2O_3 (C), all calcined at 550°C <u>except</u> 110°C dried $CoO-MoO_3/Al_2O_3$ (C').

REFERENCES

1. P.C.H. Mitchell, in 'Catalysis', C. Kemball (Ed.), Chemical Society, London, Vol. 1, 1977, p.204, and references therein.
2. (a) J.B. McKinley, in 'Catalysis', P.H. Emmett (Ed.), Reinhold, New York, Vol. 5, 1957, p.405; (b) V.H.J. de Beer and G.C.A. Schuit, in 'Preparation of Catalysts' B. Delmon, P.A. Jacobs and G. Poncelet (Eds.), Elsevier, Amsterdam, 1976, p.343; (c) Y. Kotera, K. Ogawa, M. Oba, K. Shimomura, M. Yonemura, A. Ueno, and N. Rodo, ibid, p.371; (d) W. Ripperger and W. Saum, in 'Proceedings of the Climax Second International Conference on the Chemistry and Uses of Molybdenum', P.C.H. Mithcell (Ed.), Climax Molybdenum Co. Ltd., London, 1977, p.175; (e) M. Inoguchi, T. Mizutori, K. Tate, Y. Satomi, K. Inaba, Y. Kaneko, R. Nishiyama, and T. Nagai, Bull. Japan. Petrol. Inst. 13 (1976) 19.
3. N.P. Martinez, P.C.H. Mitchell, and P. Chiplunker, in 'Proceedings of the Climax Second International Conference on the Chemistry and Uses of Molybdenum', P.C.H. Mitchell (Ed.), Climax Molybdenum Co. Ltd., London, 1977, p.164.
4. J.T. Richardson, Ind. and Eng. Chem. (Fundamentals), 3 (1964) 154; J.H. Ashley, Thesis, University of Reading, 1969, p.104; N.P. Martinez and P.C.H. Mitchell, Paper 15, Chemical Society, University of Bath, 1977.
5. A.L. Farragher and P. Cossee, 'Proceedings of the Fifth International Congress on Catalysis', J.W. Hightower (Ed.), North-Holland, Amsterdam, 1973, p.1301.
6. A. Iannibello and F. Trifiro, Z. anorg. Chem., 413 (1975) 293; A. Iannibello, S. Marengo and F. Trifiro, Chimica e Industria 57 (1975) 676.
7. H.P. Boehm, Discuss. Faraday Soc., 52 (1971) 264.
8. J. Sonnemans and P. Mars, J. Catalysis, 31 (1973) 209.
9. J.H. Ashley and P.C.H. Mitchell, J. Chem. Soc. (A), (1968) 2821; (1969) 2730.
10. J.M.J.G. Lipsch and G.C.A. Schuit, J. Catalysis, 15 (1969) 174.
11. P. Tewari and W. Lee, J. Colloid Interface Sci., 52 (1975) 77; H.C. Yao and M. Bettman, J. Catalysis, 41 (1976) 349.

DISCUSSION

E. MATIJEVIC : I am somewhat worried about the validity of equations 1a or 1b in the preprint. While they may stoichiometrically describe the situation (if the titration curves are in agreement with the reaction shown), they cannot mechanistically represent the true process. At pH 4-5 alumina is strongly positively charged (precipitation is at $\sim$9). Thus a protonated aluminum hydroxide surface species should be involved in the condensation reaction with the molybdate io

A. IANNIBELLO and P.C.H. MITCHELL : We agree with the comment. The following equations describe the interaction of metamolybdate with alumina under acidic conditions :

$$(Mo_7O_{24})^{6-} + 4H_2O \rightleftharpoons 7(MoO_4)^{2-} + 8H^+$$

$$Al-OH + H_3O^+ \rightarrow AlOH_2^+ + H_2O$$

$$Al-OH_2^+ + [MoO_4]^{2-} \rightarrow AlOMoO_3^- + H_2O$$

The observed rise of pH is then correctly described as due to removal of protons from the solution rather than to release of hydroxide ions.

THE INFLUENCE OF THE SUPPORT ON Co-Mo HYDRODESULFURIZATION CATALYSTS

H. TOPSØE[x], B.S. CLAUSEN[+], N. BURRIESCI[+▫], R. CANDIA[x] and S. MØRUP[+]

x Haldor Topsøe Research Laboratories, DK-2800, Lyngby, Denmark.
+ Laboratory of Applied Physics II, Technical University of Denmark, DK-2800, Lyngby, Denmark.

SUMMARY

The effect of support on the structural properties of cobalt-molybdenum hydrodesulfurization catalysts has been investigated by means of Mössbauer spectroscopy.

The study shows that the interaction of Co and Mo with the support plays an essential role in determining the resulting phases, both in the oxidic and in the sulfided state.

1 INTRODUCTION

The continuous scientific attention paid to hydrodesulfurization (HDS) catalysts is undoubtedly inspired by the need for preparing more efficient catalysts for sulfur removal especially from difficult feedstocks. The great research effort has yielded a better understanding of this difficult catalyst system. There still exist, however, diverging views regarding the structure of the catalysts, the relation between the structural and catalytic properties, and the changes resulting from different preparation conditions. It is clear that the nature of the support plays an important role for the catalyst properties. (For recent reviews, the reader is referred to [1,2]). Most of the published works deal with alumina-supported catalysts. Less attention has been paid to unsupported catalysts and to carbon- and silica-supported catalysts.

In the present work we have studied the influence of the sup-

---

▫Permanent address: Montedison Research Laboratories, Novara, Italy.

port on the structure and properties of the catalyst. For this purpose we have used Mössbauer spectroscopy, which earlier has been shown to be very useful for the study of Co-Mo HDS catalysts [3-5] and other catalysts [6].

In this study, we have applied Mössbauer spectroscopy mainly as a "fingerprint" technique. By comparing Mössbauer spectra of catalysts on different supports with those of model compounds and also by studying the shapes of the individual spectra one can get structural and chemical information which is not necessarily limited to the Mössbauer atoms. The fact that the studies can be carried out *in situ* and that the degree of crystallinity does not play any role, makes this technique especially useful for studies of HDS catalysts.

2 EXPERIMENTAL
2.1 Preparation of samples.

All samples were doped with ^{57}Co.

2.1.1 $CoMoO_4$ was prepared by precipitation at boiling point using stoichiometric solutions of $Co(NO_3)_2 \cdot 6H_2O$ and Na_2MoO_4, followed by washing, drying and calcination at 500°C in air. X-ray diffraction analysis showed that the resulting phase was pure $CoMoO_4$.

2.1.2 $Co:MoS_2$ was prepared by impregnation of MoS_2 powder (Riedel-De Haen AG) with a solution of $Co(NO_3)_2 \cdot 6H_2O$, followed by heat treatment in H_2/H_2S (2% H_2S) at 325°C. The sample contained approximately 1 ppm Co.

2.1.3 Co_9S_8 was prepared by sulfiding $Co(NO_3)_2 \cdot 6H_2O$ at 600°C in H_2/H_2S (2% H_2S).

2.1.4 The catalyst samples were prepared by impregnating the supports with an ammoniacal solution of cobalt nitrate and ammonium paramolybdate. The supports used were: γ-Al_2O_3 (230 m^2/g); SiO_2 Davison Grade 950 (950 m^2/g); and active carbon (900 m^2/g). Samples containing only Co (without Mo) were also prepared.

The impregnated samples were dried in air at room temperature and afterwards calcined in air for 24 hours at 500°C (for the carbon-supported samples the calcination temperature was 230°C).

2.1.5 Another sample was prepared by evaporation of the impregnation liquor to dryness at room temperature, followed by calcination in air at 500°C for 24 hours.

Sulfided catalyst samples were prepared by treating the oxidic samples at 325°C in H_2/H_2S (2% H_2S) for 24 hours.

The nomenclature used in this paper for the samples will become clear from the following example: Co1-Mo6/Al_2O_3 denotes an alumina-supported sample containing 1% Co and 6% Mo.

2.2 Mössbauer spectroscopy

The samples were pressed into thin wafers and placed in an *in situ* Mössbauer Pyrex cell connected to a gas handling system [5]. The samples were used as sources with a moving single-line absorber of $K_4Fe(CN)_6 \cdot 3H_2O$ enriched with ^{57}Fe. Zero velocity corresponds to the centroid of the spectrum obtained at room temperature with a source of ^{57}Co in natural iron. Positive velocity corresponds to the source moving away from the absorber.

3 RESULTS AND DISCUSSION

Figure 1 shows room temperature Mössbauer spectra of model compounds and of the calcined catalysts. Figure 2 shows the room temperature spectra after sulfidation at 325°C. The spectra of some model compounds are also included here. In the following we will give a brief interpretation of the various spectra. The main emphasis will be on the great differences observed among the catalysts on different supports. A quantitative analysis of the Mössbauer parameters will be published elsewhere.

3.1 $CoMoO_4$

Figure 1a shows the Mössbauer spectrum of bulk $CoMoO_4$. An analysis of this spectrum has been published earlier [7].

3.2 Co_3O_4

Co_3O_4 was not prepared, but its Mössbauer spectrum is well known [8] and looks quite similar to Fig. 1e, except for the small shoulders appearing to the right and left of the two absorption lines.

3.3 Co_9S_8

The Mössbauer spectrum of this model compound is shown in Fig. 2a. This spectrum has been discussed in a previous publication [9].

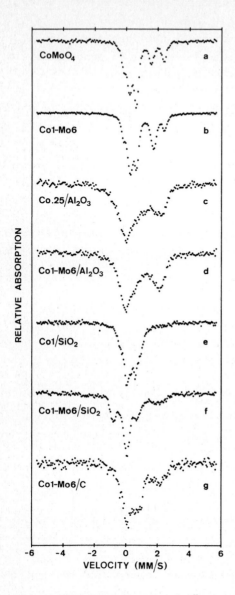

Fig. 1. Room temperature Mössbauer spectra of samples in the oxidic state.

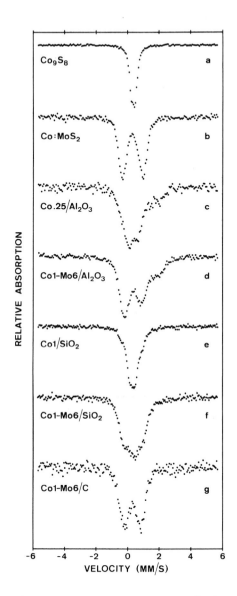

Fig. 2. Room temperature Mössbauer spectra of samples in the sulfided state.

3.4 Co:MoS$_2$

Co:MoS$_2$ gives the two-line spectrum shown in Fig. 2b [4]. This phase will be further discussed in connection with the sulfided catalysts.

3.5 Co1-Mo6 (unsupported oxidic catalyst)

The sample prepared by evaporation of the impregnation solution followed by calcination gives a Mössbauer spectrum (Fig. 1b) which is quite similar to the stoichiometric CoMoO$_4$ (Fig.1a). This result is expected, since CoMoO$_4$ is the thermodynamically stable compound under the calcination conditions used for the catalysts.

3.6 Alumina-supported samples

3.6.1 Oxidic form

Samples: Co 0.25/Al$_2$O$_3$ and Co1-Mo6/Al$_2$O$_3$. The spectra of Co 0.25/Al$_2$O$_3$ (Fig. 1c) and Co1-Mo6/Al$_2$O$_3$ (Fig. 1d) in the oxidic state are both typical for cobalt diffused into alumina [4,10]. The broad lines indicate that the cobalt is not present in a single type of site; rather, a distribution of Co surroundings exists. It is interesting that the cobalt ions in the two samples are present in quite similar surroundings. Thus, in the calcined catalyst, Co1-Mo6/Al$_2$O$_3$, the presence of molybdenum does not influence to any large extent the structure of the cobalt ions. Certainly, a cobalt molybdate phase does not form. This phase has previously [11] been proposed to be present for catalysts calcined at relatively low temperatures such as those (500°C) used in the present study. If the support interactions with Co and Mo are weak, cobalt molybdate would be expected to form upon calcination. This was in fact observed for the unsupported oxidic catalyst (Co1-Mo6) (Fig. 1b) as described above.

In studies of other Co/Al$_2$O$_3$ samples with higher cobalt concentrations it was found that, above a certain level, part of the cobalt did not interact with alumina. Instead, a separate Co$_3$O$_4$ phase was formed in accordance with the findings of Lo Jacono et al. [12]. The presence of molybdenum influenced the concentration level of cobalt necessary for the Co$_3$O$_4$ formation, but not the nature of this phase. It is interesting to note that under these circumstances where the Co$_3$O$_4$ phase is formed, the cobalt molybdate formation did not occur. This suggests a high stability of the molybdenum on the alumina surface.

With respect to the various models proposed for the structure of the oxidic catalyst, the monolayer model as originally proposed by Schuit and Gates [13] seems to represent quite closely the actual catalyst structure for catalysts with a composition typical of that used industrially. It should be stressed that the catalyst structure is dependent on many parameters (surface area of the alumina; calcination temperature; metal loadings; impregnation procedure, etc.), and other models may describe the structure encountered when the parameters are changed.

3.6.2 Sulfided form

Samples: Co 0.25/Al_2O_3 and Co1-Mo6/Al_2O_3. Upon sulfidation, the Co1-Mo6/Al_2O_3 catalyst (Fig. 2d) behaved quite differently from the sample containing no molybdenum; in this sample, part of the cobalt is transformed into a Co_9S_8 phase (compare spectra 2c and 2a), whereas in the other sample at least two cobalt positions can be distinguished, none of which is a Co_9S_8 phase. On the other hand, one of these positions gives a Mössbauer spectrum quite similar to that of Co:MoS_2 (Fig. 2b). In fact, a detailed investigation of the spectra of Co:MoS_2 and Co1-Mo6/Al_2O_3 *versus* temperature strongly suggests that cobalt in Co1-Mo6/Al_2O_3 is present in a MoS_2-like phase [4].

These results are interesting in view of the recent research on alumina-supported catalysts which has been directed toward the possible presence or absence of cobalt in MoS_2.

The Co:MoS_2 structure in the Al_2O_3-supported catalyst has most likely a two-dimensional character. This is seen from the accessibility of the cobalt atoms to gas exposure [4].

It should be noted that the Co1-Mo6/Al_2O_3 catalysts studied here have a composition typical of those used industrially. For such catalysts it appears, therefore, that a separate phase of Co_9S_8 is not present; rather, the structure resembles that proposed in the "pseudo-intercalation" model [14]. Co_9S_8 was, however present when the concentration of cobalt was high. It is interesting that previous Mössbauer spectroscopy results indicate that the Co:MoS_2 phase is catalytically active [4].

3.7 Silica-supported samples

3.7.1 Oxidic form

Samples: Co1/SiO_2 and Co1-Mo6/SiO_2. The Co1/SiO_2 sample (Fig. 1e) shows that cobalt is mainly present as Co_3O_4. In addition, a small part of the cobalt seems to have formed a compound with the silica. This shows that the interaction of cobalt with silica is present, but is much weaker than in the case of alumina. The spectrum of the Co1-Mo6/SiO_2 sample in the oxidic state (Fig. 1f) is quite complicated and very different from the spectrum of the sample without molybdenum. This indicates the formation of a Co-Mo-O-containing surface compound, the exact nature of which is not clear. The presence of a poorly crystallized cobalt molybdate phase cannot be excluded. It is, however, not the only phase formed on the silica-supported catalyst. The results for the silica-supported samples are significantly different from those of the alumina-supported samples, because neither the cobalt nor the molybdenum seems to interact appreciably with silica.

3.7.2 Sulfided form

Samples: Co1/SiO_2 and Co1-Mo6/SiO_2. Upon sulfidation, the Co_3O_4 present in the oxidic Co1/SiO_2 sample is observed to transform into Co_9S_8, whereas both Co_9S_8 and the Co in the MoS_2-like phase are present in the sulfided Co1-Mo6/SiO_2 catalyst (Fig.2d). The Co:MoS_2 type phase can form upon sulfidation of a Co-Mo-O precursor structure. This is observed also for the carbon-supported catalyst as seen below. Formation of the Co:MoS_2 structure does not seem, therefore, to require the type of precursor structures which were observed in the alumina-supported catalyst.

3.8 Carbon-supported catalysts

A cobalt molybdate-like phase is observed when the carbon-supported catalyst is calcined at 230°C (Fig. 1d). The crystallinity of this phase is probably quite poor as seen from the broad lines and from the intensity ratios, which are somewhat different from those observed in bulk $CoMoO_4$ (Fig.1a). After sulfidation at 325°C the spectrum (Fig. 2e) shows that almost all the cobalt ends up in a MoS_2-like phase, whilst sulfidation of bulk $CoMoO_4$ gives also Co_9S_8 [1]. The above discussion of the silica-supported catalysts with respect to the support interactions for the cobalt and the molybdenum holds also in the case of carbon-supported catalysts.

4 CONCLUSIONS

The present results show that the support plays an essential role for the formation of the cobalt-containing phases, both in the oxidic and in the sulfided states of the catalyst.

In the alumina-supported catalysts the support interaction is quite strong. This is in particular evident from the Mössbauer spectra of the oxidic catalyst which indicate that cobalt is present in the alumina lattice. In the sulfided state, some of the cobalt atoms are present in a MoS_2-like surface structure which is probably the catalytically active phase. At high cobalt contents (higher than that corresponding to maximum catalytic activity) a Co_9S_8 phase is also found.

In the silica-supported catalysts, the weaker support interaction leads to the formation of new phases both in the oxidic state and in the sulfided state. In the latter, cobalt is present both as Co_9S_8 and in a MoS_2-like structure.

The Mössbauer spectra of the oxidic carbon-supported catalysts are similar to those of the unsupported catalysts. This indicates a very weak support interaction. In the sulfided state, the cobalt seems to be present mainly in a MoS_2 phase.

REFERENCES

1 V.H.J. de Beer and G.C.A. Schuit, in Preparation of Catalysts, B. Delmon, P.A. Jacobs and G. Poncelet (Eds.), Elsevier, Amsterdam, 1976, p.343.
2 For a recent review of the "contact synergy model", see B.Delmon, American Chem.Soc.Petroleum Div. Preprints, 22(1977)503.
3 H. Topsøe and S. Mørup, in A.Z. Hrynkiewich and J.A. Sawicki (Eds.) Proc. Int. Conf. Mössbauer Spectrosc., Akademia Górniczo-Hutnicza Im. S. Staszica W. Krakowie, Cracow, Poland, 1(1975)305.
4 B.S. Clausen, S. Mørup, H. Topsøe and R. Candia, J.Physique, 37(1976)C6-249.
5 B.S.Clausen, Thesis, LTF II, Technical University of Denmark, 1976, unpublished.
6 J.A. Dumesic and H. Topsøe, Adv.Catal., 26(1977)121.
7 B.S. Clausen, H. Topsøe, J. Villadsen and S. Mørup, in D.Barb and D. Tarina (Eds.), Proc. Int. Conf.Mössbauer Spectrosc., Bucharest, Romania, 1977, p.155.
8 C.D. Spencer and D. Schroeer, Phys.Rev.B,9(1974)3658.
9 B.S.Clausen, H. Topsøe, J. Villadsen, S.Mørup and R. Candia, in D. Barb and D. Tarina (Eds.), Proc. Int. Conf. Mössbauer Spectrosc., Bucharest, Romania, 1977, p.177.
10 G.K.Wertheim and D.N.E.Buchanan, in A.H.Schoen and D.M.J. Coryston (Eds.), Proc. 2nd Int. Conf. Mössbauer Effect, Saclay, France, 1961, John Wiley & Sons, Inc. New York, 1962, p.130.

11 R. Moné and L. Moscou, Am. Chem. Soc., Symp.series, 20(1975)150.
12 M. Lo Jacono, K. Cimino and G.C.A. Schuit, Gazz. Chim. Ital. 103(1978)1281.
13 G.C.A. Schuit and B.C. Gates, AIChE J. 19(1973)417.
14 A.L. Farragher and P. Cossee, in J.W.Hightower (Ed.), Proc.5th Int. Congr. on Catalysis,North-Holland, Amsterdam ,1973,p.1301.

DISCUSSION

M. CARBUCICCHIO : You affirm that neither Co nor Mo interact with SiO_2. This is based on the Mössbauer spectrum for Co(1)-Mo(6)/SiO_2 (Fig. 1f) - a quite complicated spectrum - which you do not interpret. On the other hand from the spectrum for Co(1)/SiO_2 (Fig. 1c), a contribution appears which could be not negligible due to the product of the reaction between Co and SiO_2.

H. TOPSØE : We do not state that there is no interaction of the active component with SiO_2. We do, however, say that this interaction is not so important as for alumina-supported catalysts. This is seen from the large fraction of the Co that is present as Co_3O_4 in the calcined, and as Co_9S_8 in the sulfided Co(1)/SiO_2 sample. Besides the above Co phases we observe, as discussed in the paper, small "shoulders" which most likely, as you suggest, are owing to Co reacting with the SiO_2 surface.
For calcined silica-supported catalysts containing both Co and Mo, a Co-Mo-containing compound is observed. This is in contrast to the alumina-supported catalysts and shows that for silica-supported catalysts a large fraction of the Mo is interacting only weakly with the silica.

J. SCHEVE : How can you be sure that the strong gamma-Co-source in your samples does not influence solid state reactions between the oxides during calcination ? (remember the Tamman temperature of the oxides !).

H. TOPSØE : We have studied with other physical techniques samples with and without radioactive Co and have found no evidence that the solid state reactions occurring during calcination are affected by the rather weak gamma ray source ($\leqslant$1 mCi). The main problem in ^{57}Co Mössbauer source experiments is associated with the electron capture process. The influence of such processes on the Mössbauer spectra of Co-Mo catalysts has recently been dealt with (S. MØrup, B.S. Clausen and H. Topsøe, to appear in J. Physique).

E.V. LUDENA : I would like to make a comment with regard to some molecular orbital calculations we have performed in order to determine the effect of cobalt on the reaction of thiophene over a MoO catalyst. The main result is that for given configurations, there is a narrowing of the gap between the highest occupied and the lowest unoccupied molecular orbitals. Hence, it seems that there is a strong electronic effect of cobalt on molybdenum oxide favoring the hydrodesulfurization reaction.

H. TOPSØE : In the active state of our catalysts the molybdenum is sulfided. We observe a valence change of the promotor atoms located in the Co-Mo-S structure during the catalytic reaction. Since Mo and Co are closely associated in the Co-Mo-S structure, this would undoubtedly affect the electronic structure of the Mo atoms. It would indeed be interesting to compare the present Mössbauer results with the type of calculations you have performed.

B. DELMON : 1) With respect to the spectrum of Co(1)-Mo(6)/SiO_2 in Fig. 1 (spectrum f) have you any interpretation for the line at -0.7 mm s^{-1}, which shows up in no other spectrum ?
2) Concerning the sulphided catalysts, your results quite impressively show that a new cobalt species is formed. But this finding raises problems !
With unsupported systems Farragher (1), as well as ourselves (2) put a very low limit to the maximum number of group VIII metal atoms which can be intercalated or pseudo-intercalated in MoS_2 or WS_2. The totality of the Co signals in b and g (Fig. 2) must be attributed to the cobalt associated with MoS_2, as no other cobalt signal is visible. In contradiction to the above recalled results and calculations, one should thus admit that we have one intercalated cobalt atom for six molybdenum atoms: this is really much more than it ever has been assumed. What explanation could you propose ? Is there an extremely high dispersion of MoS_2 ? What do you think about some special feature related to the model of Schuit for sulfided catalysts (3) ?
3) Also in Fig. 2, curves c and d: it is clear that there is much less $CoAl_2O_4$ in catalysts containing only Co than in those containing also Mo. This agrees with our recent results (4).

(1) A.L. Farragher, P. Cossee, in "Catalysis", Proc. 5th Int.Cong. Catalysis (J.W. Hightower, ed.), North Holland, Amsterdam, 1973, 1316-1317.

(2) F. Delannay, D.S. Thakur, B. Delmon, J. Less Common Metals, in press.

(3) W.H.J. de Boer, G.C.A. Schuit, in "Preparation of Catalysts" (B. Delmon, P.A. Jacobs, G. Poncelet, eds.).Elsevier, Amsterdam, 1976, 343.

(4) B. Delmon, P. Grange, M.A. Apecetche, P. Gajardo, F. Delannay, C.R. Acad. Sci., Ser. C, $\underline{287}$ (1978), 401.

H. TOPSØE : 1) We have at the moment no detailed explanation of the line at about -0.7 mm s^{-1} in Fig. 1 f. It seems to correspond to a component with very low isomer shift which may have formed as a result of the Auger Cascade following the electron capture process. 2) We do not believe that the present results are in conflict with the results you refer to, obtained on unsupported systems. These systems will have a low degree of dispersion of the MoS_2 phase. Our results on supported catalysts point to Co being present at the surface of a MoS_2-like structure with a high degree of dispersion. With respect to Fig. 2 b we recall that the unsupported Co:MoS_2 sample contained only 1 ppm Co.

P.G. MENON : The structure of virgin, reduced sulfided catalyst may change still further on exposure of the catalyst to actual reaction conditions. For silica-supported multi-component molybdate catalysts containing Bi, Mo, Fe, Co and Ni, we have found recently (1) that exposure of the catalyst to ammoxidation of propylene results in a substantial enrichment of the surface with Bi and Fe and, consequently, 'burying" of Mo, Co, Si, etc. The Mössbauer spectra of the catalyst also changes significantly. No such changes occur on an identical thermal treatment of the catalyst without reactants. Have you found any surface enrichment or depletion on exposing your catalyst to actual hydrodesulfurization reaction, instead of sulfiding them only ?

(1) T.S.R. Prasada Rao and P.G. Menon, J. Catalysis, $\underline{51}$, (Jan. 1978).

H. TOPSØE : First, it should be mentioned that the state of the catalyst did not change significantly going from H_2/H_2S to H_2/thiophene reaction mixtures (see ref. 4 in the paper). However, very large changes were observed when the sulfided catalysts were exposed to air. It is therefore essential that sulfided catalysts are examined in situ. The changes you mention occur going from calcined to used catalysts. In the case of hydrodesulfurization catalysts you would expect from thermodynamics great changes between the structures of calcined and sulfided catalysts. Indeed, the expected changes were observed for the unsupported catalysts. The changes we observe for supported catalysts depend very much on the interaction of the Co and Mo with the support. In general, the changes may alter the fraction of various atoms exposed at the surface. For example, in the case of the Co-Mo catalysts supported on alumina, we did observe a surface enrichment of the Co atoms upon sulfidation, since Co atoms moved from positions in the alumina to positions at the surface of a Co-Mo-S phase.

C.J. WRIGHT : From your Mössbauer spectra it is presumably possible to estimate the ratio of molybdenum atoms to "intercalated" cobalt atoms in the sample of sulphided $Co(1)-Mo(6)/Al_2O_3$. What is the magnitude of this ratio, and do you have any measurements of the crystallite size of the molybdenum sulphide in your sample which would enable you to rationalize this ratio in terms of the "intercalation" model ?

H. TOPSØE : For the sulfided Co(1)-Mo(6)/alumina catalyst more than 30% of the Co atoms are present in a MoS_2-like phase. Our results point to the Co atoms being present at the surface of this phase. As a consequence the MoS_2-like sample must be present in a high state of dispersion and may have rather two- than three-dimensional character.

P.C.H. MITCHELL : Regarding the presence of Co_3O_4 in the catalyst, I would like to suggest that for "good" catalysts the formation of Co_3O_4 should be avoided or minimized and that this can be done by preparing the catalysts in such a way that cobalt and molybdenum do interact in the oxidic catalyst. I notice also that the authors prepared their catalyst by a co-impregnation method. Possibly with this procedure interaction of molybdenum and cobalt separately with the support rather than with each other is most likely. We prepared our catalysts by adding cobalt(II) solution to MoO_3/Al_2O_3 and then found evidence of a $CoO-MoO_3$ interaction.

H. TOPSØE : We agree that in order to make good catalyst, Co_3O_4 formation should be avoided in calcined catalysts, since this will be converted into Co_9S_8 upon sulfiding. Co_3O_4 formation can be avoided if Co interacts with the alumina or if, as you suggest, Co and Mo interact. From the present results it does, however, appear that the interaction of Co with alumina is more important than the interaction between Co and Mo. Co is therefore located in the alumina but, as recent Mössbauer results show, in immediate vicinity of the Mo surface layers. We have found that changing the preparation procedures (e.g. changing the order of impregnation) changes the structure of the catalyst.

STUDY OF SOME VARIABLES INVOLVED IN THE PREPARATION OF IMPREGNATED
CATALYSTS FOR THE HYDROTREATMENT OF HEAVY OILS

O. OCHOA, R. GALIASSO[x] and P. ANDREU[x]
Instituto Venezolano de Investigaciones Cientificas, Caracas, Venesuela

SUMMARY

In the present investigation it was found that the most significant variables for the hydrotreatment of heavy oils were the following: use of additives, drying and activating conditions and active metal content. The metals deposited over alumina base supports were oxides of molybdenum and nickel. The activities were tested with deasphalted venezuelan heavy oils diluted with gasoil.

The best additives modified the solute-support interactions and gave homogeneous distribution profiles for Mo through the catalyst pellets, as evidenced by electron microprobe analysis. The effect of the drying rate was found to be critical in attaining a good molybdenum distribution and activity. Low heating rates and the use of a stream of hot air during calcination yielded better catalysts. The activity increased with MoO_3 concentration up to 15%. A maximum was obtained for a NiO/MoO_3 relation of about 0,3. When impregnating nickel first, the hydrodemetallation activity was improved.

INTRODUCTION

During the hydrodesulfurization (HDS) and hydrodemetallation (HDM) reactions on $NiMo/Al_2O_3$ catalysts, a deactivation phenomenon takes place due to carbon and vanadium deposition (10% of C and 1% V) during the first 24 hours. The kinetics of these reactions is not well established for resines and asphaltenes present in the heavy ends. Furthermore, strong diffusion control of big molecules into the narrow pores takes place. The reactions which occur are the following :

---

[x] Present address: Instituto Tecnologico Venezolano del Petroleo, Caracas, Venezuela.

organosulfur compound + H_2 ⟶ desulfurized organic compound + H_2S

$$\begin{array}{c} R \quad \quad R \\ \diagdown \quad \diagup \\ V \\ \diagup \quad \diagdown \\ R \quad \quad R \end{array} + H_2 \longrightarrow \text{V on the catalyst} + RH$$

There are few systematic studies about the influence of the multiple variables involved in the preparation of catalysts by impregnation of porous alumina base supports for light hydrocarbon and for heavy oils desulfurization (1-4). There is no open literature mentioning hydrodemetallation catalysts preparation.

This work attempts to clarify which of the parameters found as important through the routes of support impregnation are important for the activity and selectivity and specially for the stability, when working with heavy molecules. The mean objective is the control of the homogeneity in the metal distribution profile into the catalyst particle. Crust catalysts have been previously prepared (5).

EXPERIMENTAL

Catalysts were prepared using commercial γ-alumina as support and ammonium heptamolybdate (AHM) and nickel nitrate (NN) as impregnating salts. The dry techniques was generally used. Additives employed were an ammonium salt (AS), a peroxide compound (P) and an amine (A), among others. Chromatographic plates were prepared with suspensions of the support in a pulverized form. Adsorption isotherms studies were carried out using powders and pellets.

The activity was tested in a trickle bed reactor, the charge being fed through a pump, and preheated and mixed with a H_2 stream. Some tests were performed in a batch reactor. The feeds employed were diluted Boscan and Morichal oils with 500 and 130 ppm of V respectively, 2.5% S and 22-24° API.

For the drying experiments a McBain balance was employed when working with single pellets. For bed drying a thermostated reactor was used. The hydrothermal treatment of the support was carried out in a tubular reactor also employed in the calcination of the catalysts. The support was subjected to a stream of steam diluted with nitrogen.

The metal contents on the catalyst were determined by neutron activation analysis and X ray fluorescence. The Mo distribution along the catalyst pellets was performed by microprobe analysis. Vanadium in oil was determined by neutron activation and sulfur by titulation of SO_2 in a LECO apparatus. Physicochemical characterization of the catalysts was achieved by standard methods. For thermogravimetric studies a Du Pont instrument and the McBain thermobalance were employed.

RESULTS AND DISCUSSION
IMPREGNATION
Solute-support interactions

To prevent the formation of a white precipitate when AHM solutions are placed in contact with the support, several additives were assayed. Ammonium salts, peroxides and amines insure stable solutions. Other additives lead to poor results. Yamagata et al. (6) have explained the appearance of the precipitate as due to the interaction between p-molybdic acid and acidic OH of the support.

A comparative technique by modified thin layer chromatography was used to assay different additives. Further analysis of impregnated pellets by electron microprobe gave the distribution profiles shown in fig. 1.

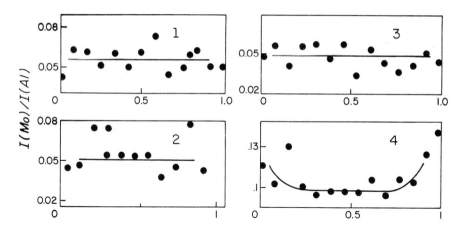

Fig. 1. Electron microprobe analyses of Mo. 1) P; 2) As; 3) A
4) without additive

Additive A gave the best result by TLC (greater R_f of molybdenum). The mechanical strength of the support was damaged by additive P. In order to select among A and AS, catalysts were prepared with the same concentration of MoO_3 (18%) and either of the two additives. Since the activities were almost equal, the final selection was done on the basis of the favourable effect of modifying the pore radius distribution which was exhibited more pronouncely by AS, and on the basis of the toxicity of A, which was higher than that of AS. Nevertheless, the mechanical strength is a function of the concentration of AS.

By employing the above mentioned TLC, elution experiments were performed to obtain qualitative information of the minimum amount of AS required. Fig. 2 is a photograph corresponding to a plate on which solutions of different concentrations of AHM and of a fixed concentration of AHM and different ones of AS were deposited. Addition of increasing volumes of water allowed detection of different types of interactions. In the presence of AS, the R_f varied with concentration, whereas for different concentrations of the salt in the absence of additive, the R_f was the same. These results suggest that, above some concentration of AS, most of the interaction is eliminated and more salt is displaced by the solvent. These results are in agreement with electron microprobe determinations.

In order to study the adsorption isotherms, the pulverized support was placed in contact with excess solution of AHM during 12 hours, in the presence of A or AS. Fig. 3 shows the concentrations retained by the support. Three curves are seen. The first corresponds to the total concentration. The second is obtained by substracting from the first the initial porous volume equilibrium concentration. This result must be corrected taking into account the decrease in volume with concentration of the salt. In this way, a third curve is obtained which exhibits a saturation concentration of 6-7% MoO_3. This correcting procedure is an approximation, which does not take into account other possible effects such as pore plugging and exclusion of molybdenum species from the narrower pores.

Using the Langmuir isotherm, the relation among the concentration adsorbed in the equilibrium ($c_a(eq)$) and the external concentration (c_e) can be expressed as (7) :

$$\frac{c_a(eq)}{c_e} = - K c_a(eq) + sK$$

where K is the adsorption-desorption equilibrium constant and s is the number of active centers. The values obtained for AHM were the following:
with additive AS: $K = 0.53$ lt/Mol and $s = 1.71$ Mol/Kg
with additive A: $K = 0.61$ lt/Mol and $s = 1.23$ Mol/Kg

These results show that when using A, the number of active centers available for adsorption is reduced, supporting the TLC observations. When using AS, the amount of Mo retained as MoO_3, after the impregnate support was washed for 24 hours, was near 2%. With A as additive, this value dropped to 0.5%. This also implies in this case a lesser degree of irreversible adsorption.

For NN the results also fit the Langmuir isotherm. The values of K and s where, respectively, of 1.7 lt/Mol and 2.5 Mol/Kg. These values are lower than those reported previously (7). Irreversible adsorption was not detected by washing.

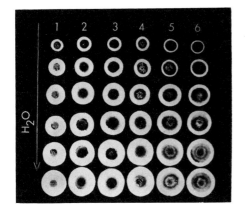

Fig.2. TLC at different AHM and AS concentrations.

1.- 5% AHM   4.- 10% AHM; 0.5% SA
2.- 10% AHM 5.- 10% AHM; 2.5% SA
3.- 15% AHM 6.- 10% AHM; 5.0% SA

Fig.3. Adsorption isotherm of AHM on γ-alumina. AS as additive.

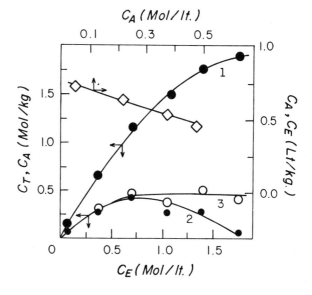

Using only 120% of pore volume for the solution, the metals were carried into the pores of the dry support by capillary forces and the concentration in the catalyst was quickly raised to 90-95 % of the total concentration attainable. The remaining 5-10% was reached by diffusion following partial adsorption of the salt initially present in the pores,

Using a mathematical model similar to that developed by Vincent and Merrill (8), it was found that adsorption of AHM was the limiting step during the establishment of the equilibrium. The values of the constants were: $K_1 = 0.15$ $K_{11} = 0.35$ $K_{111} = 3.0$ (see addendum).

DRYING

Kinetics: These experiments were performed with single pellets at vario conditions, as summarized in table 1. The drying curves are shown in fi 4. They show two typical periods, the first being the so called constan rate and the second the falling rate periods. Two models were assayed i order to find out the possible mechanism of drying: the diffusion model and the capillary flow model (9). The best fit was obtained with the latter.

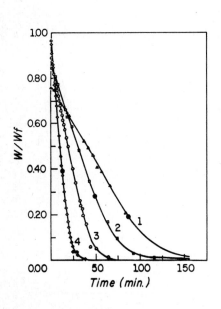

TABLE 1.
Drying conditions for Mo/Al_2O_3 catalysts

CAT	TEMPERATURE	GAS FLOW LT/H
SO-1	30	2
SO-2	30	10
SO-3	50	10
SO-4	100	10

Fig. 4 . Moisture content of impregnated pellets as a function of time

Influence of drying conditions on the molybdenum distribution and cata lyst activity: In this case three drying conditions were used and ex periences were performed with catalyst beds. Fig. 5 gives the distribu tion profiles obtained; the activities for these catalysts can be foun in table 2.

The mildest drying conditions obviously give better macroscale distribution and activity. The remarkable fact is that with reduction in size the differences in activity persist, suggesting that not only the macroscale molybdenum distribution is significant, but also its distribution across the microstructures along the pellet. It seems that the demetallation reaction is more sensitive than the desulfurization one to the catalyst size, a fact which could be explained on the basis of diffusional limitations.

TABLE 2. Activity (T=380°C, VHA=0.32 h^{-1} atm.) as a function of drying conditions

CATALYST	Particle diameter			
	5 mm		0.36 mm	
	HDS %	HDM %	HDS %	HDM %
T0-1	38	30	50	47
T0-2	30	27	45	43
T0-3	27	22	40	38

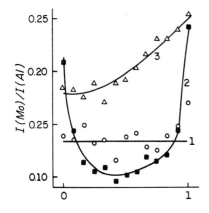

Fig. 5. Molybdenum distribution as a function of drying conditions.
1. T0-1: 30°C, 10 lt/h
2. T0-2: 50°C, 10 lt/h
3. T0-3: 50°C, 100 lt/h

When there is no interaction between solute and support, the solution flows from the macropores suctioned by the micropores, originating a lack of solute in the former, when dry (10,11). When drying begins, all the pores are filled with liquid, and the temperature of the solid corresponds to that of the wet bulb. As drying proceeds, the macropores will partially empty, and the surface temperature will rise. The concentration will grow in the narrower micropores. When reaching the critical suction radius (r_{cs}) suction will no longer take place. The solute will precipitate when attaining the supersaturation of the solution or by surface effects. The final result will be a marked heterogeneity in the micro and macropore distribution along all the solid, in discrete microregions. Nothing can be said at this stage about the radial distribution.

When there exists reversible and irreversible interactions, part of the material will stay adsorbed on the walls when drying proceeds. This material corresponds to the irreversible concentration, and a equilibrium will be established at any moment between the solute remaining in solution and that adsorbed. In this way, besides the micro-macropore heterogeneity, heterogeneity among the micropores from the external and the internal regions of the pellet may exist.

If the drying rate is much larger than the desorption rate, the final result would be lesser heterogeneity as there will be no significant migration of the adsorbed species. When an additive is present whose vapour tension is different from that of the solvent, their evaporation rates will differ, and this situation could change the intensity of the adsorption and the mechanism of migration inside the pores. The elimination of stabilizer would produce early precipitation, or concomitant change in pH modifications in the solute-support interactions.

With respect to the influence of the distribution on the activity, in the case of a homogeneous radial distribution with heterogeneity among micro and macropores, two situations can be possible :

    a. The high molecular weight molecules are excluded from the narrower pores, as is the case for the resins and asphaltenes. The activity will be low since most of the Mo is not available.

    b. Light molecules like tiophene would reach much more active Mo since diffusional restrictions are smaller. Microscale heterogeneity is not significant.

ACTIVE METALS CONCENTRATION

MoO_3 content : The relative activity of Mo/Al_2O_3 catalysts was plotted against the trioxide concentration and depicted in fig. 6. Activity was referred to a commercial NiMoAl catalyst (MoO_3 14%, NiO 3%). HDS and H reached a maximum about 15% MoO_3. Higher concentration lead to a loss HDS activity. For HDM a flat maximum is apparently obtained, and due to the limited range of concentration studied, it is not clear whether over 20% the activity decreases. The decrease in activity has been explained by other authors and by us (12,2) as due to the detrimental effect of clusters formed over the initial active centers. In addition, another explanation valid for heavy molecules is given by the micropore plugging during drying stages (shorter time for oversaturation and precipitation

As can be seen, nickel affected the activity for the HDS reaction, because the relative activities for this reaction were less than one.

The HDM values were, on the other hand, approximately 1.

NiO content : Changing the NiO content at a fixed MoO_3 concentration (14%), a maximum was obtained for the ratio of about 0.3 NiO/MoO_3. This is in agreement with the reported values (13,14). These maxima have been explained by different theories which in general, are based on the probability of having molybdenum surrounded by accesible nickel (15).

Total metal content : When the total amount of metals was changed from 0.12×10^{-2} to 0.17×10^{-2} Mol/g support, but for a fixed oxides relation, the activity remained almost constant. At the higher concentration it slightly increased. In order to prepare industrial catalysts the minimum content in this range should be used since after 24 hours on stream, the equilibrium value of the activity was the same for all these catalysts.

Impregnation order : When molybdenum was impregnated first or coimpregnated with nickel the HDS and HDM activities were almost the same. When impregnating Ni first, the activity was still the same for the HDS but higher for HDM (see fig. 8). The explanation could lie in the possibility that NN could adsorb on the active centers excluding Mo and giving a different adsorption for the latter (5). A similar effect was obtained by Mitchell and coworkers (16) for Co. Ni/Al_2O_3 has been claimed as a good demetallation catalyst (17,5). Probably accesible nickel sulfide compounds are more acidic than those of Mo and crack the heavy molecules.

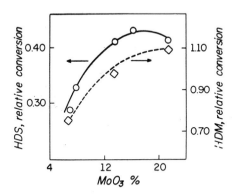

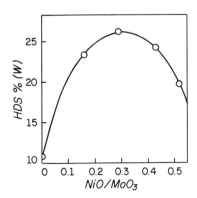

Fig. 6. Relative activity as a function of MoO_3 content.

Fig. 7. HDS activity as a function of NiO/MoO_3 ratio.

A few experiments were carried out in this field to confirm previous results obtained with thiophene as testing molecule (3,4).

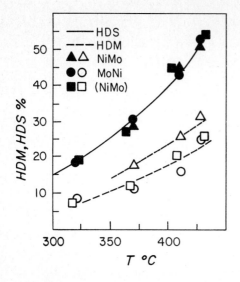

Fig. 8. Activation as a function of impregnation order.

Catalysts were activated from a drying temperature of 50°C to a calcining one of 500-600°C, using two rates for the increase in temperature (50 and 100°C/h). The activities can be found in table 3.

TABLE 3.
Activity of catalysts calcined at various conditions

Catalyst	Temp. (rate)	Air g/cm^2 h	HDS %	HDM %	T_{calc}
R 107	50	3.11	70	48	550
R 103	100	3.11	65	46	550
R 106	50	3.11	62	43	500
R 105	50	3.11	60	45	600
Commercial A			70	49	

Higher activity was observed with the lower rate of increase of temperature. To study this phenomen, thermogravimetric experiments were performed with AHM and NN on alumina. The slow thermogram (50°C/h) is similar to those published (18), and shows for the AHM two well defined peaks due to water loss (100/120°C) and ammonia loss (200/250°C). Complete decomposition occurs at higher temperatures. This suggests that polymeric molybdates are absent. The higher rate gives less defined peaks (see figs. 9 and 10).

DSC measurements demonstrated exothermic processes taking place at

200/250°C. The narrowest peaks were obtained for the lower rate. Finally, XRD showed slightly lower cristallinity for the higher rate. More detailed studies on crystal configuration with rate of decomposition of the salt are needed to explain the activity results.

Calcining temperature is important. The activity has a smooth maximum at 550°C. Moné and coworkers (19) explain this effect as due to reaction of Ni with alumina to give inactive Ni aluminate. ESCA measurements are in progress and could confirm this theory.

In the same way, using a stream of hot air through the catalyst during calcination, more controlled heat transfer, oxidation of ammonia and sweeping of the gaseous products were achieved. Better activity was also observed.

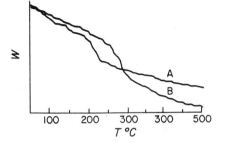

Fig. 9. Weight decrease as a function of temperature for a Mo/Al_2O_3 catalyst.

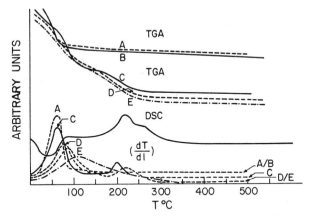

Fig. 10. TGA, DSC of: A) Al_2O_3 B) NN/Al_2O_3 C) AHM/Al_2O_3 0.5°C/min
D) AHM/Al_2O_3 1°C/min E) AHM/Al_2O_3 5°C/min.

Hydrothermal treatment of the support changed the micropore distribution. At 620°C the average micropore radius was raised from 66 to about 200Å. Surface loss was of 50% at the 24 hours, and a exponential time dependence was found for this reduction. Lower temperatures did not modify micropore distribution appreciably, and higher ones started thermal sintering with pore volume loss. The activity did not change in the HDS and HDM reactions with these modifications. Mass transfer control rate for heavy molecules (n_{eff} = 0.2-0.6) originated that only pores of diameter higher than 400 Å were used. These results support the conclusions drawn from the drying stage results.

REFERENCES

1. Y. Kotera, N. Todo, K. Muramatsu, M. Kurita, T. Sato, M. Ogawa and T. Kabe, Int. Chem. Eng. 11, 752 (1971).
2. M. Inoguchi, R. Nishiyama, Y. Satori, T. Misutori, N. Inaba, H. Kagaya, K. Tate, H. Hosaka, K. Niume and T. Ota, Bull. Japan Petrol. Inst. 3, 13 (1971).
3. R. Galiasso, 5th Iberoam. Congress of Catalysis, Lisboa (1976).
4. J. Laine, Doctoral Thesis, Imperial College, London (1971).
5. A. Cuenca, R. Galiasso and P. Andreu, PACHEC'77, Denver, Colorado (1977).
6. N. Yamagata, Y. Owada, S. Okazaki and K. Tanabe, J. Cat. 47,358 (1977).
7. J. Cervello, Thesis, Universidad Complutense, Madrid (1974).
8. R.C. Vincent and R.P. Merrill, J. Cat. 35, 206 (1974).
9. R. Galiasso, O. Ochoa and P. Andreu, to be presented in the 6th Iberoamerican Congress on Catalysis, Rio de Janeiro, Brasil 1978.
10. R.W. Maatman and C.D. Prater, Ind. Eng. Chem., 49, 253 (1957).
11. Ch. Chen and R.B. Anderson, Ind. Eng. Chem. Prod. Res. Dev., 12, (1973).
12. N. Todo, K. Muramatsu, M. Kurita, K. Ogawa, T. Sato, M. Ogawa and Y. Kotera, Bull. Japan Petrol. Inst., 14, 89 (1972).
13. S.P. Ahuja, M.L. Derrien and J.F. Le Page, Ind. Eng. Chem.Prod.Res Dev., 9, 272 (1970).
14. J.T. Richardson, Ind. Eng. Chem. Fund., 3, 154 (1964).
15. V.H.J. De Beer, in preparation of Catalysts (B. Delmon Ed.) 1976, p. 343.
16. M. Martinez, P.C.H. Mitchell and P. Chiplunker, Climax 2nd Conf., Oxford 1976.
17. A. Vargas, M. Sc. Thesis, IVIC 1977.
18. E. Ma, Bull. Japan Chem. Soc., 37, 171 (1964).
19. R. Moné, ref. 15 p. 381.

ADDENDUM :

$-df/dt + 0.5t^{-0.5} df/dR = -k_1 f(1-\theta) + k_{11}\theta;$ $f = C/C_o;$ $t = $ time $R = r/R_i$

$d(\theta C_s)/dt = k_1 f(1-\theta)/k_{111} + k_{11}\theta/k_{111}$ k dimensionless

AKNOWLEDGEMENT. To Prof. B. Delmon, Dr.Sc. and Analytical Service of IVIC for analyses.

DISCUSSION

J.F. LEPAGE : As far as the impregnation of your catalyst is concerned, what do you recommend if demetallation is aimed at : first Mo then Co (or Ni) or first Co then Mo or Co and Mo together.

R.E. GALIASSO : After 10 or 12 hours on stream the catalysts you mentioned have the same activity due to carbon and vanadium deposition. Mo or Ni/Al_2O_3 catalysts have better initial activity and thermal stability.

R. PRICE : A substantial variation in hydrodemetallization activity has been observed for different catalyst preparations. Would the authors care to speculate on the nature of the active sites, and support properties involved in the hydrodemetallization reaction ?

R.E. GALIASSO : With the present status of our knowledge of the hydrodemetallization process it is very difficult to speculate about active sites, because it is not known if there is a catalytic or adsorption process. Metals and support, both play an important role.

L. MOSCOU : Fig. 5 of your paper shows molybdenum concentration profiles over the particle as a function of drying conditions. Can you clarify the very different average Mo-contents between experiments 1 and 3 ? Did you suggest that this difference in average molybdenum content is also due to changed drying conditions ?

R.E. GALIASSO : The total amount of metals is the same for all drying experiments. The difference in the Mo content from sample TO 1 to TO 3 is due mainly to the gravity forces during static drying and the heterogeneity between particles. To obtain qualitative results of Mo distributions a great deal of sample needed to be considered.

R. BADILLA-OHLBAUM : Can you give the characteristics of your trickle bed reactor and the full conditions of operation ? It appears, from your results (Fig. 8), that you see a difference in catalysts for HDM but not for HDS. One would expect the reverse considering that HDM is more affected by diffusion in this kind of support.

R.E. GALIASSO : The small trickle bed reactor used has 416 cm^3 of catalyst diluted with ceramics (1:1), and the operation conditions were : T = 380°C, VVH = 0,33 h^{-1}, P_{H_2} = 80 atm. Some results were obtained in a 100 cm^3 reactor in similar conditions. The difference in HDM activity could be due to the presence of accessible SNi adsorbed on Al_2O_3 in addition to "conventional" Mo active sites.

PREPARATION AND PROPERTIES OF THIOMOLYBDATE GRAPHITE CATALYSTS

G.C. STEVENS and T. EDMONDS
The British Petroleum Company Limited, BP Research Centre,
Sunbury-on-Thames, Middlesex, England

ABSTRACT

This study aims to take fundamental information determined by ESCA following the activation of a CoMo/Al_2O_3 catalyst and use it to prepare a more active form of that catalyst system. The requirements to be met in the preparation of a more active catalyst were isolated as:-
1 Molybdenum must be deposited exclusively as sulphide.
2 The deposition of MoS_2 must maximise the area of basal plane to favour hydrogenolysis.
3 A promoter for the basal plane of MoS_2.
4 Elimination of Al_2O_3 to avoid the formation of the MoO_3-Al_2O_3 monolayer.

The development of the preparation of this catalyst will be described, and will include monitoring its surface states, elemental distribution and activity. The final result was a highly active catalyst.

INTRODUCTION

A series of publications has described our fundamental studies on the CoMo-Al_2O_3 system (refs 1 - 3). ESCA analysis showed that the active CoMo catalyst contained a state similar to Mo^{4+}-sulphide in MoS_2 on the surface (ref 2). Although this was the predominant state, detailed analysis of the spectra also revealed the presence of a Mo^{4+}-oxide component and the magnitude of the total Mo^{4+} signal was related to the rates of reaction of the catalyst with thiophene and carbon disulphide. In recent publications other authors using different techniques have subsequently reported results which are consistent with our interpretation (refs 4 & 5).

These results drew attention to the activity of MoS_2 itself and we examined in particular some of the structural influences of the basal and edge planes of the disulphide (ref 3). The results suggested that with thiophene the basal plane of MoS_2 was the reaction site giving mainly desulphurisation selectivity.

In this paper we describe work which set out to develop a new formulation of the CoMo system designed to exploit the results we have reported. We aimed to produce an entirely sulphidic system in which the basal plane of MoS$_2$ predominated without influence of other surfaces, particularly catalyst support surfaces. We were especially concerned to avoid formation of an oxidic MoO$_3$-Al$_2$O$_3$ monolayer at any stage (ref 6). Finally we wanted to have a supported system with as large a surface area as possible to encourage molybdenum sulphide distribution as well as to maximise the possible interaction between the molybdenum and the promoter.

SYSTEM CHOSEN FOR STUDY

A high purity carbon black was chosen as support. The material was graphitised at 2700°C to remove any oxygen containing functions such as hydroxyl or carbonyl groups leaving a high surface area material, predominantly basal plane graphite. Ammonium thiomolybdate was used as the source of molybdenum and this was impregnated onto the support prior to controlled decomposition to produce supported MoS$_2$. In other preparations cobalt ions were also impregnated to promote MoS$_2$.

In this paper we describe:-
- the conditions for decomposition of thiomolybdate
- development of impregnation method and physical properties of catalysts produced
- catalytic properties of the promoted catalyst system

PREPARATION AND DECOMPOSITION OF AMMONIUM THIOMOLYBDATE

Ammonium thiomolybdate was prepared by passing H$_2$S into ammonium molybdate. MoS$_3$ is formed from the thiomolybdate by decomposition at 200°C under vacuum (ref 7) and further decomposition to MoS$_2$ occurs under hydrogen (ref 8).

The decomposition of ammonium thiomolybdate was studied in an atmospheric pressure microreactor equipped with an automatic sampling valve and gas chromatograph on the reactor effluent. As the salt thermally decomposed hydrogen sulphide was liberated. The rate of decomposition as a function of temperature could therefore be examined recording the change in H$_2$S GC peak height as the furnace temperature was raised.

Under hydrogen no H$_2$S peaks were observed below 150°C in the samples taken by the GC system. At higher temperatures H$_2$S began to appear in the effluent samples which were taken at one minute intervals throughout the experiment. At 180°C there was steady evolution of H$_2$S and this reached a maximum rate at about 200°C.

These experiments were repeated using ammonium thiomolybdate under nitrogen and also with a sample of graphitised carbon impregnated with ammonium thiomolybdate solution so as to deposit about 8 per cent weight molybdenum on the support. Table summarises the temperatures at which H$_2$S first appeared and the temperatures at which maximum H$_2$S evolved. With the supported thiomolybdate sample, H$_2$S appeared only after standing at 225°C for 30 minutes and all H$_2$S was given off at this temperature

TABLE 1

Decomposition of ammonium thiomolybdate

Sample	Gas	Temperature °C for	
		H$_2$S Appearance	Max H$_2$S Evolution
$(NH_4)_2MoS_4$	N$_2$	160	200
	H$_2$	160	200
$(NH_4)_2MoS_4$/graphitised C	H$_2$	225	225

The chemical activities of the samples of MoS$_2$ prepared by thermal decomposition were compared with those obtained by grinding commercially available MoS$_2$ (ref 3). The conversion of thiophene vapour in a stream of hydrogen gas was measured at atmospheric pressure and 325°C. 0.2 g samples of MoS$_2$ were used with hydrogen flowing at 30 ml min^{-1} and thiophene being injected continuously at the rate of 0.05 ml h^{-1}. The conversions of thiophene were steady after 30 minutes at reaction temperature and are listed in Table 2. The table also includes the ratio of butane to mixed butenes in the microreactor effluent.

TABLE 2

Conversion of thiophene vapour in atmospheric pressure microreactor at 325°C

Sample	Stable % Thiophene Conversion	Ratio $\dfrac{\text{Butane}}{\text{Butenes}}$
MoS$_2$ n-heptane ground	80	0.8
air ground	76	2.9
by $(NH_4)_2MoS_4$ decomposition	41	1.3
MoS$_2$/graphitised C (8% wt Mo from $(NH_4)_2MoS_4$)	50	0.7

Table 2 shows that the samples of ground MoS$_2$, prepared as described previously (ref 3), were more active and had characteristically different product ratios. Of the materials derived by decomposition of ammonium thiomolybdate, the supported sample was the more active though having only 8 per cent weight Mo. The supported material also gave less butane analogous to the MoS$_2$ sample ground in heptane which contains a high proportion of basal plane surface area (ref 3).

These results show that supported and unsupported ammonium thiomolybdate thermally decomposes above 225°C. When the thiomolybdate is supported on graphitised carbon, the desulphurisation activity is improved, probably because the total surface is greater. The product pattern suggests that the MoS$_2$ produced by decomposition on graphitised carbon has a higher proportion of basal plane surface area than that produced by decomposition of unsupported ammonium thiomolybdate.

DEVELOPMENT OF IMPREGNATION TECHNIQUE FOR THIOMOLYBDATE GRAPHITE CATALYSTS

Our first attempts at making an all sulphide, uniformly distributed catalyst system employed direct evaporation. Catalyst samples were prepared by impregnation of the graphitised carbon support using aqueous solutions of cobalt nitrate and ammonium thiomolybdate. The cobalt nitrate solution was made up to contain sufficient cobalt to give about 3 per cent weight on the support and was dissolved in a similar volume of water to that of the support. Water was removed under vacuum on a steam bath and the deposited salt was decomposed by heating to 150°C for about 2 hours in air. The cobalt containing support was then impregnated with a series of ammonium thiomolybdate solutions to bring the molybdenum content to about 8 per cent weight.

ESCA spectra from samples prepared by this procedure were compared with spectra from $(NH_4)_2MoS_4$, MoS_2 and MoO_3. Table 3 lists the binding energies of the main peaks. The standard for MoS_2 and MoO_3 was $Au(4f_{7/2}) = 83.8$ eV. The carbon support was calibrated against gold and the measured value $C(1s) = 284.3$ eV was used as a secondary standard for the catalyst and thiomolybdate. No charging effects were observed with the catalyst samples.

TABLE 3

ESCA data based on $Au(4f_{7/2}) = 83.8$ eV or $C(1s) = 284.3$ eV

Sample	Binding energies				
	$C(1s)$	$O(1s)$	$Mo(3d_{5/2})$	$Co(2p_{3/2})$	$S(2p_{3/2})$
$(NH_4)_2MoS_4$	284.3	531.8	228.4		162.0
MoS_2	284.1	531.5	228.9		161.7
MoO_3	284.5	531.6	232.5		-
3% Co/8% Mo/graphitised C	284.3	531.4	232.3		-

The results in Table 3 show, surprisingly, the complete absence of molybdenum disulphide on the catalyst surface. Although the Mo(3d) signals were simple giving no indication of more than one state of molybdenum, the binding energies clearly show the presence of the fully oxidised Mo^{6+}-oxide state as in MoO_3. An unexpectedly small sulphur (2p) signal was detected on the freshly prepared catalyst by ESCA On the other hand X-ray fluorescence, which has a greater sampling depth, did show the presence of sulphur. Taken together these results indicate almost complete oxidation of the surface of the catalyst during preparation.

Samples containing thiomolybdate on graphitised carbon pellets were examined by electron microprobe to confirm that cobalt and molybdenum were distributed evenly on both the surface and over the cross-section of the pellet. Catalysts containing 3 per cent weight cobalt showed a slightly higher cobalt concentration at the surface of the pellet but at 2 per cent weight there was an even distribution over the cross-section. However, despite good distribution across individual pellets, the intensities of cobalt varied widely between pellets.

To examine this problem further, samples of graphitised carbon pellets were impregnated with 2 per cent weight cobalt using the technique described and the distribution of cobalt counts measured by the microprobe on 20 pellets selected by chance. The results showed that cobalt loadings on individual pellets could vary by up to tenfold. Subsequent impregnation of ammonium thiomolybdate did not affect the distribution.

In an attempt to overcome the problems of oxidation and uneven distribution of active components the preparation method was modified by using a nitrogen swept rotary evaporator to remove water from solutions used for impregnation. As before, cobalt nitrate was impregnated first, decomposed and then ammonium thiomolybdate was deposited.

Figure 1 shows the ESCA spectra from a sample prepared using this alternative method. It can be seen that the Mo(3d) signals (Figure 1(a)) are still complicated and suggest a significant amount of Mo^{6+}-oxide is still present although the larger signal can be assigned to an '$(NH_4)_2MoS_4$' like state. The oxidation of the catalyst surface has been substantially reduced.

The distribution of cobalt was also checked with the electron microprobe. The sample prepared using rotary evaporation showed a much more uniform distribution of active component among the twenty pellets examined. In catalysts containing up to 2 per cent weight cobalt and 6 per cent weight molybdenum there was no sign of shell impregnation.

As a final check on the composition of the surface of the active catalyst, ESCA spectra were measured after reaction with thiophen in hydrogen at 325°C. The Mo(3d) spectrum in Figure 1(b) shows the presence of just one molybdenum state at binding energies very similar to those from MoS_2. This result confirms that the thiomolybdate on graphitised carbon decomposes under the experimental conditions to form, exclusively, a surface similar to that of MoS_2. Full binding energies from both the fresh and used catalysts are listed in Table 4.

TABLE 4

ESCA data from catalysts prepared by rotary evaporation

Sample	Binding Energies (eV)*				I O(1s) / I Mo($3d_{5/2}$)
	Mo($3d_{5/2}$)	S($2p_{3/2}$)	Co($2p_{3/2}$)	O(1s)	
Fresh catalyst (2 wt % Co; 6 wt % Mo)	232.5 229.3	162.9	778.6	531.0	
Catalyst after reaction	228.8	162.0	778.1	531.5	0.2

* Standard C(1s) = 284.3 eV

Apart from changes in the Mo(3d) spectrum already described, the S(2p) signal is simplified to give the single sulphide sulphur spectrum found in MoS_2 itself.

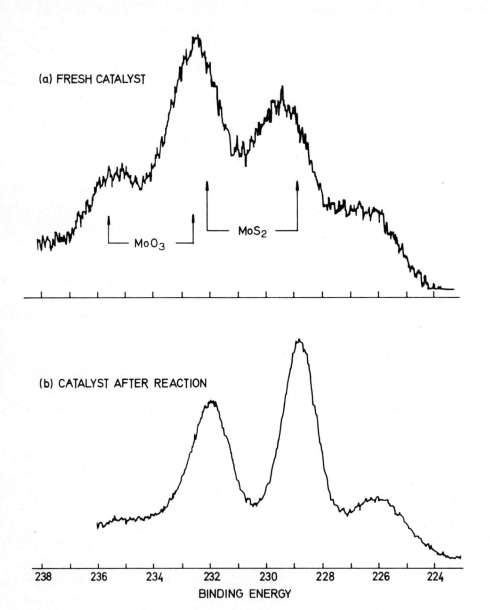

FIG 1. ESCA ANALYSIS OF CoMo CATALYST SUPPORTED ON GRAPHITIZED CARBON - Mo (3d) SPECTRA

The oxygen (1s) signal is reduced to very low levels and the ratio $O(1s):Mo(3d_{5/2})$ approaches the very low value found in n-heptane ground MoS_2 (cf Table 5, ref 3). Oxygen containing species preferentially absorb on the MoS_2 edge plane (ref 3) so the low value of the relative O(1s) intensity indicates that on the graphitised carbon catalyst there is a high proportion of MoS_2 basal plane surface area.

On both fresh catalyst and sulphided catalyst only one major $Co(2p_{3/2})$ signal is observed at a binding energy similar to that in a cobalt sulphide (cf cobalt oxide 781.0 eV (ref 3)). This indicates a close interaction between cobalt ions and MoS_2 and an exchange of ligands during preparation.

These results have shown that care is needed in preparing an evenly distributed fully sulphided system. Given adequate technique, ammonium thiomolybdate and cobalt ions can be evenly distributed by impregnation onto graphitised carbon pellets. There is ready ligand exchange with Co^{2+} ions and, after activation, the MoS_2 basal plane makes up a large part of the catalyst surface area.

CATALYTIC PROPERTIES OF THE PROMOTED MoS_2-GRAPHITISED CARBON SYSTEM

Cobalt promoted MoS_2 on graphitised carbon prepared by the rotary evaporation method was tested for conversion of thiophene vapour using the atmospheric pressure microreactor test described for use with MoS_2 samples. Table 5 shows data for thiophene conversion and compares the novel catalyst with a conventional commercial CoMo Al_2O_3 catalyst.

TABLE 5

Conversion of thiophene vapour in atmosphere pressure microreactor at 325°C

Sample	Stable % Thiophene Conversion	Ratio Butane/Butene
CoMo Al_2O_3 (2.4% wt Co 7.5% wt Mo)	42	0.5
MoS_2/graphitised C (8% wt Mo)	50	0.7
Co/MoS_2/graphitised C (3% wt Co, 8% wt Mo)	100	All butane

It is clear that cobalt promotion has a large effect on the MoS_2/graphitised carbon catalyst and produces a catalyst of exceptional activity.

Four compositions of cobalt promoted MoS_2 on graphitised carbon catalyst were prepared as indicated in Table 6 and thiophene conversions were measured at 260, 280, 300 and 320°C using 0.13 g of catalyst in each case. In this test hydrogen flow was at 30 ml/minute and thiophene was injected at 0.05 ml/h as before. The results are plotted in Figure 2 and temperatures for 50 per cent thiophene conversion were estimated and recorded in Table 6. Comparison with earlier results in Table 2 at 325°C emphasises the improvement in activity obtained.

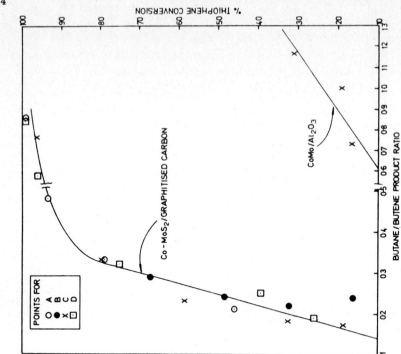

FIG 3. ACTIVITY AND SELECTIVITY OF THIOMOLYBDATE-GRAPHITE CATALYSTS

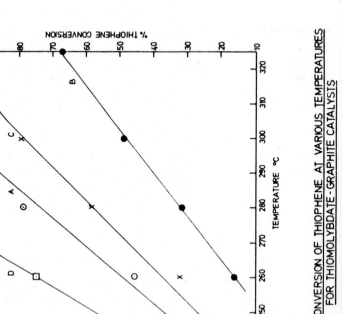

FIG 2 CONVERSION OF THIOPHENE AT VARIOUS TEMPERATURES FOR THIOMOLYBDATE-GRAPHITE CATALYSTS

TABLE 6
Effect of catalyst composition on thiophene conversion

Sample	% wt Cobalt	% wt Molybdenum	Temp °C (50% conversion)
A	1	3	261
B	2	3	303
C	1	6	272
D	2	6	246

In all experiments the ratio of butane to butenes in the product was measured and in Figure 3 these ratios are plotted against per cent conversion for thiophene. As can be seen all the points lie on the same straight line up to 80 per cent desulphurisation and the very low value of butane/butenes at this conversion (0.33) emphasises the very selective behaviour of the catalysts. The figure includes a line for a typical commercial catalyst for comparison.

The results have shown the very strong promotional power of cobalt on a supported MoS_2 catalyst which has mainly basal plane surface exposed. The promotional effect varies with Co:Mo ratio and appears to be largest at a weight ratio of about 1:3.

RELEVANCE TO MODELS OF THE CoMo SYSTEM

Three main models of the CoMo system have been described - the monolayer model argues for cobalt promotion via incorporation in the alumina surface. In the graphite supported system the promotion caused by cobalt in the absence of alumina shows that this model is not the only way in which cobalt promotion can occur. In the intercalation model promotion occurs via the intercalation of the cobalt in the edge sites of MoS_2. This model, however, requires very low loadings of cobalt much lower than the 1:3 weight ratio of cobalt to molybdenum observed here and would not be favoured in systems with a high proportion of basal plane surface area. The synergy model assumes a very close interaction between cobalt and molybdenum sulphides. There is clear evidence for this in the present study from the exchange of ligands between cobalt and molybdenum at the preparation stage and the ESCA data showing the active phases to be MoS_2 and a sulphide of cobalt. Thus, the present results favour the synergy model.

SUMMARY OF CONCLUSIONS

A very active, highly selective desulphurisation catalyst can be prepared by cobalt promotion of the MoS_2 basal plane. The preparation of such a catalyst has been described and the activity shown to depend on the ratio of cobalt to molybdenum.

ACKNOWLEDGEMENT

Permission to publish this paper has been given by The British Petroleum Company Limited.

REFERENCES

1 G.C. Stevens and T. Edmonds, J Catal, 44(1976)488
2 G.C. Stevens and T. Edmonds, J Catal, 37(1975)544
3 G.C. Stevens and T. Edmonds, J Less Common Metals, 54(1977)321-330
4 P. Grange, B. Delmon et al, J Less Common Metals, 54(1977)311-320
5 M.J.M. van der Aalst and V.H.J. de Beer, J Catal, 49(1977)247-253
6 G.C.A. Schuit and B.C. Gates, AIChEJ, 19(1973)417
7 J.C. Wildervanck and F. Jellinck, Z. Anorg Allg Cjem, 328(1964)309
8 E. Furimsky and C.H. Amberg, Can J Chem, 53(1975) 3567
9 A.L. Farragher and P. Cossee, Proc 5th International Cong Catalysis, Miami Beach, (1972)1301
10 G. Hagenbach, P.H. Courty and B. Delmon, J Catal, 31(1973)264

DISCUSSION

F. KERKHOF : Are the MoS_2 crystallites smaller when cobalt is used ? We suggest that this may be concluded from the Mo/C XPS intensity ratio, which should be higher in case of smaller crystallites (or a "monolayer" of MoS_2).

T. EDMONDS : We did not study the unpromoted catalyst used in Table 2 by XPS. Our XPS studies were confined to the promoted catalysts. Consequently we cannot give an answer directly to this interesting question. However, if the butane : butene ratio for the unpromoted catalyst in Table 2 is compared with this ratio for the promoted catalyst at the same thiophene conversion (50 per cent), given in Figure 3, a marked decrease is noted following promotion. This could indicate that the promotor increases the ratio of basal plane edge sites.

B. DELMON : Your approach is quite original and fruitful. Therefore, I would like to ask you some clarification of your sentence, p. 7, 2nd paragraph of your paper : "This indicates a close interaction between cobalt ions and MoS_2 and an exchange of ligands during preparation".

T. EDMONDS : The argument is as follows : cobalt is deposited as nitrate and decomposed to oxide. Molybdenum is deposited as thiomolybdate (and ultimately decomposed to sulphide). Yet following catalyst preparation we find that cobalt has a binding energy which can only be interpreted by the cobalt cation being in close proximity to sulphide anions; the molybdenum spectrum indicates association with both oxygen and sulphur anions. Consequently, even if some oxidation has occurred during preparation, anion exchange must explain the presence of cobalt in association with sulphur anions. This exchange can only occur if the cobalt ions and MoS_2, or the thiomolybdate anion, have interacted.

L. MOSCOU : I have two questions on this paper.
1) Did you make any attempt to look for the presence of separate MoS_2 particles under the high resolution electron microscope ?
2) What happens if these catalysts are contacted with air at room temperature and then contacted with thiophene under test conditions ? Will the structure and catalytic activity of the catalyst be restored ?

T. EDMONDS : 1. The catalyst was not studied by high resolution electron microscopy.
2. Only minor changes occur when these catalysts are exposed to air at room temperature over a time scale of weeks or even months. The major change which is noted (by ESCA) is the conversion of S^{2-} to SO_4^{2-} but our evidence suggests that the sulphate anion is readily re-reduced to sulphide in hydrogen. There may be an extremely limited oxidation of MoS_2 to MoO_3. Consequently there is little change when the catalyst is reacted with thiophene under test conditions again since the chemical form of the catalyst is essentially unchanged.

CATALYST STABILIZATION AND DEACTIVATION COMPARED WITH CATALYST
PREPARATION.

R.E. MONTARNAL
Institut Français du Pétrole, Rueil-Malmaison, France.

ABSTRACT

Once a catalyst has been prepared and placed in its reactor, a
crucial problem concerns its different ways of evolution. This
paper attempts to make as rational an analysis as possible, rather
than an exhaustive study of the possible transformations, which can
be classified under two essential headings : i) stabilization, to
achieve a steady state performance and ii) deactivation during the
working time of the catalyst.

INTRODUCTION

Once inside a catalytic reactor, a heterogeneous catalyst does not
remain in the same fixed state, as assumed by the academic definition
of catalysis, but may go through various, extreme transformations.
1. First of all, the final stages of preparation are sometimes preformed within the reactor, following the introduction of a catalyst
precursor.
2. At the beginning of catalysis the fresh catalyst undergoes an
evolution leading more or less quickly to performance stabilization.
3. Once the stabilization period has been achieved, performances
generally decrease slowly under the influence of deactivation
phenomena : aging period.
4. The deactivated catalyst can be regenerated inside or outside
the reactor, or may be discarded if regeneration is impossible or
uneconomic.

Until a few years ago researches undertaken in these areas were
of a rather empirical nature due to the complexity of the phenomena.
More progress has been made recently, e.g. by the Berkeley Symposium
(ref. 1), and the present paper aims to demonstrate some general
parameters governing these transformations. Of course, due to the

vastness and complexity of the topic, some over-simplifications have been necessary, but hopefully excessive divergence from reality has been avoided. Some classifications and some interpretations certainly may be debatable and some cases will also be omitted. Moreover the material of the text and the examples chosen will belong rather to the area of Refining and their generalization to cover all heterogeneous catalysis may require some variations.

1. FINAL STEPS OF PREPARATION WITHIN THE REACTOR

These last stages of preparation are indeed of the same type as those performed outside the reactor. We indicate here the main reasons for finishing the preparation inside the reactor.

The main reason is the risk of degradation of some of the final catalyst in contact with the atmosphere during transportation or handling. Thus many nickel catalysts are pyrophoric and must be introduced into the reactor as the oxides (or massic alloys, for Raney nickel). The reduction step may lead to troublesome complications if, for example, the catalytic reaction can be performed at low temperature (hydrogenation at 150°C) while the reduction must be at 400°C. Hence the use of passivation techniques aids performance of the reduction in milder conditions. For a reforming catalyst, platinum compounds are also reduced in situ, although it would be possible to handle a reduced catalyst. Hydrodesulfurization catalysts are introduced in the form of Co and Mo oxides and are sulfided inside the reactor by a combination $H_2S + H_2$. Some degradation of the sulfide could occur by exposure to air, leading to a longer transient period of stabilization or even to a lower steady state of performances. Moreover we must note that, for catalysts subjected to regeneration by burning coke with oxygen, a reduction step has to be performed inside the reactor to obtain the desired zero valency state of the metallic catalyst.

A more fundamental reason concerns the usefulness of performing the final stages of preparation by means of the reactions themselves, which can prepare the catalyst surface to produce the best performance. The classic example is of Fe_3O_4 reduction by $N_2 + H_2$, to obtain iron for NH_3 synthesis. But such a process belongs to both the final stages of preparation and to catalyst stabilization by reconstruction phenomena, as seen below. Indeed this raises the problem of knowing the nature of the fresh catalyst.

The answer to this question is complicated and can only be considered after catalyst stabilization has been discussed.

2. GENERALITIES ON CATALYST EVOLUTION DURING CATALYSIS

This paper essentially concerns the case of an open flow reactor, either for gradient concentration technology (piston flow), or uniform concentration technology. As a simplification, it can sometimes be useful to consider a differential reactor for which concentrations and temperature around the catalyst are perfectly defined. Generally, the evolution of activity as a function of time has the form presented in Fig. 1. During a relatively short transient period (a few hours, a few days) the activity decreases or increases rapidly, until a stationary level is reached. Indeed a true steady state corresponding to the dashed line in the figure is not obtained. During the second period, the activity decreases slowly due to deactivation (also called aging).

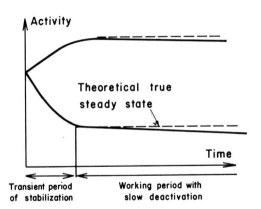

The distinction between the two periods, which lies essentially in their length results from the actions for each different processes that we shall try to characterize according to the following classification.

Fig. 1 : General evolution of activity during catalysis.

Stabilization period

We shall distinguish :
- An intrinsic stabilization, e.g. inherent to the nature of the catalyst-reaction system itself.
- A stabilization due to the transient influence of a parasite reaction which will essentially be coke formation.
- A stabilization due to a reversible selective poisoning, in other words due to a kinetic inhibition by some compounds in the feed that are irrelevant to the catalytic reaction.

Aging period or deactivation period during working

The slow downward drift of performances can result :
- from the continuation of coke formation, but according to kinetic characteristics and even mechanisms different from those which act

during stabilization;
- from the evolution of the catalyst itself independently of any catalytic reaction;
- from irreversible poisoning by some compounds in the feed;
- from parasite reactions due to a bad catalyst conception or manufacture;
- from various other causes : attrition, breaking, etc.

The problem of regeneration will also be mentioned.

3. STABILIZATION PERIOD

The general concept of a catalyst evolving when placed in contact with reactants is rather old. Many authors have invoked the **action of reactants on the catalyst** as a counterpart of the action of the catalyst on the reactants, according to schematic reaction (1) occuring during the transient period of stabilization

$$\text{Fresh catalyst } (S_0) + \text{Reactants} \rightleftarrows \text{Stabilized Catalyst } (S_1) + \text{Products (during stabilization)} \quad (1)$$

The problem relevant to stabilization is first to determine the nature and kinetics of the reactions leading from S_0 to S_1, and secondly (and mainly) to characterize the final state of the catalyst and the final performance, for these govern the reactor design.

The possibility of reaching the final state either by a **reversible** or an **irreversible** process should first be considered. The steady state will be reached by a reversible path if, once this state has been obtained following transient modification of the operating conditions, by returning to the original conditions, we can again obtain the original performance. In another form, the steady state is reversibly obtained if it depends only on operating conditions and not on the path followed to reach it.

Remark : The steady state of equilibrium for the catalyst - reactant system does not correspond to a thermodynamic equilibrium between reactants, as appears evident for differential reactors. Equilibrium results from the dynamic play of antogonistic reactions between the homogeneous phase and the solid.

3.1. Intrinsic stabilization

We shall try to move from simplicity to complexity, or rather, from well established to more controversial.

3.1.1. <u>Stabilization</u> related only to the catalytic mechanism. By this we mean the case where stabilization is attained by the involvement of elementary steps of the catalysis itself and thus has a reversible character. Different examples can be given.

The classical mechanism of LANGMUIR-HINSHELWOOD involves, as does any mechanism, a transient period evolution, since the stationary surface coverage in different species cannot be established immediately. But this period is often very short, due to the high values of adsorption - desorption rate constants, and so can be neglected. It is interesting to consider simple cases for which this period is sufficiently long to be observed. Let us, for example, look at the two reaction schemes in Figure 2, with the indicated values of rate constants, which lead to a slow accumulation of the inhibitor σ_I. At the initial time, the adsorption - desorption equilibrium is quite rapidly established for A and gives for σ_{AO} the value indicated in the Figure, because σ_I is still equal to zero. After an infinite period, the different values of $\sigma_{A\infty}$ and $\sigma_{I\infty}$ can be easily obtained from the stationary-state equations, and are also indicated in the Figure. In other words, the σ_A fraction decreases from σ_{AO} to $\sigma_{A\infty}$, while the σ_I fraction increases from zero to $\sigma_{I\infty}$. Hence the formation rate of B decreases from $k_s \sigma_{AO}$ to $k_s \sigma_{A\infty}$. The period of stabilization is greater as k_i is weak, and it is easy to understand that the stabilization time can be measured.

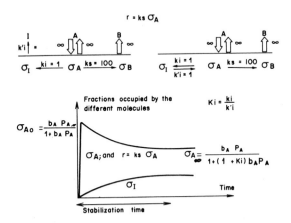

Fig. 2 : Transient period for a LANGMUIR-HINSHELWOOD mechanism (∞ represents reaction rate constants of quite high values).

left $A \genfrac{}{}{0pt}{}{\rightarrow B}{\rightarrow I}$

right $A \longrightarrow B$

σ represents the surface fraction occupied by a given compound.

Another characteristic example is the case of a protonic acid catalyst used for the isomerization of n-paraffins : $RH \rightleftarrows iRH$ (ref. 2). The complexity of the phenomena leads us to give only the essential aspects. Beginning with the protonic form of the catalysts SH^+, its superficial structure goes towards a steady state under the influence of several reactions, which are elementary steps of the catalysis and can be expressed overall as (2).

$$SH^+ + \text{Reactants (RH)} \underset{}{\overset{k_i}{\rightleftharpoons}} SiR^+ + \text{Products} (H_2, iRH) \quad (2)$$

In the steady state the catalyst surface is shared between SH^+ and SiR^+ (SR^+ being much less stable than SiR^+ appears only in a very weak proportion). Such an evolution is presented in Fig. 3a. On the other hand, the catalytic transformation of RH is governed by the following two elementary reactions :

$$RH + SH^+ \xrightarrow{k_1} iRH \quad \left(\begin{array}{c} \text{generally with} \\ k_3 >> k_1 \end{array} \right)$$

$$RH + SiR^+ \xrightarrow{k_3} iRH$$

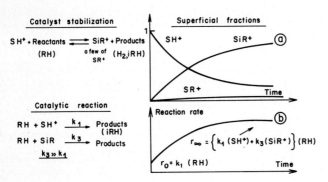

Fig. 3 : Stabilization process for a protonic acid catalyst, during n paraffin isomerization $nRH \rightleftharpoons iRH$ (ref. 2) (ref. 33).

Thus the RH transformation rate goes from the initial equation :

$$r_o = k_1 [RH][SH^+]_o = k_1 [RH], \text{ since } [SH^+]_o = 1$$

to the final equation :

$$r_\infty = \{k_1 [SH^+]_\infty + k_3 [SiR^+]_\infty\} [RH]$$

Since $k_3 >> k_1$, the reaction rate increases from r_o to r_∞, as shown in Fig. 3b. Indeed, in the steady state SiR^+ is practically the active form of the catalyst. The length of the transient period increases as k_1 weakens relative to k_3. The relevant experimental results have been established for liquid super-acid catalysts, but the same phenomena can occur for example with zeolites. For xylene isomerization on Y zeolite such a transient period of one hour has been observed (ref. 3).

A third example is the oxidation of hydrocarbons proceeding by a redox mechanism on oxide catalysts. The two elementary reactions govern both the superficial evolution of the solid, and the conversion of reactants leading to a steady state where the solid surface remains in an intermediary state of oxidation (ref. 4). In fact, the real mechanism is more complex, for the bulk evolution of the solid goes in parallel with the superficial evolution.

3.1.2. Complex stabilization

a) NH_3 synthesis or decomposition.

Some authors have shown the existence of transient periods of several hours before reaching a new steady state, after modification of experimental conditons, for NH_3 synthesis (ref. 5). Thus on Fe at 250°C, by increasing N_2 from 5 to 25%, the NH_3 formation rate increases, but it takes more than 10 hours to reach the new steady state. By returning to 5% N_2, we again obtain the initial performance, again after a long transition period. The reversibility of the phenomenon permits invoking the intervention of one or several slow steps of the catalytic mechanism. It is well-known that the nitrogen activation by chemisorption is a slow step in NH_3 synthesis.

Besides, others experiments have resulted in some superficial structural modifications of the catalyst by the reactants (ref. 6). Thus fig. 4 shows that Fe_3O_4 reduced by the mixture $N_2 + H_2$ produces an iron which is immediately active when catalytic conditions are established. On the contrary, Fe_3O_4 reduced by H_2 is initially inactive for synthesis and only becomes active after a prolonged period in contact with $N_2 + H_2$. The two final steady-states are not very different. The interpretation of these results stemmed from the need for a restructuring, by chemisorbed N_2, of the iron surface produced by H_2 reduction, to create the crystallographic plane (111) which is active in catalysis. This plane is obtained directly in the case of reduction by $N_2 + H_2$. Such a reconstruction

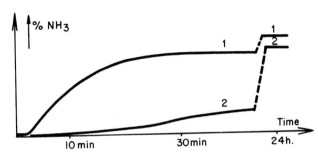

Fig. 4 : Time dependence of ammonia yield at 275°C 1) catalyst reduced in N_2-H_2 mixture 2) catalyst reduced in pure H_2 (ref. 6).

is intrinsic, since it is created by the reactant N_2. It may or may not be reversible. Other authors also invoke reconstruction phenomena to interpret the effect of pretreatments on the transient period and on the final state for NH_3 synthesis on Fe/Mg O (ref. 7). For NH_3 decomposition also, the catalyst Ru/Al_2O_3 obtained by reduction with H_2, is not immediately active and must be activated by NH_3 at 500°C, which can also be interpreted as requiring a surface reconstruction (ref. 8). In conclusion, the superficial modifications occurring during the stabilization of catalysts for NH_3 synthesis can involve the elementary steps of the catalysis, but also some superficial reconstruction phenomena induced by the reactants. Moreover these reconstructions can be of a purely structural nature, or imply strongly bonded nitrogen species, going from chemisorbed nitrogen to nitride, i.e. the incorporation of N in a crystallographic lattice. The reversibility of transformations is questionable.

   b) Alloys or complex catalysts.

The reconstruction concept used for a pure metal must be extended for the case of more complex catalysts. For alloys, the superficial enrichment with metal having a weaker heat of sublimation is well known, by thermal treatment in a vacuum or inert atmosphere (ref. 9). Besides, in the presence of reactants the superficial enrichment concerns the metal giving the stronger bonding with these reactants, if of course, the segregation is kinetically possible (sufficient temperature). Thus while for Pd-Au, this enrichment concerns gold in a vacuum, it concerns palladium in the presence of oxygen (ref. 10). Hence, during the catalytic reaction, the superficial composition (and structure of course) can change either by a purely thermic effect or under the influence of strongly bonded reactants. The two effects may be opposed. The reversibility of the final state depends on the experimental conditions. Almost similar reconstructions have been observed for other complex catalysts. They can be used, for example, to interpret the evolution of nickel sulfide properties during the selective hydrogenation of acetylene (ref. 11). Under the influence of acetelyne, the authors invoke the migration of sulfur atoms according to the reversible superficial reaction (3).

$$2 \begin{array}{c} S \\ \\ S \end{array}\!\!\!\!>\!\!Ni\!\!-\!\!S \rightleftharpoons \begin{array}{c} S \\ \\ S \end{array}\!\!\!\!>\!\!Ni\!\!<\!\!\!\begin{array}{c} S \\ \\ S \end{array} + S\!-\!Ni\!\cdots \quad (3)$$

These migrations induce activity modifications through transient periods of stabilization. In a paper presented in this Symposium (ref. 12), the catalyst stabilization also seems to be governed by a reconstruction phenomenon, in the general sense. For the ethylbenzene dehydrogenation into styrene on Fe_3O_3 promoted by Cr-K, the decrease in a few hours, of the parasitic reaction of dealkylation can be interpreted by some structural catalyst modification; possibly the migration of the promotors.

In conclusion, the intrinsic catalyst stabilization due to the interaction between the solid and the reactants is an unavoidable phenomenon which leads to the true steady state of the catalyst-reactant system, at least if it is permitted by the kinetics. We can consider that the fresh catalyst is defined by this steady state and that its characterization would have to be performed in this state.

3.2. Pseudo-stabilization under the influence of transient parasite coke formation

We must first recall that coke is generally formed of polycondensed aromatics with a wide variety of possible structures ranging from undesorbable "liquid" products to graphite. The rapid deactivation induced by parasite coke formation can lead to a practical stabilization of performances. The decrease of performances stops, for a coke level which remains constant and depends on the severity of the operating conditions. Once this stabilization is achieved, if the severity of the operating conditions is increased, the coke level is also increased for the new state of stabilization which is reached. On returning to the previous conditions, the performances are lower than previously. This evolution is then of an irreversible nature. It is indeed a pseudo-stabilization, because the nature of the catalyst is greatly modified by parasitic irreversible coke deposition. Due to this irreversibility it is impossible to invoke a stabilization process such as in Figure 2, for which product, σ_I, would represent the coke. The coke results from an irreversible reaction (4) such as (for hydrocarbons reactions):

$$\text{Reactants + Fresh catalyst} \xrightarrow[n_c]{k_c} \text{Coked catalyst + Products} \quad (4)$$
$$(H_2 \text{ and light products})$$

The rate of coke formation can be written in the general form:

$$r_c = k_c \; n_c \; (C^{ons}) = k_c \; n_c \; (C_{HUns})$$

C_{HUns} represents the concentration of highly unsaturated products, which are coke precursors. The rate of coke formation declines until it is cancelled-out by the decomposition of this coke itself on the coking sites, so as to decrease their number (n_c) or their activity (k_c). This schematic process is generally accepted, but can operate according to different kinetic forms. The simplest form of decrease of n_c and r_c is an exponential curve. But the decrease can be linear, or even non-existent during a prolonged period (ref. 13). Beyond, it must be explained that in spite of the total self poisoning for coke formation, there is only partial self poisoning for the catalytic reaction, whose activity decreases to a finite value.

-We can invoke a surface heterogeneity with two types of site. Only the more active sites (n_c) would be able to perform coking, while the total sites (coking + non coking = n_c + n) would promote the catalytic reaction. The coke deposition would eliminate only the coking sites, n_c. This over-simplified hypothesis can be improved by invoking an activity site distribution.

-We can consider a single type of highly active site on the fresh catalyst, able to perform both the catalytic reaction and the more difficult coking. The coke deposition would decrease the site activity, which is relevant to the decrease in k_c instead of n_c. The activity decrease of a site can result from the coke deposition on the site itself, but also and even more probably from the ligand effect (or electronic effect) of the coke on a neighbouring site. Thus there only remains the reaction activity.

-We can also consider a demanding nature for the coking, which is a bimolecular reaction, since it involves a polycondensation, as opposed to a non demanding nature for the catalytic reaction. Coke formation would require many more sites than the reaction. So, the sole effect of site dilution, due to coke deposition, could decrease coking drastically, until it is eliminated, while the catalytic reaction would be reduced only in proportion to dilution.

A demonstrative example is the evolution of n-heptene on a protonic acid heterogeneous catalyst according to the reaction path of Figure 5 (ref. 14).

The coke percentage stabilizes itself at a given level where its formation is cancelled out, while n-heptene transformation decreases to a finite level. Why ? Coking is caused by the hydrogen transfer reaction between heptenes and isobutene, producing isobutane, and heptadienes which are coke precursors. This reaction is difficult in the conventional sense, for it involves a difficult

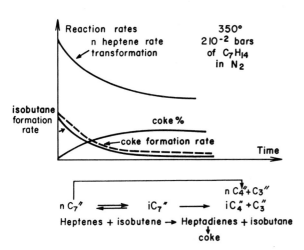

Fig. 5 : n-heptene reaction on acid silica-alumina catalyst. Pseudo-stabilization by coke formation (ref. 14).

hydride abstraction and is demanding in the BOUDART sense for it involves at least two sites. This explains why self poisoning coke formation leads drastically to its own cancellation. On the contrary catalytic isomerization and cracking are performed through carbonium ions, i.e. intermediates which are easily obtained by proton addition to n-heptene, and isomerization involves only one site for the skeletal rearrangement. An important point in this example is that the coke formation rate here is particularly easy to characterize from the parallel isobutane formation rate.

The self poisoning is particularly clear and well known for the reaction of paraffins and H_2 on transition metals at around 200° to 400°C. Conversion decreases quickly to a stable level. Moreover, because the reaction is complex (with hydrogenolysis, isomerization, cyclization, dehydrogenation), the different paths are selectively reduced. Basically, the most drastic decrease concerns the most demanding reactions; either they need special sites (corner, edges...) or a collection of several sites. Very little research has been done to characterize the final state of the pseudo stabilized catalyst.

Hydrogenation reactions at low temperature can themselves be self poisoned by parasitic coke deposition. It is known that in the absence of hydrogen, an olefin is chemisorbed on a transition metal and undergoes a self-hydrogenation, giving both the relevant paraffin in the gas phase, and a highly dehydrogenated chemisorbed complex which is a coke precursor. Of course the absence of H_2

corresponds to extreme conditions. With the olefin + H_2 mixture, as feed, for a catalytic hydrogenation, the presence of H_2 reduces coke deposition drastically. However, the fact remains that the steady-state of the working catalyst can be different from the state of the fresh catalyst and to a certain extent may depend on the experimental conditions. A good characterization of the final catalyst would also be very useful.

Catalytic reforming is a more complex case due to the bi-functionality of the catalyst and to the complexity of the feed. Bearing this in mind, with the example of n-heptane as a model reactant, Figure 6 shows the deactivation observed for a fresh catalyst (ref. 15).

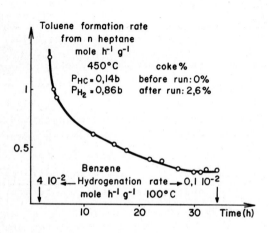

Fig. 6 : Pseudo-stabilization by coke formation, for a reforming catalyst (ref. 15).

The coke level goes to 2.6%, while the chloride is maintained at around 1%. We observe in correlation a decrease from 4 to 0.1 in the hydrogenation activity of benzene. All this illustrates the modification of the catalyst.

<u>Conclusions</u> : Catalysts can undergo rapid, profound and irreversible transformations by coke deposition during the first hours of operation. Hence several consequences must be emphasized :

- The only well-defined catalyst, for both fundamental investigation and industrial uses, is the stabilized catalyst.

- The catalyst characterization should be done on this final catalyst. This is hardly ever the case because of the difficulties encountered. It would be very useful to adapt physical methods for characterizing a coked catalyst by measurement of free metallic sites, acidity, etc... This is an indispensable requirement for the correct definition of the T.O.N. and for all the consequent deductions. In the same way, it would be fruitful to use systematically, catalytic test reactions capable of characterizing individual functions of the coked catalyst, i.e. hydrogenation reaction, reaction connected with a given type of acidity, etc...

3.3. Stabilization by selective reversible poisoning

Among the infinite variety of cases of poisoning, it is possible to introduce two types of differentiation.

-Differentiation between two extreme poles which are : (1) a selective poisoning concerning the catalytic species themselves, and (2) a non selective poisoning concerning poison deposition on any point of the catalyst surface. Indeed, the term poison should be applied only to the first case, but it will be extended here to all types of activity modifications by deposition of foreign compounds present in the feed and not involved in the catalytic reaction.

-Differentiation between reversible or non-reversible poisoning. Non-selective poisoning is generally irreversible and will be examined in the section concerning deactivation. Selective poisoning can be reversible in which case it leads to a true stabilization, and the poison is then an inhibitor. It can be irreversible, in which case it leads rather to a progressive deactivation.

Considering the example of the influence of H_2S on Group VIII metals, the problem of selective poisoning will be examined for the case of reversibility, but we will also see how it can be compared with the case of irreversibility.

a) <u>Selective poisoning of group VIII metals by H_2S</u>

<u>Isotherms of chemisorption</u>. The strong but variable affinity of group VIII metals for sulfide compounds is clearly shown in the thermodynamic diagrams of Figure 7 for two extreme metals : Ni and Pt, respectively hard and soft (ref. 16).

The sulfide area is reached for a much weaker H_2S/H_2 ratio in the case of Ni than in the case of Pt. For both cases, H_2S chemisorption on the metal corresponds to the line AB in Figure 7. This chemisorption proceeds according to the equations in Figure 9.

If the thermodynamic equilibrium is considered for H_2 chemisorption, we easily obtain the equations for fractions σ_V, σ_H, σ_S. From the σ_S equation reported in Figure 9, we see that the K_S/K_H ratio governs the surface coverage by S. For a given temperature, K_S/K_H is much higher for Ni than for Pt. Beyond, K_S decreases when the temperature increases, for (5) is exothermic due to the easy H_2S

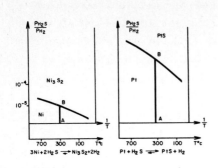

Fig. 7 : Thermodynamic diagrams, for the <u>bulk</u> of the solids.

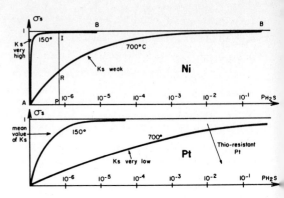

Fig. 8 : Isotherms for the dissociative <u>chemisorption</u> of H_2S on Ni and Pt (reversibility and apparent irreversibility).

$$3\,Ni + H_2S \underset{k's}{\overset{k_s\ \ K_s}{\rightleftarrows}} NiS + 2\,NiH + Q \quad (5)$$
$$(3\sigma v) \qquad\qquad (\sigma s)\ (2\sigma H)$$

(With 2 steps : NiSH ⟶ NiS)

$$2\,NiH \underset{K_H}{\rightleftarrows} 2\,Ni + H_2 \quad (6)$$
$$(2\sigma H) \qquad\qquad (2\sigma v)$$

$$Ni + H_2S \overset{K_G}{\rightleftarrows} NiS + H_2 \quad (7)$$

$$K_G = \frac{K_s}{K_H}$$

$$\sigma_s = \frac{\left(\frac{K_s}{K_H}\right)\frac{P_{H_2S}}{P_{H_2}}}{1 + (K_H P_{H_2})^{1/2} + \left(\frac{K_s}{K_H}\right)\frac{P_{H_2S}}{P_{H_2}}}$$

Fig. 9 : Dissociative chemisorption of H_2S and H_2 on Ni. σ_V free fraction of metal; σ_H fraction covered by hydrogen; σ_S fraction covered by sulfu

$$K_S = \frac{k_S}{k'_S} \qquad K_H = \frac{k_H}{k'_H}$$

dissociation. Hence the isotherms giving σ_S versus P_{H2S} provide the forms presented in Figure 8, for Ni or for Pt, and at low or high temperature.

b) <u>Reversibility and practical irreversibility</u>

Strictly speaking, H_2S chemisorption, then poisoning, is reversible for both Ni and Pt since (5) (6) (7) are reversible. Let us consider however the H_2S pressure at point P in Figure 8 for Ni, which at points R and I respectively, cuts the isotherms for 700° or 150°C. The temperature of 150° can correspond to industrial benzene hydrogenation on Ni, and 700° to methane Steam Reforming. Introducing H_2S at this pressure gives, for both reactions, the evolutions presented in Figure 10. After a certain time H_2S is discarded from the feed and the evolution of the activity is again given in Figure 10.

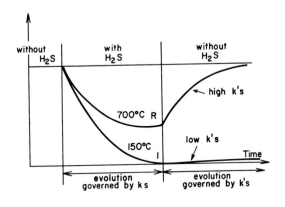

Fig. 10 : Ni poisoning by H_2S :
- reversible at 700°C, for $k'_S \simeq k_S$
- apparently irreversible at 150°, for $k'_S \ll k_S$.

The stationary level is established at R or I, according to the values of K_S, while the time required for sulfur elimination depends mainly on the rate constant of desorption k'_S. For steam-reforming at 700°, stabilization corresponds to an inhibition and partial poisoning. By eliminating H_2S, the initial activity can be regained after a time having the same degree of magnitude as for deactivation. For benzene hydrogenation at 150°C, stabilization corresponds to practically complete poisoning due to the huge value of K_S. By eliminating H_2S, the initial activity would be regained but only after a time of infinite length due to the quite small value of k'_S. Of course the high value of $K_S = k_S/k'_S$ is connected with the low value of k'_S. To speed up sulfur elimination, k'_S can be increased by using high temperatures, but the risk of Ni sintering then appears. To conclude on Ni :

-For steam-reforming, selective poisoning is partial and reversible, and a reversible stabilization is attained.

-For benzene hydrogenation, selective poisoning is practically complete and irreversible. So just very weak quantities of sulfur compounds are tolerated in the feed. Anyway this leads to a progressive and complete deactivation as described in section 4-2.

In the case of platinum all these phenomena tend to produce less poisoning and more reversibility. At 500°C, in the case of Catalytic Reforming, the reversibility makes it possible to eliminate any accidental excess of sulfur. Of course reality is more complex, and for example, the last traces of sulfur can be quite difficult to eliminate due to the site heterogeneity of the metal. By the use of so-called thio-resistant platinum we again move in the direction of less poisoning and more reversibility due to the decrease in K_S. Thio-resistance is obtained by deposition of highly dispersed Pt on an acid support (ref. 17). The accepting power of the support decreases the electronic density of Pt, which is then unable to produce such strong bonding with S, for Pt acts as a donor in the formation of PtS (ref. 18).

3.4. Stabilization of the catalyst in its efficient structure

By this we mean the addition to the feed, of chemical species, which do not contribute to the stoichiometry of the catalysis, but which are necessary to maintain the efficient composition and structure of the catalyst. Two illustrative examples concern an intrinsic effect on the catalyst itself. The first is the addition of HCl to the feed of a catalytic reforming unit in order to maintain the acidity of the support. Under the influence of the temperature and water in the feed, this acidity decreases but is restored by this addition of HCl. The second is the addition of H_2S, to the feed of a hydrodesulfurization unit in order to produce a P_{H_2S}/P_{H_2} ratio capable of maintaining the cobalt molybdenum in the sulfidation state (see Figure 7 for a comparative illustration).

On the other hand hydrogen is often added, not to modify the catalyst itself, but to modify the selectivity of the reaction by hydrogenating the coke precursor. In this way this stabilization process belongs rather to the section on stabilization by transient parasitic coke formation.

4. DEACTIVATION DURING WORKING

While the final state is obviously the important point for stabilization, it is the kinetic evolution which is important for deactivation phenomena, in order ot define the variation of operating conditions needed to maintain the performances and to determine the moment when the catalyst has to be regenerated or discarded. Among the causes of deactivation described above, we shall not consider coke formation, which is the continuation of phenomena leading in a first phase to a pseudo-stabilization; a more or less similar mechanism can operate but is considerably slower on the catalyst which is deeply modified by the first and rapid coke deposition.

4.1. Deactivation, relevant to the catalyst itself (the case of sintering)

The underlying cause of this deactivation is the thermodynamic metastable state of a catalyst, if we consider its possible evolution, independently from catalysis, which, fortunately is greatly slowed down by kinetic factors. The simplest example concerns the large surface area of porous catalysts which tends to decrease, diminishing the overall free energy of the system. Among the numerous methods of textural or structural deactivation we shall consider only the sintering of metal deposited on a support. This deactivation process becomes significant for high temperature reactions such as steam reforming or post-combustion of exhaust gases. During catalyst regeneration by coke combustion, the high local temperatures can also lead to significant sintering.

Two essential sintering mechanisms are generally invoked (ref. 19) : i) sintering by crystallite migration, followed by their coalescence after collision ii) sintering by migration, towards large crystallites, of atomic or molecular compounds extracted from small crystallites.

For the first mechanism (ref. 20) it seems that once crystallites collide the coalescence is rapid, if we consider the ease of sintering of bulky metals at temperature as low as 200°C. Hence the limiting step will usually be the diffusion step. The diffusion rate depends mainly on the bonding strength, between crystallite and carrier. Since the value of the overall bonding increases with crystallite size, it is only for small crystallites that this mechanism appears likely to cause sintering, even if we consider the surface migration of a particle by its change in shape, through a type of creeping process (ref. 21). Moreover, the redispersion phenomenon is difficult to explain by this mechanism.

The second mechanism (ref. 22), called Ostwald Ripening, proceeds according to the three successive steps shown in Figure 11.

i) From crystallites, atomic or molecular species escape to produce, in the gas or in the adsorbed phase, highly dispersed atoms, not necessarily in a metallic form.

ii) The migration from small to large crystallites proceeds under the influence of a concentration gradient, which will be justified below.

iii) The capture of diffusing species can occur in two ways :
either by collision and in incorporation into a large crystallite, hence leading to crystallite growing and sintering

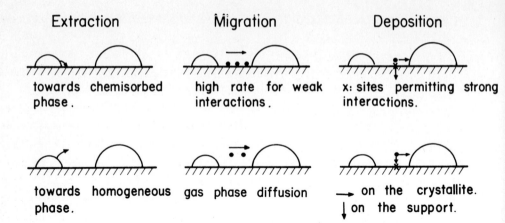

Fig. 11 : Ostwald Ripening mechanism (growing of large crystallites at the expense of small crystallites).

or by settling on some free support sites, with which the bonding strength may be quite high. Such a phenomenon is the possible source of redispersion.

The concentration gradient from a surrounding of small crystallites towards a surrounding of large crystallites has been explained by thermodynamic considerations (ref. 19). If the thermodynamic equilibrium is achieved, large crytallites are in equilibrium with a lower concentration of dispersed atoms than small crystallites (this is the two-dimensional analog of Ostwald Ripening, by the application of Kelvin's law). Of course thermodynamic equilibrium may be far from being attained during sintering, but it remains that the rate of loss of atoms is smaller than the rate of capture for large crystallites, while for small crystallites the rate of loss is greater than the rate of capture. The Ostwald Ripening mechanism can be used to interpret more experimental data than the Crystallite Migration mechanism. We shall try to show its potential by examining some examples. An important general parameter which governs the sintering rate is also the interaction strength of catalytic atoms, with the bulk of the metal crystallite, with the support, or with the homogeneous phase surrounding the crystallite. But depending on whether this parameter affects one or the other of the three above mentioned steps, it can enhance or hinder the sintering rate, or even induce a redispersion.

From such considerations, it is possible to attempt interpretation for example, of the apparently contradictory influences of an oxygen partial pressure on the metal dispersion. The most generally observed result is that oxygen enhances sintering rate as compared with the case of a reducing medium (ref. 19). However it has been possible to observe a crystallite redispersion by using an oxygen atmosphere (ref. 23). Besides, a redispersion is claimed, although without decisive proof, when using oxygen and chloride compounds for the platinum of a reforming catalyst for example. It is possible to reconcile these phenomena, using ideas developed by different workers (ref. 24). The extraction step is relatively difficult in a reducing medium, due to the high bonding strength of a metallic atom with the crystallite (the heat of sublimation can be higher than 100 kcal/mole is many cases), and to the weak bonding strength of zero metal with a carrier such as alumina. In an oxygen medium, the extraction step is made much easier, both superficial crystallite oxidation and by the formation of more stable compounds after extraction. For extraction by volatilization towards the gas phase, we obtain PtO for example (or platinum oxychloride with $O_2 + Cl_2$). For extraction towards a chemisorbed phase, we obtain Pt^{2+}/Al_2O_3, for example. Then, after the extraction and migration steps :

 - either the diffusing species are captured by a large crystallite; sintering occurs, and so is explained its enhancement by oxygen,
 - or, the diffusing species can be strongly held by some particular sites of the support, and this induces redispersion. For species diffusing in the adsorbed phase such scavenging could result from the intervention of particular sites of the support : corner, edges, various defects, responsible for the well known "decorating effect". For diffusion in the gas phase, a mechanism of interaction with the support must be invoked.

Of course the reacting molecules themselves can contribute to crystallite evolution. Strong bonding between reactant and metal can promote surface reconstruction, as seen previously but can also facilitate the atom extraction towards the adsorbed or homogeneous phase, thus leading to sintering, to metal elimination, or to redispersion. Examples are given of sintering enhanced by reactants (ref. 25), and the possibility of metal loss by reaction with reactant is well known (RuO_2 in post combustion catalysts).

These few examples illustrate the interpreting power of the Ostwald Ripening mechanism. However, according to the authors themselves, the hypotheses have not been proved directly, and choice between the two basic theories is still the subject of discussion. The kinetic evolution itself does not make discrimination possible (ref. 19).

In conclusion, a huge amount of experimentation has been performed which has led to practical results and to the establishment of intellectual concepts of the sintering mechanism. A more systematic experimental approach now seems needed to progress our understanding of these phenomena.

4.2. Irreversible poisoning

For selective poisoning, an intrinsic modification of the catalytic species occurs. For non-selective poisoning the decrease in activity results mainly from the covering of the catalytic sites and sometimes from pore plugging. In any case, location of the poison has a great influence on the downward drift of performances and can be discussed from Figure 12. The phenomenon depends on the competition between the penetration rate of poisons (or poison precursors) and their deposition rate on the catalyst surface. Two extreme cases can be considered.

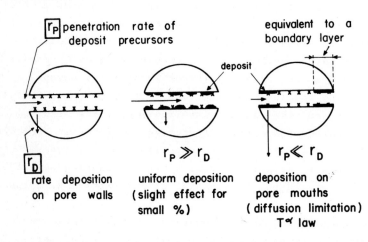

Fig. 12 : Different aspects of irreversible poisoning (X : active sites)

$r_P \gg r_D$. Deposition takes place uniformly all along the pore. If the poisoning is non-selective the rate decrease results from statistic covering and may be weak. Let us consider e.g. that 1% of metal is placed on a support and covers 1% of the surface area; fo 1% of statistic poison deposition (having the same textural characteristics as the metal crystallites), only a small fraction of these crystallites will be covered. If the poisoning is selective, its eff will be greater and could be more complex, considering mainly selecti vities, exactly as in the case of reversible selective poisoning.

$r_P \ll r_D$. Deposition takes place on the pore mouths, with quite different consequences from the previous case. The bead periphery

becomes inactive. For fast reactions a diffusion - limitation
appears (or is greatly increased), and this limitation is of the
same nature as that created by a larger immobile boundary around
a catalyst bead. Hence, the activity varies as T^α with temperature,
which produces apparently very low activation energies. Of course
the metal distribution through the catalyst bead has a great influence
on such poisoning. If the metal has been deposited on the periphery
to avoid diffusion-limitation (in the case of a fast reaction), the
decrease in activity may be drastic. So a compromise has to be sought.
Several authors have published theoretical or experimental papers
for achieving such a compromise (ref. 26).

The poisoning of nickel hydrogenation catalyst by sulfide compounds
(at low temperature) is an example of selective irreversible
poisoning (see Figure 10). Consideration of the influence of lead
on the activity of post-combustion catalysts is particularly
interesting, although it would be difficult to distinguish between
selective or non-selective poisoning. The poison precursors consist
mainly of more or less volatile lead halides probably diffusing
in both the gaseous and adsorbed phase. Their penetration rate
into the pore increases extensively with the temperature. If
poisoning has taken place at low temperature it mainly affects the
pore mouth. Hence due to the apparently weak activation energy,
conversion will increase very slowly with temperature, for such a
catalyst, and may remain inadequate, as illustrated by Figure 13
(ref. 27).

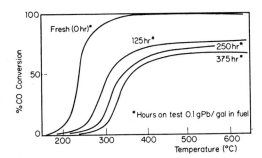

Fig. 13 : Laboratory evaluation of a lead-poisoned $Pt-Al_2O_3$ catalyst GHSV = $90,000/hr^{-1}$ lead accumulation = 0.1 g Pb/gal; converter = 2300 cm^2 radial flow at 590°C (poisoning at low temperature) (ref. 27a), and for different times.

If poisoning occurs at high temperature, the lead compounds penetrate
statistically into the entire pore, and deactivation is less, even
for a higher total lead percentage (1.5% of uniformly dispersed lead
can be less harmful than 1% on pore mouths).

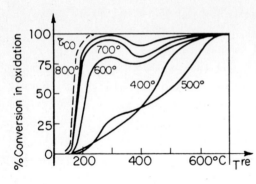

Fig. 14 : Catalyst with 0.08% Pd on alumina carrier. Activity after aging during 30 hr in a burner at different temperatures, (0.25 g/l Pb; 1 Cl - 0.5 Br). Laboratory activity test for GHSV = 30,000 h^{-1} (ref. 27b).

---------Activity for the fresh catalyst.

———————Activity after aging.

Figure 14 shows that, for poisoning at 700°C, deactivation may be much less than for poisoning at 400°C. Moreover, it the catalyst poisoned at 400°C is heated to 700°C, for example, the activity may increase as the result of both a better distribution of the lead and of the elimination of some lead by its desorption from the catalyst's surface. We can note that the proof of a selective contribution for lead poisoning is given by the interesting complex evolution of activity in CO oxidation versus metal coverage by lead. Until half of the metal covered the activity increases, a phenomenon which is interpreted by an intrinsic effect (electronic or ligand effect) of lead on platinum (ref. 28). More precisely, the oxidation rate can be written, for Pt or Pd catalytic species :

$$r = k \frac{b_{O_2} P_{O_2}}{1 + b_{CO} P_{CO}} \quad ; \text{ with } b_{CO} P_{CO} \gg 1 \text{ (ref. 28 et 27b)}.$$

k being the rate constant, b_{O_2} and b_{CO} the adsorption coefficients of O_2 and CO. By adding a small percentage of lead, the great decrease of b_{CO} has much more influence than the small decrease of k b_{O_2} (the reactivity) and of the active site number of Pt or Pd. By increasing the lead percentage again, the activity goes through a maximum and decreases again when the decrease of k b_{O_2} and the active site number becomes greater than that of b_{CO}.

Another example of pore mouth poisoning is observed in the case of hydrodesulfurization on Co-Mo/Al_2O_3 catalysts, of vacuum residues containing vanadium and nickel compounds. Figure 15 shows that a vanadium deposition occurs at the bead periphery (in the form of vanadium sulfide) while nickel passes (also in the form of nickel sulfide) deeper into the heart of the bead (ref. 29). The r_p/r_D ratio is different for the two poisons. The deposition involves the intervention of the catalytic species. The aromatic asphaltenic compounds in which the vanadium, for example, is combined, are hydrogenated into naphthenic structures, which are then hydrogenolysed, leading thus to vanadium sulfide deposition. Moreover geometrical factors certainly have an important influence on the penetration rate

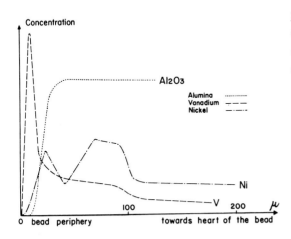

Fig. 15 : Hydrodesulfurization of vacuum residues. Ni and V concentration all along a bead diameter (microprobe analysis).

through the pore of the large molecules or micelles of asphaltenic compounds.

In the case of pore mouth poisoning, a means of couteracting deactivation consists of using a catalyst with bimodal pore size distribution (or a spread out pore size distribution). Due to the high deposition rate for this poisoning mode, the deposit occurs immediately in the large pores which operate like a filter, thus providing protection for small pores, responsible for the high catalytic surface area and activity. The large pores act as the guard case often located at the entrance of a reactor to provide protection for the catalytic bed. The general concept of competition between transportation and deposition inside the pores as described above, can be transposed for phenomena occuring in the reactant flow along the catalytic bed.

4.3. Catalyst defects

Here we shall describe some examples of parasitic reactions due to bad design or manufacture of the catalyst. "Foreign" species can be included in an "ideal" catalyst. Thus the presence of group VIII metals in a catalyst for methanol synthesis can catalyze methane formation accompanied by high heat emission. Besides a decrease in yield, this causes sintering of the catalyst. Another more general case of catalyst defects concerns the parasite activity of the support. A characteristic example is the selective hydrogenation of diolefins or acetylenic compounds or, more practically, of steam-cracking gasoline, on catalysts such as Pd or Ni/Al_2O_3. Due to the reactivity of these highly unsaturated compounds, their polymerization can occur even for a small acidity of the alumina, leading to polymer deposition on the support. The way to counteract this deactivation is of course to decrease the acidity of the carrier and also to use the best technology. By operating for example, in the liquid phase, the catalyst can be cleaned of the accumulated liquid polymers.

5. REGENERATION

A general definition of regeneration can be the recovery (restoration) of activity after irreversible damage. In this sense, the intrinsic stabilization is not, of course, affected by the regeneration concept. Reversible poisoning (inhibition), which is cancelled out by the elimination of poison from the feed, is likewise not invoked in regeneration.

Elimination of coke by burning with oxygen is a true regeneration. The well-known precaution to be taken consists in feeding in a progressively growing oxygen percentage in order to avoid high temperatures leading to sintering (and eventually to explosion). Sometimes the presence of sulfur is claimed to enhance sintering during coke burning (case of reforming catalyst), and sulfur has to be eliminated by previous H_2 treatment according to the reaction (ref. 30).

$$H_2 + PtS \rightleftharpoons Pt + H_2S$$

The damage induced by sintering is much more difficult to restore. For deposited metal however, we have tried to interpret the redispersion of platinum by oxygen or rather by the $O_2 + Cl_2$ mixture claimed by various patents. Much more systematic experimental investigation is needed, firstly to prove that redispersion does occur effectively, and secondly to establish its mechanism. Another recently proved possibility for example, for the decrease of accessible platinum for a reforming catalyst by hydrogen treatment at high temperature, is the formation of Pt-Al alloys (ref. 31). In this case, of course, regeneration by oxygen is quite different (ref. 32).

Regeneration outside of the reactor, rejuvenation, is generally achieved by chemical treatment and will not be discussed here.

6. GENERAL CONCLUSION

The variety and complexity of the phenomena involved in the stabilization, deactivation and regeneration of catalysts should be clear from what has been said. We have only been able to give a slight glimpse of the types of variations that we feel to be important and with which we are most familiar. However, we feel able to conclude that the enormous number of experimental results obtained from fundamental research or industrial achievements can now form the basis for general laws and mechanisms which are valid as an effective working hypothesis in the study of these phenomena.

Furthermore, experimental techniques are becoming more and more effective in characterizing the state of catalysts at different stages of their evolution and in following catalytic performances in the steady or transient state.

Therefore, we feel that systematic, well directed and properly exploited research should enable rapid advances in the understanding of these phenomena while at the same time providing practical applications from the industrial standpoint.

REFERENCES

1. Conference on Catalyst Deactivation and Poisoning, Berkeley (may 1978).
2. A. Roumegous, B. Torck, J.P. Franck and R. Montarnal, To be published.
   R. Bonifay, B. Torck and M. Hellin, Bull. Soc. Chim. Fr 9-10 (1977) 808-814.
3. F. Avendano Thesis Paris (1975). Editions Technip Paris.
4. P. Boutry, P. Courty, J.C. Daumas and R. Montarnal, Bull. Soc. Chim. Fr. 10 (1968) 4050-4056.
5. G. Rambeau and H. Amariglio, J. Chim. Phy. 75 (1978) 110-115.
6. R. Brill and J. Kurzidim, Colloques Internationaux du C.N.R.S. 187 Paris (Juillet 1969) 99-101. Editions du C.N.R.S. Paris (1970).
7. J.A. Dumesic, H. Topse and M. Boudart J. Catal. 37 (1975) 513-522.
8. A.G. Friedlander, Ph. Courty and R. Montarnal J. Catal. 48 (1977) 312-321.
9. W.M. Sachtler and R.A. Van Santen Advances in Catalysis 26, 69-119 Academic Press N.Y. (1977).
10. R.L. Moss and L. Whalley Advances in Catalysis 22, 115-185 Academic Press N.Y. (1972) V. Ponec Catalysis Review Sc and Eng. 11, 41-70 Marcel Dekker N.Y. (1975).
11. A. Takeuchi, K.I. Tanaka and K. Miyahara J. of Catal. 40 (1975) 101-107.
12. Ph. Courty and J.F. Le Page Second International Symposium on the Scientific Bases for the Preparation of Heterogeneous Catalysis Louvain (Sept. 1978).
13. J.R. Rostrup-Nielsen "Steam Reforming Catalyst" Danish Technical Press Copenhagen (1975).
    J.R. Rostrup-Nielsen and D.L. Trimm J. Catal. 48 (1977) 155-165.
14. C. Portier, G. Thomas, J.F. Le Page and R. Montarnal C.R. Acad. Sc. Paris Serie C, 284 (1977) 865-867.

15. C.E. Alvarez Herrera Thesis Poitiers 1977 Editions Technip Paris
16. R. Montarnal and G. Martino Franco-Sovietic Seminar on Catalysis Kiev (Sept. 1974) and Rev. Inst. Francais du Pet. Vol XXXII N°3 (1977) 367-391.
17. J.A. Rabo, V. Schumaker and P.E. Pickert Proc. 3rd Int. Congr. Catalysis North Holland Pub. CY Amsterdam (1965) Vol. 2, 1264.
18. G.D. Chukin and Al Proc. Sixth Int. Cong. on Catalysis London (July 1976). The Chemical Society London (1977) 668-676.
19. S.E. Wanke and P.C. Flynn Catal. Rev. Sci. Eng. 12 (1) (1975) 93-135 Marcel Dekker (1975).
20. E. Ruckenstein and B. Pulvermacher AICHE J. 19 (1973) 356.
21. R. Kern, A. Masson and J.J. Metois Surface Sci. 27 (1971) 483.
22. P.C. Flynn and S.E. Wanke J. Catal. 34 (1974) 390.
23. R.M.J. Fiederow and S.E. Wanke J. Catal. 43 (1976) 34-42.
    R.M.J. Fiederow, B.S. Chahar and S.E. Wanke J. Catal. 51 (1978) 193-202.
24. J.W. Geus Euchem conference on colloïdal metal particles in catalysis Namur (December 1977). Paper on Preparation and Thermostability of Supported Metal Catalyst.
25. A.K. Joesoef Thesis Lille (1977) Editions Technip Paris.
26. J.C. Summers and L.L. Hegedus J. Catal. 51 (1978) 185-192.
27. R. Montarnal Fifth Ibero American Symposium on Catalysis Lisbonn (July 1976).
    27a) R.L. Klimisch and J.M. Komarny The Catalytic Chemistry of Nitrogen Oxides Plenum Press (1975).
    27b) M. Prigent and A. Sugier Fifth Ibero American Symposium on Catalysis Lisbonne (July 1976).
28. G. Praline Thesis Nancy (1978).
29. F. Audibert and P. Duhaut Rev. Inst.Francais Pet. Vol. XXV N°5 (1970) 613-622.
30. USP 3, 481, 861 (UOP).
31. F.M. Dautzenberg Euchem Conference on colloïdal metal particles Namur (December 1977) (and in publication annonced in the paper of reference 32).
32. F.M. Dautzenberg and H.B.M. Wolters J. Catal. 51 (1978) 26-39.
33. A. Roumegous Thesis Paris 1978 Editions Technip Paris.

DISCUSSION

H. SHINGU : As to your definitions and classifications about stabilization or catalyst evolution during catalysis, particularly referring to Fig. 1, I would like to point out that :
1) the steady state terminology for the stationary level of activity (you mention even true state in Fig. 1) should be correctly referred to as stationary state, because the kinetic term steady state signifies generally that the overall rates are changing with time, i.e. $\Sigma\ d^2x/dt^2 \neq 0$, whereas in the stationary state the rates are time independent, i.e. $\Sigma\ d^2x/dt^2 = 0$, so that the theoretical true steady state in Fig. 1 may seem to be self-contradictory. The currently prevailing misuse of the term steady state, especially in reaction engineering fields, seems to originate from a confusion between stabilization and deactivation during working time, accordin to your classification and analysis.

2) the transient period of catalyst stabilization usually involves much complicated surface processes, and may be better analyzed with respect to the reversibility or reproducibility and the time required for the stabilization.

Essentially, the intrinsic stabilization should be reversible or more generally reproducible and require only short time for the transient period to be reached (e.g., less than several minutes with a high selectivity catalyst). Irreversible stabilization usually requires much longer time, and cannot be referred to as intrinsic, because the catalyst surface structure is always essentially modified by this treatment, as you have concluded about the complex stabilization.

R. MONTARNAL : As far as my paper is concerned, the concept of steady-state is presented for an open flow reactor, in the absence of any deactivation process. In this case, a steady state is attained after the time at which the concentrations in a given slice of catalyst remain constant, and so, the reaction rate remains constant in each slice. This true steady state refers to a reversible intrinsic phenomenon which is due to the mechanism of the catalytic reaction. It corresponds to what I have called intrinsic stabilization. This must be distinguished from the pseudo-stabilization due to coke formation, which is an irreversible phenomenon and leads to new catalyst different from the initial one. Once this pseudo-stabilization by deactivation is stopped, the coked catalyst works with its own steady state, depending on its own intrinsic catalytic properties.

D.J.C. YATES : I should like to draw your attention to several U.S. Patents issued in my name in the last few years, where you will find precise experimental evidence that noble metals can easily be redispersed to atomic dispersion, starting from particles in the 20-30 nm region, by the addition of halogens.

R. MONTARNAL : As you say, as early as 1954, in French patent 1,099,233 it is claimed, on the basis of X-ray diffraction results, that a redispersion occurred of platinum crystallites of about 200 Å into crystallites less than 50 Å in size. Such claims are again taken up in US patent 3,134,732 for platinum. They are also taken up, for example, in your French patents 2,240,762 and 2,257,337 "for iridium- and halogen-based catalysts on a refractory

support, using sophisticated techniques in the case of iridium".
What I meant to say was that I have not found any scientific
publication concerning this redispersion. I feel that the intimate
mechanism of this redispersion still remains to be determined,
even though mention can be made of an interaction between volatile
species and the OH sites of the support, as I suggested in my
reply to Charcosset.

J.W.E. COENEN : In the process you call Oswald ripening, often
intermediate formation of a volatile or soluble compound which
acts as the transport agent is an important step with nickel
catalysts. I shall come accross two examples, both involving
crystallite growth caused by unfavorable transient situation
during processing.
One concerns methanation. If the catalyst is brought into contact
with a carbon monoxide containing gas stream under conditions where
$Ni(CO)_4$ formation is thermodynamically favoured, a situation may
occur where small Ni crystals are "eaten" and at the same time, or
later $Ni(CO)_4$ is decomposed and Ni deposited on larger crystals.
The second example refers to fatty acid hydrogenation. Nickel drop
formation is possible and occurs at low hydrogen pressures again
with loss of the smaller crystals. When later hydrogen pressure is
built up, nickel metal is deposited but not in very active form
(autoclave walls, large nickel crystals).

R. MONTARNAL : You gave two quite nice and characteristic examples
of the general idea which I paraphrased in my paper in the following
terms : "Of course the reactional molecules themselves can contribute
to crystallite evolution. Strong bonding between reactant and metal
can promote surface reconstruction, as seen before, but also make
easier the atom extraction towards the adsorbed or homogeneous phase,
thus leading either to sintering, to metal elimination or why not
to redispersion".

H. CHARCOSSET : In your communication, you consider the redispersion
of sintered Pt by O_2 or by O_2 and Cl_2 as a rather questionable
phenomenon. Could you comment in more detail on this ?

R.E. MONTARNAL : The platinum redispersion by means of O_2 and Cl_2
finally appears to be rather well proved. It seems to proceed by
formation of volatile compounds from Pt crystallites (complexes of

Pt-Al-chlorides or Pt-oxychlorides, at the temperature used for redispersion). The migrating species are trapped by sites such as surface OH groups. The strong interaction with these sites allows deposition of these species in a highly dispersed state leading to small crystallites after reduction.

R. PRICE : Catalysts with bimodal pore size distributions were mentioned as a possible means of reducing the extent of poisoning by metals. In our experience with this type of catalyst the greatest amount of carbon and metals deposition occurs in the smaller pores (<50 Å) and the resultant loss of surface area is not recovered on regeneration. Can this be interpreted in terms of the poisoning mechanism which you describe or does it suggest, perhaps, a difference in the metal-containing poisons present in our coal-derived feedstock compared to the metallic components of petroleum based feedstocks ?

R. MONTARNAL : By using catalyst with a uniform distribution of rather small pores, we observe coke and metal deposition. This agrees with your result that these deposits can occur in the small pores of a catalyst with a bimodal distribution of pore sizes. In our experiments, the metal accumulation was, as presented in the paper, at the periphery of the pellet. This result only concerns the macro-distribution and does not provide any information on the microdistribution between large and small pores. I would like to emphasize that in our experiments, the deposition at the periphery of the pellet is indicative of a deposition rate which is faster than the penetration rate. So, if added macropores have a sufficiently high catalytic surface area to allow fast reactions to occur, they may be able to achieve a sort of "scavenging" of metal compounds (deposited as sulfides), leading then to a protection of the micropores. This explanation has been advanced by different authors, especially in patents, for petroleum based feedstocks with V and Ni. The possibility that such a protection occurs, depends on many factors as the nature of the charge, of the catalyst, and of the operating conditions. So it is possible that, for coal derived feedstocks containing for example Fe and Ti compounds and minerals, the characteristics of the deposition could be different.

N.   : I agree that a more systematic experimental approach to thermal sintering would be beneficial. These experiments should cover a wide range of catalytic metals, metal concentrations and support types with different atmospheres to give a more complete picture. However, it seems likely that both Oswald ripening and crystallite growth migration may operate together, one phenomenon predominating under a given set of experimental conditions. Could you comment on the usefulness of crystallite particle size measurements, particularly at the early stage of sintering experiments attempting to differentiate between the possible mechanisms ?

R. MONTARNAL : The answer to your question is given in the review of Wanke and Flynn (1). These authors state that if support surfaces were energetically homogeneous, a number of tests could distinguish between crystallite and atomic migration. The crystallite migration model predicts that metals on catalysts with unisized or narrow pore size distribution will sinter readily, while the atomic migration model predicts such catalysts would sinter extremely slowly. However, real surfaces are energetically heterogeneous; if one includes strong trapping sites and preferred adsorption areas, either model can account for a wide variety of phenomena. Model discrimination will be extremely difficult for this case. Moreover specific mathematical kinetic laws have been developed by different authors for the evolution of crystallite sizes, in order to be able to differentiate between the involved mechanisms. But due to the complexity of a real catalyst, it seems that till now "the experimental identification of the sintering mechanism(s) is a problem that awaits solution". However, in spite of this negative conclusion by the authors mentioned, it seems to us that crystallite migration is able to explain sintering of small crystallites, but only in the initial stages and cannot explain redispersion.
1. S. Wanke and P.C. Flynn, Catal. Rev. Sci. Eng. (1) $\underline{12}$, 93 (1975).

J.W. HIGHTOWER : In your lecture you mentioned the surface migration and the gas phase migration models in the sintering of supported metal catalysts. Could you comment on the relative importance of these two effects in the case of supported Pt catalysts either in reforming or in auto exhaust gas purification ?

R. MONTARNAL : In the case of supported Pt for catalysts of auto exhaust purification, sintering is observed in an oxidative atmosphere and by using unleaded gasoline. In this case, no chloride compounds need to be added, and the formation of rather volatile compounds such as platinum oxychlorides or complexes of platinum-aluminium chlorides cannot be invoked. It is then generally admitted that the migration of Pt species occurs in the adsorbed phase, by diffusion of ionic platinum compounds. But the detailed mechanism of extraction from the Pt crystallite and of the migration remains to be established. Considering now the case of a reforming catalyst, two recent publications (1) (2) have shown that in the presence of O_2, redispersion first occurs from 500 to 600°C, at still higher temperatures sintering occurs again. The authors give a detailed discussion of the different modes of transfer for platinum compounds. They conclude (1) that "migration of $PtO_{2(s)}$ (adsorbed) is restricted to individual support grains and vapor phase transport is required for transfer of Pt from one grain to another". However for reforming catalysts, we must also consider that chloride compounds are present and that, at the catalytic temperature, the phenomena are more complex. In conclusion, in spite of the possibility of migration in the adsorbed phase, each individual case requires a specific discussion.

(1) R.M.J. Fiedorov and S.E. Wanke, J. Catal. $\underline{43}$, 34 (1976).
(2) R.M.J. Fiedorov, B.S. Chahar and S.E. Wanke, J. Catal. $\underline{51}$, 193 (1978).

E.J. NEWSON : In reference to Fig. 15 on HDS vacuum residues and deposition of Ni and V from the oil in the pores of the alumina based catalyst, it would be more appropriate to call this phenomenon pore-plugging deactivation and not pore-mouth poisoning. This is easily shown by the drastic decrease in pore volume of the spent catalyst, say from 0.5 to 0.2 cc/g.

The questions really refer to the ultimate life-time of the HDS catalyst. Using a catalyst with a bimodal pore size distribution would probably give a slight gain in initial HDS activity but a higher demetallation rate is a necessary consequence. Since some of the small pores have probably been given up to make "large pores which operate like a filter, it does not necessarily follow (remember the higher demetallation rate) that ultimate life of the catalyst will be increased. Which data do you have to substantiate the claim for increase in ultimate life, say 1000 to 2000 hours ? Which feed analyses and process conditions were applicable ?

R. MONTARNAL : Of course, pore plugging can occur but I think it often occurs after a rather long period of pore mouth poisoning as the result of the progressive growing of the volume of the deposits at the pore mouth. The drastic decrease in pore volume from 0.5 to 0.2 cc/g does not seem to be a definite proof for plugging, since the pore volume is determined at low temperature, often with insufficient washing of the catalyst on which adsorbed hydrocarbons (or micelles) of high molecular weight can cause an apparent plugging. At higher temperature, under the catalytic conditions, micelles can be "dissociated" or hydrocarbons desorbed, leading to a decrease of plugging, while pore poisoning is maintained.

I don't entirely agree with your idea that large pores are obtained at the expense of small ones, for a macroporosity can be added to a catalyst pellet while maintaining a high fraction of microporosity. Concerning the comparison between the ultimate life time of a catalyst with bimodal and monomodal pore size distribution, the answer is not simple. It is difficult to build two "identical" catalysts, differing only by the presence or the absence of macropores. It also seems true that macroporosity increases the "chemical" life-time of the catalyst, unfortunately at the expense of the mechanical strength of the catalyst pellet.

B. DELMON : You have quite authoritatively presented the major categories of deactivation processes. However, I wonder whether one additional category, which you did not mention should not be alluded to. I refer to what could be called "catastrophic" deactivation which occurs quite often at the scale of individual pellets, but sometimes also in the whole bed. Catastrophic deactivation is best examplified as follows. Imagine a catalyst for partial oxidation. Suppose that on-stream a solid state phenomenon forms another phase acting as a complete oxidation catalyst. The temperature of the pellet will increase enormously, thus triggering further solid state transformations. Since complete oxidation consumes more oxygen, a highly reducing product stream (CO, etc...) is formed, and consequently intense reduction of the solid phases may occur. In this way, some kind of autocatalytic destruction of the active phase takes place.

In my opinion, it is this kind of autocatalytic catastrophic deactivation process which could justify to invoke a distinct category of deactivation processes. Indeed, the other categories you have mentioned correspond more or less to progressive changes many times with auto-adjustment.

With catastrophic phenomena, preparation of catalysts is still more

critical than with the other categories of deactivation. Any nucleus of the unwanted phase could trigger catastrophic nucleation if it is inadvertently formed during preparation.
I believe that the above remarks are far from being mere speculation, especially in the case of partial oxidation processes.

R. MONTARNAL : Your remark illustrates that it is nearly impossible to consider all cases of such a broad topic as catalyst stabilization and deactivation. However your example of "catastrophic" deactivation could be inserted under my heading : "Catalyst Defects". In this part, I have only mentioned "Foreign and Parasite Species included in an Ideal Catalyst, due to Poor Design or Manufacture of the Catalyst, and thus leading to some Parasite Reactions". By parasite reactions, I mainly refer to catalytic parasite reactions. You are quite right in emphasizing the importance of catalyst preparation when such autocatalytic "Catastrophic Deactivation" could occur, as it is in the case of mixed oxides used for partial oxidation processes.

N. PERNICONE : You have reported some results concerning transient phenomena in kinetic runs of ammonia synthesis catalysts. I would like to stress that both Amarijlio and Brill's experiments (refs. 5-6 of your paper) have been performed at atmospheric pressure and low temperatures (below 300°C). During one activity test, carried out in the pressure and temperature ranges currently used in industrial practice, we never detected such phenomena. They are certainly interesting from a basic point of view, but not relevant to the industrial conditions.

R. MONTARNAL : My purpose was to give some examples of "intrinsic stabilization". It is possible that under industrial conditions transient periods would not exist. However, it is also possible that the response time of the performance test would be too long to allow the detection of the transient period observed by the authors invoked in my paper.

Z. PAAL : I should like to add two short comments as far as pseudo-stabilization due to coke formation is concerned.
I completely agree with you that the severity influences coke formation. In addition to the amount, the quality of coke also can be affected. For example, coking of platinum may leed to the formation

of carbonaceous species (one carbon atom per Pt atom, probably in dissociated form). In the presence of such deposits platinum exhibits higher total activity in the presence of lower hydrogen pressure (i.e. under more severe conditions) (1). Considering also selectivity changes under the combined effect of hydrogen and carbonaceous deposits we believe that the second explanation offered by you (i.e. the ligand effect) may be valid in this case.
(1) Paul Z., Dobrovolosky M., Téténijr P., J. Catal. $\underline{46}$, 65 (1977).

R. MONTARNAL : I am little surprised by your result according to which the deactivation seems highest in the case of the highest hydrogen pressure, i.e. with less severe conditions. What is the catalytic reaction used in your experiments ?

Z. PAAL : The decrease of activity in higher hydrogen pressure was observed in 3-methylpentane reactions (isomerization - hydrogenolysis - cyclisation) and was explained by competition of hydrogen and hydrocarbon for sites left after partial carbonization.

R. MONTARNAL : I agree on the explanation according to which hydrogen acts as an inhibitor (1) (2). This leads to the hydrogenolysis of n-hexane on iridium, the reaction rate being :

$$r = \frac{K_1 K_2 \, (b_P P_{HC}) \, (b_H P_{H_2})^2}{\left\{ (b_H P_{H_2})^2 + (b_H P_{H_2})^{3/2} + b_P P_{HC} \left[(b_H P_{H_2}) + K_1 (b_H P_{H_2})^{1/2} + K_1 K_2 \right] \right\}^2}$$

This equation is similar to the one you have also considered in your most recent paper (3) :

$$r = k \, \frac{\frac{K_{CH} \, P_{CH}}{\sqrt{K_H \, P_{CH}}}}{1 + \sqrt{K_H P_H} \, + \, \frac{K_{CH} P_{CH}}{\sqrt{K_H P_P}} \, + \, \frac{K \, P_{CH}}{(\sqrt{K_H P_H})^a}}$$

(1) J.P. Boitiaux, G. Martino and R. Montarnal, C.R. Acad. Sc. t 280 (23 Juin 1975), Série C, p. 145.
(2) J.P. Boitiaux, Thèse Paris 1976.
(3) Z. Paal, K. Mausek and P. Tetenyi, Acta. Chim. Acad. Sci. Hung. Tomus 94(2), p. 119 (1977).

Z. PAAL : Under more severe conditions a severe coking occurs leading to "bulk" coke formation. In this respect, n-hexane and n-hexene which may form coke via polycondensation of their **all-trans polyene** dehydrogenation products (1) or via unsaturated C_5-cyclic polymers (2), cause more severe deactivation than any C_6-cyclic species (cyclohexane, cyclohexene or benzene). This points to the importance of the structure of the reactant. A systematic study in this respect might reveal the pathways of coking.

(1) Paal Z., Fyer J.R., Thomson S.J., Mechanisms of Hydrocarbon Reactions. A symposium Akad. Kiado. Budapest, 1975, p. 137.
(2) Myers C.G., Lang W.H., Wein P.B., Ind. Eng. Chem. $\underline{53}$, 299 (1961).

R. MONTARNAL : In the case of the hydrogenolysis of n-hexane, we have observed a rather different type of result. The deactivation caused by the presence of small quantities of benzene, cyclohexene and cyclohexane was strong and of the same order of magnitude. But it was for the hydrogenolysis of n-hexane on Ir instead of on Pt, and for quite different operation conditions (1) : T = 180°C, P_{H_2} = 0.95 bar and P_{C_6} = 0.05 bar.

(1) Boitiaux J.P., Thesis, Paris 1976.

PREPARATION AND PROPERTIES OF MONODISPERSED COLLOIDAL METAL HYDROUS OXIDES*

EGON MATIJEVIĆ
Institute of Colloid and Surface Science and Department of Chemistry, Clarkson College of Technology, Potsdam, New York 13676 (U.S.A.)

ABSTRACT

The methods for preparation of colloidal dispersions consisting of exceedingly uniform particles of hydrous oxides of a number of metals are described. Some of the sols contain spherical amorphous particles, while others are well crystallized submicronic or micronic solids of different morphologies. These novel systems have been used to establish the chemical mechanisms of metal (hydrous) oxides formation and growth. Various properties of such uniform particles (surface charge, magnetism, color, etc.) are illustrated, and their use in the studies of interactions with different solutes (adsorption), with other particulate matter (heterocoagulation), and with surfaces of different materials (adhesion) are discussed.

INTRODUCTION

Few families of inorganic compounds are as significant as metal (hydrous) oxides*. These materials appear in the nature as different minerals and ores and they are produced by hydrolytic corrosion of metals. In addition, metal hydrous oxides in various forms find numerous uses, such as pigments, catalysts, catalyst carriers, fillers, coatings, etc. Thus, it comes as no surprise that much work has been done on these compounds, both in terms of their preparation and characterization; nevertheless, relatively little is still known about the mechanism of formation of any of the metal oxides or hydroxides. This lack of knowledge is easily understood if one recognizes the complexity of the processes involved in the precipitation of such solids. For example, a minor change in pH can result in the formation of entirely different species in terms of chemical composition as well as of morphology. Equally important are the roles of other parameters, primarily of the temperature and of the nature of the anions in the solution in which the precipitation takes place. The latter can exercise a pro-

---

*Metal (hydrous) oxide designation is taken here in a rather general way; it includes oxides, hydroxides, hydrated oxides, oxyhydroxides, etc.

---

* Published in Pure Appl. Chem., 50, 9-10 (1978), p. 1193.

found effect on the solid phase formed, even though the respective anions may not appear as constituent species of the precipitate. All the factors mentioned above (pH, temperature, anions) affect the complexation of the solutes which act as precursors to the nucleation of the metal hydrous oxides and which later determine their growth.

The sensitivity of the described precipitation processes explains the poor reproducibility usually encountered in the studies of metal hydrous oxide formation. It also accounts for the finding that the resulting solids are mostly ill defined in shape and polydisperse.

Over the past few years we have succeeded in the preparation of colloidal dispersions of a number of metal hydrous oxides consisting of particles exceedingly uniform in size and shape (1). Such sols can be repeatedly obtained by relatively simple procedures and, as one would expect, the morphology, chemical composition, and other characteristics of the precipitates depend strongly on the experimental conditions. The reproducibility in the generation of well defined suspensions makes the elucidation of the chemical formation mechanisms possible. In order to accomplish the latter aim, the knowledge of the composition of all species in solution, in which the solid phase is formed, and particularly of the complexes containing the metal ion which is the major constituent of the precipitate, is implied. The task in obtaining this information is by no means easy. The monodispersed sols can also be employed in studies of various phenomena, such as in adsorption, adhesion, heterocoagulation, color determination, to mention a few.

In this presentation an overview of the author's program in metal hydrous oxides will be offered with examples in the following areas of research:
  a. Preparation of monodispersed sols.
  b. Mechanism for formation.
  c. Characterization of the particulate matter.
  d. Interactions with solutes.
  e. Interactions with other particles.
  f. Adhesion.

PREPARATION OF MONODISPERSED SOLS OF METAL HYDROUS OXIDES

If uniform particles are to be obtained from a solution in which the precipitating components are continuously generated, secondary nucleation must be avoided. This condition implies that, upon reaching the critical supersaturation leading to the burst of nuclei, the rate of crystal growth must exceed the rate of nuclei formation. In the case of systems studied in this work, complexes of metal ions with hydroxyls, and often with other anions, are the precursors to the solid phase separation. Thus, hydroxylation is the controlling step, which will determine the nature of the final product. The usual procedure of adding base leads to local supersaturations and, consequently, to poorly defined stages of nucleation and particle growth. As a result the produced solids are irregular in shape and of broad size distributions.

Hydroxylation is greatly accelerated with temperature; depending on the metal and on the temperature of aging, hydrolyzed metal ion complexes may form in solutions of various degrees of acidity. Thus, it is possible to regulate the rate of complex formation in an aqueous solution of metal salts by proper adjustment of pH and temperature. Once the solution becomes supersaturated in the constituent species of a given metal hydrous oxide, the nucleation occurs. Further aging at a convenient temperature will continue to produce the precipitating complexes which, assuming a proper rate of their generation, are consumed in particle growth. Consequently, no secondary nucleation will take place, and a uniform uptake of the constituent species by the existing primary particles yields a monodispersed sol.

The chemical composition of the solid will strongly depend on the nature of the complex solutes produced on aging. In this respect the anions play a dominant role. In some cases well defined stoichiometric species form which include one or more hydroxyl and negatively charged ions, whereas in other cases the anions may cause condensation of hydrolysis products into polynuclear solutes of different degrees of complexity. As a result the precipitated particles from solutions of different salts of the same metal ion may have different composition and structure.

To illustrate the above described situations in Figure 1 are given electron micrographs of four different systems, all precipitated by aging acidified ferric salt solutions. The scanning electron micrograph 1a shows spherical particles of hematite (α-Fe_2O_3) obtained by aging a solution 0.0315 M in $FeCl_3$ and 0.005 M in HCl for 2 weeks at 100°C (2), whereas 1b is a transmission electron micrograph of β-FeOOH particles generated in solution 0.27 M in $FeCl_3$ and 0.01 M in HCl heated at 100°C for 24 hr. Figure 1c represents submicron alunite type crystals of the composition $Fe_3(OH)_5(SO_4)_2 \cdot 2 H_2O$ which formed on aging at 98°C for 2 hr a solution 0.18 M in $Fe(NO_3)_3$ and 0.32 M in Na_2SO_4 (3). Finally, Figure 1d shows particles obtained on heating for 20 min at 100°C a solution which was 0.0038 M in $FeCl_3$, 0.24 M in H_3PO_4 with NaOH added to adjust the pH to 1.86. The chemical analysis of these particles gave a composition consistent with $FePO_4$.

The four examples show that in the presence of chloride ions, under rather similar conditions, two entirely different sols are generated in terms of chemical composition and particle morphology (Figures 1a and b). The solid β-FeOOH contained considerable amounts of chloride ions when freshly prepared, but these anions could be removed by repeated washing with water without any apparent change in particle size or shape. On the other hand, solids formed on heating

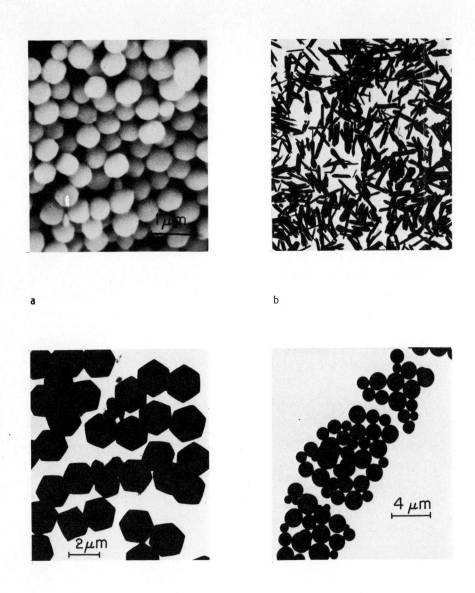

Fig. 1. Scanning and transmission electron micrographs of particles precipitated from various ferric salt solutions. (a) Hematite, α-Fe_2O_3 (solution 0.0315 M in $FeCl_3$ and 0.005 M in HCl aged for 2 weeks at 100°C). (b) β-FeOOH (solution 0.27 M in $FeCl_3$ and 0.01 M in HCl aged for 24 hr at 100°C). (c) Ferric basic sulfate $Fe_3(OH)_5(SO_4)_2 \cdot 2H_2O$ (solution 0.18 M in $Fe(NO_3)_3$ and 0.32 M in Na_2SO_4 aged for 2 hr at 98°C). (d) Ferric phosphate, $FePO_4$ (solution 0.0038 M in $FeCl_3$ and 0.24 M in H_3PO_4 at pH 1.86 aged for 20 min at 100°C).

ferric salt solutions in the presence of sulfate ions consist of stoichiometrically stable and structurally well defined basic ferric sulfates. Finally, no detectable amounts of hydroxyl ligands are found in the systems precipitated as described above from aged solutions in the presence of phosphate ions. The resulting ferric phosphate redissolves on cooling.

The four different dispersions shown in Figure 2 were obtained by aging acidified aluminum salt solutions. Again the anions play an essential role. The perfect spheres (Figure 2a) were generated on heating aluminum sulfate solutions (2×10^{-3} M) for 48 hr at 97°C (initial pH 4.0). The precipitate contained a considerable amount of sulfate ions, but these anions were readily leached out by rinsing the solids with water yielding pure amorphous aluminum hydroxide particles without change in shape (4). Heating at the same temperature for 20 hr a solution which was 0.050 M each in $Al(NO_3)_3$ and Na_2HPO_4 and 0.035 M in HNO_3 (initial pH of the mixture being 2.0) gave also spherical particles (Figure 2b) but the X-ray and chemical analysis identified the solids to be consistent with the composition of the mineral variscite, $AlPO_4$. Scanning electron micrographs (2c and d) show two unusual morphologies of boehmite, obtained by aging of aluminum chloride (0.0050 M $AlCl_3$) and aluminum perchlorate (0.0030 M $Al(ClO_4)_3$) solutions, respectively, at 125°C for 12 hr (5).

Again, it is clearly demonstrated that rather different, yet quite uniform colloidal metal oxide dispersions, can be prepared by homogeneous precipitation of different salt solutions in which different anions have a profound effect on the properties of the solid formed.

As a last example, Figure 3 gives a transmission electron micrograph and a replica of Co_3O_4 particles which have a spinel structure. The significant finding is that such systems are formed on heating at 100°C of cobalt(II) salt solutions in oxygen or air *only* in the presence of acetate ions. No precipitate was obtained when, under otherwise identical conditions, the sulfate, chloride, perchlorate, or nitrate salts of cobalt(II) were aged.

Many more monodispersed sols of different hydrous metal oxides, involving those of chromium (6,7) copper (8), titanium (9), and other metals, have now been prepared following similar procedures.

MECHANISM OF METAL HYDROUS OXIDE FORMATION

The examples in the preceding section clearly show that it is necessary to know the composition of the solution in terms of the nature and of the concentration of all complexes if one is to explain the chemical processes in homogeneous precipitation of metal hydrous oxides. Unfortunately such information is not always available, particularly for solutions at higher temperatures. Thus, one has no choice but to determine the composition of the different solutes under the actual conditions of the solid phase formation.

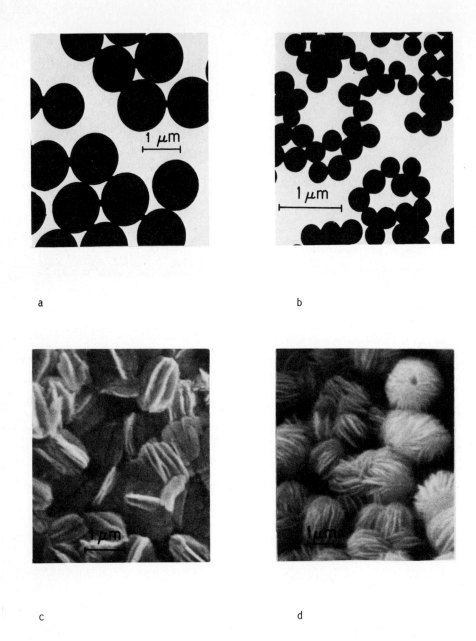

Fig. 2. Transmission and scanning electron micrographs of particles precipitated from various aluminum salt solutions. (a) Amorphous aluminum hydrous oxide (solution 0.0020 M in $Al_2(SO_4)_3$, pH 4.0 aged for 48 hr at 97°C). (b) Variscite, $AlPO_4$ (solution 0.050 M in $Al(NO_3)_3$ and Na_2HPO_4 and 0.035 M in HNO_3 aged for 20 hr at 97°C). (c) and (d) Boehmite, α-AlOOH (solutions of 0.0050 M in $AlCl_3$ and 0.0030 M $Al(ClO_4)_3$, respectively, aged for 12 hr at 125°C).

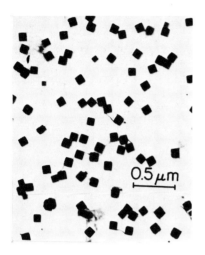

a b

Fig. 3. Electron micrograph and a replica of Co_3O_4 spinel particles obtained by heating a 0.010 M Co(II)-acetate solution for 4 hr at 100 C.

Until recently no substantiated mechanisms on metal hydrous oxide formation have been available (10). Using the monodispersed sols it was possible to develop a better understanding of the essential processes in the generation of chromium hydroxide (11,12), titanium dioxide (9), and ferric basic sulfates (13). In all these cases the particles formed in the presence of sulfate ions, but the role of these anions was quite different in each system.

Thus, in solutions containing chromium and sulfate ions solid chromium basic sulfate precursor precipitates first, which acts as a heterogeneous nucleating material for the amorphous chromium hydroxide (11). An application of the Nielsen's chronomal analysis showed that the particles grow via surface reaction which involves polynuclear layer growth (12).

In the highly acidic solutions of titanium salts, sulfate ions bind the titanium(IV) ions into solute complexes, which on prolonged aging at elevated temperatures slowly decompose. These species act as a reservoir for the titanium ions; on decomplexation they hydrolyze and precipitate.

Finally, in solutions containing ferric and sulfate ions at low pH, well defined monomeric and dimeric solute hydroxy and sulfato complexes of iron(III) form, giving crystals of fixed stoichiometric composition (13).

The role of chloride ions in the precipitation of different ferric hydrous oxides has not been explained yet, but the work on the chloro-ferric complexes in acidic solutions, which is presently carried out in our laboratory, may shed light on the findings that a relatively small variation in conditions not only yield particles of different morphologies, but also of entirely different chemical compositions.

CHARACTERIZATION OF MONODISPERSED METAL HYDROUS OXIDE SOLS

Depending on application, the characterization of colloidal particles may take many forms. In this review only two properties will be discussed: surface charge and magnetism.

Surface Charge

Owing to the nature of the stabilizing species (OH^-, H^+) the surface charge of any metal hydrous oxide is strongly pH dependent. Furthermore, the electrokinetic point of zero charge (isoelectric point, i.e.p.) varies not only with the constituent metal, but also with the composition of its hydrous oxides. As a matter of fact the reported values for the i.e.p. for apparently the same materials differed by several units in pH (14). For example, the literature data for the i.e.p. of synthetic titanium dioxide range from pH 2.6 to 7.3, whereas for a very pure sample of TiO_2 it was found to be between 4.5 and 5.2 (15). In another example, the cubic Co_3O_4 particles illustrated in Figure 3 show an i.e.p. at pH 5.4 (16); a considerably higher value of 11.4 was reported for a Co(II,III) oxide by Tewari and Campbell (17).

Since the isoelectric point defines the pH below which the particles are positively charged and above which they are negatively charged, this property of a metal hydrous oxide is essential in considering the sol stability, particle adhesion, heterocoagulation with other particles, and interactions with different solutes.

Figure 4 gives the electrokinetic mobility data of four aluminum hydroxide sols as a function of pH. Three of these curves refer to systems illustrated in Figure 2. The interesting finding is that the largest difference in the mobility curves is for two amorphous spherical particles prepared by aging aluminum sulfate solutions. The lower values of the i.e.p. is for the sol, as directly generated (diamonds), whereas the higher value is for the same sol from which the sulfate ions were removed by repeated rinsing with water (triangles). This example shows that anionic "impurities" have a pronounced effect on the surface charge characteristics of metal hydrous oxides. Obviously, the discrepancies in many reported data may be due to similar causes.

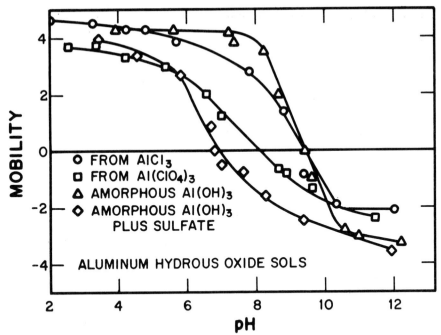

Fig. 4. Electrokinetic mobilities (μm/sec/V/cm) as a function of pH of four different colloidal aluminum hydrous oxides as illustrated in Figure 1c (O), 1d (□), 1a (◇), and the same particles as shown in 1a from which sulfate ions were removed by leaching (△).

Magnetism

It is well known that magnetic properties of different (hydrous) oxides of a given metal strongly depend on their composition. Much less well understood is the relationship between various types of magnetism and the shape of the particles of the same chemical composition or of the materials having the same chemical composition and shape but different particle size. The monodispersed metal hydrous oxides lend themselves exceedingly well for the investigation of magnetic properties as a function of various parameters.

Figure 5 shows the magnetization as a function of applied magnetic field for nearly spherical α-Fe_2O_3 particles (30-40 nm modal diameter at 298°K). It was suggested that hematite particles of this size should be in the antiferromagnetic state (18). Thus, the σ vs H plot should exhibit some hysteresis. Within the experimental error no hysteresis is seen and the curve is typically S-shaped, characteristic of pure superparamagnetism (possibly in coexistence with weak ferromagnetism). This observation would suggest that the proposed relationship of the magnetic states to particle size (18) for Fe_2O_3 systems may not be correct.

Heating the same hematite particles at high temperatures (960°K) in the helium atmosphere of 500 torr drastically alters the magnetic properties of α-Fe_2O_3

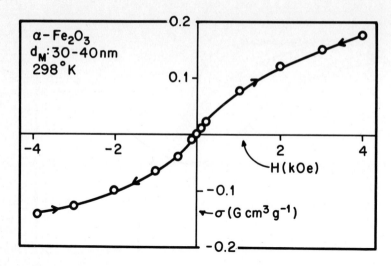

Fig. 5. Magnetization at 298°K as a function of applied field of spherical α-Fe$_2$O$_3$ particles having modal diameters of 30-40 nm.

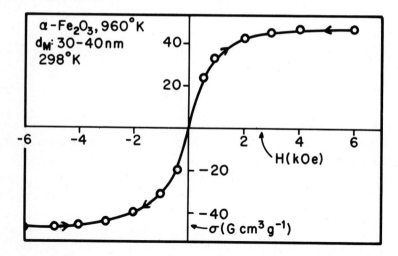

Fig. 6. Magnetization at 298°K as a function of applied magnetic field of spherical α-Fe$_2$O$_3$ particles having modal diameters of 30-40 nm after heat treatment at 960°K in a He atmosphere of 500 torr.

(Figure 6), although little change could be observed with respect to particle size and shape. The high σ values indicate the presence of ferromagnetic components, presumably Fe$_3$O$_4$ or γ-Fe$_2$O$_3$, in these particles. Obviously, the crystallinity of the solids must have been affected by the heat treatment.

INTERACTIONS WITH SOLUTES

It is well known that the surfaces of hydrous metal oxides are rather reactive, i.e. various molecular or ionic species adsorb on such solids from aqueous solutions. The interfacial processes depend on the pH, since the latter is the factor controlling the surface charge. It is important to note that the opposite charge between the adsorbent and the adsorbate is not a sufficient condition for adsorption. The electrostatic attraction will only facilitate the approach of the solute species to the particle surface; other forces are needed to keep them at the interface. These forces may be hydrogen bond or other types of bindings (such as chelation), surface precipitation etc. It is the ability of the constituent metal ions and of the potential determining species to interact with a variety of solutes that make metal hydrous oxide surfaces so reactive.

Two examples to be given here deal with the interaction of monodispersed,

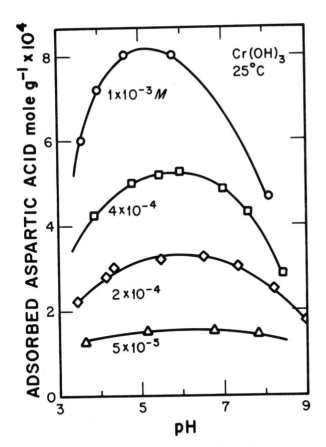

Fig. 7. The total amount of aspartic acid sorbed on spherical chromium hydroxide particles (modal diameter 350 nm) as a function of pH using different initial concentrations of aqueous solutions of the amino acid.

spherical, amorphous chromium hydroxide particles with aspartic acid and of crystalline rod-like β-FeOOH particles with ethylenediamine tetraacetic acid (EDTA).

Figure 7 gives the total amount of aspartic acid sorbed on a chromium hydroxide sol having particles of a modal diameter of 350 nm as a function of pH using different initial concentrations of the solute. Analogous measurements showed that at 90°C the adsorbed amounts are considerably higher (19). The pH effect is obviously due to the charge on the solid. The i.e.p. of the used sol is at pH ∿ 8.5 and little uptake of aspartic acid was observed above this pH value. It is evident that opposite charges of the adsorbent and adsorbate are needed to bring about the interaction.

The significant finding is that the amounts of the amino acid taken up by the metal hydroxide are approximately four orders of magnitude higher than the quantities calculated on the basis of monolayer adsorption assuming geometric surface areas of the spherical adsorbent particles. The obvious conclusion is that the major fraction of the aspartic acid molecules is *absorbed* in the interior of

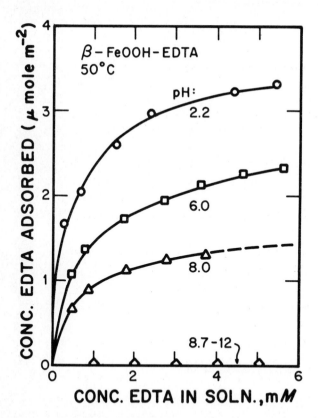

Fig. 8. Adsorption isotherms at 50°C for EDTA on β-FeOOH consisting of rod-like particles with an average length of 1 μm and a width of 0.3 μm. Each curve is for a different pH value.

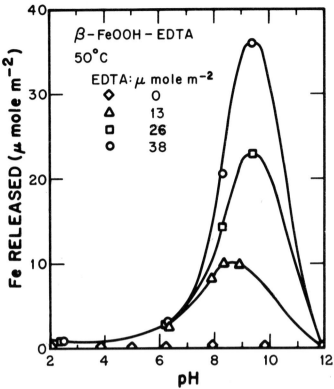

Fig. 9. Solubility of β-FeOOH (expressed as released iron) in the absence and in the presence of EDTA in different concentrations as a function of pH at 50°C.

the adsorbent, which presumes that the amorphous chromium hydroxide is permeable to this amino acid. The absorption seems to be due either to chelation or to coordination of chromium ions with the aspartic acid molecules. The much stronger uptake of these solute species at higher temperature supports the suggested chemisorption mechanism.

Contrary to the findings with aspartic acid, the same chromium hydroxide sol did not adsorb any measurable quantities of tryptophan. The latter is a larger molecule with a stronger aromatic (hydrophobic) character; its monocarboxylate nature (as distinguished from the dicarboxylic aspartic acid) precludes chelate formation by the carboxyl groups only.

The amount of EDTA adsorbed on β-FeOOH is also pH dependent and the uptake decreases strongly with increasing pH (Figure 8). The area per Fe atom in β-FeOOH, calculated from the crystal structure of this solid, is 27 $Å^2$, and the area per ligand molecule at the condition of maximum adsorption at pH 2.8 (at 25°C), as calculated from the corresponding adsorption isotherm and the BET specific surface area of the ferricoxyhydroxide, is 33 $Å^2$. The latter value is in reasonable agreement with the estimated cross-sectional area of the EDTA molecule, which indicates that at saturation a monomolecular layer is formed based on a Fe:EDTA interfacial complex of 1:1. The surface complexation is further supported by the value of the free energy of adsorption, which was found to be $\Delta G°_{ads}$ = -9 kcal/mole (20).

It was of special interest to investigate if the interaction of EDTA with the same β-FeOOH was accompanied by dissolution of some of the oxyhydroxide. Figure 9 is a plot of the amount of ferric ions released as a function of pH in the absence of EDTA and in the presence of three different concentrations of this chelating agent. A very pronounced maximum is observed. At lower pH values, at which considerable adsorption takes place, the release of ferric ions is rather small; the dissolution is inhibited by the surface complexation of lattice ferric ions with the organic ligand.

The maximum dissolution of ferric ions is over the pH range 8 - 10. Under these conditions no adsorption takes place. Partial dissolution of particles yields positively charged ferric hydrolysis products ($Fe(OH)_2^+$, $FeOH^{2+}$, $Fe_2(OH)_2^{4+}$) which react with EDTA anions and, thus, enhance further release of ferric ions from the solid β-FeOOH.

Finally, at the highest pH values (> 12), the complete inhibition of the dissolution can be understood, if one considers that the reaction

$$FeOOH + EDTA^{4-} + H_2O \rightleftharpoons FeEDTA^- + 3\ OH^-$$

is strongly shifted to the left.

The two examples offered clearly illustrate the important role of specific interactions between a solid and a solute resulting in entirely different behavior of different systems.

INTERACTIONS OF METAL HYDROUS OXIDES WITH OTHER PARTICULATE MATTER
Heterocoagulation

Most of the colloid stability studies have been carried out with single systems, preferably monodispersed sols, such as polymer latexes. Yet, by far a majority of the naturally occurring dispersions, or those used in various applications, consists of mixed type particles, which may vary in composition, size, shape, and other properties. It is, therefore, of fundamental interest to study heterocoagulation phenomena, which implies stability of systems containing dissimilar suspended solids.

In order to apply the theoretical analysis it is necessary to work with spherical particles of known size, potential, and specific attraction characteristics. In addition, the stability of such mixed systems depends on the ratio of the particle number concentrations as well as on the ionic composition of the suspending media.

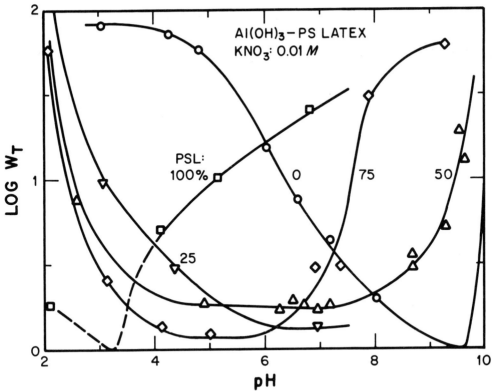

Fig. 10. Total stability ratio, W_T, as a function of pH for a pure aluminum hydroxide sol (modal diameter 570 nm) (○), a pure polystyrene latex (PSL, modal diameter 380 nm) (□), and for mixtures of the two sols containing 25% (▽), 50% (△) and 75% (◇) PSL particles (in terms of number concentration).

Monodispersed spherical metal hydrous oxide sols are particularly suitable for the study of interactions between unlike particles, because the necessary parameters can be experimentally determined. A comprehensive investigation was carried out with a binary system consisting of a polymer (polyvinyl chloride, PVC) latex stabilized with sulfate ions and spherical chromium hydroxide particles (21). The advantage of such a combination is that changing pH affects little the surface potential of the latex, whereas the metal hydroxide particles not only undergo a change in this quantity, but the sign of the charge can be reversed. The obtained data showed that excellent qualitative agreement existed between the experimental results and the theoretical calculations for dispersions of these two colloids.

To illustrate the effects of interactions in a mixed system, the stability ratios of sols consisting of spherical aluminum hydroxide particles (4) having a modal diameter of 570 nm and of a polystyrene latex (PSL, modal diameter 380 nm) stabilized by sulfate ions will be given. The rate of coagulation was followed by means of laser light scattering at a low angle ($\theta = 5°$). Figure 10 is a plot

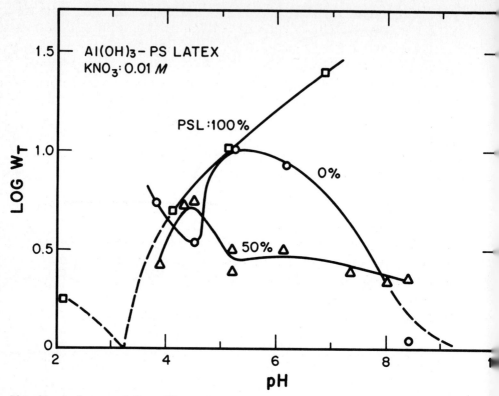

Fig. 11. Analogous plot as Figure 10 except that the aluminum hydroxide particles contained some sulfate ions.

of the total stability ratio, W_T, defined as

$$1/W_T = (n_1^2/W_{11}) + (n_2^2/W_{22}) + (2n_1n_1/W_{12}) \qquad (1)$$

where W_{11} and W_{22} are the homocoagulation stability ratios for aluminum hydroxide and the latex, respectively, and W_{12} is the heterocoagulation stability ratio. n_1 and n_2 are the primary particle number fractions of the two dissimilar particles.

As seen in Figure 10, the latex is least stable at low pH ($\sim$ 3) whereas pure aluminum hydroxide is unstable at high pH ($\sim$ 9.5). These pH values are close to the isoelectric points of the two colloidal systems. Depending on the particle number ratio and pH the binary systems may be either less or more stable than individual sols. For example, at pH 6 all mixed dispersions are less stable than the single systems. This is understood if one considers that the particles of PSL and of aluminum hydroxide carry opposite charges. At pH 8 the system containing 25% aluminum hydroxide particles is more stable than the pure aluminum hydroxide sol. In the studied case, aluminum hydroxide sol was carefully washed to eliminate all sulfate ions which are present in the particles as they are prepared (4).

Figure 11 gives an anologous plot to the previous one except that the aluminum hydroxide particles still contained some sulfate ions. The difference in the behavior of the purified and unpurified aluminum hydroxide sol is quite dramatic. In addition, the reproducibility of the results in the latter case is rather poor. This study exemplifies the sensitivity of such systems to anionic contaminations which often tend to be disregarded.

Particle Adsorption

Mixed systems which contain particles greatly divergent in size cannot be analyzed in the same manner as those having particles of comparable size. In the former case, the more finely dispersed systems may coagulate selectively, a heterofloc may form, or the small particles may adsorb on the larger ones, causing a change in the properties of the latter. All of these phenomena were observed in a binary system containing negatively charged polyvinyl chloride (PVC) latex (particle diameter 1020 nm) and silica (diameter $\sim$ 14 nm). It was shown that under certain conditions silica adsorbs on latex enhancing its stability toward electrolytes (22).

The adsorption of small particles on larger ones is of particular interest, as

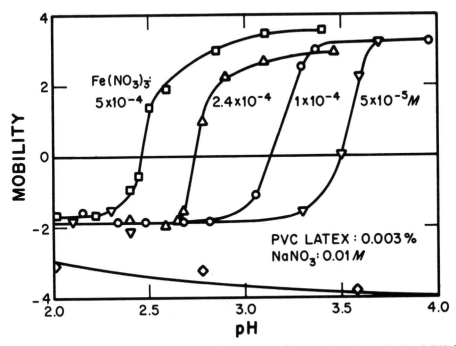

Fig. 12. Electrokinetic mobilities (μm/sec/V/cm) as a function of pH of PVC latex (0.003% by wt, modal particle diameter 320 nm) in the absence ($\Diamond$) and in the presence of Fe(NO$_3$)$_3$: 5×10^{-5} ($\triangledown$), 1×10^{-4} (O), 2.4×10^{-4} ($\triangle$), and 5×10^{-4} M ($\square$).

other properties - in addition to stability - can be altered, such as the surface charge, reactivity, wettability, adhesivity, pigment characteristics, etc.

Figure 12 shows the mobility of a (PVC) latex as a function of pH and of the same latex in the presence of four different concentrations of $Fe(NO_3)_3$. In all cases the addition of the ferric salt had a strong effect on the particle charge causing the charge reversal from negative to positive at appropriate pH values. Independent measurements showed that the sharp change in the electrokinetic mobility was associated with the precipitation of ferric hydroxide. Consequently, the modification of the latex surface was due to the deposition of the metal hydrous oxide on the polymer particles. Needless to say the so coated latex showed different stability, adhesion characteristics, etc. than the untreated material.

It is expected that reactants which affect the precipitation of ferric hydroxide would also influence its interaction with the latex. This is clearly evident in Figure 13, which gives the mobility data of the same latex in the presence of $Fe(NO_3)_3$ to which NaF was added in different concentrations. Fluoride ions are known to complex with the ferric ion and, as a result, the precipitation phenomena as well as the surface charge groups of ferric hydroxide are altered by these

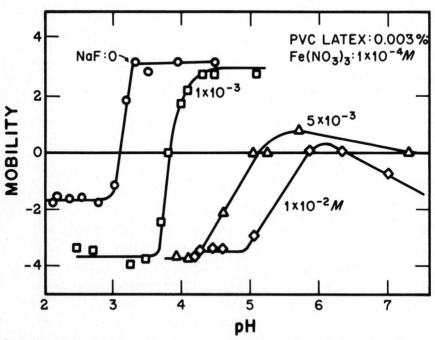

Fig. 13. Electrokinetic mobilities as a function of pH of the same PVC latex as in Figure 12 in the presence of 1×10^{-4} M $Fe(NO_3)_3$ (O) to which 1×10^{-3} (□), 5×10^{-3} (Δ), and 1×10^{-2} M (◊) NaF was added, respectively.

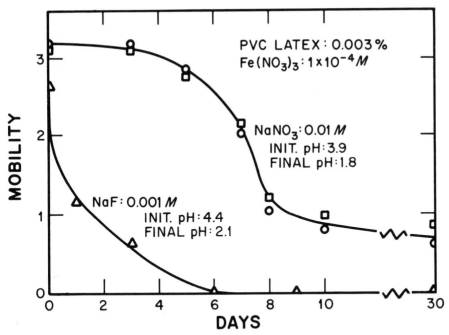

Fig. 14. The change of electrokinetic mobilities (μm/sec/V/cm) as a function of time (days) of the same PVC latex as in Figure 12 on acidification of a sol containing 1×10^{-4} M $Fe(NO_3)_3$. Circles and squares represent systems to which HCl and HNO_3, respectively, were added to lower the pH in the presence of 0.01 M $NaNO_3$; triangles represent systems the pH of which was lowered by HNO_3 in the presence of 0.0010 M NaF prior to aging.

anions. Indeed, with increasing fluoride concentration the charge reversal occurs at higher pH values and the magnitude of the positive charge decreases.

Acidification of the sols containing particles with adsorbed metal oxides should bring about dissolution of the coating and, consequently, a change in the surface charge to less positive, or even negative. Figure 14 shows that in the presence of HNO_3 and HCl very long times are needed to dissolve, at least in part, the metal oxide coating. However, even after one month the charge is not reversed back to negative; obviously, the metal hydrous oxide must be polymerized at the surface to which it adheres tightly. Addition of only 0.0010 M NaF greatly accelerates the removal of the oxide layer and the particles eventually become uncharged. The latter observation implies that the negatively charged potential determining groups of the original latex surface are neutralized by the metal counterion complexes. Apparently, the bonds formed between the stabilizing sulfate ions and the adsorbed ferric species are not broken by acidification even in the presence of F^-.

PARTICLE ADHESION

The deposition of colloidal particles on other solids and their removal from these substrates depend on the chemical and physical forces acting between the adhering surfaces. In the absence of chemical bonds the adhesion can be treated as heterocoagulation of dissimilar particles, taking the radius of one to go to infinity. In order to compare the theoretical predictions with experimental data it is necessary to have sufficiently well defined systems with known parameters. Again, monodispersed metal hydrous oxides, particularly those of spherical particles, can serve as excellent models for particle adhesion and removal studies. Using the packed column procedure (23), the interactions of such sols with glass and steel have been studied as a function of various conditions (pH, different electrolytes, temperature, etc.).

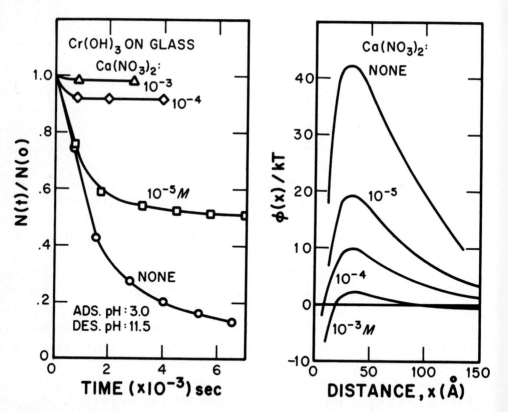

Fig. 15. Left: Fraction of monodispersed spherical chromium hydroxide particles (modal diameter 280 nm) desorbed from glass on repeated elution with rinse solution of pH 11.5 in the absence (O) and in the presence of 10^{-5} (□), 10^{-4} (◇), and 10^{-3} M $Ca(NO_3)_2$. The particles were adsorbed on the glass from a sol at pH 3. Right: Calculated potential energy curves as a function of distance using the sphere-plate model (Eqs. 2 and 4) for the same systems shown left.

Figure 15 illustrates the results obtained with spherical chromium hydroxide particles (diameter 280 nm) on glass (24). At pH 3 glass beads placed in a column rapidly and quantitatively remove these metal hydroxide particles from an aqueous suspension by adhesion. Under these conditions the sol is positively charged and stable whereas the glass beads are negatively charged. Precautions are taken that no filtration takes place in the course of the deposition process. The subsequent removal of particles depends on the composition of the rinsing solution. At pH 11.5, at which both solids are negatively charged, rapid desorption of the particles is observed (Figure 15 left). However, if the rinse solution contains $Ca(NO_3)_2$, desorption can be reduced or completely inhibited depending on the concentration of the electrolyte. The higher the charge of the added cation, the lower is the concentration needed to prevent particle removal (24).

The results can be interpreted in terms of the total interaction energies. The attractive energy, ϕ_A, as a function of distance (x) was calculated using the equation:

$$\phi_A(x) = -\frac{A_{132}}{6}\left[\frac{2a(x+a)}{x(x+2a)} - \ln\frac{x+2a}{x}\right]. \tag{2}$$

A_{132} is the overall Hamaker constant for the system sphere-medium-plate, which can be approximated by:

$$A_{132} \cong (\sqrt{A_{11}} - \sqrt{A_{33}})(\sqrt{A_{22}} - \sqrt{A_{33}}), \tag{3}$$

where subscript 1 applies to chromium hydroxide, 2 to glass, and 3 to water. The value A_{11} for chromium hydroxide was taken as 14.9 kT, A_{22} for glass as 20.9 kT, and A_{33} for water as 10.3 kT.

For double layer repulsive energy, ϕ_R, the plate/sphere expression of Hogg, Healy and Fuerstenau (25) was taken:

$$\phi_R(x) = \frac{\varepsilon a}{4}\left[(\psi_1 + \psi_2)^2 \ln(1 + e^{-\kappa x}) + (\psi_1 - \psi_2)^2 \ln(1 - e^{-\kappa x})\right] \tag{4}$$

in which ψ_1 and ψ_2 are the surface potentials (equated with the corresponding ζ-potentials), for the plate and the particles, respectively, ε is the dielectric constant, and κ the reciprocal Debye-Hückel thickness.

The right side of Figure 15 gives the calculated total interaction energy curves, $\phi(x) = \phi_A(x) + \phi_R(x)$, as a function of distance of separation for the chromium hydroxide/glass systems in the absence and in the presence of the same concentrations of $Ca(NO_3)_2$ as used in the desorption experiments. At the highest salt concentration essentially only attraction prevails and indeed no particle removal is observed. When no calcium nitrate is added, a number of particles

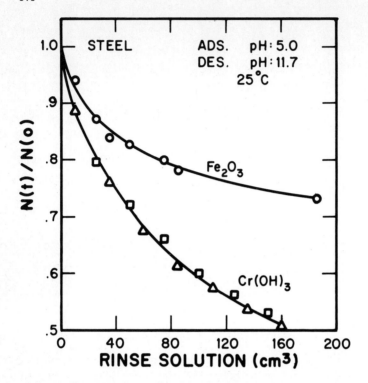

Fig. 16. Fraction of monodispersed spherical chromium hydroxide particles (modal diameter 280 nm) desorbed from steel on rinsing with an aqueous solution of pH 11.7. Number of particles originally deposited on 3 g of steel at pH 5.0 was 2.9×10^9 (□) and 1.8×10^{10} (△). Circles give the analogous data for desorption of spherical hematite particles (modal diameter 140 nm) from the same steel. The number of originally deposited hematite particles: 3.0×10^9.

appears to be at sufficient distance which enables them to overcome the energy barrier and desorb; however, particles which escaped cannot readsorb due to the high potential barrier. Thus, the double layer theory, albeit in its oversimplified form, explains at least semiquantitatively the adhesion phenomena in the described system.

Similar observations were made with chromium hydroxide on steel (Figure 16). The reproducibility of data is best shown by triangles and squares which represent two separate runs made with beds of steel beads on which the number of originally adsorbed particles differed by nearly one order of magnitude.

In the same diagram is also included the desorption curve of spherical hematite particles from the same steel under otherwise identical conditions. The large difference in the desorption rate is primarily due to the variation in the attractive forces, as determined by the Hamaker **constants**. This illustrates the sensitivity of the adhesion phenomena to the physical properties of the adsorbent/adsorbate system.

CONCLUDING REMARKS

It was not the intention of this review to give an in depth analysis of each problem discussed. Readers are referred to individually cited reports for more detailed information of the experimental techniques, the results obtained, and the theoretical analyses. However, it is hoped that the different cases described in this presentation clearly show the usefulness of the monodispersed metal hydrous oxide systems in the studies of various interfacial phenomena.

ACKNOWLEDGMENTS

This article is based on work done with the support of the NSF grants CHE77 02185 and ENG 75-08403 as well as EPRI contract RP-966-1.

The author is greatly indebted to his many collaborators, and specifically to R. Brace, B. Gray, E. Katsanis, J. Kolakowski, H. Kumanomido, R. Kuo, J. Rubio, R. Sapieszko, H. Sasaki, P. Scheiner, W. Scott, T. Sugimoto, and R. Wilhelmy, on whose contributions was based this presentation.

REFERENCES
1. E. Matijević, Progr. Colloid Polymer Sci., 61(1976)24.
2. E. Matijević and P. Scheiner, J. Colloid Interface Sci., 63(1978)509.
3. E. Matijević, R. Sapieszko and J.B. Melville, J. Colloid Interface Sci., 50 (1975)567.
4. R. Brace and E. Matijević, J. Inorg. Nucl. Chem., 35(1973)3691.
5. W.B. Scott and E. Matijević, J. Colloid Interface Sci. (in press).
6. R. Demchak and E. Matijević, J. Colloid Interface Sci., 31(1969)257.
7. E. Matijević, A.D. Lindsay, S. Kratohvil, M.E. Jones, R.I. Larson and N.W. Cayey, J. Colloid Interface Sci., 36(1971)273.
8. P. McFadyen and E. Matijević, J. Inorg. Nucl. Chem., 35(1973)1883.
9. E. Matijević, M. Budnik and L. Meites, J. Colloid Interface Sci., 61(1977)302.
10. K.H. Lieser, Angew Chem., Int. Ed., 8(1969)188.
11. A. Bell and E. Matijević, J. Inorg. Nucl. Chem., 37(1975)907.
12. A. Bell and E. Matijević, J. Phys. Chem. 78(1974)2621.
13. R.S. Sapieszko, R.C. Patel, and E. Matijević, J. Phys. Chem., 81(1977)1061.
14. G.A. Parks, Chem. Rev., 65(1965)177.
15. M. Visca and E. Matijević, J. Colloid Interface Sci. (in press).
16. T. Sugimoto and E. Matijević, J. Inorg. Nucl. Chem. (in press).
17. P.H. Tewari and A.B. Campbell, J. Colloid Interface Sci., 55(1976)531.
18. Yu.F. Krupyanskii and I.P. Suzdalev, Sov. Phys.-JETP, 38(1974)859.
19. H. Kumanomido, R.C. Patel and E. Matijević, J. Colloid Interface Sci. (in press).
20. J. Rubio and E. Matijević, J. Colloid Interface Sci. (in press).
21. A. Bleier and E. Matijević, J. Colloid Interface Sci., 55(1976)510.
22. A. Bleier and E. Matijević, J. Chem. Soc., Faraday Trans. I, 74(1978)1346.
23. E.J. Clayfield and E.C. Lumb, Disc. Faraday Soc., 42(1966)285.
24. J. Kolakowski and E. Matijević, J. Chem. Soc., Faraday Trans. I (in press).
25. R. Hogg, T.W. Healy and D.W. Fuerstenau, Trans. Faraday Soc., 62(1966)1638.

DISCUSSION

R. POISSON : Thône Poulenc has patented processes for controlling the size of zeolite, for example, between the micron and 10 microns in concentrated medium.

My questions :
1) You have been talking about polymer sulfated cations. How large they are ? What kind of polycondensation they have (Linear ?).
In general, how do you see the polycondensation of cations (geometrically speaking). What is the location of the counter ions (first or second hydration shell ?), their degree of freedom compared to a normal salt.
2) when the polycondensation is conducted at sufficiently high concentration, a slight neutralisation (homogeneous !) permits the transition from the sol to the gel state. Is this compulsory salt accompanied by a neutralisation of the ions to form an insoluble salt or is this fact independent of the sol or gel state; in other words, can we have polycations in the gel state ?)

E. MATIJEVIC : 1) There is no information available on the distribution of molecular weights or conformation of polynuclear complexes of chromium containing sulfate ions. Hall and Eyring (J. Am. Chem. Soc. 72, 782, 1950) have shown that chromium hydroxylated species undergo continuous polymerization on heating. Thus, it is not possible to define a given polymer in solution of such salts aged at elevated temperatures. In dilute solutions there should be no considerably greater restriction in the motion of such polymeric ions and their counterions than one would expect in electrolyte solutions of highly charged ionic species.
The chemical mechanism involved in the formation of spherical chromium hydroxyde particles has been discussed in greater detail by my associate, A. Bell (J. Inorg. Nucl. Chem. 37, 907, 1975; J. Phys. Chem. 78, 2621, 1974).
2) None of the work presented here has yielded gels. Gel formation has been studied in great detail, among other by Sing (Brunel University) or Teichner (Lyon). However, it is indeed possible to have polycations in the gel state.
Re comment :
My presentation dealt only with metal (hydrous) oxides. Except for some earlier work by Heller, who prepared monodispersed β-FeOOH sols, little progress has been made in the production of monodispersed metal (hydrous) oxides of well defined morphologies. In this respect our work is unique.
Monodispersed dispersions of various other materials such as organic latexes, elements (sulfur, selenium, gold, etc.), different inorganic salts, silica, tungstic acid and others have been frequently reported

in the literature. Since zeolites cannot be properly described as <u>metal</u> oxides this family of compounds has not been considered in my lecture.

B. DELMON : There are many questions that scientists interested in catalysis could ask you in relation to possible applications of your preparation techniques of catalytic materials, well defined both in structure and in texture; these questions could especially concern the possibility of making mixed oxides (compound oxides with a given stoichiometry, or solid solutions) or sulphides. But my question is related to the solid state chemistry aspect of the preparation of powders with uniform size. I agree that the control of nucleation, and, more precisely, of the number of nuclei, is crucial. But concerning the second step involved, the growth of nuclei, one could remark that there is always a factor working against uniform size. It is the lower stability and growth rate of smaller nuclei, as a consequence of the changes of the free energy of the surface with crystallite size. Larger nuclei have a **tendency** to grow at the expense of the smaller ones, thus progressively increasing the dispersion in size. The only way I know to suppress **this tendency and to restore the equality in growth rate is to introduce diffusion limitations in the supply of material with which nuclei** are fed. It has been observed, that this kind of limitation plays a crucial role in the growth of X-zeolite crystallites. In the latter case, the diffusion limitations can be made to work by dispersing the growing nuclei in a gel. In your case, the crystallites grow in a liquid. To what factor do you attribute the uniformity in size ?

E. MATIJEVIC : The procedure described apparently yields nuclei of uniform size and, consequently, the Ostwald Ripening (the growth of larger nuclei at the expense of the smaller ones) does not take place.
The most important condition for the uniform growth of particles is the proper rate of production of solute species on which nuclei feed. In the systems illustrated in this work the rate is controlled by the concentration of the metal salt, pH, temperature, and in some cases, by the concentration of specific anions. Thus, as long as the rate of formation of the constituent species which are built onto the particles (rate of crystal growth), is sufficiently high, secondary nucleation will not take place. This assures a reasonably uniform particle size. Thus, only one burst of nuclei occurs,

which then grow uniformly.

Finally, during the entire growth process particles are charged which prevents their aggregation.

O.P. KRIVORUCHKO : What is the main reason for formation of amorphous or crystal sols of metal hydrous oxides ?

What is the reason for the influence of metal concentration and the anion nature ? According to the considerations developed by us, the intermediates at the sol particles formation are the polynuclear **hydroxocomplexes of the corresponding metal.** The direction of polycondensation of metal aquo-ions and the nature of complexes drastically depend on the total metal concentration in the solution as well as on the nature of the initial salt anion. Theses circumstances determine the most important properties of sols and gels of hydroxides. Are the sols formed continuous or through some limited set of species ?

It was shown experimentally that for our systems the sol particle formation proceeded through a limited set of polymeric hydroxocomplexes. So for definite conditions, the particles of aluminum hydrous-oxide sol were formed through consecutive stages of formation of complexes, which contained 2 and 13 Al (III) ions.

E. MATIJEVIC : It depends on the system and on the conditions whether the particle formation is a result of continuous polymerization or crystal growth of well defined species. For example, in the case of the precipitation of basic ferric sulfates in acidic solutions, only well defined monomeric and dimeric complexes are involved in particle nucleation and growth. On the other hand, polymeric hydroxo (and substituted hydroxo) complexes are precursors to chromium hydroxide sol formation. Thus, it is not possible to offer a generalized scheme for the preparation of the uniform hydrous oxides. However, if continuous polymerization takes place, the resulting particles will most likely be amorphous. Well defined solute complexes of low degree of polymerization will yield crystalline materials. The role of the anions will also differ from case to case. In some systems the anions (such as sulfate or phosphate) may promote polymerization, as in the case with chromium hydroxide. In other cases well defined coordinated metal-anion species (with or without hydroxyl ions) may form (e.g. ferric sulfate solutes). Obviously in each case the resulting materials will differ for reasons given above.

O.P. KRIVORUCHKO : What do you think about the mechanism of transformation of amorphous aggregates of primary particles into the crystals ? In the mechanism of formation and growth of the metal hydroxide crystals which you discuss, it is supposed that primary amorphous particles are dissolved. We showed that the transition of hardly solved hydroxides from the amorphous to the **crystalline** phase is the result of the crystallization in the bulk of the primary particles rather than dissolution. After the achievement of the determined degree of crystallinity, formation and growth of the secondary crystals by means of oriented growth of one type of primary particles to another one at the same facet become possible.

E. MATIJEVIC : I never stated that amorphous particles must dissolve in order for crystals to form. On the contrary, we showed in the case of spherical amorphous aluminium hydroxide particles, that removal of sulfate ions and subsequent heating of the sol, resulted in direct crystallization of the particles into boehmite (Brace and Matijevic, J. Inorg. Nucl. Chem. $\underline{35}$, 3691, 1973). Obviously, removing the sulfate ions was essential to crystal formation.

V. FENELONOV : Your lecture is very interesting for catalyst preparation but the sizes of the particles which you have shown to us on the slides are very large. If I try to use your sols for support preparation, I would get a support which surface area would be only a few square meter per gram. The sols with particle size of about 30-40 Å are more useful and important for catalyst preparation. Therefore, I am interested to know whether it should be possible to use your conclusions for the preparation and the examination of small particle systems ? Are you sure that there is no difference in the genesis of small and large sized particles ?

E. MATIJEVIC : We have tried to grow larger particles because they find various applications, such as pigments or corrosion product models. Furthermore, particles of spherical shape can be analyzed for size distribution in situ by light scattering, as long as they behave as Mie scatterers.
In some instances we prepared much smaller particles, such as spherical hematite sols. It is quite conceivable that some of the

dispersions could be obtained in particles sizes as small as a few nanometers, whereas in some other cases this may not be possible. One important parameter in this respect would be the initial concentration of metal salt solution. Futhermore, heterogeneous nucleation could be employed to control the particle size.

K.S.W. SING : 1) Can you remove sulphate ions from the spherical particles of amorphous alumina without changing the particle shape ? It is possible to remove sulphate from the surface layer (eq. by ion exchange or restructuring the surface) without changing the bulk composition ? 2) Does the electrophoretic mobility curve for pure aluminium hydroxide (or oxide-hydroxide) depend on the chemical or physical structure of the surface ?

E. MATIJEVIC : 1. It is quite possible to remove all of sulfate ions from spherical colloidal amorphous aluminium hydroxide without changing the particle shape (Brace and Matijevic, J. Inorg. Nucl. Chem. $\underline{35}$, 3691 (1973). It is conceivable that incomplete washing would remove the surface sulfate groups, while leaving some of these anions in the bulk of the particle. However, it is quite likely that a redistribution of the sulfate ions throughout the particle would take place with time.
2. Different physical structures of pure aluminium hydroxide could have some effect on the **electrokinetic** properties of these systems. However, I believe that the average values of mobilities are not sufficiently sensitive to demonstrate such differences. The electrokinetic properties are determined by the charged surface groups; in the case of aluminium hydroxide these are $-AlO^-$ and $-AlOH_2^+$. Even highly charged colloidal particles carry relatively few charges (e.g. one charge per 500-1000 Å^2). This low density of charge groups would make the effects of structure differences difficult to detect. We have shown, for example, that pure aluminium hydroxide spherical particles, and latex particles coated with a monolayer of aluminium hydroxide showed essentially identical electrokinetic behaviour as a function of pH.

L. RIEKERT : What is the space-time yield for the preparation of monodisperse alumina ? In other words how many kg can be made per hour and m^3 of reaction volume ?

E. MATIJEVIC : The yield on monodispersed sols varies from case to case. Some systems have been prepared only in very small quantities in dilute dispersion. Specifically in the case of monodispersed alumina, our procedure was somewhat modified at Unilever Research Laboratories at Port Sunlight, England (Dr. D. Barby), so that kilogram quantities could be prepared in concentrated slurries over short periods of time (hours).

M.V. TWIGG : An important difference between systems containing chromium(III) and aluminium(JII), iron(III), etc. is that ligand substitution reactions of chromium(III) are relatively very slow. Is this kinetic inertness reflected in the times needed for the preparation of your products ?

E. MATIJEVIC : It is not possible to relate times necessary to generate monodispersed hydrous oxides of different metals to their liquid substitution reactions. A more important parameter seems to be the temperature; the higher the aging temperature, the shorter is the reaction time. However, one cannot expect that monodispersed sols will form at any temperature. To obtain uniform particles, the particle growth must be faster than particle nucleation (after the initial burst of nuclei). If above a given temperature secondary nucleation takes place, the sols will become polydisperse.

PROCESS FOR THE PRODUCTION OF SPHERICAL CATALYST SUPPORTS

R.M. CAHEN; J.M. ANDRE; H.R. DEBUS*
Labofina S.A., chaussée de Vilvorde 98-100, B-1120 Bruxelles

SUMMARY

Spherical alumina catalyst supports are produced by this process in which a mixture of an aluminum hydroxide hydrosol and organic monomers is introduced dropwise, through calibrated orifices, into a column of heated oil. In the course of this procedure, each drop forms a sphere; the monomers contained therein polymerise and give sufficient rigidity to the spheres so that they may be handled with ease during the ensuing steps. Excess water and the organic polymer are eliminated during the final calcination stage leaving very hard alumina spheres, the purity of which lies above 99 %. The properties of the starting mixture and of the obtained spheres are described. Examples of catalysts made from these supports are given and their desulfurization activity is compared to that of commercial catalysts.

INTRODUCTION

The shapes and sizes of the catalysts usually loaded in fixed bed reactors are :
- pellets or cylinders (diameter 3-7 mm; length 3-7 mm)
- extrudates (diameter 1.5-5 mm; length 2-8 mm)
- spheres (diameter 1.5-5 mm).

Pellets are produced by tabletting; their price is rather high as each cylinder is formed individually in a set of punches and dies and because tool wear is important. The particles produced in this way have excellent mechanical properties.

The major part of fixed bed catalysts are in the form of extrudates. Their manufacturing costs are low due to the great capacity of the extruders. The main disadvantage of these catalysts is their moderate mechanical strength. Neither pellets nor extrudates can be used in moving beds.

Alumina spheres, which are mainly used in catalytic reforming as carriers for noble metals such as platinum have the double advantage of high mechanical strength

---

*This work was carried out under the direction of the late Dr. Henri R. Debus; the present paper is dedicated to his memory.

(little losses of precious metals due to attrition and breakage) and ideal reactor packing (good contact between catalyst and reagents).

Spheronizing processes may be divided into methods which rely upon the build-up of smaller particles into spheres by means of a rolling, or "snow-ball" technique, and those which form spheres by individually shaping particles. The production costs are low for the first technique, however the spheres are irregular and their particle size is randomly distributed. The spheres manufactured by individual shaping are more expensive, however their shape and calibration are very regular. In general, catalysts of this type are produced by so called "oil-drop" i.e. by suspending drops of an aqueous gel-forming liquid in a water immiscible liquid such as mineral oil so as to form spherical drops. The coagulation of the gel is brought about by a change of pH. The spheres are then aged for a sufficient time so that the drops can be withdrawn, washed, dried and calcined.

DESCRIPTION OF THE PROCESS

The process developed by Labofina and patented in a number of countries belongs to the "oil-drop" group; its originality lies in the incorporation of an organic monomer into an aluminum hydrosol which is then introduced dropwise, through calibrated orifices, into a column of heated oil. Each drop of oil-insoluble mixture forms a sphere. Under the influence of the heated oil, the polymerisation of the monomer is initiated. Gelation by modification of the pH is not necessary as the spheres maintain their form through the effect of the polymer. The resulting spheres can be directly unloaded, washed, dried and calcined without further ageing. The organic matter is completely removed during the calcination step. A simplified flow-sheet is shown in figure 1.

Figure 1 shows that the reaction mixture is constituted of an inorganic part (alumina precursors) and an organic part (monomer, polymerisation initiators and reticulation agents).

It is beyond the scope of this paper to give technological details such as the nature of the oil, diameter and length of the calibrated orifices, length of the oil column which are key factors, but which may vary from case to case and are adapted to the individual circumstances. It is however important to give more details concerning the components of the reaction mixture, which constitutes the essential feature of the process.

As the alumina spheres will be used as catalyst supports, a number of conditions must be respected, so that the final product will have the desired properties of purity, structure and porous texture.

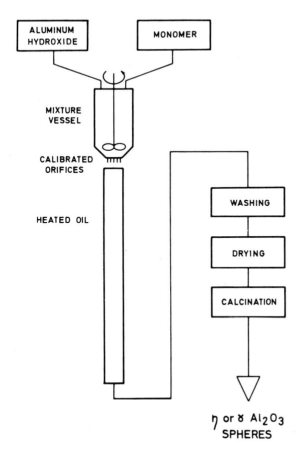

Fig. 1 Simplified flow-sheet

The inorganic compounds

The aluminas most commonly used as catalysts and catalyst supports are of the η or γ form. The preparation and characterization of "active aluminas" has been extensively described by Lippens and Steggerda (ref. 1). Such aluminas may be obtained by low temperature calcination (300-600°C) of gelatinous boehmite(pseudo-boehmite). This is the starting material used in the process described here.

The first difficulty encountered was the necessity to incorporate a gel into a reaction mixture which must remain sufficiently fluid so as to pass the calibrated orifices and take the form of droplets. A suitably fluid mixture is obtained when an alkaline monomer is used; in this case the pH is adjusted to a sufficiently high value by means of ammonia. The alumina spheres made in alkaline medium have very interesting properties; however acid type monomers give a product which is by far superior.

The introduction of an acid monomer into the reaction mixture consisting of a fluid suspension of pseudoboehmite in water brings about its immediate gelification into a stiff unusable paste.

This difficulty has been overcome by the addition of highly acid products to the reaction mixture so as to form a hydrous alumina sol. The choice of the acid constituents is however critical. Suitable hydrosols are produced by dissolving aluminum metal in aluminum trichloride solution, or by adding aluminum salt solutions, nitric acid or nitric and perchloric acid mixtures to the pseudoboehmite gel.

The second problem is to avoid inorganic constituents that remain in the alumina after calcination and that may impair the activity of the finished catalyst. Impurities may also inhibit the polymerisation of the monomer so that the spheres leaving the oil column will collapse. For these reasons, polymerisation initiators must be carefully chosen (this will be the subject of the next chapter) and, at this stage of the process, it is not always possible to introduce any type of catalytically active compound at any concentration.

It should however be stressed that fluid reaction mixtures containing variable concentrations of catalyst precursors have been made and have successfully been "oil-dropped"; among the compounds that have been added to the pseudoboehmite the following can be mentioned : salts of nickel, molybdenum, chromium, iron, cobalt, zinc, silicium, titanium, germanium, lanthanum, platinum, palladium, rhodium, gold; this list is not exhaustive. Each addition to the initial reaction mixture is a particular case and the technology must be adapted.

Catalyst precursors can more easily be introduced in the subsequent steps of the process; as the spheres are very firm and cohesive in the wet stage after leaving the hot oil column, they can be readily handled, washed, impregnated, transported to dryers and ovens.

The organic components and polymerisation initiators

The process is based on the incorporation of an easily polymerisable organic monomer to the aluminum hydrosol; this implies that the monomer as well as the polymer must be water soluble. Good results have been obtained with the principal monomers that comply with this requirement. For economical reasons it is preferable to use the cheapest monomers available because the polymer is finally burnt off during the calcination step. Furthermore, no solid residue must be left by the polymer after its decomposition. These reasons have led to the choice of acrylic acid and acrylamide; more sophisticated monomers, such as N-hydroxymethylacrylamide and N-hydroxymethylmethacrylamide for example, have been used to confirm the results.

The polymerisation of the monomer within the alumina hydrosol droplet is not sufficient to make a rigid sphere at the outlet of the oil column. Two factors, one natural, the other artificial, contribute to bring about the required rigidity. The first one is due to the tendency of the aluminum hydroxide to bond with the polymer chain; under certain conditions, very small amounts of monomer, as low as 0.5 % wt. based on Al_2O_3, are required to create the tridimensional network responsible for the good cohesion of the spheres. This can also be brought about by adding a certain amount of reticulation agents to the monomer. An amount of 5 % (based on monomer) of dihydroxyethylene-glycol-bis-acrylamide is sufficient to reticulate the straight-chain polymers. This product - made from acrylamide and ethyleneglycol - can under certain conditions be made in situ by adding both components directly to the acrylic acid monomer.

Finally, in order to have an efficiently working system, it is necessary that all of the monomer contained in each droplet of reacting mixture polymerises during the span of time it drops down the column of oil. The usual polymerisation initiators for these types of monomers are oxidation-reduction couples. As already stated above, care must be taken in chosing the components that are added to the alumina, so as to finish with a very pure product. As the quantity of polymerisation initiators are small, for certain alumina grades it is possible to use couples that leave some residue on calcination, such as persulfates-bisulfites. If the purity requirements for the alumina spheres are very stringent, oxidation-reduction couples such as hydrogen peroxide-hydroquinone or hydrazine are preferred. For certain particular reaction mixtures, the well known inhibiting effect of oxygen on polymerisation has been applied to monitor the rate of this reaction to a certain degree.

PROPERTIES OF THE ALUMINA SPHERES

The flow-sheet represented in fig. 1 shows that drying and calcination steps are required. The wet spheres contain large amounts of water and some polymer which must be eliminated. The pseudoboehmite must furthermore be transformed to "active alumina", η or γ Al_2O_3. During this final phase of the process, the spheres shrink in size and gain hardness which is brought about by the development of the Al_2O_3 structure. The range of physical properties of typical alumina spheres produced by this process are given in table 1.

It is possible to widen somewhat the range of the properties reported in table 1; however this will have an influence on several properties simultaneously. Thus for example, the lowering of apparent bulk density inevitably will bring about some loss of mechanical strength.

The alumina spheres described in table 1 are comparable and very often superior to commercially available ones. They are quite equivalent by their purity, size, density, surface area and pore volume; they are superior by their crushing strength and low attrition. Typical oil-dropped commercial alumina spheres have crushing strengths of 3-4 kgf and attritions around 0.5 % wt.; compared to extrudates and to the more irregular alumina spheres made by rolling techniques, the superiority of the spheres described here is even more striking.

TABLE 1

Physical properties of the alumina spheres

Properties	Range	Typical (a)
Alumina type		η
Purity, % wt. Al_2O_3	99-99.9	99.9
Diameter, mm	1-3	1.8
Apparent bulk density, g/ml	0.6-0.8	0.705
Crushing strength		
individual, kgf	7-10	8.3
bulk, kg/cm^2	20- >40	>40
Attrition losses, % wt.	<1	0.03
Specific surface area, m^2/g	200-300	224
Pore volume, ml/g	0.4-0.7	0.55

(a) Sample produced from commercial type pseudoboehmite and acrylic acid in aqueous solution of nitric and perchloric acids; hydrogen peroxide/hydroquinone were used as polymerisation activators.

TESTING OF CATALYSTS MADE FROM ALUMINA SPHERES

In the preceding chapter the alumina spheres were shown to have the physical properties required for a good catalyst support. The spheres were therefore impregnated with cobalt and molybdenum salts; the catalyst obtained was used to desulfurize gasoil and was compared in identical tests to the most recent commercial hydrodesulfurization catalysts.

Figure 2 represents the results of these tests which were carried out on a 60 ml catalyst reactor at 350°C, 40 kg/cm^2 pressure, hydrogen to feed ratio of 150 l (NTP)/l and liquid hourly space velocities (LHSV) comprised between 2 and 6; LHSV is defined as volume of feed per volume catalyst and hour. The gasoil feed had a boiling range of 200-350°C and a sulfur content of 1.3 % wt. It can be seen that the activity of the Labofina sample compares well with that of commercial catalysts, as both commercial samples can be rated among the best ones available.

The physical properties of the catalyst support are not so important for naphtha and gasoil desulfurization as compared to sulfur removal from petroleum residues; in this case pore-size distribution plays a major role (ref. 2). The efficiency of the spherical alumina support has also been tested for this application. A cobalt-molybdenum oxides catalyst has been prepared by impregnation of a spherical alumina support with the metal salts. Catalyst properties are reported in table 2.

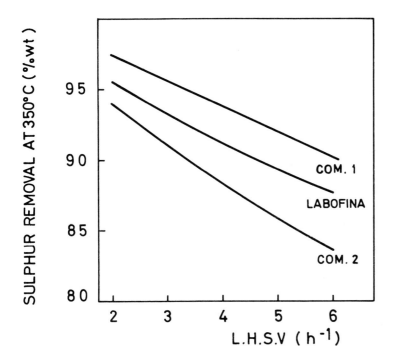

Fig. 2   Gasoil hydrodesulfurization

TABLE 2

Properties of residue desulfurization catalyst

MoO_3, % wt.	15.2
CoO, % wt.	2.8
SiO_2, % wt.	1.4
Al_2O_3	type h
Surface area, m2/g	200
Apparent bulk density, g/ml	0.768
Total pore volume, ml/g	0.56
Volume of pores smaller than 100 Å diameter, ml/g	0.42

This catalyst has been included in a testing project aimed at the selection of suitable commercial catalysts for residue desulfurization. Each test lasted two weeks and covered activity as well as selectivity with respect to metals removal, gasoil production, lowering of the viscosity and specific gravity of the 350°C + product. The feedstock used was Kuwait atmospheric residue (350°C + fraction) containing 4.15 % sulfur, 59 ppm vanadium and 17.5 ppm nickel.

The catalysts (300 ml) were first presulfided and then tested at the following conditions :

Temperature : 400°C
Pressure : 100 kg/cm2
H_2/feed : 1000 l (NTP)/l
LHSV : variable between 0.3 and 1.5

To compare the performances of the different samples, a given catalyst has been arbitrarily chosen as reference and its relative volumetric activity (RVA) set at 100; RVA is defined by the following relationship :

$$RVA = 100 \times \left[\frac{LHSV\ unknown}{LHSV\ reference} \right] \text{ at 1 \% sulfur remaining in the 350°C + product.}$$

Table 3 summarizes the results obtained. High values for RVA mean better activity. The metals removal activity is given for constant sulfur conversion. High values mean short catalyst life expectancy. The sulfur removal activity of the catalyst made from the spherical alumina support is equivalent to that of the best commercial samples tested; its metal removal selectivity needs however to be improved.

TABLE 3

Desulfurization and metals removal of Kuwait atmospheric residue

Catalyst	RVA	% metals removed at 75 % sulfur removal	
		Nickel	Vanadium
Commercial 1 (Reference)	100	48	58
Commercial 2	142	58	72
Commercial 3	99	41	59
Commercial 4	171	39	50
Commercial 5	125	52	64
Commercial 6	60	59	69
Commercial 7	215	66	77
Commercial 8	149	65	78
Labofina 10/864	175	52	59

CONCLUSIONS

The production of spherical alumina catalyst supports according to the new technique described in this paper has a number of definite technological and economical advantages. The range of finished products can be quite large. The alumina spheres have physical properties that meet the very stringent specifications required for the production of catalysts. Performance tests of hydrodesulfurization catalyst samples made from these supports showed that they ranked among the best commercial ones.

ACKNOWLEDGMENT

This work was subsidized by the "Institut pour l'Encouragement de la Recherche Scientifique dans l'Industrie et l'Agriculture (I.R.S.I.A.)".

REFERENCES
1. B.C. Lippens and J.J. Steggerda, in B.G. Linsen (Ed.), Physical and Chemical Aspects of Adsorbents and Catalysts, Academic Press, London and New York 1970, ch. 4.
2. R.L. Richardson and S.K. Alley, in J.W. Ward (Ed.) and S.A. Qader (Ed.), Hydrocracking and Hydrotreating, ACS Symposium Series 20, American Chemical Society, Washington DC, 1975, ch. 9.

DISCUSSION

R. POISSON : Why do you say η or γ Al_2O_3 ? I thought that everybody agreed that boehmite and crystalline boehmite (pseudoboehmite) conduct to the γ phase and not to η alumina ?

R.M. CAHEN : The authors cited in reference (1) of our paper indicate that pseudo-boehmite"... decomposes at about 300°C into an alumina for which a doubling of the 1.98 Å and 1.4 Å diffraction no longer observed"... Many authors refer to this form as η-alumina. However, as the 4.6 Å band is extremely broad and does not show any sign of a sharp peak, we prefer to call it a γ-alumina..." In accordance with the above authors' definition of γ-alumina, we can state that this is the product we obtain; as however, the same form is called η-alumina by other, we gave both designations in our paper.

E.R. BECKER : Does the porosity or pore size distribution vary as a function of particle radius in your supports prepared by the methods you described ?
R.M. CAHEN : Pore-size distribution has only been measured on average

samples of the alumina spheres; we do not therefore know if there is a variation in function of particle radius.

P.R. COURTY : Can you give information on the method you have used for the determination of the attrition properties of your alumina carrier (0.03% attrition). A SOCONY test measurement would be helpful for this type of carrier, because of its great severity.

R.M. CAHEN : Attrition is measured by tumbling a weighed sample of spheres in a tube which turns around an axis normal to its length for a given amount of time. The quantity of powder collected is weighed and attrition is reported as percentage powder found per initial weight. A more severe test method, as suggested by Courty, would be helpful in this case, in view of the strength of the particles.

V. FATTORE : Can you control your process in order to obtain particles with lower apparent bulk density, for instance 0.4-0.5 g/cc ? Going up with the dimensions of the particles can you still produce a product with almost spherical shape ?

R.M. CAHEN : 1. Spherical particles of low apparent bulk density have been produced, however at the cost of some mechanical strength. 2. We have not tried to increase particle diameter above 2 mm; the spherical shape of the 2 mm particles has been maintained.

METHODS OF SATURATION WITH ALKALI IONS. INFLUENCE OF THE PROPERTIES OF OXIDES

R. HOMBEK, J. KIJENSKI and S. MALINOWSKI
Institute of Organic Chemistry and Technology
Technical University (Politechnika), Warsaw (Poland)

INTRODUCTION

Many industrial catalysts, especially for dehydrogenation and cracking processes, are modified by the introduction of alkali metal ions. The problem of the influence of alkali ions on the properties of oxide catalysts has been investigated by us (ref. 1-5) since several years. Numerous papers concerning this subject have been published by other authors (ref. 6-15). Alumina is the most thoroughly investigated catalyst from this point of view. Parera and Figoli (ref. 6) and Scharme (ref. 7) reported that the addition of sodium to alumina involved a decrease of surface acidity. Chuang and cowork. (ref. 8) and Bremer and cowork.(ref. 9) have found using IR spectroscopy that sodium ions react with the most acidic surface OH groups only. According to Pines (ref. 10) it is possible that sodium hydroxide undergoes a reaction with Lewis acid centres :

$$- Al + NaOH \rightarrow - Al-OH^- Na^+$$

On the other hand our recent results (ref. 2) and those obtained by other authors (ref. 12) support the supposition that after impregnation with small amounts of alkali hydroxide the acidity of the surface increases, so the studied process is much more complicated than one might assume. The common method of introducing alkali ions onto oxide surfaces lies in the saturation of the solid oxide with an alkali hydroxide aqueous solution.

In the present work the influence of various alkali ions, introduced from ethanolic solutions of corresponding alkoxides, on the physicochemical properties of alumina was investigated. Our new impregnation procedure permits avoiding the reaction of water with dehydrated alumina surface. Water being a source of considerable changes in the properties of catalysts prepared by the hydroxide addition method. We have taken into consideration the higher reactivity of the alkoxides than that of the respective hydroxides, assuming that the reaction with the Al_2O_3 surface will be more effective and cover a greater number of surface sites.

EXPERIMENTAL

Alumina has been obtained by precipitation of $Al(OH)_3$ with water from a benzene solution $Al(iso-C_3H_7)_3$. The hydroxide was then washed with distilled water and dried at 120°C for 24 hr. The resulting preparation was calcined at 550 or 750°C for 24 hr in a stream of dry and oxygen-free nitrogen. Lithium, sodium, potassium and caesium ethoxides were obtained by dissolving the corresponding alkali metals in absolute alcohol at room temperature.

Alkali cations were introduced onto the Al_2O_3 surface during adsorption of alkoxides from absolute alcohol solution. Impregnation procedure was as follows : alumina calcined at 550 or 750°C was cooled down to room temperature and was soaked in a known volume of corresponding alkali metal alkoxide. After 20 hr the catalyst was filtered-off, washed with absolute alcohol and then calcined at 100, 150, 200, 220, 250, 350 and 550°C in a stream of dry and de-oxidized nitrogen for 3 hr. The number of alkali cations deposited on the Al_2O_3 surface was determined by titration of the filtered-off alkoxide excess with a hydrochloric acid solution in the presence of phenolphtalein. The titer of the alkoxide used for the impregnation was determined by the same method.

Surface basicity and acidity of the catalysts were determined by the titration method with benzoic acid or n-butylamine using a series of Hammett indicators (ref. 14,15). The specific surfaces areas of the catalysts were measured by the BET method.

Hydroxyl group concentrations were determined by titration with sodium naphthenide (ref. 16) and by chromatography using the reaction with $Zn(CH_3)_2 \cdot 2THF$ (ref. 17).

One-electron donor and one-electron acceptor properties were determined on the basis of adsorption on the catalyst surface of appropriate acceptor and donor and by recording the signals of newly-formed anion- or cation-radicals using ESR spectroscopy (ref. 18,19). Perylene was used as the electron donor and tetracyanoethylene was used as the electron acceptor. Quantitative results were obtained by comparing the signal intensity of the investigated sample with that of the standard - DPPH solution in NaCl. Adsorption of perylene was carried out in the presence of molecular oxygen. ESR measurements were performed on an X-band spectrometer (modulation 100 kHz) at room temperature.

RESULTS

Basic and acidic properties

The values of maximum basic strength (H_{-max}) of catalysts under study are given in Table 1. For comparison the values of basic strength of sodium hydroxide doped alumina were investigated in the same conditions.

TABLE 1.
Basic and acidic strength (H_{-max} and H_{omax}) of alkali alkoxides doped alumina

Alkali cation	Na^+ [a]		Li^+		Na^+		K^+		Cs^+	
Calcination temperature (°C)	H_{-max}	H_{omax}	H_{-max}	H_{omax}	H_{-max}	H_{omax}	H_{-max}	H_{omax}	H_{-max}	H_{omax}
20	18.4		17.2		18.4		18.4		18.4	
100	18.4	∞	17.2	∞	18.4	∞	22.3	∞	22.3	∞
150	18.4	<-4	18.4	<-4	22.3	<-4	26.5	<-4	26.5	<-4
200	18.4	<-4	18.4	<-4	26.5	<-4	26.5	2.8	26.5	2.8
220	22.3		22.3		26.5	2.8	26.4	-3.0	26.5	-3.0
250	22.3	-3.0	22.3	-3.0	26.5	-5.6	26.5	-5.6	26.5	-5.6
350	24.6	-3.0	24.6	-5.6	26.5	-5.6	26.5	-5.6	26.5	-5.6
550	24.6	-3.0	26.5	-5.6	26.5	-5.6	26.5	-5.6	26.5	-5.6

[a] Na^+ = from NaOH

The basic strength of catalysts doped with alkoxides and sodium hydroxide increases with the rise of calcination temperature after alumina impregnation. H_{-max} of preparations containing alkali alkoxide is in all cases smaller or equal to H_{-} of pure alumina. Caesium ethoxide doped alumina reaches the basic strength $26.5 < H_{-} < 33$ after calcination at 150°C, and Al_2O_3 doped with lithium ethoxide reaches the same value after treatment at 550°C.

None of the catalysts reaches the acid strength corresponding to pure alumina ($H_{omax} < -8.2$). Calcination below 200°C resulted in the lack of acidic properties for each examined catalyst. Treatment at higher temperatures caused an increase of the H_{omax} value to -5.6.

Since the highest H_{-max} and H_{omax} values of the investigated catalysts were observed for catalysts calcined at 550°C the quantitative measurements of basic and acidic centres at various strengths were made for those catalysts only (Table 2). Generally catalysts obtained from alumina pretreated at 750°C have a smaller concentration of basic centres than those with Al_2O_3 precalcined at 550°C. The highest concentration of basic sites is achieved on the alumina pretreated at 550°C and then doped with potassium ethoxide. On the surface of this catalyst two types of basic centres (at strengths of $26.5 < H < 33$ and $12.2 < H_{-} < 15.0$, respectively) were found. The smallest concentration of basic centres was found on alumina precalcined at 750°C and doped with lithium ethoxide. All catalysts in this series (precalcination Al_2O_3 temperature 750°C) have one type of basic centre only at $26.5 < H_{-} < 33$. Catalysts of both series have stronger basic properties than pure alumina. Alumina precalcined at 550°C and doped with potassium ethoxide has the greatest concentration of acidic sites. In the case of all the catalysts under study higher concentrations of acidic

598

TABLE 2

Basic and acidic sites concentrations on investigated catalysts surfaces (calcination after impregnation at 550°C).

Catalyst	Basic centres concentration at various strength (H_-) (mmol/g)								Acidic centres concentration at various strength (H_o)(mmol/g)				
	12.2	15.0	17.2	18.4	22.3	24.6	26.5	33.0	+4.8	+1.5	−3.0	−5.6	−8.2
Al_2O_3 (550)[a]	1.17	0.99	0.99	0.99	0.91	0.91	0.91	0	2.73	1.92	0.96	0.63	0.56
Al_2O_3 (750)	1.02	1.02	1.02	1.02	1.02	0.93	0.81	0	2.79	2.79	0.81	0.40	0.40
Al_2O_3 (550) − Li	1.60	1.60	1.60	1.31	1.31	1.00	1.00	0	2.77	2.11	2.11	0.23	0
Al_2O_3 (750) − Li	0.93	0.93	0.93	0.93	0.93	0.93	0.93	0	0.59	0.42	0.21	0.09	0
Al_2O_3 (550) − Na	1.82	1.82	1.82	1.82	1.19	1.19	1.19	0	7.64	7.64	7.64	6.47	0
Al_2O_3 (750) − Na	1.42	1.42	1.42	1.42	1.42	1.42	1.42	0	2.89	2.89	2.89	2.02	0
Al_2O_3 (550) − K	2.22	2.03	2.03	2.03	2.03	2.03	2.03	0	7.00	7.00	7.00	6.65	0
Al_2O_3 (760) − K	1.69	1.69	1.69	1.69	1.69	1.69	1.69	0	3.49	3.49	2.99	2.52	0
Al_2O_3 (550) − Cs	2.42	2.22	2.22	2.02	1.81	1.81	1.81	0	4.50	4.50	4.19	3.27	0
Al_2O_3 (750) − Cs	1.86	1.86	1.86	1.86	1.86	1.86	1.86	0	3.97	3.97	3.09	3.09	0

[a] Al_2O_3 (550) = pretreatment temperature of alumina 550°C

centres were observed for systems obtained from alumina calcined at lower temperatures, i.e. at 550°C.

Specific surface area

Specific surface areas of alumina doped with lithium, sodium, potassium and caesium ethoxides are presented in Table 3. Introduction of alkali alkoxides on Al_2O_3 surface resulted in an increase of specific surface area (about 10%), the differences between preparations doped with various ethoxides are insignificant.

Hydroxyl group concentration

Surface hydroxyl group concentrations found for the investigated catalysts are given in Table 3. Introduction of alkali ethoxides on alumina surface causes a decrease of the amount of OH groups. Only alumina doped with caesium alkoxide has a concentration of surface hydroxyl groups similar to that of the pure alumina.

Alkali ion content

The quantity of alkali cations deposited on alumina surface was measured by titration with HCl (Table 3). In the series of catalysts investigated the number of lithium ions deposited during the reaction of the alkoxide with Al_2O_3 was the greatest among alkali cations. The decrease of hydroxyl group concentration, as compared with pure alumina, was the greatest in the case of the lithium ions doped alumina. The number of Cs^+ ions deposited on alumina surface was the smallest in the series and simultaneously after reaction of caesium alkoxide the minimal change in OH groups concentration was observed.

All alumina preparations have a γ-Al_2O_3 structure, which has been checked by X-ray analysis. Assuming that 100 faces dominate on the surface, the ratio of the number of alkali cations per 100 $\overset{\circ}{A}^2$ of catalyst to the number of surface oxygen anions on the same area of surface was calculated.

One-electron donor and one-electron acceptor properties

The concentrations of one-electron donor centres on the surfaces increase with an increase of the calcination temperature after impregnation, reaching a maximum in the case of preparations calcined at 250°C. Calcination at higher temperatures causes a rapid fall of the one-electron donor centre concentration. The greatest number of centres of this type occurs on the surface of Al_2O_3 doped with caesium alkoxide (Fig. 1).

Catalysts which have not been calcined after impregnation do not exhibit any one-electron acceptor properties at all. Among the preparations calcined at 550°C only the catalyst doped with lithium ions had one-electron acceptor properties comparable to pure alumina. Minute concentrations of one-electron

TABLE 3

Specific surface area, alkali ion content and hydroxyl group concentration on investigated systems surfaces (calcination after impregnation at 550°C)

Catalyst	Specific surface area (m^2/g)	OH group, concentration (mmol/g) a	b	Alk. ion content (mmol/g)	c	d
Al_2O_3 (550)[e]	169	0.29	0.30	0		
Al_2O_3 (750)	107	0.15	0.17	0		
Al_2O_3 (550) - Li	188	0.19	0.17	1.41	5.02	2.49
Al_2O_3 (750) - Li	117	0.04	0.031	1.17		
Al_2O_3 (550) - Na	193	0.21	0.20	1.07	3.81	3.28
Al_2O_3 (750) - Na	114	0.055	0.047	0.76		
Al_2O_3 (550) - K	194	0.26	0.23	0.79	2.81	4.45
Al_2O_3 (750) - K	116	0.09	0.078	0.64		
Al_2O_3 (550) - Cs	196	0.28	0.25	0.58	2.07	6.04
Al_2O_3 (750) - Cs	120	0.15	0.14	0.43		

a = alkali naphthenide titration
b = chromatographic method
c = alkali ion number per 100$Å^2$ of catalyst surface
d = oxygen ions number to alkali ions number ratio on 100$Å^2$ of catalyst surface

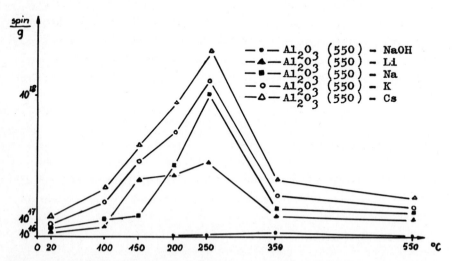

Fig. 1. One-electron donor sites concentration as a function of the calcination temperature after alumina impregnation

acceptor centres occurred on the surfaces of all the other catalysts (Table 4).

TABLE 4

One-electron donor and one-electron acceptor centres concentration on investigated catalysts surfaces (calcination after impregnation at 550°C)

Catalyst	One-electron donor centres concentration (spin/g)	One-electron acceptor centres concentration (spin/g)
Al_2O_3 (550)	1.53×10^{17}	2.1×10^{17}
Al_2O_3 (750)	1.48×10^{17}	4.4×10^{16}
Al_2O_3 (550) - Li	1.31×10^{17}	1.1×10^{17}
Al_2O_3 (750) - Li	1.61×10^{17}	2.2×10^{16}
Al_2O_3 (550) - Na	1.74×10^{17}	1.2×10^{15}
Al_2O_3 (750) - Na	3.32×10^{17}	4.5×10^{15}
Al_2O_3 (550) - K	1.97×10^{17}	5.2×10^{14}
Al_2O_3 (750) - K	3.28×10^{17}	2.7×10^{15}
Al_2O_3 (550) - Cs	2.79×10^{17}	$<5 \times 10^{14}$
Al_2O_3 (750) - Cs	3.20×10^{17}	1.7×10^{15}

DISCUSSION

The addition of alkali metal ethoxides to alumina causes considerable changes in the physicochemical properties of the oxide. The type of apparent change depends on the pretreatment temperature of alumina and on the calcination temperature of the catalyst after impregnation. It is quite clear that the properties of the catalytic systems obtained depend mainly on the pathway of the reaction between alkoxide and surface active sites on Al_2O_3. At room temperature it is possible that alkali metal ethoxides undergo reactions with surface hydroxyl groups and with Lewis type acidic centres, presumably with the naked surface aluminum cations. As a result of the reaction with hydroxyl groups the new bond between alkali metal and surface oxygen would be formed:

$$\equiv Al-OH + RO^-M^+ \rightarrow -Al-O^-M^+ + ROH \qquad (1)$$

In the case of reaction with Lewis sites the equation may be envisaged as follows:

$$\underset{Al}{O\diagdown}\underset{}{\diagup O\diagdown}\underset{Al}{\diagup O} + RO^-M^+ \rightarrow \underset{Al}{O\diagdown}\overset{OR \quad M}{\underset{|}{\diagup O\diagdown}}\underset{Al}{\diagup O} \qquad (2)$$

Reaction (1) causes a decrease of the surface OH group concentration. As a result of reaction (2) the blocking of most acidic surface centres (Lewis type centres with strength $H_o^- < -8.2$) occurs.

The basic strength of catalysts not calcined after the introduction of alkoxide is much smaller than for pure alumina. This fact seems to be a result of suppressing the surface oxygen anions O^{2-} with alkali cations (Eq. 2). The surface alkoxide anions may act as electron pair donors and are probably responsible for the basicity in the range of $H_- = 17.2 - 18.2$. According to McEwen (ref. 20) the ethoxide anions in solution have the same basic strength.

The relatively low concentration of one-electron donor sites on the surface of catalytic systems not calcined after impregnation may be a result of the destruction of part of the surface OH groups and the suppression of surface oxygen anions. It is well known that both types of active sites can act as one-electron donors. In this same reaction pathway the one-electron acceptor sites (aluminum cations) (ref. 21) are blocked by alkoxide anions.

The calcination of the impregnated catalysts causes considerable changes in the properties and structure of the surfaces.

$$\begin{array}{c} M \\ | \\ O \end{array} \begin{array}{c} C_2H_5 \\ | \\ O \end{array} \qquad \begin{array}{c} M \\ | \\ O \end{array} \begin{array}{c} H \\ | \\ O \end{array}$$
$$\text{O}{-}\text{Al}{-}\text{O}{-}\text{Al}{-}\text{O} \;\rightarrow\; \text{O}{-}\text{Al}{-}\text{O}{-}\text{Al}{-}\text{O} \;+\; C_2H_4 \qquad (3)$$

As a result of reaction (3) the formation of new hydroxyl groups and new naked oxygen anions may take place. The two surface sites formed are responsible for the increase of the concentration and strength of the basic sites. An increase in the concentration of basic and one-electron donor centres is probably due to the formation of new naked oxygen anions after the evolution of ethylene. Secondary hydroxyl groups formed in reaction (3) may also act as new one-electron donor centres.

Differences in the properties of alumina doped with alkoxides of different alkali metals point to the role played by the properties of the cation introduced (electron affinity and ionic radius). The electron affinity and ionic radius of the cation in the respective alkali metal alkoxide determine the basicity and the reactivity of the anion. It seems likely that the reaction of the alkoxide molecule with the surface hydroxyl group of alumina begins with the attack of the anion at the acidic proton of the hydroxyl group as in analogous homophase reactions. The second step is the addition of the cation to the oxygen anion O^{2-} released. Such a mechanism is supported by the observed differences in the reaction of alkoxides and sodium hydroxide with the Al_2O_3 surface. Obviously, the structure of the anion is important so far as it determines the type of product formed during the calcination of the impregnated material.

Larger cations such as potassium and caesium possess greater or comparable radii in relation to oxygen anions. It cannot be ruled out that their coordination number is higher than that of Li^+ or Na^+. The presumption is supported by the ratio of the alkali metal ions to the oxygen anions per 100 Å^2 of alumina surface (Table 3). One must presume that the ratios close to the value of the coordination number in the sequence of the cations under discussion may be an explanation of the fact that various quantities of different cations are retained on the same alumina surface under unaltered conditions. Moreover the ionic radius determine the hardness of the cation. The small lithium cation is much harder than the other alkali metal cations. As the oxygen anion is one of the hardest anions, its affinity for a cation will grow with an increase of cation hardness and should be highest for the lithium cation.

It is interesting to note that catalysts doped with alkali metal alkoxides (calcination after impregnation at 550°C) possess higher overall concentrations of acidic centres than the initial alumina preparations. We think that the alkali cations are responsible for the formation of new acidic sites on the surface. It is possible that the titration with n-butylamine leads to side reactions in which complexes of the amine with surface alkali cations are formed. The number of coordinated titre molecules may be greater than one. A similar phenomenon is observed during the formation of stable complexes of polycyclic amines with alkali metal cations in solutions (ref. 22). Thus, Benesi's method would provide results allowing a qualitative comparison of acidity rather than the obtention of true, absolute values of the acid sites concentration.

REFERENCES

1. S. Malinowski, W. Grabowski,W.J. Palion and B. Zielinski, Bull. Acad. Polon. Sci., Ser. Sci. Chim., 21 (1973) 737.
2. M. Marczewski, S. Malinowski, Bull. Acad. Polon. Sci., Ser. Sci. Chim., 24 (1976) 1.
3. M. Marczewski, S. Malinowski, Bull. Acad. Polon. Sci., Ser. Sci. Chim., 24 (1976) 187.
4. J. Kijenski, S. Malinowski, Bull. Acad. Polon., Sci., Ser. Sci. Chim., 25 (1977) 329.
5. J. Kijenski, S. Malinowski, J.C.S., Faraday I, 74 (1978) 250.
6. N.S. Figoli, S.A. Hillar, M. Parera, J. Catal., 20 (1971) 230.
7. L.D. Scharme, J. Phys. Chem., 20 (1974) 2070.
8. T.T. Chuang, I.G. Dalla Lona, J.C.S., Faraday I, 68 (1972) 777.
9. H. Bremer, K.H. Steinberg, K.D. Wendlant, Z. Anorg. Allgem. Chem., 366 (1969) 130.
10. H. Pines, W.O. Haag, J. Am. Chem. Soc., 82 (1960) 2471.
11. K.V. Topchieva, K. Yun Pin, I.V. Smirnova, Adv. Catal., 9 (1957) 799.
12. S. Santhangopalan, C.N. Pillai, Indian J. Chem., 11 (1973) 957.
13. M. Iato, T. Kanbayashi, N. Kobayashi, Y. Shima, J. Catal., 7 (1967) 342.
14. H.A. Benesi, J. Phys. Chem., 61 (1957) 970.
15. K. Tanabe, Solid Acids and Bases, Academic Press, New York, 1970.
16. J. Kijenski, R. Hombek, S. Malinowski, J. Catal., 50 (1977) 186.
17. L. Nondek, React. Kinet. Catal. Lett., 2 (1976) 283.
18. M. Che, C. Naccache, B. Imelik, Trans. Faraday Soc., 63 (1967) 2254.
19. B.D. Flockhart, J.A.N. Scott, R.C. Pink, Trans. Faraday Soc., 62 (1966) 730.
20. W.K. McEwen, J. Am.Chem.Soc., 58 (1936) 1124.
21. J.B. Peri, J. Phys. Chem., 69 (1965) 220.
22. J.J. Christensen, D.J. Eatough, R.M. Izatt, Chem.Rev. 74 (1974) 351.

MECHANISM OF FORMATION OF A CATALYTICALLY ACTIVE PHASE IN THE REACTION OF CrO_2Cl_2 WITH SILICA GEL

D. Mehandjiev[x], S. Angelov[x] and D. Damyanov[xx]
[x]Institute of General and Inorganic Chemistry, Bulgarian Academy of Sciences, 1113 Sofia, Bulgaria
[xx]High School for Chemical Technology, Bourgas, Bulgaria

ABSTRACT

A new method for the preparation of supported chromium oxide catalysts for ethylene polymerization has been proposed. It is based on the reaction of chromium dioxydichloride vapour with the silanols on the silica gel surface. Hydrolysis of the surface compound obtained leads to the formation of "secondary" hydroxyl groups, which can again interact with the volatile halide. The chromium content in the surface phase and its properties were controlled by the initial concentration and distribution of silanol groups and the number of "halogenation-hydrolysis" cycles. The structure and composition of the active phase on completion of the halogenation or of the cycle "halogenation-hydrolysis[x]" was studied. The catalytic activity of the samples with respect to the ethylene polymerization was studied in a system which contained molecularly deposited chromium oxide, diethyl aluminium chloride and diphenyl magnesium.

---

1. INTRODUCTION

In the industrial production of polyethylene under low and moderate pressures, Philips Petroleum & Co chromium oxide catalysts containing up to 5 wt.% Cr on various carriers are widely applied. The usual technology of preparation of such catalysts consists of three stages: 1) impregnation of the carrier (mostly, SiO_2 or SiO_2/Al_2O_3) with a water solution of CrO_3 or a salt of Cr(VI); 2) drying, and 3) activation under fluidized conditions in a stream of dry air at temperatures of 400° to 800°C. The catalysts obtained are kept in an atmosphere of dry air or inert gases at room temperature (ref. 1)

The active phase deposition from a water solution has a number of disadvantages. E.g.,on removal of the water (drying) and heat-treatment (activation), the metal oxide particles are aggregated, which means that part of the active surface and perhaps some useful properties of the isolated metal ions are lost. For this reason, in a recent investigation attempts have been made to develop new methods for depositing the active phase, which would ensure maximum utilization of the catalytic properties of each deposited transition-metal ion, e.g. (ref. 2).

The purpose of the present paper was to show the possibility of preparation of catalysts for polymerization of ethylene using a new method of chromium oxide deposition on silica gel surfaces. The method is based on the so-called reactions of "molecular" deposition proposed by Kol'tsov and Aleskovskii (ref. 3) for modifying the silica gel surface. In our case, the active phase was deposited using the reaction of chromium dioxydichloride (chromyl chloride) with the silanols on the silica gel surface.

$$\equiv Si - OH + CrO_2Cl_2 \rightarrow \equiv Si - O - CrO_2Cl + HCl \quad (1^a)$$

or

$$\begin{matrix} \equiv Si - OH \\ + CrO_2Cl_2 \\ \equiv Si - OH \end{matrix} \rightarrow \begin{matrix} \equiv Si - O \\ \\ \equiv Si - O \end{matrix} \!\!\!> CrO_2 + 2HCl \quad (1^b)$$

Hydrolysis of the surface compound obtained

$$\equiv Si - O - CrO_2Cl + H_2O \rightarrow \equiv Si - O - CrO_2OH + HCl \quad (2)$$

leads to the formation of "secondary" hydroxyl groups which can again interact with the volatile halide. Reactions (1) and (2) form a cycle of molecular deposition during which one layer of hydrated chromium oxide is formed on the surface. Bienert et al. (ref. 4), Smirnov et al. (ref. 5) and D.Damyanov et al. (ref. 6) pointed out the possibility of preparing catalytically active phases on the surface of silica gel using reactions with some transition metal halides and oxyhalides.

2. EXPERIMENTAL

2.1. Samples

Commercial silica gel with a specific surface area of 257 m^2/g and pore volume of 0.85 cm^3/g at a relative pressure $P/P_o=0.95$ was used as a carrier. The iron impurities were preliminary removed by repeated treatment of the silica gel with hydrochloric acid and washing until negative reaction for chlorine ions. The chromyl chloride was obtained by a laboratory method described in ref. 7, then distilled several times. Two series of catalysts were obtained.

The silica gel of the first series of a samples was heated under high vacuum /about 10^{-5} Torr/ at 470 K for 8 h. After that, chromyl chloride vapours were allowed to come into contact with the carrier for

12 h at 440 K. On completion of the halogenation, the hydrogen chloride evolved and the residual chromyl chloride were removed by heat-treatment under high vacuum. The hydrolysis was also carried out at 440 K by water vapours in a flow of spectrographically pure nitrogen for 12 h. Samples symbolized as MD n, where n=1,2,3,4,5 denotes the number of the cycles "halogenation-hydrolysis", were obtained.

By carrying out, only once, the cycle of molecular deposition on silica gel previously dehydroxylated in air, the samples MD 300, MD 400, MD 500, MD 600 and MD 700 were obtained. The numbers denote the temperatures of preliminary dehydroxylation in °C.

2.2. Sample characterization

The catalysts were analyzed for chromium and chlorine as described in ref. 8. The amount of structural water (surface OH groups and intraglobular water) (ref. 9) in silica gel was determined thermogravimetrically before the heat treatment. The specific surface areas and the pore size distributions of the samples were obtained from the adsorption isotherms of nitrogen at its boiling temperature.

2.3. IR studies

The IR spectra of the samples were recorded with a UR-20 apparatus (Karl Zeiss, DDR). The powdered silica gel was pressed into thin plates of about 30 mg with dimensions 10 x 30 mm. In order to avoid contact with air and water vapours, the heat-treatment under vaccum and the reactions of molecular deposition were carried out in the same quartz cell where the spectra were recorded (ref. 10).

2.4. Investigation of the magnetic properties

The magnetic susceptibility was measured by the Faraday method. The apparatus constructed in the Institute of General and Inorganic Chemistry of the Bulgarian Academy of Sciences enabled one to carry out the measurements in vacuo or in the atmosphere of various gases. Before each measurement the samples were heat-treated in this apparatus for 1 h at 440 K. Spectrographically pure nitrogen under a pressure of 1-2 Torr was introduced and the measurement was carried out. The susceptibility change depending on temperature was studied between 77.4 and 440 K. The magnetic susceptibility of three samples (MD 1, MD 5 and MD 600) was measured within the range 4-300 K by collaborators of Prof. M. Pouchard in Laboratoire de Chimie du Solide du C.N.R.S., Bordeaux.

The EPR spectra of the samples were recorded within the range of 77.4 400 K with a commercially available spectrometer JES-3BX.

2.5. Determination of the catalytic activity

The sample activity towards ethylene polymerization was studied in a three component Ziegler-Natta system containing, along with the chromium oxide molecularly deposited on silica gel, diethyl aluminium chloride (Al $(C_2H_5)_2$ Cl, 1.2 g/l) and diphenyl magnesium ($(C_6H_5)_2$Mg, 0.3 g/l). The first co-catalyst was a preparation of the firm Schering-AG the second one was synthetized as a 2% solution in chloro-benzene. The ethylene has a purity of 99.85%. Isooctane pre-dried on a molecular sieve 3A was used as polymerization medium. The reaction proceeded at a constant ethylene pressure of 10 atm and a temperature of 360 K. The activity was judged by the yield of polyethylene 30 min after the beginning of the process.

3. RESULTS AND DISCUSSION
3.1. Composition and structure of the chromium-oxide layers

Table 1 contains the results of the analytical determinations of the catalyst compositions. It is evident that with increasing number of the cycles, the chromium amount and the Cr/Cl ratio increase whereas the specific surface area and the pore volume decrease. This may be explained with the gradual filling of the pores during the formation of each subsequent layer. With the increase of the temperature of pre-dehydroxylation (samples MD 300 to MD 700) the structural water amount decreases

TABLE 1

Sample	Cr mg at/g	OH m mol/g	Cr/Cl	surface area m^2/g before*	after**	mean pore radius Å before	after	pore volume cm^3/g before	af
MD-1	1.26	3.84	1.97	257	203	0.85	0.70	54	5
MD-2	2.22	-	3.96	203	187	0.70	0.67	58	5
MD-3	2.78	-	4.15	187	150	0.67	0.51	58	5
MD-4	3.50	-	5.83	150	119	0.51	0.43	58	5
MD-5	4.40	-	5.79	119	105	0.43	0.33	58	5
MD300	1.37	3.55	1.87	263	236	0.89	0.69	54	5
MD400	1.39	3.13	1.82	256	233	0.88	0.69	53	5
MD500	1.50	1.69	2.08	275	223	0.88	0.75	51	4
MD600	1.24	1.37	1.48	268	225	0.88	0.73	48	4
MD700	0.40	1.00	1.41	256	227	0.88	0.84	45	5

*before the cycle of molecular deposition
** after the cycle of molecular deposition

whereas that of chromium, at first negligible, increases and shows a sharp decrease at dehydroxylation temperature of 700°C. This is probably due to the dehydroxylation mainly in the narrower pores and the scarifying of the surface as a result of the removal of the intraglobular water. This is confirmed by the relatively small change of initial specific surface area.

The table shows also, in the samples pre-heated up to 500°C and treated after, that only once with chromyl chloride, the Cr/Cl ratio is almost the same (∿2) and does not correspond to the stoichiometry of halogenation according to the reaction (1^a) in which one CrO_2Cl_2 molecule reacts with one OH group of the surface. This "anomaly" can be explained by assuming that nearly half of the CrO_2Cl_2 molecules reacts with two hydroxyl groups each (reaction 1^b). The proceeding of a reaction of the type 1^a and/or 1^b is confirmed by the IR spectra. Fig. 1 shows the IR spectrum of the initial silica gel and the spectra after various treatments in the process of molecular deposition. As it is known (ref.9), the silica gel spectrum consists of a band at 3745 cm^{-1} which is ascribed to the "free" surface hydroxyl groups, a wide band between 3500 and 3700 cm^{-1} which is due to hydroxyl groups connected by hydrogen bonds (bound groups) and a band at 3680 cm^{-1} caused by the intraglobular water. These bands can be seen in the figure (spectra 1^a and 1^b). The band at 3680 cm^{-1} appeared after heat-treatment under vacuum for 6 h at 470 K (spectrum 1^b). On treatment with chromyl chloride vapour, the band at 3745 cm^{-1} disappeared and the intensity of that between 3500 and 3700 cm^{-1} decreased. However, this is obscured by the band of the intraglobular water. With a view to obtaining more precise information, the samples were subjected to preliminary deuteration by treatment with heavy water vapour at 298 K for 12 h. This led to the formation of OD groups on the surface. The silica gel spectrum after deuteration is shown in Fig. 1^c. The free OD groups gave a band at 2760 cm^{-1}, whereas a band with a maximum at 2600 cm^{-1} corresponded to the bound OD groups. The band at 3680 cm^{-1} corresponding to the intraglobular water did not change. The reaction of molecular deposition (treatment for 12 h at 440 K with chromyl chloride vapour) led to a substantial change in the IR spectrum. The band of the free hydroxyl groups, OD, at 2670 cm^{-1} disappeared and the band around 2600 cm^{-1} strongly decreased. The fact that the latter band did not completely disappear indicates that chromyl chlorida cannot react with the bound OD groups in the narrowest pores. A new band at 1750 cm^{-1} appeared in the spectrum after the reaction. A second deuteration and treatment under high vacuum produced no change in the spectrum (spectra 1^e and 1^f). It is evident that during the treatment with chromyl chloride vapour, (1^a) and (1^b) type reactions proceed and cause the disappearance of the bands for the OD groups. The appearance of the new band at 1750 cm^{-1} is due to a surface complex containing chromium, oxygen and silicon.

The study of the sample porous structures showed that the specific surface area and the pore volume decreased as a result of deposition (Table 1). The distribution curves of the pores with respect to their radii preserved their character, a small shift of the maxima towards pores with larger radii being observed.

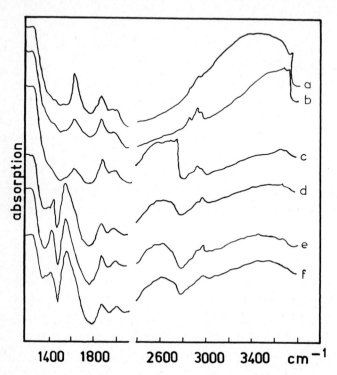

Fig. 1. IR spectra of the silica gel used as support after various treatments: a, initial sample; b, after high vacuum treatment at 470 K for 6 h; c, after deuteration at 298 K for 12 h and a subsequent treatment under high vacuum at 470 K for 4 h; after contact with chromium dioxydichloride vapour at 440 K for 3 h and post-treatment under high vacuum at the same temperature; e, after contact with D_2O at 298 K for 12 h and treatment under high vacuum at 440 K for 3 h; f, after contact with chromyl chloride at 440 K and pumping at the same temperature for 25 h.

If all the chromyl chloride molecules react with the silica gel surface according only to equations (1^a) and (1^b), the valency of the chromium anchored to the surface should remain the same as in CrO_2Cl. However, within the range 77- 400 K, the samples showed EPR signals. They were analogous to the signals of the so-called γ- and β- phases (ref.11) observed with chromium oxide deposited by the traditional technology of impregnation on silica and alumina-silica carriers. In our case, the γ- phase signal ascribed to Cr^V ions predominates immediately after halogenation. This signal was rather narrow (∼20G) and slightly asymmetric. The maximum difference between g_I and g_{II} was found to be 0.045 with g_I = 1.975 ± 0.002 and g_{II} = 1.928 ± 0.004. A signal of this type was also observed in the initial chromyl chloride even after distillation. Here it was completely symmetric with g = 1.986 ± 0.002 and indicated a resolved hyperfine structure due to the nuclear spin of Cr^{53}. The fact that the signal is preserved after distillation shows that the molecules containing Cr^V ions are stable and can react with the surface hydroxyl groups, the valency state of chromium in them remaining the same. A comparison of the signal intensities of the γ- phase in the chromyl chloride and on the surface after halogenation showed, however, that the Cr^V amount per gram chromium on the surface was 20 - 40 times as large as that in chromyl chloride. This indicates that during halogenation part (about 0.1%) of the chromium ions are reduced to Cr^V probably by the HCl evolved.

The γ- phase signals described in literature for silica gel samples impregnated with Cr^{VI} - containing solutions are rather asymmetric (ref.11). Thus, there is a greater similarity between the signals obtained by us and those observed by L.L. van Reijen et al.(ref.12) and Yu.I.Pecherskaya and V.B.Kazanskii (ref.13) with chromium oxide deposited on alumina supports. These authors ascribed the more symmetrical γ- phase signal to chromyl ions with Cr^V in strongly distorted square pyramidal coordination. By analogy one can assume that in our case the Cr^V ions obtained during the halogenation have higher coordination numbers than the Cr^V ions stabilized on a silica gel surface by the traditional technology usually have.

After hydrolysis of the samples, a considerable decrease of the γ- phase signal and a substantial increase of the β- phase signal, especially of the $β_N$ - type one, is observed. According to the nomenclature adopted in ref.11, $β_N$ is a phase containing Cr^{III} - ions which are aggregated and exchange coupled. Table 2 contains the widths of the $β_N$ - phase lines measured at room temperature. It is evident that they are smaller than those usually obtained with chromium oxyde layers deposited by impregnation. This can be explained by a higher homogeneity of distribution of the Cr^{III} ions on sites with different symmetry of the crystal field. The amount of chromium reduced after the hydrolysis to Cr^{III} was established on the basis of the experimentally determined effective magnetic moments of the samples. As evidenced from Table 2, the magnetic moments ($μ_{eff}$) calculated from the atomic magnetic

TABLE 2

Samples	ΔH G	μ_{eff} BM	$\dfrac{Cr^{6+}}{Cr^{3+}}$	Cr content (wt %)	Yield (g PE / g catalyst)	Yield (g PE / g Cr)
MD-1	330	2.00	2.7	6.1	91	1441
MD-2	270	2.13	2.3	10.3	136	1214
MD-3	300	2.26	1.9	12.6	104	745
MD-4	310	2.39	1.6	15.4	63	357
MD-5	330	2.42	1.6	18.6	55	250
MD300	267	2.47	1.4	6.6	31	418
MD400	280	2.45	1.5	6.7	42	601
MD500	330	2.30	1.9	7.2	77	926
MD600	310	2.30	1.9	6.1	91	1477
MD700	330	2.23	2.0	2.0	82	3832

PE = polyethylene

susceptibilities of deposited chromium are smaller than the spin only value typical of Cr^{III} (μ_{III} = 3.87 BM). This shows that in addition to the paramagnetic ion species established with EPR in the hydrolyze samples, there remains also some unreduced Cr^{VI}. Of course, it is not excluded that some quantities of Cr^{IV} and even Cr^{II} could also be present, but one can assume that the amounts of these intermediate forms are negligibly small, as in the amount of Cr^{V} estimated from the EPR spectrum of the γ-type. On this basis, we determined the Cr^{VI}/Cr^{III} ratio using the formula:

$$\mu^2_{eff} = \Delta \mu^2_{III} + (1 - \Delta) \mu^2_{VI}$$

where Δ is the relative amount of Cr^{3+} in the sample. As $\mu_{VI}=0$, $\Delta = (\mu/3.87)^2$. Thus, the ratio between the two predominating paramagnetic ion species was calculated (Table 2). It is evident that with the increase of the number of the layers, the relative amount of Cr^{III} increases. On a single treatment with chromyl chloride with increasing degree of pre-dehydroxylation, the relative amount of Cr^{VI} increases.

Since the chromium amount in the samples is rather high, the possibility of formation of sufficiently large clusters of exchange coupled Cr^{III} ions is not excluded. It was interesting to see whether some kind of magnetic ordering could appear under these conditions. As established by measurements of the magnetic susceptibility at temperatures up to 4 K, the Curie-Weiss law is obeyed up to 50 K; around 50 K, the χ (T) curve shows a small maximum; after that the susceptibility sharply increases with further decrease of temperature. This result is in agreement with the conclusion made on the basis of the EPR studies: the main part of the Cr^{III} produced in the hydrolysis reaction is included in a β_N - phase in which chromium (III) ions are antiferromagnetically coupled. The antiferromagnetic exchange interaction leads to some magnetic ordering at around 50 K. Only below this temperature one observes paramagnetism of isolated centres which are considerably less numerous.

3.2. Catalytic properties

Table 2 shows the characteristics of the catalytic activity with respect to polymerisation of ethylene in g polyethylene per g catalyst, and g polyethylene per g Cr, 30 min after the beginning of the process. It is evident that there is a correlation between the amount of chromium reduced on hydrolysis to Cr^{III}, and the activity: the larger the relative amount of Cr^{III}, the lower the catalyst activity. Since the number of reduced ions is larger when the density of the surface hydroxyl groups is higher, one may conclude that samples with larger mean distances between the chromium ions are more active and less able to form clusters of exchange coupled Cr (III) ions.

The treatment of a freshly halogenated sample with diethyl aluminium chloride at 360 K leads to a monotonous decrease of the intensity of the γ- phase signal.

Simultaneously, the catalytic activity increases and reaches its maximum. This shows that the Cr^V accumulated during the halogenation is not catalytically active in the reaction of polymerisation. In addition to this, no other kind of the γ- type signal appear during the treatment described, i.e. no other kind of Cr^V are produced.

The present study showed that by the method of the molecular deposition of chromium oxide on silica gel one could prepare catalysts for ethylene polymerization. Its advantages consist of the possibility of regulating the thickness, structure and composition of the active phase by varying the number of cycles of molecular deposition and the density of hydroxyl groups on the carrier surface.

ACKNOWLEDGEMENTS

Thanks are due to Prof.M.Pouchard (Laboratoire de Chimie du Solide du C.N.R.S., Bordeaux, France) for the low temperature measurements of the magnetic susceptibility. We are also indebited to Mr.D.Panayotov for his assistance in running the IR spectra.

REFERENCES

1 A.Clark, Catalysis Rev. $\underline{3}$(2), 145 (1969)
2 Yu.I. Yermakov, Catalysis Rev.-Sci.Eng. $\underline{13}$(1), 77 (1976)
3 S.I. Kol'tzov, G.N. Kuznezova and V.B. Aleskovskii, Zh.prikl. Khimii $\underline{40}$, 2774 (1967)
4 R.Bienert, W.Hanke, U.Illgen, H.-G.Jerschkewitz, G.Lischke, G.Öhlman and I.W. Schulz, USSR Conference on the Mechanism of the Heterogeneous - Catalytical Reactions, September 1974, Moscow, Rep.№ 78
5 V.M. Smirnov, S.I. Kol'tzov and V.B. Aleskovskii, Izvestiya Visshykh Uchebnykh Zavedenii Khimiya i Khim.Techn., $\underline{17}$, 1759 (1974)
6 D.Damyanov, D.Mehandzhiev and Tzv.Obretenov, 3rd International Symp. on Heterogeneous Cat., October 1975, Varna, Rep.№ 95
7 G.Brauer, Hdb.d.Präparat.Anorg.Chem., B.II, Ferdinand Enke Verlag, Stuttgart, 1962, p.1211
8 D.Damyanov and D.Mehandjiev, Commun.Dept.Chem.Bulg.Acad.Sci., $\underline{9}$, 294 (1976)
9 V.Ya. Davydov, L.T. Zhuravlev and A.V. Kiselev, Trans.Faraday Soc., $\underline{60}$, 2254 (1964)
10 D.Damyanov, D.Panayotov, D.Mehandjiev and Tzv.Obretenov, Commun.Dep Chem.Bulg.Acad.Sci., $\underline{9}$, 394 (1976)
11 Ch.P.Poole, Jr. and D.C.Mac Iver, in D.D.Eley, J.H. de Boer, M.Boudart (eds.) Adv.Catalysis, Vol.17, Academic Press, N.-Y. and London 1967, p.223
12 L.L. van Reijen, P.Cossee and H.J. van Haren, J.Phis.Chem., 38 572 (1963)
13 Yu.I.Pecherskaya and V.B. Kazanskii in O.V. Krylov (Ed.) Problemy kinetiki i kataliza, Vol.13, I-vo "Nauka", Moskva, 1968, p.236

THE USE OF TRANSPORT REACTIONS FOR DISPERSING SUPPORTED
SPECIES

R. HAASE, H.-G. JERSCHKEWITZ, G. ÖHLMANN, J. RICHTER-MENDAU and
J. SCHEVE
VEB Chemiekombinat Bitterfeld, DDR-44 Bitterfeld
Zentralinstitut für physikalische Chemie der Akademie der
Wissenschaften der DDR, DDR-1199 Berlin-Adlershof

ABSTRACT

For manufacturing catalysts a main object is to disperse the active material on the support. Methods of fixing the catalytically active substances only on the surface and not to place them in the bulk are favourable. By using methods of mass transport a dispersing in the reaction vessel before and during the reaction is imaginable. Therefore we investigated the spreading of "molten" V_2O_5 on and the reacting of vanadiumoxichloride with the surface of SiO_2, TiO_2, and mordenite at various conditions. These processes are similar to those in industrial kilns. It could be demonstrated that the spreading rate is predominantly determined by the surface roughness while the effect itself is strongly influenced by the stoichiometry of the vanadium oxide. According to the melting point V_2O_4 does not disperse on these conditions. Because the host lattices of the support and the ambient atmosphere influence the V^{5+}/V^{4+} relation they also determine the dispersing of the initial vanadium oxide.

These findings enable us to choose appropriate conditions for getting highly dispersed coats of catalytically active material.

INTRODUCTION

From an economic point of view the preparation of industrial catalysts is favoured by dry mixing of the support and the catalytically active components. By means of this only one operation is necessary for getting the green pellets of the catalyst. But a special problem arises in this case, the homogeneous and high dispersion of the active material. There are three methods of mass transport for meeting this goal; diffusion, viscous flow or spreading and evaporation and nucleation.

An additional amount of HCl in the firing gas or wetting the green pellets with liquid hydrochloric acid can form volatile or liquid oxichlorides of transition metals, which can easily diffuse, or spread over the support. The redox conditions of the system determine the reaction and transport rates. This method is the only one for dispersing a molten layer.

We modelled the industrial process using V_2O_5 and V_2O_4 and different supports.

EXPERIMENTAL

Small cylinders of pressed powder of V_2O_5 or V_2O_4 2 mm in diameter and 2 mm high are formed and annealed. These pieces are put on to the surface of the support which has been pressed as a wet paste in a combustion boat and dried. This sample is placed in a quartz tube and heated in a special furnace. One half of this tube furnace can be lifted and the sample is visible during the heating operation. Gas, one liter per hour, flows through the quartz tube, consisting of air, air saturated with HCl or air loaded with a mixture of HCl and CH_3OH. Diatomite, anatase and mordenite, pure and impregnated with K_2SO_4 serve as supports.

In a second set of runs, discs 5 mm thick are cut from a fused silica rod, 16 mm in diameter. These discs are ground with sandpaper of different mesh to obtain varying roughness of the surface. Then they are used as supports for the spreading experiments of V_2O_5. The roughness is measured with the stylus method.

RESULTS

The experiments describe the basic processes used in manufacturing the DS-type catalysts of the VEB Chemiekombinat Bitterfeld Chemical Company. Therefore diatomite is preferred as support but nevertheless anatase and mordenite are also investigated because they are mixed or impregnated with V_2O_5 in industrial catalysts, too.

The first surprising observation is that V_2O_5 spreads over the surface of fused silica at 710 K, that means 250 K lower than its melting point. The roughness of the three quartz discs is demonstrated in fig. 1 by the depth of the grooves, formed by grinding with sandpaper. The depth should be smaller than 3×10^{-1} cm for promoting the spreading of powder (ref. 1).

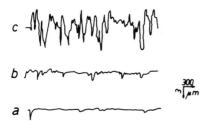

Fig. 1. Roughness of discs cut from a fused silica rod, ground with sandpaper of a) 280 mesh, b) 150 mesh, c) 100 mesh.

The covered area strongly depends on the roughness of the surface. The surface of the disc a in fig. 2 has been ground with 280 mesh silica sand, disc b with 150 mesh and disc c with 100 mesh.

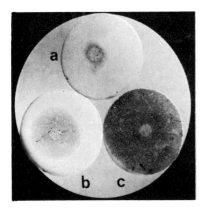

Fig. 2. V_2O_5 coats on the surfaces of different rough silica surfaces, spreaded from 2 mm pellets in dry air at 710 K.

The effect is identified as spreading because the increase of the area with time mainly follows a square root law, which is only valid for a spreading mechanism (ref. 1) (fig. 3).

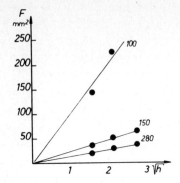

Fig. 3. Time dependence of the spreading of V_2O_5 over silica surfaces ground with sandpaper of different mesh.

Continuing our experiments we measured the dispersing of the starting V_2O_5 or V_2O_4 pellets in various gas atmospheres.

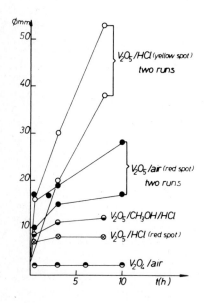

Fig. 4. Dispersing of vanadium oxides on the surface of diatomite.

Similar to the manufacturing conditions, where gas-fired kilns are used, a stream of gas flows over the sample. For getting uniform flat surfaces of the different samples, we smoothed the surface with a spatula. Because the combustion boat was to small no circular coat of vanadium oxide was available. Therefore the diameter was used for comparing the different effects. In the case of diatomite they are demonstrated in fig. 4. We got the highest dispersion when air was bubbled through conc. HCl. We observed two different forms of vanadium oxide coats on the surface: a red non-transparent spot and a yellow one which is transparent. The diameter of the latter one was 5 times larger. After three hours the red spot did not grow further. The differences of two runs are also drawn in fig. 4. In air V_2O_5 only yielded the red spot. The first three hours its diameter was equal to those resulting in an air/HCl mixture, and then it further increased slower. V_2O_4 did not coat the surface of diatomite when using air as ambient atmosphere. The spot of V_2O_5 on diatomite increased in a stronger reducing atmosphere, i.e. CH_3OH/HCl/air, a little bit quicker than in HCl/air.

The variation of spreading with temperature is shown in fig. 5.

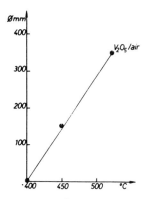

Fig. 5. Temperature dependence of the spreading of V_2O_5 over a diatomite surface at 7 h runs.

At 673 K no dispersion was observed but already at 723 K V_2O_5 crept over the silica surface, that means 240 K below the melting point. This agrees with other observations, for example at 723 K the electrical conductivity changes. This is discussed to be due to surface melting (ref. 2).

In order to meet manufacturing conditions as real as possible we used a diatomite, mixed with K_2SO_4 in the same mass ratio as in SO_2 converting catalysts.

The results are given in fig. 6. It is to be seen, that V_2O_5 and V_2O_4 are not spreading on the surface in a non oxidizing atmosphere.

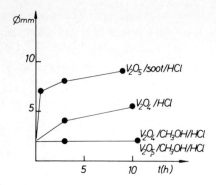

Fig. 6. Variation of the surface area of V_2O_5, solved in K_2SO_4, which has been soaked in diatomite.

Ambient air/HCl partly oxidizes V_2O_4 and it is solved in K_2SO_4, yielding a red coloured spot of V_2O_5. The diameter of V_2O_5 mixed with carbon black is a little bit larger than that of the V_2O_4 spot.

Fig. 7 demonstrates the dispersion of vanadium-V-oxide and vanadium-IV-oxide on the surface of anatase at 873 K. The runs with V_2O_5 on anatase and a mixture of air and HCl yielded the same results as on diatomite. We observed also two different sorts of vanadium oxide spots on the surface of TiO_2. A red spot, which was non-transparent and a yellow one, nearly ten times larger than the red spot and transparent. V_2O_4 did not coat the surface neither in air nor in the mixture of gaseous HCl and air. Nevertheless on a mixed TiO_2/K_2SO_4 sample vanadium-IV-oxide is actually better solved than V_2O_5.

The third material investigated as core material for vanadium oxide coated catalyst was mordenite. This support exhibited a special behaviour. V_2O_5 covered a three times larger part of the surface if mordenite was mixed with K_2SO_4 than on the pure powder. In both cases, the spots decreased after three hours. In this set of runs the effect of the concentration of the transport medium was observed. The mass dispersed per unit time increased with the concentration of HCl (fig. 8).

From the preceding results we assumed the supports to control strongly the dispersing of the vanadium oxide. This was confirmed by the experiments shown in fig. 9. The three different supports, anatase, guhr, and mordenite dispersed V_2O_5 using standard conditions, in significantly various extents. Anatase was most active

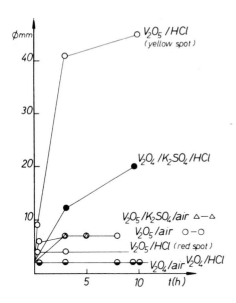

Fig. 7. Spreading and dissolving of vanadium oxides on anatase surface.

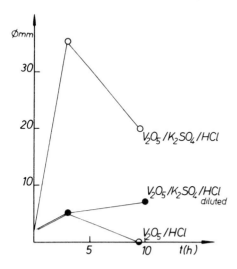

Fig. 8. The effect of mordenite/K_2SO_4 and the concentration of HCl on the dispersion of V_2O_5.

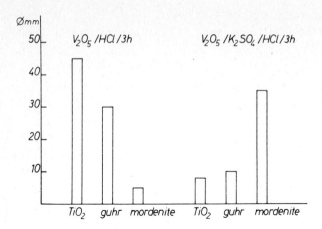

Fig. 9. Comparision of TiO_2, SiO_2 and alumosilicate with regard to the mass transport of V_2O_5.

and mordenite was less effective. Guhr took a position between the other two.

A reverse sequence was observed when the powders of the support material were mixed with K_2SO_4.

DISCUSSION

The two forms of V_2O_5 coatings observed on the surface of diatomite demonstrate, that when firing a mixture of vanadium oxide with a support in a stream of air, containing hydrochloric acid, two different processes of mass transport take place (fig. 10).

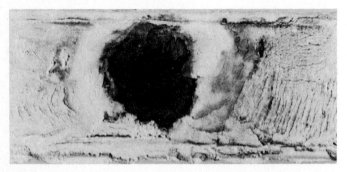

Fig. 10. Diatomite covered with transparent and non-transparent V_2O_5 coats.

The first one occured in ambient air and is due to a spreading of V_2O_5 (fig. 1 and 2). According to the melting point of the different VO_x phases only V_2O_5 pure or better slightly reduced is expected to exhibit such behaviour (ref. 3). V_2O_4 (mp 2240 K) should not spread. This prediction is confirmed (fig. 3 and 6). Because a slight reduktion decreases the melting point this effect should increase the spreading (fig. 5). Furthermore the surface is more coated by a spreading process if the annealing temperature is raised (fig. 4).

The transparent yellow spot observed in an air/HCl gas stream in addition to the non-transparent red one results from transporting V_2O_5 by means of $VOCl_3$. This is supported by the dependence of this effect from the concentration of the transport medium (fig. 7). A second argument is the decrease of the red spot with time and the observation of condensed V_2O_5 on the cold wall of the silica tube. The dispersion via gas phase is also confirmed by the asymmetrical form of the transparent spot, resulting from the dragging of gaseous $VOCl_3$ in the direction of the gas stream.

The two forms of mass transport differ also in the way of penetration of the bulk. The volatile $VOCl_3$ diffuses through the sample till to the bottom of the combustion boat. The molten part of the V_2O_5 or V_2O_{5-x} spreads only across the surface if the viscosity remains high i.e. at moderate temperatures. At high temperatures (about 1000 K) the versatile fluid penetrates the bulk too. This effect depends from the diameter of the pores. At a large-grained support only bulk impregnation is obtained. The pellet drops into the support.

The results obtained from mixtures of the oxides with K_2SO_4 can be explained by assuming the corresponding vanadium oxide to form a melt with K_2SO_4. If pure V_2O_4 or V_2O_5 are used in a reducing atmosphere, no dispersing on the surface is observed (fig. 5). Under slightly oxidizing conditions or in air V_2O_5 is quasi extracted by the K_2SO_4. According to ref. 4 a V_2O_4/K_2SO_4 mixture melts at 728 K.

The great effect of K_2SO_4 mixed with mordenite can be due to the great capillary tension operating on the melt in the small pores of the mordenite.

Resuming our results we can state that dispersing of vanadium oxide on supports is enhanced by using hydrochloric acid. A slight reduction is also favourable. Both processes are already used for manufacturing catalysts.

REFERENCES

1 Ja. E. Geguzin, Physik des Sinterns, VEB Deutscher Verlag für Grundstoffindustrie, Leipzig, 1973
2 H. Clark and D.J. Berets, Advances in Catalysis 9 (1957) 204
3 H. Oppermann, W. Reichelt, G. Krabbes and E. Wolf, Kristall und Technik 12 (1) (1977) 717-728
4 S. Hähle and A. Meisel, Kinetika i Kataliz 12 (5) 1276 (1971)

DISCUSSION

B. DELMON : In the field of ceramics we have observed that the impregnation of alumina-silicates (e.g. kaolin) with nitrates gives a substantially better dispersion of the mineralizers (e.g. alkali metals, Mg, Ca, Cu, Zn as oxides) than other salts. Should this effect be related to the phenomena you describe ? Nitrates, while decomposing, might provide a more oxidizing atmosphere, thus explaining the difference. Or should additional factors be taken into account ?

J. SCHEVE : Of course the impregnation is improved by using nitrates but in this case you get also penetration of the bulk. Using a gas phase process such as oxychlorination you get only surface contamination. The dispersion of a coating on a surface is enhanced by a slight reduction. Therefore a more oxidizing atmosphere, when using nitrates is not a favourable one. The dropping of the melting point, due to nonstoichiometry is the major factor to spread the coating material on the surface of the support.

S.P.S. ANDREW : The subject of the spreading of one oxide over a metal or over another oxide has been much studied by technologists interested in enamelling. What has been found is that spreading is enhanced when there is a chemical likeness or affinity between the two materials or an intermediate compound forms between the two layers. The oxidative power of the atmosphere is most important. Have you compared your findings with this theory ?

J. SCHEVE : According to our knowledge (Op. cit. (1)) spreading theory tells us that transport on the surface is favoured by two phenomena : physically by the bending of the surface and chemically by decreasing the melting point. Chemical reaction with the support material prevents such spreading as shown in the case of V_2O_4 on

anatase. Chemical affinity is only necessary if you use a flux
material, but this has not been the aim of our experiments. The
influence of the oxidative power of the atmosphere has been
demonstrated.
The basic model we used is that of Thomson i.e. decreasing the
vapour pressure by bending of the surface. According to this theory
lattice defects created by the ambient atmosphere also influence
the melting point. Our experiments are not analogous to enamelling
operations, because in the case flux material is used for enhanced
spreading, and the coated surface is fused. Chemical likeness
should exist between flux material and the powder to be spread.
If the porous support can react with the coating sample, it does
not spread but is soaked into the pores or the space between the
grains as shown in our paper.

F. ROOZEBOOM : From your experiments concerning the spreading
behaviour as a function of surface roughness, I get the impression
that you used the geometric surface, instead of the actual surface
area. The latter surface will be larger than the former by some
roughness factor. Did you take into account the roughness factors
of your samples in plotting the surface areas in Fig. 3 ?
If, for example, the roughness factors of the 280 mesh and 150 mesh
sandpaper-treated samples are 10 and 5 times larger than the 100 mesh
sample, respectively, the lines in Fig. 3 would even coincide. This
might suggest that, by sandpaper-treating, you mainly introduced
a macroscopic surface change, which will have no effect on the
interaction between the vanadium oxide and the support.

J. SCHEVE : The surface area in Fig. 3 is indeed a geometrical one,
because we used fused silica rod. According to theory (Op. cit. (1))
even macroscopic roughness influences strongly the spreading. This
is what we proved in this experiment. The lines of Fig. 3 would
be much more disconnected it we take into account the roughness
factor. The 280 mesh sample is smoother and the real surface area
therefore lower than in the case of the 100 mesh sample, which
even from a geometrical point of view is much more covered by the
V_2O_5 (Compare Figs. 1 and 2 with Fig. 3 in the paper).

SYNTHESIS OF NEW CATALYTIC MATERIALS: METAL CARBIDES OF THE GROUP VI B ELEMENTS

L. LECLERCQ[*], K. IMURA, S. YOSHIDA[x], T. BARBEE and M. BOUDART
Department of Chemical Engineering, Stanford University, Stanford, California 94305 (U.S.A.)

ABSTRACT

In a study of the surface reactivity of tungsten carbide powders, the role of surface carbon and oxygen has been pointed out. The presence of graphitic carbon on the surface as revealed by Auger electron spectroscopy and electron spin resonance, suppresses hydrogen chemisorption. By contrast, the latter can be enhanced by hydrogen spillover occurring to the oxide surface region from the surface carbide.

To avoid free carbon contamination, molybdenum oxycarbide samples, with a deficiency in carbon content compared to the theoretical value for MoOC or Mo_2C, have been synthesized from molybdenum hexacarbonyl $Mo(CO)_6$ by condensation of vapors in a low pressure gas. The resulting high specific area powders exhibit a higher activity in the synthesis of hydrocarbons from $CO-H_2$ mixtures than molybdenum metal. Moreover a further carburization of the initial oxycarbide powder increases its catalytic activity which becomes closely comparable to that reported by different authors for ruthenium catalysts.

INTRODUCTION

The incentive to find substitute materials to replace platinum group metals is growing rapidly with the number and amplitude of industrial processes which require their use. What are the essential properties that new catalytic materials should possess to simulate the surface reactivity of Group VIII metals of the last two transition series?

Perhaps the answer could be formulated by Sabatier's dictum: a most important property of an excellent catalyst is its ability to bind many organic and inorganic molecules strongly enough but not too strongly. This leads to the simple idea that, by dissolving a non-metallic element, e.g. carbon, in a metal of the Group VI B (tungsten or molybdenum which are strong adsorbers of molecules like CO) it is possible to temper the surface reactivity of the host metal by increasing the electron to atom ratio of the host. The net result might be that carbides of

[*]On leave from University of Poitiers.
[x]On leave from Kyoto University.

Group VI B metals behave like Group VIII metals in their surface reactivity. This has been observed by Levy and Boudart (ref. 1) who reported that tungsten carbide exhibits platinum-like behavior for hydrogen chemisorption and isomerization of 2,2-dimethylpropane. This latter isomerization is carried out at appreciable rates only by platinum and iridium (ref. 2).

There are not only chemical but also structural analogies between platinum and tungsten carbide surfaces (refs. 3,4). Numerous studies deal with tungsten carbide since tungsten carbide electrodes have been reported to function well and durably in a methanol fuel cell (ref. 5). Another desirable property of WC surfaces is their possible resistance to sulfur (refs. 6,7). But the nature of active sites in WC is still a matter of controversy (refs. 8,9). Thus we have re-examined the tungsten carbide surface by Auger electron spectroscopy (AES) and electron paramagnetic resonance (EPR). The influence of carbon and oxygen on surface reactivity for hydrogen chemisorption has been clarified.

It seems that the Golden Rule of Catalysis formulated by Sabatier is quite generally applicable to carbides. Thus, carbided nickel surfaces behave like copper in their selectivity for the decomposition of formic acid (ref. 10). In addition, Sinfelt (ref. 11) reported an increase of the activity of molybdenum for the hydrogenolysis of ethane, due to the formation of a bulk molybdenum carbide Mo_2C. The analogy of activity between Mo_2C and Ru has been noted previously (ref. 1). Such an analogy might be interesting in the Fischer-Tropsch (FT) synthesis of hydrocarbons from CO and H_2 mixtures since ruthenium is an active FT catalyst. Thus, we have synthesized molybdenum carbides of high specific surface areas and investigated their catalytic activity and selectivity in the FT reactions, in relation to their chemisorptive properties. The latter have been improved by an activation treatment consisting of further carburization of the original preparation.

PREPARATION OF CARBIDE CATALYSTS

Traditional methods of preparation (ref. 12) do not look very promising in the case of refractory metals such as tungsten (ref. 13). Besides in the case of molybdenum on alumina, it was shown (ref. 14) that the metal interacts strongly with the support and have its properties modified. During the impregnation of the support, Al-O-Mo complexes occur which are very difficult to reduce (ref. 15).

The preparation of metal carbides has been thoroughly reviewed (refs. 16-19) and it can be summarized in the following Table, from reference 16, used to describe the nomenclature of the different samples studied in this paper.

TABLE I

General Preparation of Metal Carbides

Method	Reaction	Symbols Studied
(1) Direct reaction either by melting or sintering of the elements or metal hydrides in a protected atmosphere or in vacuum.	Me + C → MeC	WC-AEG commercially obtained at AEG Telefunken (preparation not available).
(2) Direct reaction of the metal oxide and excess carbon in a protective or reducing atmosphere.	MeO + C → MeC + CO	WC-PWA from Pratt and Whitney Aircraft (MeO=WO_3)
(3) Reaction of the metal with a carburizing gas	Me + CO → MeC + CO_2 Me + C_xH_y → MeC + H_2	Sample S 3, Me = Mo $C_xH_y = C_4H_{10}$ Sample S 2 (Mo_2C) Me = Mo, $C_xH_y = CH_4$
(4) Precipitation from the gas phase by reacting the metal halide or metal carbonyl in hydrogen or inert gas.	$MeCl_4$ + C_xH_y + H_2 → MeC + HCl + C_mH_n Me carbonyl + H_2 → MeC + (CO, CO_2, H_2, H_2O)	WC-CMR, $MeCl_4 = WCl_6$ $C_xH_y = C_4H_{10}$ Mo oxycarbides S 208 ... 220 Me carbonyl = $Mo(CO)_6$ MeC = (MoOC, Mo_2C)

By necessity, catalytic materials must exhibit specific surface areas in the 50 - 100 m^2g^{-1} range corresponding to particle sizes in the vicinity of 100 Å. Although tungsten carbide is now available with a surface area of 30 m^2g^{-1} from highly dispersed tungstic acid (refs. 20,21), no one had reported the preparation of carbide particles in the vicinity of 100 Å until our recent work at Stanford in the central facilities of Stanford's Center for Material Research (CMR).

The effectiveness of fine particles synthesis by vapor condensation in reactive gases, has been demonstrated at CMR in a series of experiments in which a range of compounds including tungsten and molybdenum carbides have been produced. In this process, metal compounds (tungsten chloride WCl_6 or molybdenum hexacarbonyl $Mo(CO)_6$ are evaporated into hydrogen or CO at low pressure (10-50 Torr). A resistance heated baffle box vapor source was used at low pressure of hydrogen and at a controlled temperature of about 623-773 K depending upon the metal compounds to be vaporized. The vapor of the metal compound passed over a hot tungsten filament at 1400 to 2100 K where it reacted with the ambient gas (isobutane and hydrogen, or simply hydrogen or CO for molybdenum carbonyl, with or without argon). This technique has been available since the work of Pfund in 1933 (ref. 22), though it has not been widely used. However, several investigators (refs. 23-27) in Japan have studied this process in more details.

TUNGSTEN CARBIDE

Characterization

In addition to the WC-CMR sample, different kinds of tungsten carbide had been obtained from various sources (Table I). Specific surface area was measured by the nitrogen BET method. The crystal structures of these sample were obtained by X-ray diffraction.

TABLE 2

Characteristics of the tungsten carbide samples studied

WC Samples	BET Surface Area $m^2 g^{-1}$	Hydrogen Uptake (I) μ mol m^{-2}	Crystal Structure Crystal System	Lattice Constant a_o	c_o (nm)
AEG	6.3	5.0	hexagonal	0.2901	0.2884
PWA	5.5	0.4	hexagonal	0.2905	0.2836
CMR	15	0.1	cubic	0.2901	0.2830

(I) The Sample was pre-exposed to oxygen and then evacuated at 423 K for 1 h.

In order to investigate the surface properties of WC, EPR and AES studies were carried out at room temperature. In the EPR work, the magnetic field was calibrated with DPPH powder. All WC samples gave symmetrical singlet signals whose g value were between 2.002 and 2.003 with width between 5 and 7 Gauss. The line shape and parameters indicated that the signals could be assigned to radicals associated with carbonaceous compounds similar to carbon black, whose origin of spins is believed to be unpaired electrons stabilized in the condensed aromatic rings (ref. 28). A quantitative analysis was not attempted in this study, but spin density for the CMR sample could be estimated to be ten times longer than in the other WC samples. It was thus obvious that the WC-CMR sample contained a large amount of carbonaceous compounds.

In the AES work, the most interesting feature is the line shape of the carbon KLL transitions (270 eV). The WC-AEG sample gave a triplet signal and the PWA and CMR samples gave a singlet signal. Many AES studies for metal carbides (refs. 9, 29) agree that carbidic carbon gives a characteristic signal similar to the one of WC-AEG, while it is well known that graphitic carbon gives a singlet signal like the one found on PWA and CMR samples with no significant difference in line shape between the carbon KLL signals. Thus the latter two are covered with graphitic carbon.

Moreover, a AES triplet signal due to surface oxygen was found for all tungsten carbides and the signal was very stable. Hydrogen treatment at room temperature and a pressure of 20 kPa followed by evacuation at 423 K caused little change in the oxygen peak of the AEG sample. In addition, hydrogen treatment at 573 K hardly affected hydrogen adsorption in this sample. These results suggest that the main part of the oxygen species existed in an oxide layer on the surface.

Hydrogen adsorption on WC-AEG

Hydrogen adsorption was carried out at room temperature on the WC-AEG sample which exhibited the highest surface reactivity. Water vapor was introduced in various amounts with the hydrogen onto the AEG sample after the latter had been exposed to air and evacuated at 423 K. It was shown (ref. 30) that the adsorption rates of hydrogen increased greatly with the amount of preadsorbed water. But water accelerates the reduction of WO_3 to H_xWO_3 at room temperature in the presence of platinum by hydrogen spillover from platinum to WO_3 (refs. 31,32). Besides, Levy and Boudart (ref. 1) found that WC mixed with WO_3 also catalyzes the formation of H_xWO_3 at room temperature. Similarly, the surface oxide of WC-AEG appears to react at room temperature with spillover hydrogen from the WC part of the surface, with the formation of H_xWO_3 increasing the total hydrogen uptake on the AEG sample in the presence of water.

Conclusion

This study pointed out the influence of graphitic carbon which can make a surface inactive for hydrogen chemisorption while the carbide region could adsorb hydrogen in the dissociative form to cause spillover to the oxide region and the formation of H_xWO_3 at room temperature. Therefore in discussing the catalytic activity of tungsten carbide one should be careful to consider the actual surface composition consisting not only of tungsten and carbidic carbon but also of oxygen and graphitic carbon. Surface composition influences considerably the surface reactivity as will be illustrated again in the next part in the case of molybdenum carbides.

MOLYBDENUM CARBIDES

Preparation and Characterization

As has been seen, the synthesis of fine particles by vapor condensation in reactive gases was not successful for WC-CMR samples, giving uncontrollable carbon deposits. The above described preparation of molybdenum carbides by decomposition of $Mo(CO)_6$ has the advantage that the carbon supply is already included in the starting material. The conditions of temperatures (source and spiral) and pressures were sought to give lower content of carbon than the stoichiometric value for Mo_2C.

In addition, a molybdenum carbide (sample S 2) has been obtained by carburizing Mo in a flowing mixture of $CH_4:H_2$ (3:2), with an excess of CH_4 during 3 h at 973 K. The Mo was prepared in situ by successive reductions of MoO_3 in H_2. The method of Sinfelt et al. (ref. 11) was also used to make a sample to be compared with the different molybdenum carbides.

By decomposition of $Mo(CO)_6$ at low pressure and high temperature followed by vapor condensation, several molybdenum oxycarbide samples (S 202 to S 220) have been obtained exhibiting a f.c.c. lattice as shown by X-ray diffraction.

TABLE 3

Composition and Particle Size of Molybdenum Oxycarbides as a Function of Experimental Conditions

Samples	Gas Composition Pressure (Torr)			Source T K	Spiral T K	Atom Ratio C : Mo : O	X-ray Crystallite Size (Å)		Particle Size (Å) from BET surface
	Ar	H_2	CO				from 1/2 height	integral width	
S 202	0	20	0	573	1373	0.40: 1 : 2	60	48	–
S 203	20	20	0	573	1373	0.33: 1 : 1.8	72	73	–
S 208	0	10	0	773	1373	0.32: 1 : 1.7	86	–	160
S 209	0	10	0	773	1373*	0.31: 1 : 1.1	90	–	113
S 214	0	150	0	873	2073	0.35: 1 : 1.1	70	–	–
S 215	50	45	0	873	2073	0.31: 1 : 0.5	68	–	–
S 217	50	0	50	873	1773	0.33: 1 : 0.6	130	–	–
S 220	20	0	80	873	2073	0.35: 1 : 1.7	140	–	–

*Change of the shape of the spiral.

The molybdenum carbide structure when first reported (ref. 33) was prepared by treating metallic molybdenum with a mixture of CO and H_2. It had a f.c.c. lattice (4.14 Å) with 30.35 atom % of carbon. The formula Mo_2C was assigned to this carbide. But other authors (ref. 34) prepared molybdenum oxycarbides by pyrolysis of molybdenum hexacarbonyl $Mo(CO)_6$ under a low pressure of hydrogen, to give at 573 K, a range of oxycarbides isostructural with Mo_2N, with also f.c.c. lattice (4.17-4.18 Å). Molybdenum oxycarbide is an interstitial compound where oxygen and carbon occupy interstitial sites according to Hägg's rule (ref. 35), in a complex arrangement varying between the two limiting formulae Mo_2C and MoOC. It is well known that transition metal carbides, particularly of the Group VI metals, can dissolve considerable amounts of oxygen by substitution for carbon (ref. 17).

Only the sample S 2 "overcarburized" by an excess of methane in hydrogen, exhibits a hexagonal structure corresponding to Mo_2C. Its carbon content (9.81 weight %) is much larger than the theoretical value for Mo_2C (6.25 weight %). The presence of graphitic carbon on its surface has been evidenced by AES.

Particle Size

As shown in Table 3, the crystallite sizes were determined by the X-ray diffraction line broadening method from the measurement of half height peaks or from the integral widths. They have been qualitatively verified by electron microscopy. The results, though in acceptable agreement, have to be taken as rough values in the vicinity of 100 Å, as also shown by values determined from the BET surface area. Note that X-ray line broadening gives a volume average diameter while gas adsorption gives a surface average diameter, the two values generally not agreeing with each other.

Adsorption Measurements

Adsorption measurements were performed with a conventional Pyrex glass constant volume adsorption system provided with a Texas Instrument quartz spiral Bourdon gauge. Prior to adsorption measurements, the samples were heated in hydrogen at 623 K for 2 h and then evacuated at 623 K for another 2 h or sometimes overnight. Adsorption isotherms of CO were measured at room temperature (except for Mo metal at 195 K in a Dry Ice and acetone bath to avoid CO dissociation) between 5.4×10^3 Pa and 4.3×10^4 Pa. The linear portion of the isotherm was extrapolated to zero pressure to obtain CO the chemisorption value. Adsorption of CO was measured on new or used samples (i.e. after the FT catalytic tests). Chemisorption of CO was taken as the measure of active sites on the catalytic surface and all activity data were based on the CO uptake value with the only assumption that one CO molecule per site. Surface areas by BET were also determined with N_2 or Ar at liquid nitrogen temperature.

The hydrogen uptake at room temperature is very small on the different molybdenum carbides studied. This seems to be a general property of carbides including tungsten carbides (refs. 36,37). It was found that hydrogen adsorption is greatly reduced on carburized tungsten because the hydrogen molecules are not dissociated at room temperature. Adsorption of CO and H_2 is much more abundant on molybdenum metal than on molybdenum carbides (Table 4).

TABLE 4

Comparative Adsorption Measurement and Activity in Methanation Reaction of CO, H_2 Mixtures on Molybdenum Oxycarbides, Mo_2C and Mo.

Samples	Bulk Composition (1) atomic C/Mo (2) atomic O/Mo New	Used	H_2 Uptake (RT) $\mu mol\ g^{-1}$ New Used	CO Uptake $\mu mol\ g^{-1}$ New Used	BET Sur. Area $m^2\ g^{-1}$ New Used	E kcal mol^{-1}	(c) CH_4 Formation N $\times 10^3\ s^{-1}$
Oxycarbides S 208	(1) 0.32 (2) 1.7	0.65 1.3	− 1.4	(a) 31.7 21.7	42 36.5	14	(d) 38.5
S 215 S 217	(1) 0.32 (2) 0.55	0.5 0.9	− 0.6	(a) − 13.3	− 18	13	(d) 24
S 220	(1) 0.35 (2) 1.65	− −	0.35 1.7	(a) 25 15.2	20 −	11	(d) 41.5
S 209	(1) 0.31 (2) 1.10	− −	0.25 −	(a) 39 12.6	59 33	−	(e) 22.5
Mo_2C S 2	(1) 0.9	−	0 0.4	(a) − 1.75	− 6	11	(d) 19.5
Mo S 4	−	−	− 10	(b) − 105	− 24	16	(d) 2.5

(a) Co uptake at room temperature (RT)
(b) Co uptake at 195 K
(c) Turnover number in methane formation measured at $P_t = 1$ atm., $P_{H_2}/P_{CO} = 3$
(d) Idem, at 573 K obtained from Arrhenius plots.
(e) Idem, at 571.5 K.

Catalytic Tests: Synthesis of Hydrocarbons from CO and H_2

After the same pretreatment of the catalysts used for adsorption measurements, reaction rates were measured either in a flow reactor or in a differential batch recycle reactor. The specific rate is expressed as a turnover frequency (N) in units of reciprocal second (s^{-1}), i.e. molecules of product formed per unit surface active site per second. As shown in Table 4, molybdenum carbides are more active by a factor of 10 to 20 than molybdenum metal in methane formation from CO and H_2. The FT synthesis seems to modify the surface properties of the oxycarbide catalysts. The carbon content determined by chemical analysis is significantly increased while specific surface areas and CO adsorption decreased.

Activation Treatments: Carburization

A further carburization, conceived as an activation treatment has been performed on both samples S 209 (oxycarbide) and S 4 (Mo metal) by a recycling mixture of n-butane (nC_4) and hydrogen ($P_{H_2}/P_{C_4}=9$) during 3 h at 773 K, controlled by analyzing the products formed in the hydrogenolysis of n-butane, until only methane and hydrogen were left in the gas phase. Moreover, S 209 was activated in another way by recycling an original mixture of CO and H_2 ($P_{H_2}/P_{CO}=3$) over the catalyst during 2 h at 773 K. The results are shown in Table 5.

TABLE 5

Comparison of Activity of Mo Catalysts in the CO + H_2 Reaction with that of Ru Catalysts

Catalysts	Activation Treatment	B.E.T. Sur. Area (used) $m^2 g^{-1}$	CO uptake $\mu mol\, g^{-1}$	Experimental Conditions P_t (Torr) P_{H_2}/P_{CO} T K			Turnover Frequency N_{CH_4} $\times 10^3 s^{-1}$	N_{C_n} (3) $\times 10^3 s^{-1}$
Oxycarbide S 209	H_2, 2 h 623 K	59 (new) 33	39 (new) 12.6	760	3	572	22.5	29
S 209 after carburization	$nC_4 + H_2$ 773 K	31.5	33	760	3	578	186	221
S 209 after carburization	$CO + H_2$ 773 K	34.5	40	760	0.6	575	39.5	59
S 209 after (1) carburization	$CO + H_2$ 773 K	33.5	29	750	3	564	45	55
S 209 after (2) carburization	$CO + H_2$ 773 K	25	21	750	3	546	28.5	31
Mo Powder S 4	MoO_2 + H_2, 16 h 873 K	24	105	760	3	582	2.5	3.3

S 4 after carburization	$nC_4 + H_2$ 773 K	12.5	7.5	760	3	584	179	221
5% Ru/Al$_2$O$_3$	H$_2$ 723 K	–	35(new) 28.5(used)	760	3	548	181	(ref. 38)
0.5 to 5% Ru/Al$_2$O$_3$	H$_2$, 2h 723 K	–	–	750	3	553	25.4	(ref. 39)
1% Ru/SiO$_2$	H$_2$/N$_2$ 4h 673 K	–	20.4(used)	80	3	600	57.3	(ref. 40)

(1) Carburization of a new sample during 3 h.
(2) Extensive carburization following (1) during 6 h after adsorption
(3) N_{C_n} = turnover frequency for total hydrocarbons formation.

The activity for the reaction of methanation (N_{CH_4}) is greatly increased over that of Mo metal by about two orders of magnitude, after carburization by n-butane and hydrogen, on the carburized samples S 209 and S 4 which then exhibit identical activity. These improved activities are of the same order of magnitude as those found by some authors for Ru catalysts (refs. 38-40). It can be noted from Table 5, that the carburization of molybdenum metal (sample S 4) decreases by a factor of fifteen the number of active sites while the specific surface area of the material is only halved. By contrast, the number of active sites for CO adsorption does not change very much after carburization either by nC_4-H_2 or CO-H_2 at 773 K. In a certain way, it seems that the activation treatment stabilized the active surface of the catalysts. But the carburization must be controlled because extensive carburization with CO and H_2 at 773 K leads to a sensible decrease of the number of active sites and of the specific surface area. The activation treatment by CO and H_2 at 773 K is less efficient than the one by nC_4 and H_2 at 773 K, probably due to the influence of surface oxygen after breaking of the CO bond.

Selectivity

Selectivity is defined as the ratio of the quantity of a given product (extrapolated to zero conversion) to the total quantity of hydrocarbons formed, also at zero conversion. All the molybdenum carbides have a selectivity resembling that of ruthenium in favor of methane formation. In the same way, this selectivity can be monitored, in a short range, by variation of pressures ratio (P_{H_2}/P_{CO}) or temperature. For instance, with S 209 sample carburized by CO and H_2, the selectivity in methane varies from 66.3% ($P_{H_2}/P_{CO}=0.6$, at 574 K) to 92.4% ($P_{H_2}/P_{CO}=3$, T=546 K). The starting material (S 209) has a selectivity pattern very close to that of molybdenum metal. Both give hydrocarbons up to C_8. This suggest that the molybdenum oxycarbide (S 209) could have some molybdenum sites favoring chain growth because of its carbon deficiency. The carburization could correct this deficiency and change the selectivity toward methane.

CONCLUSION

Molybdenum carbide powders with high specific surface area exhibit favorable alteration of their chemisorptive properties in CO adsorption as compared to those of Mo, W and WC. Consequently they can be classified among the best catalysts for hydrocarbons synthesis from $CO-H_2$ mixtures. Such catalytic materials could compete favorably with ruthenium which is rare and more expensive than molybdenum by a factor of more than twenty five.

ACKNOWLEDGMENT

This work was initiated with support from the Center for Materials Research at Stanford University under the NSF-MRL Program.

REFERENCES

1 M. Boudart and R. Levy, Science, 181 (1973) 547.
2 M. Boudart and L. D. Ptak, J. Catal., 16 (1970) 90.
3 D. F. Ollis and M. Boudart, The Structure and Chemistry of Solid Surfaces, G. A. Somorjai, in J. Wiley (ed.), New York, 1969, paper 63.
4 M. Boudart and D. F. Ollis, Surface Sci., 23 (1970) 320.
5 L. Baudendistel, H. Böhm, J. Leggler, L. Louis and F. A. Pohl, 7th Intersoc. Energy Conversion Eng. Cong., 1972, p. 20.
6 G. Schulz-Ekloff, D. Baresel and W. Sarholz, Conference on Catalysis, Roermond, 1974.
7 G. A. Mills and F. W. Steffgen, Catalysis Reviews 8 (1973) 159-210.
8 P. N. Ross and P. Stonehart, J. Catal., 39 (1975) 298-301.
9 P. N. Ross and P. Stonehart, J. Catal., 48 (1977) 42.
10 R. J. Madix, Catal. Rev.-Sci. and Engrg., 15 (1977) 293.
11 J. H. Sinfelt and D. J. C. Yates, Nature Physical Science, 229 (1971) 27.
12 J. H. Sinfelt, Ann. Rev. Materials Science, 2 (1972) 641.
13 P. Biloen and G. T. Pott, J. Catal., 30 (1973) 169.
14 A. Cossi Faustin, J. Grimblot, M. Guelton and J. P. Beaufils, C. R. Acad. Sci., Paris, serie C t. 281 (1975) 487.
15 F. E. Massoth, J. Catal., 30 (1973) 204.
16 P. Schwarzkopf and R. Kieffer in collaboration with W. Leszynski and F. Benesovsky, Refractory Hard Metals, Macmillan (ed.), New York, 1953.
17 E. K. Storms, The Refractory Carbides, Academic Press, New York, 1967.
18 L. E. Toth, Transition Metal Carbides and Nitrides, Academic Press, New York, 1971.
19 T. Ya. Kosolapova, Carbides, Plenum Press, New York, 1971.
20 V. Sh. Palanker, D. V. Sokolsky, E. A. Mazulevsky and N. Baibatyrov, J. of Power Sources, 1 (1976/77) 169.
21 E. A. Mazulevsky, V. Sh. Palanker, E. N. Baibatyrov, A. M. Khisametdinov and E. I. Domanovskaya, Kinetika i Kataliz, 18 (1977) 767-775.
22 A. H. Pfund, J. Opt. Soc. Amer., 23 (1933) 375.
23 K. Kimoto, Y. Kamiya, M. Nonoyama and R. Uyeda, Japan J. of Appl. Phys., 2 (1963) 702
24 A. Tasaki, S. Tomiyama, S. Iida, N. Wada and R. Uyeda, Japan J. of Appl. Phys., 4 (1965) 707.
25 S. Yatsuya, R. Uyeda and Y. Fubano, Japan J. of Appl. Phys., 11 (1972) 408.
26 N. Wada, Japan J. of Appl. Phys., 7 (1968) 1287.
27 N. Wada, Japan J. of Appl. Phys., 6 (1967) 553.
28 R. L. Collins, M. D. Bell and G. Kraus, J. of Appl. Physics, 30 (1959) 56.
29 T. W. Haas, J. T. Grant and G. J. Dooley, J. Appl. Physics, 43 (1972) 1853.
30 S. Yoshida, K. Imura and M. Boudart, Communication presented at the Fall Meeting of the California Catalysis Society, 1977.
31 J. E. Benson and M. Boudart, J. Catal., 4 (1965) 704.

32 M. A. Vannice, J. E. Benson and M. Boudart, J. Catal., 16 (1970) 348.
33 J. J. Lander and L. H. Germer, Amer. Inst. Mining and Metallurgical Engineers Tech. Pub., 2259 (1947).
34 I. F. Ferguson, J. B. Ainscough, D. Morse and A. W. Miller, Nature, 202 (1964) 1327.
35 G. Hagg, Z. Phys. Chem., 12 (1931) 33.
36 H. Froitzheim, H. Ibach and S. Lehwald, Surface Sci., 63 (1977) 56.
37 J. B. Benziger, E. I. Ko and R. J. Madix, J. Catal., in press, 1978.
38 M. A. Vannice, J. Catal., 37 (1975) 449.
39 R. A. Dalla Betta, A. G. Piken and M. S. Shelef, J. Catal., 35 (1974) 54.
40 G. C. Bond and B. D. Turnham, J. Catal., 45 (1976) 128.

DISCUSSION

MONTARNAL : 1) It seems to me that for Fischer-Tropsch synthesis there is an inhibition of the reaction by CO. Your results show that the CO chemisorption is much lower on molybdenum carbide than on the metal (Tables 4 and 5). So could it not be possible that the activity increase you observe when going from metal to carbide, could be due to the decrease of inhibition by CO ?
In other words, the rate increase would be due to a decrease of the adsorption of CO, rather than to an increase of the reaction rate constant.
2) It also seems to me that Ru gives high molecular weights in F.T. synthesis, when used under high pressure. Do you think that your analogy between molybdenum carbide and Ru could be extended also to this characteristic property of Ru ?

L. LECLERCQ : 1) I think your suggestion is very interesting. It will need additional kinetic studies to determine the influence of CO partial pressure before and after the carburization of our catalysts.
2) It was our intention to verify if, under high pressure, the analogy between molybdenum carbide and ruthenium could be extended to high molecular weight hydrocarbon formation in F.T. synthesis. We have still to put our catalyst in suitable texture and to solve some more technological problems.

K. KOCHLOEFL : Can you give more details concerning methanation activity of molybdenum oxycarbide (comparison with usual methanation catalysts) ?

L. LECLERCQ : Nickel is the most usual industrial catalyst for methanation reaction although Ruthenium has a higher activity. According to Vannice's paper (ref. 38), the turnover frequency of

Ni/Al$_2$O$_3$ for methanation is 32 10^{-3}s^{-1} compared to Ruthenium 181 10^{-3}s^{-1} (at 275°C, H$_2$/CO = 3). Molybdenum oxycarbide can be ranked within the activity range of metals such as Co, Rh, Pd between 12 to 20 10^{-3}s^{-1}. MoOC S 209 has initially a turnover frequency of 23 10^{-3}s^{-1} (299°C, H$_2$/CO = 3).

After the activation treatment via a further carburization, the resulting Mo carbide has a turnover of 221 10^{-3}s^{-1}, quite identical to the Ruthenium activity.

J. LEMAITRE : You mentioned that carbon deposited on the surface of carbides was responsible for their poor catalytic properties. Do you know some specific treatment that could activate a carbide catalyst poisoned with carbon ?

L. LECLERCQ : Classical regeneration treatments with oxygen on WC-CMR exhibiting carbon deposition on the surface were attempted but gave oxide formation like WO$_3$. As shown by different authors (Somorjai et al. and Wise et al.) carbonaceous compounds can react with hydrogen to produce methane. Reductive regeneration seems more appropriate for such materials with W or Mo.

P.A. SERMON : Since butane and C_1-C_8 hydrocarbons in the Fischer-Tropsch synthesis almost double the carbon content of your molybdenum oxycarbides, is it possible that you are seeing retention of hydrocarbon-carbonaceous residues on the carbide surface. The catalytic activity of such residues on metals (e.g. A.S.A/Ammar and G. Webb, J. Chem. Soc Faraday Trans. I., 74, 195 (1978)) might well explain the substantial improvement in activity you observe after hydrocarbon treatment. Have you examined the structure of the additional carbonaceous material deposited by hydrocarbons ?

L. LECLERCQ : A retention of hydrocarbon-carbonaceous residues on the catalyst surface is certainly a possible explanation to the improvement in activity observed after carburization by hydrocarbon treatment. Another explanation could be found. For instance, the carburization could result in a change of the surface composition leading to surface carbon content approaching the theoretical value for Mo$_2$C. Our Auger experiments based on the carbon peak has not made it possible to determine the nature of carbonaceous species before or after the carburization treatment.

P.C.H. MITCHELL : One method which was used to prepare molybdenum carbide was to expose molybdenum powder, prepared by reducing MoO_3 in H_2, to CH_4 at 973 K. Do you consider it probable that whenever Mo metal is exposed to a hydrocarbon above a certain temperature, a layer of molybdenum carbide will form and that the real catalyst in reactions involving H_2/hydrocarbon mixtures will be molybdenum carbide rather than Mo metal ?

L. LECLERCQ : Yes, it is quite probable. Whenever Mo metal is exposed to hydrocarbons in the presence of hydrogen above a certain temperature, molybdenum carbide is formed as demonstrated by Sinfelt (ref. 11). He has shown that the specific activity of molybdenum for ethane hydrogenolysis significantly increases during carbiding. The carburization seems to involve more than carbiding of the surface since the catalytic activity does not develop fully untill bulk carbide is formed. We have also shown that carburization could simultaneously lead to carbon deposit. Anyway, at low temperature and low conversion Mo metal can be tested since it exhibits a stable activity.

A NOVEL METHOD FOR THE PREPARATION AND PRODUCTION OF SKELETON
CATALYSTS

J. PETRÓ
Organic Chemical Technology Department, Technical University
Budapest /Hungary/

ABSTRACT

A new method has been developed for the preparation of skeleton catalysts, the essence of which is that the alloy is decomposed in the solid phase and the aluminium oxide hydrate formed remains partly as Ni-support in the catalyst. In this way metal catalysts on oxide support can also be prepared.

Fundamental research work by metal microscopy, X-ray diffraction, microprobe analysis, SIMS, crystallite size distribution investigations in conjunction with the development of the catalyst are illustrated by examples.

1. INTRODUCTION

More than 50 years have elapsed since the discovery of skeleton catalysts /1925/, but their importance and significance did not decrease. This is indicated also by several hundred publications and patents relevant to the theme.

However, in this period the essential steps of technology, suitable for industrial production, remained the same, and the starting alloy of commercial skeleton-Ni was also mostly of the same composition, consisting in about 50 : 50 % of Al and Ni.

2. New skeleton-Ni catalyst

In recent years we developed a skeleton-Ni catalyst, the activity of which is considerably higher /in a given case two- to threefold/ than that of the catalyst used up to the present, particularly in the reduction of the carbonyl nitrile and nitro groups of organic molecules, and can replace the said commercial catalyst. The stability of the new catalysts is also high. In their development two factors played an important role: the elaboration of a new process for the preparation of the catalysts and the promotion of nickel.

2.1 New process for the preparation of skeleton catalysts (ref. 1)

Conventional skeleton-nickel preparation involves the feeding of the nickel-aluminium alloy into a large quantity of alkaline solution, where the dissolution of aluminium from the alloy proceeds in an exothermal process, accompanied by vigorous hydrogen evolution and foaming. After the termination of the introduction of the alloy and of foaming, after-leaching or leachings follow, which can take another 3 to 12 hours. The feeding of the alloy powder into the alkaline solution is time consuming and cumbersome process. Moreover the residential time of the first batches of the alloy in the alkaline solution is considerably longer than that of the last ones, which involves the hazard of a non-uniform product.

The essence of the process developed by us is that the alloy powder is decomposed in solid phase as follows. The alloy powder is homogenized in a kneading device with a small quantity of NaOH, and, under continuous stirring of the powder, water is added at a rate to moisten uniformly the substance without formation of a liquid phase. Decomposition is performed by water, NaOH works as a catalyst, so from aluminium mainly aluminium hydroxide is formed at the end of the process. When decomposition is terminated, giving about equal volume of water to the mass, two solid phases are separated, the lower of which contains the catalyst, while aluminium hydroxide in the form of a white precipitate is in the upper phase, and can be easily separated.

The operation is terminated by a brief treatment in a hot alkaline solution where hydrogen evolution is already insubstantial.

The aluminium hydroxide partly remains in the catalyst prepared as described above, and serves as a support for the nickel. Accordingly, the specific weight of the catalysts is lower than that of the conventional catalyst, so by mixing it can be easily dispersed in all the volume of the reaction mixture and even it is easily settled and filtered.

The preparation process is generally applicable for skeleton catalysts and, referred to the metal content, the activity of these catalysts is usually higher than that of prepared in conventional way. Since the whole quantity of the alloy reacts for an identical period, the substance is uniform.

Technical advantages are as follows: There is no foaming during the leaching process. The volume of the necessary equipment and the time of operation are about half of that needed for conventional catalyst preparation. The specific amount of NaOH is also lower. On introducing the new process, existing equipment can be used, only an additional kneading device is needed.

In conjunction with the process discussed above, another one has also been developed for preparation of catalysts.

Catalysts on oxide support are generally prepared by the joint precipitation of the hydroxides. This is followed by the washing and calcination of the hydroxides and the reduction of the metal oxide. Another possibility is the impregnation of the support, where the insolubilization in water of the substance applied, washing, drying and repeated reduction at high temperature are the usual steps.

Whichever way is chosen, both involve several steps, have a high volume demand and comprise various difficult operations.

The process developed by us has the following essential features: From the given metal or metals an alloy is prepared in the way usual for skeleton catalysts e.g. with aluminium or other alkali-soluble metal or metal combination, depending on the kind of support to be prepared . The alloy is decomposed in solid phase, as described in 2.1, essentially by water. Thus a metal on oxide support is obtained in a single step. In addition to the simplicity of this process the catalyst prepared in this way is considerably more active than that of prepared by high-temperature reduction generally in the case of nickel e.g. at 620-670 K , as in the process mentioned above the highest temperature is about 370-390 K (ref. 2).

On the basis of the new processes, a family of catalysts can be built up from alloys. Starting possibly from the same alloy, either an active "supported" skeleton catalyst for liquid-phase hydrogenation terminating the preparation with alkaline solution treatment , or an oxide supported metal catalyst for high-temperature gaseous phase reactions can be prepared according to requirement.

Fig. 1 summarizes what has been said above.

2.2 Increasing the activity of skeleton-Ni by promotion (ref. 3)

It has already been established that the activity of a nickel catalyst is enhanced by certain promotors e.g. chromium, cobalt, manganese, molybdenum, etc. . Actually, skeleton nickel catalysts have been prepared with some of the promotors, but these did not find wide industrial application.

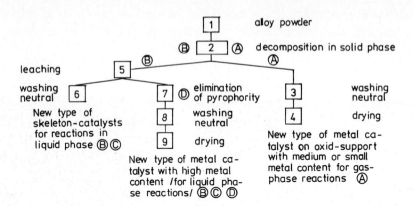

Fig. 1. New ways for catalysts preparation. A, B, C, D see ref.

In recent years we have developed various skeleton-nickel types containing several promotors, which are adequately stable, while their activity is under identical conditions 2-3-fold of that of skeleton-nickel of good quality, particularly in the hydrogenation of carbonyl and nitrile groups. This has been proved also by the results of plant-scale experiments in Hungary.

3. Some results of fundamental research work received during catalysts development.

The alloy of skeleton catalysts includes the potential catalytic properties, so that research work will begin with the investigation of these alloys.

For simple optical investigation of metal alloys metal microscopy is used. Etching of the polished surface with an alkaline or acidic solution, various contours can be developed. Fig. 2 shows the micrograph of an Al-Ni alloy, containing Mo and Cr promotors, taken in polarized light after alkaline etching. The featured structure clearly shows the separation of phases of different composition. Light spots on the picture presumably represent γ-phases, rich in nickel (Ni_2Al_3); the darker is the colour, the richer is the phase in Al. Such phases may be e.g. the β-phase $(NiAl_3)$, the Al-alloys of the promotors, etc.

Fig. 2. Optical micrograph of skeleton-Ni alloy containing Mo and Cr promotors after alkaline etching.

X-ray diffraction can give information on the metal or alloy phases present in the substance. Fig. 3/a shows the diffractograms of a skeleton-Ni alloy containing three promotors Mo, Cr, Co and Fig. 3/b of the catalyst prepared from it by the process described in Section 2.1. It can be seen that the diffractogram of the alloy is very rich in lines. Only a few phases, mainly those present in larger quantities can be unequivocally identified. The interference of the bands of a series of possible alloy phases does not permit a detailed analysis.

On the diffractogram of the catalyst prepared from the given alloy (Fig. 3/b) the broadening of the Ni-peak is indicative of the small size of the crystallites, and there is presumably also an interference between this peak and those of the phases containing promotors. Moreover, a non-identificable peak appears on the picture.

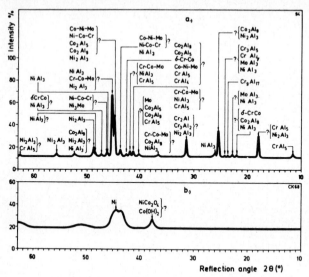

Fig. 3. X-ray diffractograms of skeleton-Ni alloy containing Mo, Cr and Co promotors **a** and of the catalyst made from it according to (ref. 1) **b**.

More detailed but still qualitative information can be obtained by microprobe analyser.

Fig 4 shows the pictures of an alloy containing 50 wt % of Al, 50 wt % of Ni and no promotor, and of the catalyst prepared from it by the process described in 2.1. The lighter is a spot on the "electron" picture the higher of the atomic number or/and the concentration of the given element, and vice versa. The γ-phases of light colour are separated from one another by β- and α-phases, the latter containing only a few % of Ni. The quantitative distribution of the phases rich in Al (β and α) and rich in Ni (γ) can be estimated on the basis of the microprobe scans showing the Al and Ni distribution, respectively, to about 50-50 %.

The microprobe scans of the catalyst prepared from this alloy were taken on pills. It can be seen on the "electron" picture that the predominant part of the grains contains Al besides Ni (mainly in the form of the hydroxide or oxide hydrate, see sect. 2.1), and this is supported also by the microprobe scans of Al and Ni distribution.

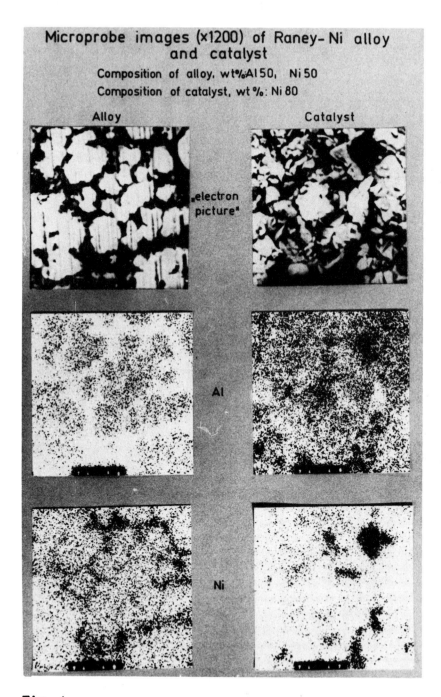

Fig. 4.

Using Mo and Cr promotors the number of phases increases, as can be observed on the "electron" picture in Fig. 5. The analysis of the scans shows that promotors have at least double effects. On one hand, they shift the "equilibrium" of the Al-Ni phases to the debit of the γ-phase towards the β-phase, and on the other hand, new phases are formed with the promotors. The distribution of Mo is uneven, it is alloyed by Cr, and mainly with aluminium. A Mo-Cr-Ni phase can also be observed, in which Al is present in a small quantity or it is totally absent. The Cr fraction unalloyed with Mo is uniformly distributed in the Al and Ni.

The structure of the alloy mentioned above becomes more complicated consisting of a small amount of silicon (Fig. 6). Almost the total quantity of Mo is alloyed with Si, a fraction of Cr is also present in this alloy, and this Mo-Si-Cr alloy is to be found in the phase rich in Al, in which the quantity of Ni is minimal. Some of Cr and Si is present in the nickel-aluminium phase.

Fig. 7 shows the curves of a microprobe line analysis for a skeleton catalyst containing Mo and Cr promotors where Mo-Cr phase can be detected.

In a catalyst containing Mo-Cr-Co promotors the changes in composition going from the surface into the bulk of the grains has been investigated by SIMS. Results in Fig. 8 show that in the external zone of about 5-15 nm of the grains the concentration of Ni, Cr and Co is somewhat lower, while that of Mo about the same as in the bulk. This result can be explained by the peritectic formation of the grains.

The crystallite size distribution and the thermal stability of catalysts containing several promotors and prepared according to section 2.1 have been investigated on the basis of X-ray line broadening. The results of these investigations are shown in Fig. 9. Accordingly, the average diameter of the crystallites is about 3 nm, and does not exceed even after heat treatment for two hours at 870 K 6-7 nm, thus, thermal stability is also good, which presumably can be attributed to the residual aluminium oxide hydrate contained in the catalyst.

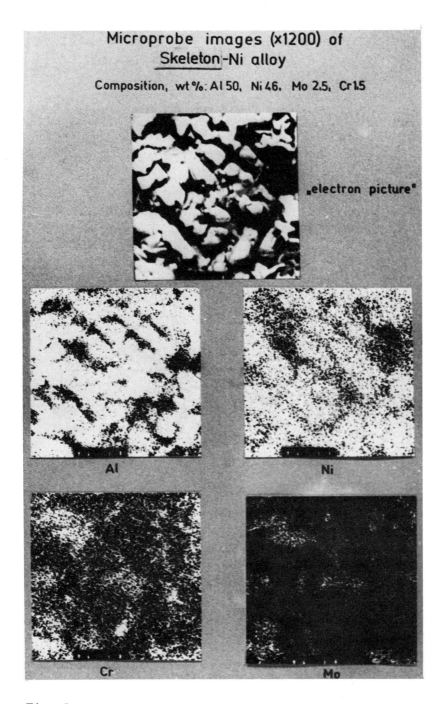

Fig. 5.

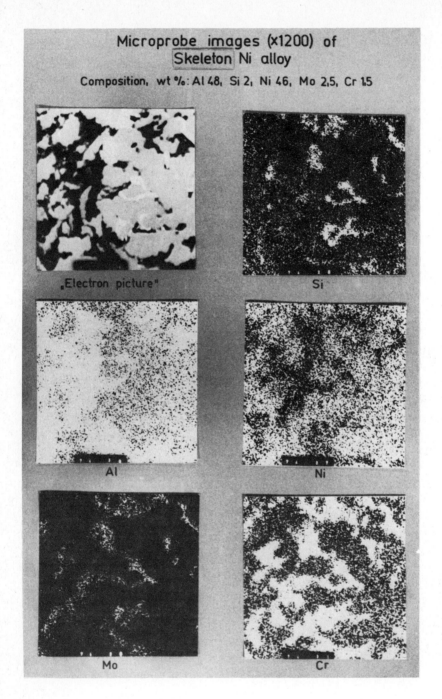

Fig. 6.

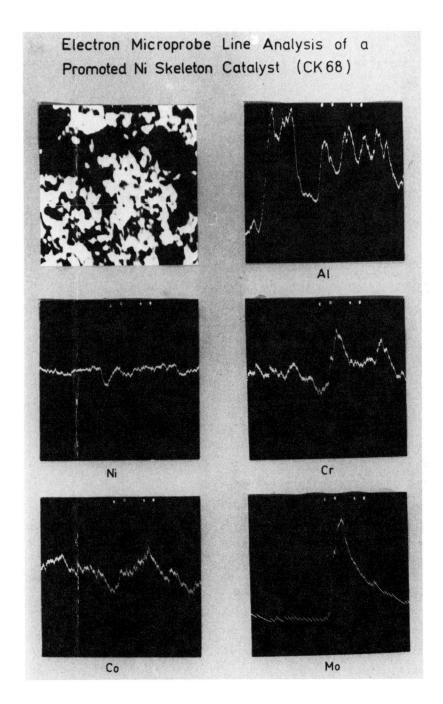

Fig. 7.

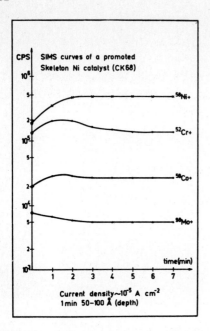

Fig. 8. SIMS curves of skeleton Ni containing Mo, Cr and Co promotors. The catalyst has been prepared according to (ref. 1).

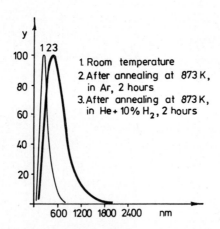

Fig. 9. Crystallite size distribution and thermal stability of skeleton-Ni catalyst containing Mo, Cr and Co promotors, prepared according to (ref 1).

ACKNOWLEDGEMENT

We thank Dr. A. Kálmán, Central Chemical Research Institute of the Hungarian Academy of Sciences for the X-ray diffraction analysis, Dr. J. Giber, Physical Department of the Technical University of Budapest for the SIMS investigations, and the Company BASF for the determination of crystallite size distributions.

REFERENCES

1	Hung. Pat.	162 281	B	see Fig. 1
	US pat.	3 809 658		
	GFR Pat.	2 209 004		
2	Hung. Pat.	169 494	A	see Fig. 1
	US Pat.	4 021 371		
	GFR Off.	2 553 660		
3	Hung. Pat.	170 253	C	see Fig. 1
	US Pat.	applied for		
	GFR Off.	2 544 761.		
D	Hung. Pat.	161 915		see Fig. 1
	GFR Pat.	22 09 000		

DISCUSSION

C.S. BROOKS : This paper is further indication of the unexploited potential for devising innovative variations on Raney and skeletal type catalysts even though it has been 51 years since their introduction by Marray-Raney. I have a specific comment on the nature of the distribution of the promotor elements Mo and Cr, as shown in the electron microprobe figures in the paper. There would appear to be considerable segregation and heterogeneity in the distribution of the promotor elements in the nickel-aluminum alloy. Equilibrium phase diagrams are not immediately available to me, but I would inquire whether the concentration of these promotor elements is not exceeding the solubility limit in nickel for the conditions of solidification. In our laboratory we have had considerable success in obtaining high orders of homogeneity in the nickel phase by using controlled solidification of eutectic alloys. We have produced pure phases of $NiAl_3$ in fibrous configurations from directional solifidication of nickel-aluminum alloys.

J. PETRO : Phase diagrams for systems like Al-Ni-Cr-Mo are not available, but it is known that in solid phase Ni dissolves 37.5 wt% Mo and (at least at higher temperatures, e.g. 1250°C) there exists a phase with composition of 35 wt% Mo + 42 wt% Cr + 23 wt% Ni (no Al !). This phase is in equilibrium with some Co-Mo phase. So, at least in simple alloys, the solubility of Mo and Cr in Ni seems to be much higher than the concentration used in our alloys.

J.W.E. COENEN : Could you give some indications on the method of obtaining crystallite size distributions from X-ray diffraction lines ?

J. PETRO : The ideal line profile of a (200) reflection has been calculated by Fourier analysis (and Stokes method). The measured and calculated curves have been fitted using different crystallite size distributions until a good agreement for the two curves has been obtained.

J.F. LE PAGE : Are your promotors Mo, Cr, Co efficient to improve the thermal stability of your catalysts in the liquid phase in presence of small amounts of water ?

J. PETRO : Generally speaking, yes. But in order to give a more precise answer, you ought to tell the temperature and H_2 pressure these types of catalysts can be used until about 150-180°C if the pressure is at least several times 10 atm.

B. DELMON : There are two ways of considering aluminium (whichever its chemical form) remaining in skeleton catalysts. As alumina, it could play essentially the role of a support or, more specifically of a dispersing substance and of an inhibitor towards sintering. Aluminium could also play the role of an electronic promotor, with micro-domains of alumina or aluminium atoms being the electronically active agents.

The picture one can also make of the ideal preparation procedure, i.e. preparation of the starting alloy and alkali leaching, depends a lot on the real reason of the activity of skeleton catalysts.

If the first role is predominant, the best is probably to allow an incipient segregation of aluminium in the alloy thus "preshaping" the spacers, by a process similar to dispersion hardening, at the boundaries of the individual nickel crystallites.

If the second role is predominant the maximum electronic effect is obtained for the maximum dispersion of aluminium.
In relation to your preparation method, what do you think is the role of aluminium or alumina ?

J. PETRO : After our investigations the predominant part of the Al remaining in the catalyst is as alumina and serves as a support for the Ni (metallic Al as free metal could not be in the catalyst, at most in Ni-rich phases it can only be present in very small quantities).

MINISYMPOSIUM ON CATALYST NORMALIZATION

NORMALIZATION OF CATALYST TEST METHODS

L. MOSCOU
Akzo Chemie Nederland bv, Research Centre Amsterdam, P.O. Box 15,
1000 AA Amsterdam, The Netherlands.

ABSTRACT
A report is given of organizational developments in the field of normalization of catalyst test methods and catalyst reference materials in Europe since 1975. The following topics are discussed: (I) results of the inquiry in 1975 on the need in Europe for normalized catalyst test methods, (II) a round-table meeting in 1976 and the formation of an ad-hoc committee on catalyst testing, (III) coordination between the various catalyst groups by the IUPAC commission I.6 on Surface and Colloid Chemistry.

1. Results of 1975 inquiry

During the 1st International Symposium on "Scientific Bases for the Preparation of Heterogeneous Catalysts", that was held in Brussels 3 years ago, a special session on the standardization of catalyst testing was organized by Dr. R. Cahen. The two major events during that session were a progress report on catalyst standardization activities as organized in the U.S.A. by "ASTM committee D-32 on Catalysts"[1] and in the U.S.S.R. and the issue of a questionnaire to those present at the session. The purpose of that inquiry was to find out how much need is felt in Europe for developing standardized catalyst testmethods.
The conclusions drawn are based on 29 inquiry forms returned out of 100:
a. A definite incentive exists to develop standardized testmethods for catalysts.
b. A number of people are willing to contribute personally to develop these methods either by submitting their own method, by testing proposed methods or by attending technical meetings.
c. No financial support is offered.
d. The most urgently needed tests are those to measure physicochemical properties of catalysts.
e. Collaboration should be established with ASTM committee D-32 on a collective and not on an individual basis.

2. The Round Table Meeting and the formation of "ad-hoc committee"

The next step was the organization of a Round Table Meeting. Hereto, Dr. Cahen contacted the BCR to advise on such organization; the BCR is the Community Bureau of Reference, being part of the Commission of the European Communities.

As a result, the "Round Table Meeting on Catalyst Normalization" was held in the BCR-Offices in Brussels on June 9th, 1976, under the chairmanship of the late Dr. K.F. Lauer; here several experts gave brief accounts on catalyst normalization activities that were in progress already in many European Countries.

The following experts accepted nomination in an "ad-hoc committee on catalyst testing": M. Baerns, R. Cahen, E.G. Derouane, W. van der Eijk, M. Michel, L. Moscou, K.S.W. Sing and F. Trifirò.

During a first meeting of this committee on October 14th, 1976, it became clear how diversified the actual normalization activities in Europe were already; it became clear also that the "committees primary task should be to coordinate the existing activities in this field, more than organizing new activities".

The following aims of the ad-hoc committee were specified:
The "ad-hoc committee should provide a basis
a. for the cooperation between the various bodies concerned with the normalization of catalyst testing and for the provision of reference materials;
b. for the critical assessment of testing procedures and for the selection of appropriate reference materials;
c. for the collection and dissemination of information gained in these activities.

It was also concluded that, to support these objectives, it is desirable that an international organization will promote, coordinate or patronize further activities, and that these objectives are beyond the scope of the BCR.

3. IUPAC

The International Union of Pure and Applied Chemistry (IUPAC) appears to be an organization that is willing and able to play such a coordinating role.

During its general assembly in August 1977, "IUPAC commission I.6. on colloid- and surface chemistry" has shown general interest to become active in the area of heterogeneous catalysis"[2]. During this session two members of the ad-hoc committee (R. Cahen and L. Moscou) reported on the activities of normalization bodies. A formal request from the ad-hoc committee to IUPAC lead to the decision that IUPAC commission I.6. will start to coordinate the normalization activities by dessimination of information between all bodies, active in the field. This will be achieved by receiving minutes and progress reports from each body and by submitting copies of them to all other bodies.
For address of the secretary of IUPAC commission I.6 see ref. 3).

Each group which is active in the field of normalization of catalyst methods
or on catalyst reference materials, and which is not yet contacted by IUPAC,
is invited to contact the secretary.

4. Present symposium

The present International Symposium is a good opportunity to present the progress
made both in the organizational area as given in the above paragraphs, and in
the development of normalized catalyst methods.
We expect representatives of normalization bodies in Belgium, Great Britain,
Italy, U.S.A. and U.S.S.R. to report on their aims, activities and results obtained
so far.

Scientific aspects of normalization will be presented in papers by Broekhoff
and by Scholten; they will discuss backgrounds that play a role in normalization
of pore size distribution and chemisorption methods respectively.

We hope and expect that this symposium will contribute to a rapid development
of normalized catalyst methods and that it will stimulate further international
cooperation and encourage liaison between the various groups.

REFERENCES

1) J.R. Kiovsky, "Progress Towards Standardized Catalyst Testing in the U.S.A. ASTM-committee D-32 on Catalysts". Symp. on the Scientific Bases for the Preparation of Heterogeneous Catalysts, Brussels, October 1975.
2) IUPAC commission I-6, Subcommittee on catalyst activity, report of meeting held 2.15 p.m., 12th August 1977 in the Palac Kultury i Nauki, Warsaw.
3) Prof. Dr. J. Lyklema, Secretary of IUPAC I.6, address: Lab. for Physical and Colloid Chemistry, State Agricultural University, De Dreijen 6, Wageningen, The Netherlands.

MESOPORE DETERMINATION FROM NITROGEN SORPTION ISOTHERMS: FUNDAMENTALS, SCOPE, LIMITATIONS

J.C.P. BROEKHOFF
Unilever Research Vlaardingen, P.O. Box 114, 3130 AC Vlaardingen
The Netherlands

SUMMARY

The thermodynamic basis for obtaining pore-size distributions from nitrogen sorption isotherms is reviewed. It is shown that pore volume and pore surface area can be determined from sorption isotherms, but only in the simplest of cases. The basic irreversibility of sorption in the hysteresis region is even then a complicating factor. In actual practise, irregularity of pore shape and interdependence of pore domains in the porous structure necessitates the use of rather drastic simplifications. As a result, the outcome of the analysis is greatly dependent upon the set of approximations used, and this should be borne in mind in drawing conclusions from the obtained data.

INTRODUCTION

Nitrogen sorption isotherms are widely used for obtaining information about the porous structure of heterogeneous catalysts in the mesopore-size region. The relative simplicity of the experimental determination and the apparent ease of interpretation of the experimental data account sufficiently for the popularity of the method. In this paper we will not discuss the experimental methods as these have been described in detail elsewhere [1], but we will restrict ourselves to a general discussion of the scope and limitations of the information on porous structures which can be obtained from the experimental data.

A porous structure can be considered as a collection of elements in pore space, each characterised by shape, size and relative allocation with respect to other elements in pore space. In practice, our information on pore structure will seldom be as detailed as that. In most cases we will have to be content with information concerning size and number of elements in pore space. The shape of the pores for actual porous structures is often difficult to define and the information concerning pore shape that can be obtained from nitrogen sorption data, is rather limited. It is common practice to assume a predominant pore shape rather than to establish it independently. Yet, it is to realised that the quantitative data on pore size obtained from nitrogen sorption data are often critically dependent upon the assumed pore shape. Information on the relative location of different elements in pore space can only be obtained in exceptional cases. Again, it is common practice to assume independent pores; i.e. the pores are assumed to be in direct communication with the external environment. If this condition is not obeyed in an actual case, it will render the common procedures for obtaining pore size data from nitrogen sorption data invalid. Even in the simplest case of independent pores of simple well-defined shape, there are difficulties in interpreting nitrogen sorption data in terms of porous structure. Pore size analysis is traditionally based upon the application of Kelvin's relation between vapour pressure of a capillary condensed phase and pore size. Kelvin's relation was not intended for pores in the mesosize range, as it basically ignores physical sorption of the adsorbate on the pore walls. The latter phenomenon necessitates additional correction terms to Kelvin's relation, as was first pointed out by Deryagin [2] as early as 1940 on the basis of thermodynamic reasoning. This has largely been ignored by subsequent authors in the field. Evidently, there is ample reason to examine the basis for obtaining pore size data from nitrogen sorption isotherms.

DEFINITION OF PORE SIZE

Pore size is an intuitive concept, with obvious meaning for simple geometric pore shapes such as slits, cylinders and spherical cavities. Actual pores may have highly irregular forms with "sizes" quite different in different directions, best to be described by the stereological method [3]. Unfortunately, the concepts of stereology are difficult to translate to the analysis of nitrogen sorption isotherms. As will be shown subsequently, nitrogen sorption data basically yield information on volume and surface area of elements in pore space. Thus, it is most convenient in the present context to define pore size R_h as the ratio between volume and surface area of a certain element in pore space:

$$R_h = V/S \qquad (1)$$

The R_h thus defined would equal $d/2$ for a slit of width d, $r/2$ for a cylinder of radius r, and $r/3$ for a spherical cavity of radius r. For an arbitrary pore shape, R_h may be taken as a characteristic size. One may consider this type of information to be insufficient for certain purposes, but one cannot hope to obtain more detailed information from nitrogen sorption isotherms.

The BET-theory will readily yield an estimate of the total surface area of the porous system from the nitrogen sorption isotherm, S_{BET}. One may argue over the theoretical significance of the BET-equation [4], but its practical validity for obtaining a reliable estimate of the total surface area is only debated in cases where micropores (with sizes <1nm) are present. If the nitrogen sorption isotherm shows a clearly defined plateau around saturation (Fig. 1), then an estimate of the total mesopore volume V_p is readily obtained from the Gurvitch rule [5]. In a number of cases, however, nitrogen uptake by the porous system near saturation becomes asymptotical, and V_p lacks clear definition. Common practice of defining V_p as the volume adsorbate taken up at a certain relative pressure (e.g. $p/p_o = 0.96$) is then potentially dangerous and should be exercised with care. A somewhat better criterion may be the uptake at the upper closure point of the sorption hysteresis loop, but one should then be certain that this volume does not depend upon the final uptake around saturation reached experimentally. In the simplest case of a clearly defined V_p, one may define a mean pore size R_m:

$$R_m = V_p/S_{BET} \qquad (2)$$

It should be realized that S_{BET} in principle also contains, the surface area present in macropores ($R_h > 100$ nm), pores which in practice will not be filled around saturation and will not contribute to V_p. In many cases, the contribution of macropore surface area will be small, and thus will not lead to serious errors in the estimation of R_m, but in the absence of a clearly defined mesopore system, R_m as derived from nitrogen sorption data becomes meaningless, (for an illustration, see Fig. 2). In some cases, e.g. that of pelletised catalysts, there is a very simple way of obtaining V_p. The total catalyst volume can then be obtained either by measuring accurately the dimensions of the catalyst or by determining its apparent density by imbibition in mercury at reduced hydrostatic pressure [6]. A determination of the volume of the solids forming the skeleton of the pore system, e.g. by helium or by imbibition into a wetting liquid, then leads to an estimate of the total V_p, including macropores. The method again is liable to errors: helium adsorption at room temperature may not always be neglected, imbibition liquids such as water may chemically react with the porous solids. Thus, a measure of the mean pores size R_m is obtained with relative case under certain restrictions. For some purposes, R_m will suffice as a characterisation of the porous system, e.g. in routine quality control in the production of heterogeneous catalysts. Often, more detailed information is required on the distribution of pore sizes. Then, one has to take recourse to interpretation of the complete nitrogen sorption isotherm by means of thermodynamic methods.

THERMODYNAMICS OF VAPOUR SORPTION IN CAPILLARIES

Nitrogen sorption in mesopores around 78 °K generally shows hysteresis: it

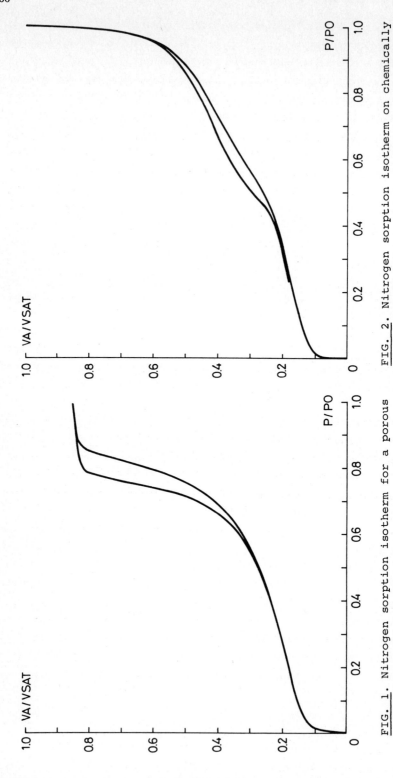

FIG. 1. Nitrogen sorption isotherm for a porous silica: well-defined mesopore system.

FIG. 2. Nitrogen sorption isotherm on chemically corroded glass spheres of around 50 μm diameter : no clearly defined mesopore system.

is found that less nitrogen is taken up at a certain nitrogen relative pressure after increase in relative pressure from a low value than after a decrease in pressure from saturation. This hysteresis is both a fundamental difficulty in interpreting sorption measurement and a source of information on the size and even shape of the porous structure elements. It is usual practice to interpret hysteresis in terms of Kelvin's relation for the capillary condensate in the pores [7]. As stated before, this is not permissible, as Kelvin's equation ignores physical adsorption in the mesopore system. Yet, without physical adsorption capillary condensation below saturation can be shown to be extremely unlikely [8], at variance with general experimental evidence. Kelvin's relation is most easily derived from thermodynamic considerations. It is now clearly necessary to include physical adsorption into the thermodynamic treatment of capillary condensation and evaporation, in order to see in what sense Kelvin's relation should be modified. Different lines of reasoning are possible, but as long as they start from the same assumptions concerning the properties of the porous system, they should, and indeed do, lead to the same final relations, as only the first two laws of thermodynamics are invoked. A general treatment can be based upon the Gibbs Maximum Work principle, as follows:

Consider the experimental set-up for measuring nitrogen sorption isotherms depicted in Fig. 3:

An adsorbent at temperature T is in equilibrium with a vapour phase at a certain (low) pressure p_g in a volume V, composed of a vapour phase volume V_g and a volume of the solid adsorbent V_s. The complete nitrogen sorption isotherm can now be traced by compressing the vapour phase with a frictionless piston, thereby increasing the pressure and reducing V_g, until the saturation pressure p_s is reached. In the absence of adsorption or capillary condensation, the volume V_g at each pressure p_g can be calculated, e.g. in the case of an ideal vapour from:

$$p_g V_g = N_t \bar{R} T \qquad (3)$$

where N_t is the total number of moles vapour in the system and $\bar{R}$ is the gas constant. In practice, V_g is found to be smaller, as N_a moles are adsorbed in the porous system:

$$p_g V_g = (N_t - N_a) \bar{R} T \qquad (4)$$

A simultaneous measurement of p_g and V_g not only yields the complete sorption isotherm, but also the work done on the system W, in reaching saturation:

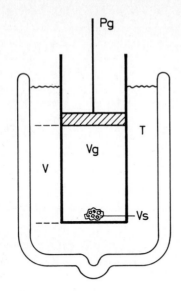

FIG. 3. Illustration of the relation between Work of Adsorption and the adsorption isotherm.

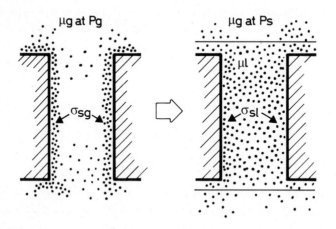

FIG. 4. Change in state of a porous system accompanying capillary condensation. The decrease in interfacial tension at the pore wall from σ_{sg} to ultimately σ_{sl} is the driving force for capillary condensation.

$$W = -\int_{p_i}^{p_s} p_g \, dV_g = N_t \bar{R} T \ln (p_s/p_i) + \bar{R} T (N_s - N_i) - \int_{p_i}^{p_s} N_a \bar{R} T \, d\ln p_g \quad (5)$$

where N_i and N_s signify the values of N_a at p_i and p_s. For simplicity, the molar volume of the adsorbate has been neglected in comparion with that of the vapour. Eq. (5) illustrates that an adsorption measurement in essence is a measurement of the work of adsorption, represented by the last term of Eq. (5). The latter quantity is readily determined from any experimental isotherm.

Gibbs Maximum Work principle now simply states that W is expended in increasing the free energy F of the system:

$$W \geqslant F_s - F_i \quad (6)$$

As illustrated in Fig. 4, F_s and F_i may simply be enumerated in terms of the fundamental thermodynamic properties of the porous system and that of the vapour phase. At p_i, the porous system contributes an amount $\sigma_{sg} S$ to F_i, at saturation this is $\sigma_{sl} S$ towards F_s. A detailed calculation of F from first principles leads to:

$$F_s - F_i = N_t \bar{R} T \ln (p_s/p_i) + \bar{R} T (N_s - N_i) - S(\sigma_{sg} - \sigma_{sl}) \quad (7)$$

Here, σ stands for the interfacial free energy per unit surface area. Its magnitude depends upon the amount adsorbed at the interface, and thus upon p_g. Combination of Eqs. (5), (6) and (7) yields the completely general relation:

$$S \geqslant \frac{\bar{R} T \int_{p_i}^{p_s} N_a \, d\ln p_g}{\sigma_{sg} - \sigma_{sl}} \quad \text{during adsorption} \quad (8)$$

Conversely, during desorption gradual lowering of the pressure causes the system to perform work $W \leqslant F_s - F_i$. This work stems from the decrease in free energy of the system, and an integration along the desorption branch of the hysteresis loop results in:

$$S \leqslant \frac{\bar{R} T \int_{p_i}^{p_s} N_a \, d\ln p_g}{\sigma_{sg} - \sigma_{sl}} \quad \text{during desorption} \quad (9)$$

As S is the true mesopore surface area, it appears that the adsorption branch of the hysteresis loop leads to estimates of the pore surface area which tend to be low, whereas the desorption branch of the hysteresis loop results in estimates of S which tend to be high. Only in the (exceptional) case where there is no hysteresis do the two estimates coincide with the true mesopore surface area.

Eqs (8) and (9) may be compared to the well-known relationship which is due to Kiselev [9] and applied to the calculation of mesopore size and surface area by Brunauer et al. [10]:

$$S = \frac{\bar{R}T \int_{p_i}^{p_s} N_a \, d \ln p_g}{\sigma_{lg}} \qquad (10)$$

It is then clear that (10) is a special case of (8), (9) if <u>a</u> sorption in the mesopore system is completely reversible and <u>b</u> $\sigma_{sg} - \sigma_{sl}$ equals σ_{lg}, the liquid surface tension under all conditions. As stated before, reversibility is exceptional. It can further be demonstrated that condition b is only satisfied if physical adsorption in the multilayer region is negligible. This is certainly not the case for physical adsorption of nitrogen at 78 °K.

For a single pore emptying at a well-defined pressure p_d, and of pore volume V and pore surface area S, equation (9) may be simply integrated to:

$$V/S \geqslant \frac{(\sigma_{sg} - \sigma_{sl}) v_1}{\bar{R}T \ln (p_s/p_d)} \qquad (9a)$$

Obviously, again Eq. (9a) reduces to the wellknown Kelvin-equation if one substitutes σ_{lg} for $\sigma_{sg} - \sigma_{sl}$. Eqs. (8) and (9a) may thus be viewed as "corrected" Kelvin-relations.

The practical application of Eqs. (8) or (9) requires a quantitative knowledge of $\sigma_{sg} - \sigma_{sl}$. We will not indulge in attempts to calculate this quantity from molecular models of physical adsorption (e.g. in Ref.[11]), as $\sigma_{sg} - \sigma_{sl}$ can be obtained directly from experimental data for solid surfaces with small curvature. The t-curve of multimolecular nitrogen sorption for oxides [12], measured on non-porous particles, essentially is a direct determination of the number of moles adsorbed per unit surface area N_a/S as a function of p_g/p_s in the case that no pore filling takes place even at high relative pressures. A line of reasoning similar to the one given above then leads to the expression:

$$\sigma_{sg} - \sigma_{sl} = \sigma_{lg} + \int_{p_i}^{p_s} \bar{R}T \frac{N_a}{S} \, d \ln p_g \qquad (11)$$

Thus, $\sigma_{sg} - \sigma_{sl}$ at any pressure p_i clearly is a function of p_i and can be directly found from a t-curve by integration along the isotherm. This integration is facilitated by using a numerical expression for the t-curve, such as the empirical one [13,14], (see Fig. 5):

$$\ln (p_g/p_s) = -0.371/t^2 + 0.387 \exp(-1.14\, t) \qquad (12)$$

An expression of a somewhat simpler form was proposed by a number of authors, of the general appearance: [15,16]

$$\ln (p_g/p_s) = -C/t^n \qquad (13)$$

In these equations t represents the statistical thickness of the adsorbed layer, defined as:

$$t = N_a v_1/S \qquad (14)$$

where v_1 is the volume of one mole of liquid adsorbate.

Substituting (13) and (14) into (11) leads to:

$$\sigma_{sg} - \sigma_{sl} = \sigma_{lg} + \frac{C\, n\, \bar{R}T}{v_1\, (n-1)\, t_i^{(n-1)}} \qquad (15)$$

where t_i is calculated from (13) at the pressure p_i. Eq. (15) demonstrates that at high pressures p_i (and thus at large values of t_i or for large values of n), $\sigma_{sg} - \sigma_{sl}$ tends to σ_{lg} and thus the thermodynamically correct expressions (8) and (9) tend to the Kiselev-Brunauer form of Kelvin's relation (10). For nitrogen sorption on oxides at 77.6 °K, reasonable estimates for C and n are 0.25 nm^3 and 3 respectively. This means that e.g. at a relative pressure of 0.5 the last term in Eq. (15) exceeds the value of σ_{lg} by approximately a factor of 2, and thus serious errors are incurred when using Eq. (10) instead of the rigorous Eqs. (8) or (9).

CALCULATION OF PORE SIZE FROM RELATIVE PRESSURE OF CAPILLARY DESORPTION

The treatment of the preceding paragraph gives the formal thermodynamic framework for estimating pore size from the experimental pressure of capillary desorption, and shows how the key thermodynamic quantity $\sigma_{sg} - \sigma_{sl}$ can be calculated from the experimentally determined t-curve of multimolecular nitrogen sorption.

It is essential to point out, however, that negligible curvature of the pore walls forms the basis for the existence of the t-curve. It seems reasonable to assume that the same values of $\sigma_{sg} - \sigma_{sl}$ will hold for slit-shaped pores with

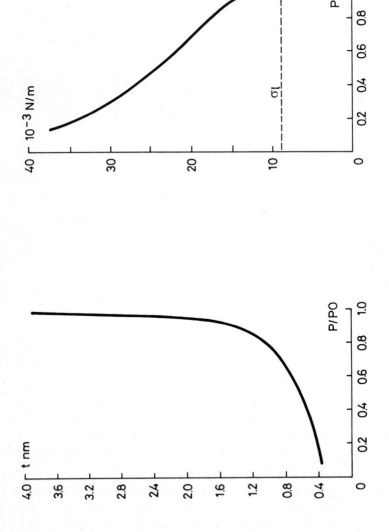

FIG. 5 a. t-curve of multimolecular nitrogen adsorption on oxides at 78°K.

FIG. 5 b. Change of $\sigma_{sg} - \sigma_{sl}$ with pressure, calculated from a. with the aid of Eq. (11).

sufficient width, but there is no a priori reason that the same values will hold for cases where the pore wall curvature is of the same order of magnitude as the reciprocal of the characteristic dimension of the pore. Thus, while Eqs. (8) and (9) are still rigorously valid, lack of knowledge of the key quantitity $\sigma_{sg} - \sigma_{sl}$ will bar any simple direct application. It is at this point that any modelless method will break down, and we will have to adopt lines of reasoning which are semi-thermodynamic at best, the use of which can only be justified by their ultimate practical success.

Three assumptions have to be made:
- Pore geometry can simply be described by either of the following geometrical models: slits, cylinders, spheroidal cavities.
- Desorption from any individual pore takes place at a single well-defined pressure p_d, leaving behind an adsorbed layer of thickness t_d.
- The adsorbed layer can be treated as an adsorbed layer of the same thickness t as that on a non-curved pore surface, except for the influence of the curvature of the interface between adsorbed layer and vapour, which is supposed to possess an interfacial tension σ_{lg}.

For the sake of generality, the experimental t-curve will be represented by:

$$\ln (p_g/p_s) = -F(t) \qquad (16)$$

It can then be shown that for the pressure of desorption the following relations will hold: [17,18]

$$R = k\, V/S \qquad (17)$$

$$R - t_d = \frac{(k-1)\, \sigma_{lg}\, v_1}{RT\, \{\ln (p_s/p_d) - F(t_d)\}} \qquad (18)$$

$$R - t_d \geqslant k \left[\frac{\sigma_{lg}\, v_1}{RT\, \ln (p_s/p_d)} + \frac{t_d \int^R F(t)\, (R - t)^{k-1}\, dt}{(R - t_d)^{k-1}\, \ln (p_s/p_d)} \right] \qquad (19)$$

where k = 1 for slits, k = 2 for cylinders, k = 3 for spheroidal cavities.

Uninviting as these equations may appear, they readily demonstrate that:
- the relation between pore size and pressure of capillary desorption is a complicated function of the t-curve of multimolecular nitrogen adsorption as well as of the pore geometry, in this case the pore wall curvature.
- even if the equations are correctly solved (e.g. by numerical methods) then the estimates of R will tend to be lower limits of the actual values.

This is a consequence of Eq. (9) which states that any intrinsic irreversibility

in the process of capillary evaporation will lead to estimates of the pore surface area which are too high.
- for all but the slit geometry, even the thickness of the adsorbed layer in an otherwise empty pore will deviate from its experimental value, in this case the t-curve on non-porous surfaces.

IRREVERSIBILITY: ORIGINS AND CONSEQUENCES OF HYSTERESIS

The existance of a hysteresis loop in the nitrogen sorption isotherm, even for the most careful and accurate determinations, points to the fact that either the adsorption branch or the desorption branch or both are essentially irreversible in nature. We have already discussed the consequence of irreversibility for the accuracy of determination of pore sizes from vapour sorption measurements. A general treatment of the origin of hysteresis has been given by Everett [19]. Here we are faced with the problem of asserting which of the branches of the hysteresis loop is reversible and thus is best used for the determination of pore sizes. It has since long been realised that scanning of the hysteresis loop, i.e. reversal of the sorption direction at any point on either the adsorption or the desorption branch leads to valuable insight into the origins of hysteresis. The theoretical framework for interpretation of scanning behaviour has been impressively developed by Everett et al. [20]. Here it suffices to recall that most of the evidence available [21] strongly points to the view that in general both adsorption and desorption branch are irreversible in nature, and thus that application of equilibrium thermodynamics to either of these branches will lead to over- or underestimates of pore sizes for fundamental rather than experimental reasons.

Again, further progress can only be made by adopting specific simple pore models. For uniform cylinders open at both ends, irreversibility during filling by capillary condensation, may be viewed as the difficulty associated with bridging the gap between a situation where the pore is empty apart from the presence of an adsorbed film, and a completely filled pore. The filling process entails a number of intermediate stages with higher free energies than either the initial or the final stage, and thus spontaneous filling is retarded. It may be shown that ultimately these free energy barriers vanish once the thickness of the adsorbed film exceeds a critical value t_{cr}, formally given by: [17]

$$- d\, F(t)/dt \big|_{t=t_{cr}} - (k-1)\, \sigma_{lg}\, v_1/(R - t_{cr})^2 = 0 \qquad (20)$$

The pressure at which this occurs again depends upon the pore size and can be calculated from (18). For slits ($k = 1$), there is obviously no critical thickness of the adsorbed layer, and pore-filling is viewed as the process of merging of opposite adsorbed layers once their thickness approaches half the pore width, or at a somewhat earlier stage [22]. For cylinders (k = 2) and spheroidal cavities (k = 3) pore-filling may be described by Eqs. (20) and (18). Everett and Haynes [23] have described hydrodynamic rather than thermodynamic criteria for spontaneous pore-filling. As they did not explicitly account for adsorption phenomena, their treatment seems to be less suitable for the analysis of mesopore sizes.

A more detailed analysis of the desorption process for uniform cylinders leads to the concept that desorption essentially is a reversible process,[21] except for the very last stage where the two menisci approaching each other from opposite sides of the cylinder finally merge, resulting in breaking of the meniscus. If this picture is correct, then Eqs. (19) and (18), after omission of the inequality sign, can be used for calculating pore distributions from the desorption branch in the case of open cylinders. The same holds for slits (k = 1), but for spheroidal cavities (k = 3) the status of Eq. (18) is uncertain to say the least. No picture has as yet been proposed for the reversible evaporation from spheroidal cavities.

For pores of irregular shape, one can seldom expect evaporation to be a reversible process. Therefore, according to Eq. (9), application of equilibrium thermodynamics to the calculation of pore sizes from the desorption branch fundamentally leads to underestimates of pore size. Conversely, the adsorption branch will lead to overestimates, unless a relation of the type of Eq. (20) is employed.

The next step is to remove the constraint that all pores are independent, viz. that they communicate independently with the external environment. One may e.g. simply view the porous system as a mutually interconnected network of pores of different sizes. In such a network, emptying of an individual element in pore space can only occur if there is at least one connecting pathway to its external environment containing pores which have already emptied. According to Ksenzhek [24], emptying of e.g. a porous grain or a tablet of catalyst, will take place by first emptying the outermost shells of the network before evaporation from the interior of the catalyst can take place. The wider the range of pore sizes and the larger the size of the network, the larger will be deviations from reversibility for evaporation from any pore in the interior of the network. In actual pore systems, this latter situation may be rule rather than exception. Relatively little work seems to have been done yet on this aspect of determining pore sizes in real systems. Notice that the actual position of the elements in pore space with respect to each other now becomes of predominant importance, and that not much is known about that in actual systems [25]. There is reason to believe that for pore filling during adsorption, network effects are much less important.

Next comes the question of the lower limit in hysteresis. It was noticed long ago that for all adsorbates there seems to be a critical relative pressure below which hysteresis in a mechanically rigid and chemically inert pore system is not observed. For nitrogen at 78 °K this seems to be around p_g/p_s = 0.42. It is thus generally accepted that hysteresis does not occur in pores below a certain size. [26]. But the real situation seems to be less simple than that. There is good reason to believe that a true capillary condensed state cannot exist below a critical relative pressure, and that around that relative pressure, all pores, independent of their size, are emptied of capillary condensate, although adsorbed films on the pore walls remain stable. Around that relative pressure the equilibrium thermodynamics relations for capillary condensation and evaporation do break down. The fundamental cause does seem to be the assumption of an incompressible fluid for the capillary condensed phase [27,28]. Real liquids are compressible and can be brought in a state of hydrodynamic tension up to a certain limit. Below p_g/p_s = 0.42 the tension acting on the capillary condensed state is believed to have exceeded its limiting value.

Thus, around that relative pressure a steep descent in the desorption branch is erroneously attributed to a peak in the pore size distribution around 2.5 nm. Actually, pores wider as well as smaller than that size may be show to contribute to the desorption phenomenon [22], and assignment on the basis of equilibrium thermodynamical equations is fundamentally invalid.

OBTAINING THE PORE SIZE DISTRIBUTION

From the foregoing discussion, the necessity of proceeding with the utmost caution in any attempt to obtain pore size distributions from nitrogen sorption data, will be obvious.

The earliest, and still most frequently used, method is that of Barret, Joyner and Halenda [29], based on the cylindrical pore model. It is in essence a numerical approximation of the basic integral equation of Wheeler [30]. Let us call L_k the total length of pores with radius between $R_k - \Delta R_k$ and $R_k + \Delta R_k$, mean size R_k, empting over any interval $p_{k-\Delta k}$ to $p_{k+\Delta k}$. We may then write:

$$V_p - V_i = \sum_{k=1}^{i} \pi (R_k - t_i)^2 L_k \qquad (21)$$

where V_i is the total volume of adsorbate at a point on the isotherm p_i/p_s, the lower bound of the i-th pressure interval, and t_i the thickness of the adsorbed film at that pressure. Eq. (21) may be solved for L_i:

$$L_i = \frac{1}{\pi (R_i - t_i)^2} \{V_p - V_i - \sum_{k=1}^{i-1} \pi (R_k - t_i)^2 L_k\} \qquad (22)$$

By dividing the isotherm into suitable pressure intervals and working downwards from saturation, the complete distribution of $L_i (R_i)$ and thus of pore volume and surface area may be obtained, provided pressure intervals are taken small enough (usually ranging from 0.005 to 0.02 relative pressure units).

The equation is specific for cylinderical pores, although the equation may be easily reworked for slit-geometry [31]. A number of remarks can be made concerning its application:

- The value of R_k is the size of pores emptying over an assigned pressure interval. Its magnitude is usually calculated from the Kelvin-equation. As we have seen, this is not permitted for mesopores. For the desorption branch of the isotherm, a solution of R_k must be found from Eqs. (18) and (19). For adsorption, the value of R_k from Eqs. (18) and (20) may be adopted.
- According to Eq. (18), the value of t_i is not only determined by p_i/p_s but also by R_k, and is thus different for each group of pores under consideration. This means that in Eq. (22) t_i has to be replaced by t_{i,R_k}, as calculated from Eq. (18), involving a considerable amount of additional work. (This may be done once and the results used subsequently, if a standard division of the isotherm in intervals is adopted. A computing device is indispensable).
- An accurate estimate of V_p is essential. As discussed before, this may pose problems in cases where the isotherm approaches asymptotically towards saturation.
- Evaporation has to be basically reversible, otherwise incorrect values of R_k will be used and incorrect values of L_k will be obtained. Any incorrect L_k assignment will influence the value of L_i calculated in a subsequent step! This is a basic (and somewhat disturbing) property of the pore size distribution calculation. If for some reason the values found for L_k from preceding steps were e.g. too low, then Eq. (22) will "generate" in subsequent steps values for L_i which are too high and thus pore volume which is totally unrealistically assigned to the smaller pore sizes. It has not been uncommon to sum

$$S_{cum} = \sum_{k=1}^{i} \pi R_k L_k \qquad (23)$$

and to proceed from saturation downward until an interval is found for which $S_{cum} = S_{BET}$. This is a dangerous procedure, as in all cases where L_i initially was underestimated, extending the pore size distribution will finally lead to equality between S_{cum} and S_{BET} by the said "generation" of pore surface area, but to physically unrealistic results. It is better to end the calculation at the lower closing point of the hysteresis loop, but in any case around values of $p_s/p_o = 0.5$ due to the final breakdown of the capillary condensed state.

An even more fundamental problem is the proper assignment of pore shape. All the equations depend heavily upon the assignment of regularly shaped cylindrical pores, and actually are rather sensitive to the choice of geometry. Some authors have argued that a model of regular sphere packing will be more realistic in many cases [32,33]. This does not seem to solve the problem, as we are still faced with the problem of accounting for the influence of pore geometry on capillary equilibria, a problem which seems rather more difficult to solve than in the case of simple pore geometry. Therefore, we prefer to adopt the latter approach for the time being.

ANALYSING THE NITROGEN SORPTION ISOTHERM

In practice, evaluation of nitrogen sorption data in our laboratory proceeds roughly as follows:

1. General inspection of the isotherm. We assume that the measured isotherm is smooth and shows no unexpected irregularities. Otherwise, a repeat determination is imperative. Permanent hysteresis at relative pressures below $p_g/p_o = 0.4$ is highly suspect of experimental error and renders further interpretation hazardous.

2. Determination of S_{BET} from a BET-plot, ideally over the pressure range $0.05 < p_g/p_o < 0.3$. Restricted linearity over this pressure range is a warning to proceed with caution and may e.g. be due to the presence of micropores (see under 4b).

3. Determination of the BET C-value.
For inorganic oxides this should be around 100. Values far in excess of 100 often point to the presence of micropores, which render the physical significance of S_{BET} doubtful. Values substantially lower than 100 may point to non-oxidic surfaces. This means that the t-curve of De Boer et al. [12] may not be valid for the system under consideration and t-curves should be determined separately for the class of materials under investigation [34].

4. Making a t-plot of the adsorption branch of the isotherm [35], Fig. 6. This should be done over the pressure range $0.08 < p_g/p_o < 0.86$. In principle, three different types of t-plot should be distinguished (see Fig. 6).
Type a., the straight line, either points to macropores only (see also under 5), or to the presence of slitshaped pores (see under 4b).
Type b., sloping downwards, points to micropores. In fortunate cases, a second straight line is found at higher relative pressures. This may be taken as representing adsorption in slit-shaped mesopores or in macropores. The surface area present in these pores, S_w, may be obtained from the slope of this second straight line:

$S_w = 1.547 \times 10^6 \, dV_a/dt$ (t in nm, Va in Nm^3/kg, S in m^2/kg).

The t-value belonging to the intercepts of the extrapolation of the two linear parts of the t-plots, gives a rough, theoretically incorrect, measure of half

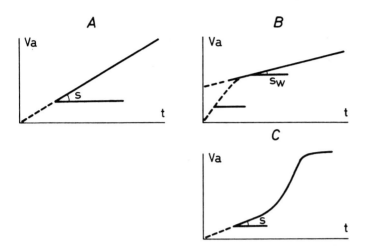

FIG. 6. T-plots of the adsorption branch of nitrogen sorption isotherms; three fundamental cases:
A. Macropores only or wide slit shaped pores
B. Narrow slit shaped pores or micropores
C. "Cylindrical" mesopores.

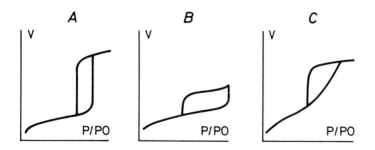

FIG. 7. Classification of hysteresis loop shapes according to de Boer:
A. Type A behaviour; "cylindrical pores"
B. Type B behaviour; slit shaped pores
C. Type E behaviour; indication of bottleneck constrictions.

the mean diameter of the micropores filling during the adsorption process. If sloping down of the t-curve is discernible below $p_g/p_o = 0.3$ then there is reason to suspect the value for S_{BET} obtained under 2.

Type c., shows an upward curvature in the t-plot with ascending t-values. This is a strong indication of the presence of curved pore surfaces, and may, with due caution, be taken as an indication of the presence of cylindrical or spheroidal pores. The pressure corresponding to the steepest slope in the t-plot may be used to obtain an indication of the radius of the most frequently occurring mesopore size from Eqs. (17,18 and 20).

5. Assessing the mesopore volume. If the isotherm shows saturation adsorption in the vincinity of $p_g/p_o = 1$ then complete mesopore filling may safely be assumed If the isotherm shows asymptotical behaviour around $p_g/p_o = 1$, then the upper closing point of the hysteresis loop may be used for assessing V_p, but one should be aware of the possibility of the presence of a non-negligible macropore surface area, interfering with subsequent attempts to calculate pore size distributions. If no clearly developed hysteresis loop is found, then the macropore system may be predominant and no mesopore volume can be calculated. Attempts to obtain a mesopore distribution will lead to false conclusions, and the analysis ends here.

6. Analysis of the shape and the width of the hysteresis loop. Here the shape analysis of De Boer may be adopted [36], Fig. 7.
A-type hysteresis loop points to "cylindershaped" pores, then the t-plot should show type c. behaviour. It should be borne in mind that this is only a model description.
B-type hysteresis may point to the presence of slit-shaped pores: the t-plot should show b- or a-type behaviour.
E-type hysteresis: often associated with "ink-bottle" pores, more generally suspect of severe irreversibility during desorption. The t-plot can be used to obtain information on the curvature of the pore walls.

7. Adoption of a "working model" for the pore shape. For the sake of pore size distribution calculation one has to choose between either "cylindrical" or "slitshaped" pores. It is wise to substantiate the choice by independent information, e.g. from electron microscopy (EM), bearing in mind that we are dealing with a model description. In cases where we are tempted to choose the slit pore model, EM often points to sheet- or layerlike structures. "Cylindrical pore" models are often adopted in cases where EM really points to agglomeration of approximately spherical particles (see under 8).

8. Selection of a pore size distribution calculation procedure. If we have chosen, under 7., a "cylinder pore" model, then the adsorption branch of the hysteresis loop should be analysed on the basis of Eqs. (18) and (20), followed by an analysis of the desorption branch on the basis of Eqs. (18) and (19).

Ideally both distribution curves should coincide. Substantial discrepancies
may point to misinterpretations in an earlier stage of the analysis and thus
are a diagnostic tool in their own right.

In both cases the total cumulative pore surface area S_{cum} should be about
equal to S_{BET}, unless micropores are present, in which case $S_{cum} < S_{BET}$ should
hold. If $S_{cum} \gg S_{BET}$ for the desorption branch but not for the adsorption
branch, then this may well point to severe irreversibility of the desorption
branch, and thus to bottle-neck effects.

If the slit-pore model is adopted, then only the desorption branch can be
used for the analysis on the basis of Eq. (19) with k = 1, which mathematically
very much simplifies the analysis. In this case S_{cum} should be about equal to
S_{BET} or S_w, depending on where the downward deviation in the t-plot of the
adsorption branch is noted. Use of the cylinder pore model in these cases
should lead to $S_{cum} > S_{BET}$ or S_w.

If part of the desorption takes place in the region $0.42 \leqslant p_g/p_o < 0.5$, then
the analysis is of restricted applicability only.

9. Finally, a balanced judgement should be made on the basis of all the information
   acquired during the analysis described, leading to a qualified picture of the
   pore structure of the analysed sample, bearing in mind the approximate nature
   of the analysis method itself.

CONCLUSIONS

Mesopore Structure determination from nitrogen sorption isotherms, is based
upon a series of simplifications of reality. It is inevitable, to adopt simplified
geometric models for the pore shape. Pore interconnectivity has largely to be
ignored. It is difficult to account for irreversibility of sorption in the
hysteresis region in a rigorous manner. The relations governing capillary
condensation as well as capillary desorption are complicated functions of
capillary effects, sorption effects and pore shape effect, and again can only be
formulated quantitatively on the basis of simplifying assumptions. An elaborate
analysis of the sorption isotherm in favourable cases can lead to a block of
reasonably consistent information, leading to a semi-quantified picture of the
mesopore structure. The method is hardly suitable for standardisation and routine
analysis. If one is not primarily interested in the physical significance of the
data obtained, as e.g. for comparative purposes in production quality control,
any convenient simplified routine for analysis can be adopted, dictated by
considerations of convenience.

REFERENCES

1. S.J. Gregg and K.S.W. Sing, Adsorption, Surface Area and Porosity, London 1967, Chapter 8.
2. B.V. Deryagin, Acta Physicochim U.R.S.S. 12, (1940) 139; Proc. Intern. Congress Surface Activity, 2 nd, London 1957, Vol II, 112.
3. F.A. Dullien, in S. Modry (Ed.), Proc. Int. Symp. on Pore Structure and Properties of Materials, Prague 1973, Vol. I, C 173.
4. J.H. de Boer, in Surface Area Determination, D.H. Everett and Ottewill (Eds.), Butterworth & Co, Ltd., London 1970.
5. L. Gurvitch, J. Phys. Chem. Soc. Russ. 47, (1915) 805.
6. H.W. Haynes, Thesis Univ. of Colorado, Boulder, Colo., 1969.
7. S.J. Gregg and K.S.W. Sing, Adsorption, Surface Area and Porosity, London 1967, Chapter 3.
8. J.C.P. Broekhoff, Thesis Delft 1969, Ch. I.
9. A.V. Kiselev, The Structure and Properties of Porous Materials, D.H. Everett and F.S. Stone (Eds.), Butterworth, London, 1957, p. 130.
10. S. Brunauer, R.Sh. Mikhail and E.E. Bodor, J. Colloid Interface Sci., 25, (1967) 353.
11. B.V. Deryagin, J. Colloid Interface Sci. 54, (1977) 157.
12. J.H. de Boer, B.G. Linsen and Th.J. Osinga, J. Catalysis 4, (1965) 643.
13. J.C.P. Broekhoff and J.H. de Boer, J. Catalysis 9, (1967) 15.
14. J.C.P. Broekhoff and B.G. Linsen, Physical and Chemical Aspects of Adsorbents and Catalysts, B.G. Linden (Ed.), Academic Press, London 1970, Ch. I.
15. J. Frenkel, Kinetic Theory of Liquids, Oxford 1946.
16. T.L. Hill, J. Phys. Colloid Chem. 54, (1950) 1186.
17. J.C.P. Broekhoff and J.H. de Boer, J. Catalysis, 9, (1967) 8; 10, (1968) 391.
18. A. Lecloux, in S. Modry (Ed.), Proc. Int. Symp. on Pore Structure and Properties of Materials, Prague 1973, Vol. 4, p. C 43
19. D.H. Everett, in E.A. Flood (Ed.), The Solid-Gas Interface, Marcel Dekker, New York 1967, Vol. II, 1055.
20. D.H. Everett and F.S. Smith, Trans. Fara. Soc. 50, (1954) 187, D.H. Everett, Trans. Fara. Soc. 50, (1954) 1077.
21. J.C.P. Broekhoff, L.F. Brown and W.P. van Beek, in S. Modry (Ed.), Proc. Iupac-Rilem Int. Symp. Pore Struct. and Prop. Mat., Prague 1973, Part IV, C 85.
22. J.C.P. Broekhoff and W.P. van Beek, submitted to Trans. Fara. Soc.
23. D.H. Everett and J.M. Haynes, J. Colloid Interface Sci. 38, (1972) 125.
24. O. Ksenzhek, Russ. J. Phys. Chem. 37, (1963) 691.
25. D.H. Everett, The Structure and Properties of Porous Materials, D.H. Everett and F.S. Stone (Eds.), Butterworth, London 1958, p. 118.
26. R.M. Harris and G. Whitaker, J. Appl. Chem. 13, (1963) 349.
27. C.G.V. Burgess and D.H. Everett, J. Colloid Interface Sci. 33, (1970) 611.
28. R.G. Avery and J.D.F. Ramsay, J. Colloid Interface Sci. 42, (1963) 597.
29. E.P. Barret, L.G. Joyner and P.H. Halenda, J. Amer. Chem. Soc. 73, (1951) 373.
30. A. Wheeler, Catalysis, Vol. II., Rheinhold, New York 1955, p. 118.
31. W.B. Innes, Anal. Chem. 29, (1957) 1069.
32. D.R. Dollimore and G.R. Heal, in S. Modry (Ed.) Proc. Intern. Symp. on Pore Structure and Properties of Materials, Prague 1973, Vol. I, A-73.
33. D.C. Havard and R. Wilson, J. Colloid Interface Sci. 57, (1976) 276.
34. A. Lecloux, in Surface Area Determination, D.H. Everett and Ottewill (Eds.), Butterworth & Co, London 1970.
35. B.C. Lippens and J.H. de Boer, J. Catalysis 4, (1965) 319.
36. J.H. de Boer, in D.H. Everett and F.S. Stone (Eds.), The Structure and Properties of Porous Materials, Butterworth, London 1958, p. 68.

DISCUSSION

K.S.W. SING : In this excellent survey of the problems involved in the determination of mesopore size distribution, you have pointed out that it is important to take into account the effect of any micropore filling contribution on the shape of the nitrogen absorption isotherm. It is useful to recall that the following classification of pore size was put forward by the IUPAC in 1972: micropores, of width less than about 2 nm; mesopores, of width between ∿ 2 nm and ∿ 50 nm; macropores of width greater than ∿ 50 nm. One may regard micropore filling as a primary stage of physisorption, i.e. taking place over the small "monolayer" range of the adsorption isotherm, and capillary condensation as a secondary process, which occurs only at higher relative pressure and after the formation of an adsorbed layer on the pore walls.

In practice, micropore filling is usually associated with the appearance of a reversible Type I isotherm, whereas capillary condensation is observed as a Type IV isotherm, which generally exhibits a hysteresis loop. You have drawn attention to the difficulty created when the steep part of the desorption branch is located at the characteristic minimum relative pressure. I would agree that in this case the usual analysis of the desorption branch provides a misleading picture of the pore size distribution. In our view, this behaviour (i.e. the type E hysteresis loop) is indicative of a wide distribution of pore size, which extends from the micropore into the mesopore range. Unfortunately, it is then impossible to obtain a unique solution for the pore size distribution from a single adsorption isotherm and a range of adsorptions must be used. To identify these and other problems in the interpretation of physisorption isotherms, it is strongly recommended that the adsorption isotherm data should always be given in any publication dealing with reference materials, i.e. standard catalysts or adsorbents.

J.C.P. BROEKHOFF : Thank you for your valuable contribution to the discussion.

N. BAERNS : In various cases it is necessary to use low temperature nitrogen adsorption as well as mercury porosimetry to obtain the pore size distribution of a porous material. When the pore sizes range from 1.5 to 1000 nm, can you make any recommendation with respect to the method of combining the two methods ? Should one use

a) the adsorption or desorption branchs of the nitrogen isotherms
or b) the penetration or retraction data of the mercury porosimetry
respectively ?

J.C.P. BROEKHOFF : Calculated pore size distributions from nitrogen
sorption as well as mercury penetration are seldom identical to
the "true" pore size distributions, but are nearly always deformed.
This deformation is dependent on the method of determination, so it
will not always be possible to join smoothly pore size distributions
from mercury penetration to those from nitrogen sorption. There is
no simple solution for this problem. For A and E type isotherms an
analysis of the adsorption branch of the nitrogen isotherm is probably to be preferred, but compatibility with mercury penetration is
only to be expected in exceptional cases (e.g. for smooth cylinders
open at both ends).

METAL SURFACE AREA AND METAL DISPERSION IN CATALYSTS

J.J.F. SCHOLTEN

Central Laboratory, DSM, P.O. Box 18, Geleen, The Netherlands
and
Department of Chemical Technology, Delft University of Technology,
Julianalaan 136, Delft-2208, The Netherlands

ABSTRACT

A survey is presented of a number of methods for the determination of free-metal surface areas.

The following metals are dealt with: iron, copper, silver, gold, nickel, cobalt, palladium and platinum.
Alloy catalysts are not considered.

Recommendations are made with respect to the choice of the adsorbate, the measuring technique and data interpretation.

1. INTRODUCTION

Besides the well known methods developed for the determination of the total surface area and the pore volume distribution of heterogeneous catalysts and other porous technical materials, special procedures are devised for measuring the extent of the <u>metallic part</u> of the surface ("free-metal surface area"). In industrial catalysis research these methods play a role in the following areas:
- In the development of new metal-on-carrier catalysts.

In most (not in all) cases, our final goal is to reach a dispersion as high as possible, and this dispersion depends i.a. on the metal-load, the time and temperature of reduction, the type of carrier material and its texture. Determination of the free-metal surface area, after and in between various preparation steps, has proved to provide a good guide for optimalization of the catalyst preparation method.

- In laboratory process development work.

It is often very instructive to measure the free-metal surface area of a catalyst before, during and after a laboratory test run. Sometimes it appears that the decline of catalytic activity with time is related to the decline of the free-metal surface area with time, and this might be caused by poisoning or by sintering of the metal crystallites. By combining the measurements with the determination of the mean metal particle size from X-ray line broadening, poisoning effects can be separated from sintering effects. Of course, the change of the total surface area and the pore size distribution of the carrier have also to be taken into account.

- In industrial practice

Here, determination of the free-metal surface area plays a role in trouble shooting. Furthermore, it is a good precaution to control, by free-metal surface area determinations, the quality and reproducibility of new catalyst batches, before loading the reactor.

In this article we first present a short survey of various receipts recommended in the literature, without striving for completeness. After that, the various pitfalls encountered in practice will be reviewed. Finally a number of recommendations will be made with respect to the choice of the adsorbate, the measuring technique and data interpretation.

One general statement, however, has to be made already at the very start of this survey: determination of the free-metal surface area is only one aspect of a more full characterization of the catalyst and of its metallic part of the surface. The catalytic behaviour of the metallic part of the surface depends on at least three variables, viz. the texture, the structure and the surface composition. Some methods available to study these aspects are shortly summarized in table 1.

TABLE 1

Survey of some methods necessary for a full characterization of a metal-on-carrier catalyst, besides free-metal surface area determination.

TEXTURE	BET method. Total accesible surface area.	
	Hg penetration. Pore size distribution.	
	Capillary condensation. Pore size distribution from Kelvin law.	
	EM. Electron microscopy for i.a. structural details.	
	X ray line broadening. Mean metal particle size.	
	SAXS. Small angle X ray scattering. Pore size distribution and metal particle size distribution.	
STRUCTURE	X ray diffraction. Bulk crystallographic structure.	
	LEED. Surface crystallography. Not applicable to technical samples.	
SURFACE ANALYSIS		
	XPS	Qualitative and semi-quantitative elemental analysis.
	AES	of the surface. Analysis 'in depth'.

An introductory treatment of these techniques may, for instance, be found in Anderson's book, 'Structure of Metallic Catalysts' (1).

2. METHODS FOR FREE-METAL SURFACE AREA DETERMINATION

2.1 Iron

To our best knowledge, Emmett and Brunauer were the pioneers in the field. In 1937 they published a paper (2) in which for the first time a very elegant method was introduced for measuring the free-iron surface area of industrial ammonia synthesis catalysts. It has not yet losed its value, and is still daily practice in many laboratories (Fig. 1).

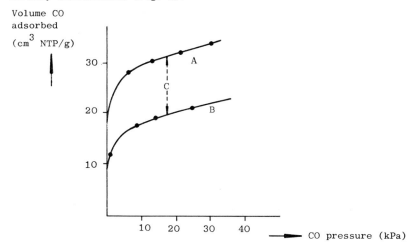

Fig. 1. Adsorption of carbon monoxide on a singly promoted ammonia synthesis catalyst. From lit. 3.
A. Total adsorption at 90 K.   C. Volume CO chemisorbed.
B. Physical adsorption, after evacuation at 195 K.

First carbon monoxide is adsorbed at 90 K; at this temperature the gas is adsorbed both physically and chemically. At 195 K the physisorbed part is pumped away very easily. When, after pumping at 195 K, a second isotherm is measured, this isotherm proves to run at a constant distance under the first. After repeated pumping at 195 K this second isotherm is reproducible; it obviously represents the physical adsorbed carbon monoxide. From the distance between the two isotherms the volume of chemically chemisorbed carbon monoxide may be derived, and this monolayer volume has to be "translated" in the free-iron surface area.

The pumping time chosen for desorption of the physisorbed part may, however, influence the final result. This is due to the fact that on pumping at 195 K also a small part of the weakly chemisorbed carbon monoxide is removed (3). This difficulty may be circumvented by subtracting the physical nitrogen adsorption isotherm from the total carbon monoxide adsorption isotherm A, instead of subtracting the poorly defined physical CO adsorption isotherm B (4). In this case we have to multiply the adsorbed quantities of nitrogen by a factor of 1.05, as appears from the almost constant ratio of the physical adsorption of nitrogen and carbon monoxide of 1.05 on different inert adsorbents (3). The factor 1.05 corresponds with the difference of 5 % in the surface area of a nitrogen and a carbon monoxide molecule.

The "translation" of the extent of the amount of carbon monoxide, chemisorbed in a monolayer, in a free-metal surface area, is hampered by a number of still unresolved fundamental questions, and this difficulty is encountered in all other free-metal surface area determinations as well. First, the distribution of the various crystallographic planes at the iron surface is unknown, and different alpha-iron planes contain different numbers of iron atoms per cm^2, n_{Fe}. For instance:

(111) face: $n_{Fe} = 0.71 \times 10^{15}$
(110) face: $n_{Fe} = 1.7 \times 10^{15}$
(100) face: $n_{Fe} = 1.2 \times 10^{15}$

Second, the number of carbon monoxide molecules that can be chemisorbed per iron atom on the various planes (the surface stoichiometry) is unknown. Various binding forms of carbon monoxide have to be considered, and the ratio of these binding forms may be a function of the iron crystallite diameter.

The best way of circumventing these difficulties would be a calibration of the method by using pure iron samples, the BET surface area of which is determined beforehand from the physical adsorption of methane or nitrogen. For iron crystallites larger than 100 nm in diameter, both Brunauer and Emmett (5) and Westerik and Zwietering (6) found that the ratio of the volume of CO chemisorbed in a monolayer to the volume of nitrogen physisorbed in a monolayer is about 1.20 to 1.25. A reastablishment of this ratio for various iron samples, applying modern ulta high vacuum techniques would be highly desirable.

2.2 Copper

The first publication dealing with the determination of the free-copper surface area of technical catalysts was by Emmett and Skau (7); exactly the same method was applied as advocated for iron samples. Owing to the low heat of carbon monoxide chemisorption on copper (ca. 71.2 kJ/mole at low coverage) it is difficult to differentiate between physically and chemically bonded gas. Therefore, the separate determination of the nitrogen adsorption isotherm (the trick also applied in the iron case) is here an absolute necessity.

In many cases the metallic part of the catalyst surface is small as compared to the total surface area. Than the above method fails, as the physical isotherm B for these samples lies very close to the total adsorption isotherm A (see Fig. 1), so that the result is given by the difference between two large numbers and, in consequence, is not very accurate.

An alternative method is the adsorptive decomposition of nitrous oxide on the metallic part of the catalyst (8):

$$N_2O \text{ (g)} \longrightarrow O \text{ (ads)} + N_2 \text{ (g)} \qquad (1)$$

From a calorimetric determination of the reaction enthalpy, Dell, Stone and Tilly (9) found that, for various values of the oxygen coverage, this enthalpy is equal to half the heat of chemisorption of oxygen plus the heat of decomposition of nitrous oxide, which amounts to -83.74 kJ/mole N_2O. From this it follows that the oxygen atoms are adsorbed on the same sites and in the same form from nitrous oxide as from pure oxygen. Hence there is thermodynamic correspondence between the two surface reactions; their kinetics are different, however.

The lower reactivity in both surface and bulk reactions of nitrous oxide as compared with that of oxygen (9, 10, 11) is related, we believe, to the difference in electronic structure between the two molecules. The linear nitrous oxide molecule, being stabilized by resonance, has the lower reactivity, whereas oxygen is the more reactive molecule, owing to its pseudo-radical character, caused by the presence of two unpaired electrons in different p-levels.

As no pressure change is involved in reaction (1), progress of the reaction has to be measured from the nitrogen enrichment as a function of time, either mass-spectrometrically or by freezing out the nitrous oxide at regular time intervals (12).

Oxygen chemisorption cannot be used, notwithstanding the lower heat of reaction; the rate of gas uptake is relatively insensitive to temperature and pressure (Rhodin (13)), and extremely fast, even at very low temperature. The applicability of nitrous oxide is based on the activated nature of the process, the activation energy being a function of coverage and increasing with coverage (8). This can be seen from Fig. 2, in which results are plotted for a copper-on-magnesium oxide catalyst, used for the dehydrogenation of cyclohexanol.

Figure 2 shows that the higher the temperature chosen, the higher the coverage reached, in accordance with the activation energy increasing with coverage. Finally, around 363 K, full coverage is reached. From experiments with pure copper samples, the BET surface area of which was known, it was found that one oxygen atom is adsorbed per two copper atoms at full coverage, taking the number of surface copper atoms $1.7 \times 10^{15}/cm^2$.

Care should be taken that no break-through of oxygen occurs (oxidation of the sub-surface layers), as observed at 393 K and at 413 K (see Fig. 2). Therefore it is safe to chose 293 K as the temperature of reaction (12), taking into account that no full coverage is reached in that case. The amount of N_2O required to cover 1 m^2 of copper surface area is 0.176 ± 0.010 cm^3 (NTP) at 293 K.

A method of determining the free-copper surface area from nitrous oxide adsorption-decomposition at 293 K by means of a chromatographic technique is described by Dvořák and Pašek (14).

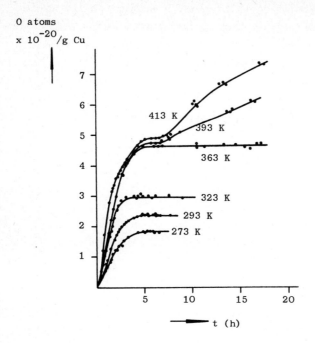

Fig. 2. Adsorptive decomposition of nitrous oxide as a function of time, on a copper-on-magnesium oxide catalyst, at various temperatures. Nitrous oxide pressure: 26.6 kPa (lit. 8).

2.3 Silver

Silver is an important catalyst for the epoxidation of ethylene and it may also be applied for oxidizing methanol to formaldehyde.

Just like with copper, the free-metal surface area can be determined from the adsorption-decomposition of nitrous oxide (15). Full oxygen coverage is arrived at around 443 K, and no break-through difficulties like with copper occur. As may be seen from Figure 3 the course of the reaction is analogous to that on copper, but in a higher temperature range. Above 443 K, correction of the total amount of nitrogen evolved is necessary, due to the start of the quasi-homogeneous decomposition of nitrous oxide, according to:

$$2 N_2O \rightleftharpoons 2 N_2 + O_2 \qquad (2)$$

It is likely that this reaction is catalyzed by the surface silver oxide.

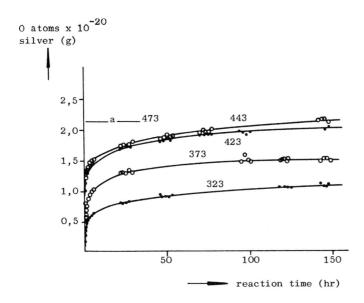

Fig. 3. Oxygen chemisorption via nitrous oxide adsorption-decomposition on a silver on alpha-alumina catalyst at various temperatures as a function of time of reaction. Initial nitrous oxide pressure 19.95 kPa. The numbers in the figure refer to the reaction temperature in K. Line a indicates full coverage adsorption, after correction for reaction (2).

Like with copper, the increase of coverage with temperature is due to the increase of activation energy with coverage.

Contrary to copper, adsorption of oxygen can very well be applied for measuring free-silver surface areas. In accordance with Kholyavenko's results (17), we found that the chemisorption isotherms at 373, 432 and 443 K coincide; the adsorption at 13.6 kPa oxygen pressure corresponds with full coverage at these temperatures (16). Interestingly, chemisorption at still higher temperature (473 K) corresponds with lower coverage, as indicated in Fig. 4.

In Fig. 5 the amount of nitrogen evolved per gram of silver during decomposition-adsorption of nitrous oxide at 423 K is plotted as a function of the BET surface area. The measurements were performed on four silver powders of high purity (15). It can be concluded from this figure that 0.4 cm^3 of evolved nitrogen corresponds to one square meter of silver. This result can be used as a basis for determining the free-silver surface area of silver-on-carrier catalysts.

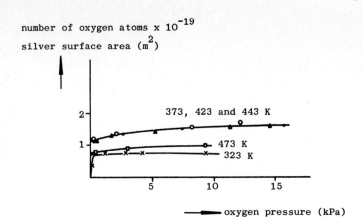

Fig. 4. Oxygen chemisorption on silver powder (lit. 16).
x: 323 K   ▲: 373 K   •: 423 K   O: 443 K   □: 473 K

It is important to note that chemisorption of oxygen leads to a result deviating from the nitrous oxide findings; more oxygen is taken up starting from O_2 (see Fig. 5). Whereas on exposure to nitrous oxide a maximum oxygen coverage of 0.2 cm^3 per square meter was arrived at, exposure to oxygen gave 0.27 cm^3 per square meter. It is known that oxygen is adsorbed on silver in both atomic and molecular forms (17), and that it is the molecular form to which the epoxidation activity is to be described. Nitrous oxide, however, gives rise to only the atomic species and, indeed, a very low epoxidation activity is found with this gas (18). From the difference of 0.07 cm^3 oxygen per square meter silver it follows that, starting from oxygen gas, about 25 % is in the molecular form.

2.4 Gold

Sometimes gold is used as alloying component in group VIII catalysts. It is noteworthy to mention the work by Schrader (19) in which chemisorption of oxygen was observed on (111)-oriented gold films in the temperature range of 473 to 873 K. Complete receipts for measuring free-gold surface areas are not worked out. In view of the low sintering temperature of gold, carefull surface oxidation via adsorptive decomposition of nitrous oxide at not too high a temperature might be a good proposal for further research.

2.5 Nickel

Probably the best method to measure the free-nickel surface area of dispersed nickel specimens is hydrogen chemisorption in the temperature range 273 - 300 K and at pressures up to about 10 - 20 kPa (1). An adsorption isotherm, measured by Sinfelt, Taylor and Yates (20) is presented in Fig. 6, and a result for hydrogen on cobalt is included.

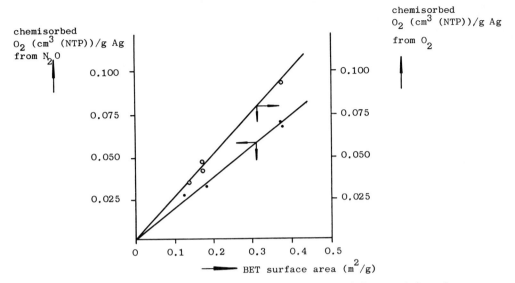

Fig. 5. Oxygen chemisorption from N_2O adsorption-decomposition, and from O_2 chemisorption, on four pure silver samples, as a function of the BET surface area. O: oxygen results   •: nitrous oxide results.

Above 5 kPa the adsorbed volume increases very little with pressure. Back-extrapolation of the isotherms in this pressure region to zero pressure, in order to find the extent of the chemisorbed monolayer, is perhaps a good standard procedure. Assuming 1.54×10^{15} nickel surface atoms per square centimeter, a surface stoichiometry of about one is found in that case.

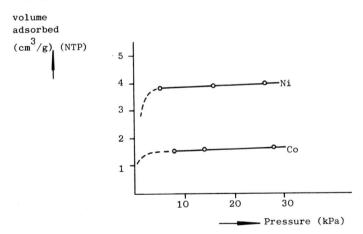

Fig. 6. Adsorption isotherms for hydrogen on silica-supported nickel and cobalt at 291 K. The samples are prereduced in flowing hydrogen, to 643 K. Evacuation at the same temperature (lit. 20).

Isosteric heats of adsorption on the three most densely packed nickel planes were calculated from isotherms by Christmann, Schober, Ertl and Neumann (21). Only small differences were found between the (111), (110) and (100) planes, the heat of chemisorption amounting to 94.2 kJ/mole and remaining constant up to about $\theta = 0.5$. At higher coverages this heat levels of to about 67 kJ/mole at full coverage, due to increasing mutual repulsion of the hydrogen atoms with increasing coverage.

It is interesting to compare the above results with those of Sweet and Rideal (22) for polycrystalline nickel films. They arrive at practically the same result, but at coverages below about 0.03 the isosteric heat of adsorption increases up to 132 kJ/mole. These high heats may be ascribed to chemisorption on edges and corners of the crystallites.

At very high coverage a weak type of hydrogen chemisorption is detected (29.3 kJ/mole). As the full-coverage definition in Sweet and Rideal's work was fully arbitrary, it might be that we are dealing here with a separate type of weakly chemisorbed hydrogen in excess of the strongly held monolayer. Such weakly bound "hydrogen in excess", the existence of which is very well described by Lynch and Flanagan (23) for the case of palladium, might play an important role in the catalytic action of nickel, and deserves separate detection.

Benndorf and Thieme (24), in a TPD study of hydrogen desorption from polycrystalline nickel, detected a small first order desorption peak at 120 K, corresponding with a heat of adsorption of about 25 kJ/mole. This peak was described by them as being due to physical hydrogen adsorption, but in our view the heat of adsorption would be 11 kJ/mole at the utmost in that case. A further study of this question is highly desirable.

In principle carbon monoxide can serve as the adsorbate gas for the determination of nickel surface areas, but the adsorption behaviour is much more complicated as found for hydrogen: at least four states of CO are detected (25). Carbon monoxide can react to $Ni(CO)_4$, especially when small nickel crystallites are present. For instance, nickel carbonyl is readily formed by passing carbon monoxide at a pressure of 101 kPa over finely divided nickel at 350-370 K (1). Of course free-nickel surface area determinations are strongly disturbed by this side-reaction.

2.6 Palladium

Hydrogen may be used as the adsorbent, but conditions have to be avoided where hydrogen absorption into the metal occurs to an undesirable extent. At 343 K and a hydrogen pressure of 133 Pa the equilibrium concentration of absorbed hydrogen does not exceed about 0.2 atom %, and these are the conditions used by Aben (26) for surface area measurements on a series of palladium/alumina, palladium/silica and palladium black specimens. Of course the seriousness of hydrogen absorption in relation to the monolayer uptake is dependent on the palladium dispersion.

For the absorption component not to exceed (say) 10 % of the monolayer uptake, we need the ratio of the total number of surface palladium atoms to the total number of palladium atoms present to be higher than 0.02.

Aben's samples were mostly reduced in hydrogen at 670 K followed by evacuation (16 h) at this temperature. It was shown that 670 K was the minimum temperature at which removal of the hydrogen was sufficiently complete, about 3 % of the surface remaining covered. This residual hydrogen can be readily removed by evacuation at 850 K, but only at the expense of significant sintering.

These conclusions are in agreement with those of Konvalinka and Scholten (27) who measured the TPD spectrum of hydrogen desorbing from 9.5 wt% Pd-on-activated carbon; indeed the last traces of hydrogen desorb above 670 K, the remaining coverage being only a few % of total coverage (see Fig. 6).

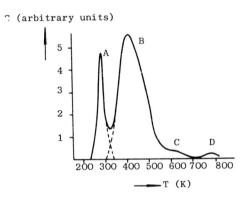

Fig. 6. TPD spectrum for hydrogen on palladium (9.4 wt% Pd-on-activated carbon). Heating rate 10 K/min. Peak maxima at 293, 407, 643 and 780 K.
A: weakly chemisorbed hydrogen ("hydrogen in excess")
B: normally dissociatively chemisorbed hydrogen.

Besides sintering and the dissolution of hydrogen in the bulk, the possibility of chemisorption of weakly bound "hydrogen in excess" has to be taken into account (23). A method to separate this weakly bound hydrogen from the normal dissociatively chemisorbed hydrogen was introduced by Scholten and Konvalinka (27). (For platinum this procedure was described earlier by Freel (33)). Hydrogen is pulsewise adsorbed from a hydrogen/argon mixture at 233 K and a partial hydrogen pressure of 0.3 kPa, and the hydrogen uptake is plotted as a function of the number of pulses (Fig. 7). In the range where the chemisorbed monolayer is formed, all hydrogen is withdrawn from the argon. After a certain number of pulses less hydrogen is withdrawn and than we enter the region of weak hydrogen chemisorption and of palladium hydride formation. By the extrapolation method indicated in Fig. 7, the kinkpoint is found which corresponds with the hydrogen monolayer.

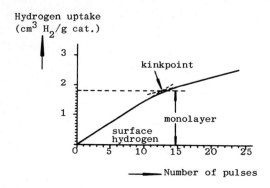

Fig. 7. Pulse-wise hydrogen chemisorption on 9.4 wt% Pd-on-activated carbon. The hydrogen uptake is catharometrically detected.

In the calculation of the free-palladium surface area from hydrogen chemisorption up to the kinkpoint it was assumed that, according to Sundquist (28), 70 % (111), 25 % (100) and 5 % (110) planes are exposed by the palladium crystallites, from which a mean surface concentration of 1.45×10^{19} Pd sites/m^2 is calculated. Furthermore, Aben's (26) surface stoichiometry of one hydrogen per surface palladium atom was accepted. In this way a surface area of 6.78 m^2 Pd/g of catalyst was arrived at.

A carbon monoxide chemisorption isotherm at 293 K was measured with a parallel sample from the same batch which underwent exactly the same pretreatment procedure. This method is described by Scholten and van Montfoort (29), and is based on a calibration with pure palladium samples, the BET surface area of which is known. A free-palladium surface area of 7.5 m^2 Pd/g was arrived at, a value much higher than found from the hydrogen pulse method. If, however, in this last method the distribution of exposed planes is taken to be 33.3 % (111), 33.3 % (110) and 33.3 % (110), the result is 8.19 m^2 Pd/g, which is in much better agreement with the CO result.

Anderson (1) rightly states that the carbon monoxide method has the difficulty that the chemisorption stoichiometry is variable because the proportion of chemisorbed species in the linear and bridged forms can vary, the former offering a chemisorption stoichiometry of one, the latter of two. Because the linear and bridged forms are bound to the surface with different energies and because their chemisorption stoichiometries are different, the relative proportions of the two forms are temperature and pressure dependent. Furthermore, the proportions appear also to depend on the metal particle size (1).

In view of the above, we agree with Anderson (1) that when dealing with transition metals which readily chemisorb hydrogen dissociatively, this should be the first choice for surface area measurement, provided complications due to hydrogen absorption are absent or can be eliminated.

We now return to Aben's method (26), where hydrogen chemisorption is used to determine the surface areas of supported samples. The samples were pretreated in hydrogen (101 kPa) and then in vacuo at 673 K. The combined amount of adsorbed and absorbed hydrogen was then determined under chosen conditions (343 K, 0.133 kPa). This was corrected for the hydrogen uptake of the support, measured from a blank run, and also for the amount of absorbed hydrogen (H/Pd = approx. 0.002) using absorption data for palladium foil. Aben's technique is not suitable for high surface area palladium blacks, because the conditions of pretreatment and measurement of hydrogen adsorption would result in a substantial loss in surface area from sample sintering. Then Sermon's technique (30) may be applied, in which the samples are contacted with hydrogen at room temperature, measuring both the amount of hydrogen chemisorbed and the amount of water generated from the reduction of surface oxygens.

2.7 Platinum

Numerous publications are devoted to the problem of the determination of free-platinum surface areas. This is caused by the fact that especially alumina-supported catalysts are frequently used in industrial processes, involving hydrogenation, dehydrogenation, oxidation and naphta reforming. A good survey of methods published up to 1974 is presented by Anderson (1). He concludes that in general hydrogen chemisorption provides the best method, but optimum conditions for various systems need to be established experimentally. When it can be done without an unreasonably large contribution from adsorption on the support, the measurement of the surface area of dispersed platinum by hydrogen chemisorption is preferably done at 273-300 K and at pressures up to about 0.2 kPa. Further Anderson believes that it is currently best to work on the basis of one hydrogen atom adsorbed per surface platinum atom for the entire platinum size range, with this assumption being regarded as tentative for sizes less than about 1.0 nm diameter.

The separation of the strong rapid chemisorption from the weak slow chemisorption is one of the main problems. We recently tried to solve this problem by applying the hydrogen pulse method (27), mentioned already in the discussion of palladium. A 1 % Pt-0.08 % Zn-on-silica catalyst, prereduced in hydrogen at 773 K and treated in vacuo at the same temperature, was pulse-loaded with hydrogen in argon at 233 K. All further conditions were equal to those mentioned in lit. 27. Results are given in Fig. 8.

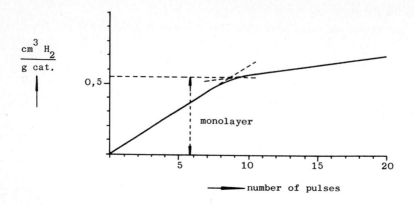

Fig. 8. Pulse-loading of 1 % Pt-on-silica catalyst at 233 K.
Unpublished results by Konvalinka and Scholten. The platinum was alloyed with 0.08 % Zn.

The first eight pulses are rapidly and totally adsorbed, but at the ninth pulse less hydrogen is taken up, and here we enter the region of weak adsorption. By the extrapolation method indicated in the figure a standard method is introduced to find a measure for the extent of the monolayer.

Of course this extrapolation method is an arbitrary one, but more information is gained on the extent and types of weak and strong chemisorption when the pulse run is immediately followed by a TPD run. The TPD diagram for the 1 % Pt-0.08 % Zn-on-silica catalyst is plotted in Fig. 9.

The first desorption trace (top of the peak at 295 K) corresponds to the desorption of weakly chemisorbed hydrogen (β-hydrogen, in the terminology of Cvetanovic and Amenomiya (31)). At least two forms of strongly chemisorbed hydrogen are found (γ- and δ-hydrogen, according to lit. 31); top of the peaks at 460 and 580 K. Above 640 K there is still a substantial amount of desorbing hydrogen. This is partly hydrogen desorbing from platinum, as follows from separate runs with platinum black as the adsorbent (ϵ-hydrogen, top of peak at 720 K).

There are, however, strong indications that above 640 K hydrogen not only desorbs from platinum but also from the silica carrier (spilled-over hydrogen). That spillover occurs indeed follows from the fact that the integrated amount of desorbed hydrogen was only 65 % of the amount pulse-loaded on the catalyst.

The foregoing illustrates the limited significance of measuring the free-metal surface area only, especially when trying to correlate this quantity with the activity of the catalyst. This also follows from the work by Aben, van der Eijk and Oelderik (32), who demonstrated that in the hydrogenation of benzene over alumina-supported-platinum it is likely that only β-hydrogen is involved in the hydrogenation, the benzene hydrogenation rate being a linear function of the population of the β-peaks.

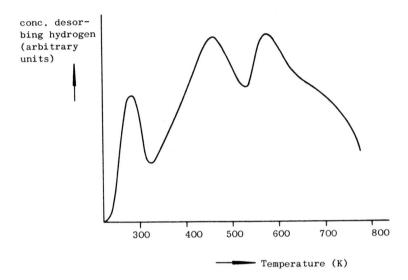

Fig. 9. Temperature programmed desorption of hydrogen from a 1 % Pt-0.08 % Zn-on-silica catalyst. Heating rate 10 K/min.
Apparatus and method are described in lit. 27.

Various other methods for the determination of free-platinum surface areas are available. Carbon monoxide chemisorption is compared with hydrogen chemisorption by J. Freel (33). The ratio between the number of carbon monoxide molecules and hydrogen atoms taken up at full coverage is about 0.87, but for platinum dispersions with H/Pt > 0.25 much lower values are found. Therefore, as argued alrady in the discussion of palladium, hydrogen chemisorption must be recommended as the best determinant of platinum surface areas.

Hydrogen-oxygen titration, the reaction between a monolayer of chemisorbed oxygen and gaseous hydrogen, is very popular as an alternative method for the estimation of surface area in dispersed platinum samples (see for instance lit. 33). This method will not be fully discussed in this paper, and only some small remarks will be made.

From the stoichiometry:

$$O_{(s)} + 3/2\ H_{2\ (g)} \longrightarrow H_{(s)} + H_2O \qquad (3)$$

it follows that the sensitivity of the method is three times as high as for hydrogen chemisorption. This increase in sensitivity can even as well be attained by taking a sample three times as large, and applying normal hydrogen chemisorption, and in doing so the complications of the titration method are avoided.

It was pointed out by various authors that the results of the titration method are strongly influenced by pretreatment. Samples given a "mild" pretreatment in hydrogen have to be distinguished from those given "prolonged" pretreatment in hydrogen (33) (34) (35). This phenomenon might be related to the fact that alumina in the immediate surrounding of the platinum crystallite is partly reduced during "prolonged" pretreatment, whereas the heat of solution of aluminum in platinum is gained, and a Pt/Al-alloy is formed:

platinum + γ-alumina + hydrogen $\longrightarrow$ less γ-alumina + water + Pt/Al-alloy

As a result, more oxygen is taken up as corresponds to the free-platinum surface area due to the re-oxidation of the aluminum, and hence the stoichiometry of the titration method is out of balance.

Another interpretation is possible as well. During "prolonged" pretreatment oxygen-deficient alumina patches are formed to which the platinum crystallites become strongly bound:

$$Pt + \gamma\text{-}Al_2O_3 + x\ H_2 \longrightarrow Pt/Al_2O_{3-x} + x\ H_2O \qquad (4)$$

Then, more oxygen is consumed during the O_2 chemisorption run as part of it is used for the re-oxidation of the alumina. If this is true, it is likely that the chemical bonds of the platinum to the carrier may act as "bridges", by which hydrogen spillover from the platinum to the alumina surface is enhanced (36).

If the above interpretation is correct, there are two reasons for a deviation from the expected stoichiometry: to much oxygen is taken up, and hydrogen is spilled over to the support.

3. GENERAL CONSIDERATIONS AND RECOMMENDATIONS

3.1 First of all, in judging the value of free-metal surface area determinations, one should be aware of the fact that in many cases the condition of the metal surface area "in situ" (during catalytic action) deviates considerably from the condition arrived at after the pretreatment procedure which always precedes the measurement. During catalytic action the metallic part of the catalyst surface is often covered with strongly chemisorbed atoms or carbonaceous deposits, and only a small part of the metal surface is catalytically active. As a result of the pretreatment (oxidation, reduction, followed by evacuation at higher temperature) the condition of the surface is seriously altered, whereas at the same time sintering and reaction of the metal particles with the carrier may occur.

This is one of the reasons that in many cases a direct relation between free-metal surface area and catalytic activity is absent.

3.2 Sometimes <u>a linear relation between free-metal surface area and catalytic activity</u> is found. For instance, Mars and coworkers (37), studying phenol hydrogenation over four different types of nickel catalysts differing in free-nickel surface area from 3 to 100 m^2/g, arrived at an equal activity per m^2 Ni in the temperature range from 308 K to 363 K for all samples investigated.

In other cases, the activity seems to be related to a special type of chemisorbed hydrogen, and it is to be expected that weakly chemisorbed groups are more likely to be involved in catalysis than the strongly chemisorbed ones. A good example may be found in the work by Aben, van der Eijk and Oelderik (see 2.8).

3.3 <u>After pretreatment</u>, including evacuation of the catalysts, <u>the hydrogen coverage of the metal is not always zero</u> (26) (29). This may influence the results in two ways: in applying hydrogen as the adsorbent too low a surface area is found, and using O_2 as the adsorbent too high a value may be arrived at, due to reaction with the remaining hydrogen.

Another unfavourable side-effect of hydrogen is that a number of metals either take up hydrogen in true solution as dissolved atoms, or form hydrides.

Metals in which hydrogen forms a true solution are, for instance, aluminium, chromium, molybdenum, tungsten, iron, cobalt, nickel, manganese, copper, silver and platinum. Solubility data are summarized by Anderson (1). It appears that, except for manganese which behaves exceptional, the hydrogen solubility in all these metals extrapolated to 273 K is less than about 10^{-3} atom% at a hydrogen pressure of 101 kPa.

There is a substantial number of transition metals which form hydrides, often of variable stoichiometry, and these include titanium, zirconium, hafnium, thorium, vanadium, niobium, tantalum, cerium, lanthanum, rare earth metals, and palladium. With these methals the amount of hydrogen absorbed can often be substantial, and it decreases with increasing temperature. Again, data are summarized by Anderson (1). In the vicinity of room temperature and at a hydrogen pressure of 101 kPa the uptake can reach the region of 30-75 atom%.

3.4 <u>Small amounts of impurity metals</u>, often undetected by normal analytical procedures, <u>may accumulate at the surface of metal crystallites.</u> As a rule, metals with the lower surface energy strongly accumulate at the surface (38). If they chemisorb hydrogen, their presence remains unnoticed; if they do not, too low a free-metal surface area is found. An extension of the measurement with temperature programmed desorption of hydrogen can, in favourable cases, provide us with further information

For instance, Scholten and Konvalinka (27), in studying hydrogen chemisorption on pure palladium, arrived at an enthalpy of adsorption of normal dissociatively adsorbed hydrogen of - (90 $\pm$ 5) kJ/mole, from an analysis of the TPD spectra. However, contaminated samples, containing zinc, lead, calcium and carbon as surface impurities (this followed from ESCA analysis) produced a decrease in the enthalpy of adsorption of about 15 %, the enthalpy being of the order of - (65 $\pm$ 5) kJ/mole. Hence, the shift in peak positions in TPD spectra may give a first indication for the presence of impurities.

A palladium-on-carbon sample, intentionally contaminated with 5 at.% germanium in the Pd, gave a shift in the peak position of - 24 K, corresponding with a decrease of the enthalpy of adsorption of about 4.6 kJ/mole. Due to the presence of germanium, the Pd(3d 5/2) line in the ESCA spectrum is shifted from 334.6 (Pd on carbon) to 335.4 eV. Hence, a chemical shift of the order of 1 eV is found, indicating an increase in the electron density on the Pd sites (27).

In this respect it is important to note that Pd/Ge-on-carbon differs essentially in its catalytic behaviour from pure Pd-on-carbon (in the reduction of HNO_3 to NH_2OH). However, from surface area measurements with hydrogen or carbon monoxide as the adsorbates, the difference between both samples can hardly be detected, and additional measurements (TPD, AES, ESCA) are necessary.

3.5 As argued already in 2.7, application of the pulse method, followed by the determination of the hydrogen TPD spectrum, has the additional advantage that information is gained about the extent of "hydrogen in excess" (weakly adsorbed hydrogen) and on the occurence of hydrogen spillover. In principle CO-TPD may be applied as well.

3.6 Important extra information is obtained by measuring X-ray diffraction line broadening (1). When the metal crystallites are partly poisoned, or partly inaccesible due to chemical fixation to the carrier surface, very large differences are found between the mean particle size calculated from chemisorption and from line broadening. However, the absolute accuracy with which the average particle diameter can be obtained by line broadening should not be overestimated, and the influence of particle shape and size distribution factors probably limits this to an accuracy of about 30 %.

Furthermore, there may be sensitivity limitations with supported catalysts if the metal loading is low, and difficulties will in general be encountered at a metal content $<$ 0.5 wt.% in the case of platinum. Since the intensity of diffracted radiation is proportional to the square of the atomic number, this problem becomes more severe for lower atomic number elements.

Depending on the type of diffractometer applied, very small crystallites (say < 2 nm) are totally missed, whereas very large crystallites (say > 50 nm) do not contribute to the line broadening.

Of course, Scanning Electron Microscopy (SEM) and Transmission Electron Microscopy (TEM) may be used to determine mean particle diameters and to discover the detailed form of the crystallites and other structural details (1), but these methods are very laborious.

3.7 Supports for metallic catalysts are not necessarily inert. The consequence of this is twofold: chemisorption may occur on the support, and hence disturb the surface area determination, and gases chemisorbed on the support may play a role in the catalytic reaction itself.

For example, the palladium crystallites in a palladium-on-γ-alumina catalyst are active for H_2-D_2 exchange. After evacuation above 723 K, γ-alumina becomes active for this reaction as well, and the support starts to chemisorb some hydrogen.

Another exmple. Recently we found that the nitrous oxide method, recommended for the determination of copper and silver free-metal surface areas in 2.2 and 2.3, may also be applied with palladium and platinum catalysts (39), and the same general reaction behaviour as outlined in Fig. 2 and 3 was found. With platinum, maximum oxygen coverage is reached at 393 K and 25 kPa pressure. However, under these conditions the silica carrier appeared to be moderately active for N_2O decomposition, and corrections for this effect had to be made by carrying out a separate run with the plain carrier.

It goes without saying that corrections have to be made for physical adsorption, both on the carrier and on the chemisorbed layer.

3.8 Free-metal surface areas may be calculated by calibration with unsupported metal powders of known BET surface area. However, such powders pose a difficult problem for surface cleaning, as they are very susceptible to sintering and particle growth. Moreover, in applying such method, it is tacitly assumed that the distribution of crystallogrphic planes at the surface of the metal powders equals that of the crystallites on the carrier, whereas in reality this distribution is influenced by particle size and by the interaction of the metal particles with the support (see J.R. Anderson, lit. 1, chapter 5).

Calibration with UHV evaporated metal films of known BET surface area (from Xe or CH_4 isotherms) is probably the best method, but this is rarely done.

3.9 An alternative for the calibration method is calculating the free-metal surface area from the extent of the chemisorbed monolayer. This requires a knowledge of the chemisorption stoichiometry at monolayer coverage, and of the number of metal atoms per unit surface area.

In many cases the chemisorption stoichiometry is unknown or variable (CO), and the number of metal atoms per unit area of surface is mostly approximated by assuming that the surface is formed from equal proportions of the main low index planes. There is no scientific basis for this last supposition (see 3.8), and from the observation of the actual equilibrium shapes of f.c.c. crystals by Sundquist (28), a distribution 70 % (111), 25 % (100) and 5 % (110) seems to be a better approximation. Of course the shape of the crystallites is dictated by the anisotropy of the surface energy, and hence will be influenced by the interaction with the support; therefore a general answer to this question can not be formulated.

Perhaps the best advice is to combine the methods indicated in 3.8 and 3.9.

3.10 <u>Dispersion and mean particle size</u>. It is convenient to define the state of subdivision of the metal in terms of the ratio of the total number of surface atoms to the total number of metal atoms present (1): $D_m = n_s/n_t$, where n_s is the number of surface atoms and n_t the total number of metal atoms.

Sometimes it is found from the extent of the chemisorbed layer that D_m is nearly one, its maximum value. This does not necessarily mean that the metal is dispersed atomically. For instance, palladium crystallites with a diameter of 1 nm and containing about 63 Pd atoms, have nearly 90 % of their atoms at the surface.

A low D_m value, on the other hand, does not always mean that the metal crystallites are large. Sometimes, a large part of the metal atoms is inaccesible due to strong binding with the carrier or reaction with the carrier. For instance, nickel catalysts, prepared by nickel precipitation, and particularly those prepared by co-precipitation, are very difficult to reduce, and sometimes only half of the nickel is present as metallic nickel. Another example: in the preparation of Pt-on-alumina catalysts, starting from H_2PtCl_6 and HCl, it appears that after calcining, part of the platinum is dissolved in the alumina surface as Pt(III).

Therefore, in the calculation of D_m from n_s and n_t, one has to differentiate between the metallic and the non-metallic part of n_t, the total number of metal atoms in the starting compound.

Otherwise n_s may be low due to irreversible surface poisoning; in such case the mean crystallite size found from X-ray line broadening points to a much higher dispersion.

It goes without saying that the dispersion D_m does not give any information about the particle size distribution: this distribution follows from electron microscopy, or, in the case of ferromagnetic particles, from a magnetochemical investigation.

<u>The mean particle size</u> may be calculated from the free-metal surface area S and the total volume of the dispersed metal, V. For spherical particles $\bar{d}_{v.s.}$, the mean volume-surface diameter is given by:

$$\bar{d}_{v.s.} = 6 \, V/S \tag{5}$$

Exactly the same expression is found for cubes; than $\bar{d}_{v.s.}$ is the mean edge of the cube.

$\bar{d}_{v.s.}$ is defined by:

$$\bar{d}_{v.s.} = \sum_i n_i d_i^3 / \sum_i n_i d_i^2 \tag{6}$$

where n_i is the number of particles of diameter d_i.

X-ray line broadening, however, gives the weight-mean diameter $\bar{d}_w$:

$$\bar{d}_w = \sum_i n_i d_i^4 / \sum_i n_i d_i^3 \tag{7}$$

From electron microscopy, both (6) and (7) may be calculated, and, of course, also other types of mean diameters.

One should be aware that such mean values are not merely comparable, and not only for mathematical reasons. The upper and lower limits of particle detection in X-ray and EM studies are different, and in the application of eq. (5) a large error is introduced if part of the surface is inaccesible.

REFERENCES

1. J.R. Anderson, "Structure of Metallic Catalysts", Academic Press, London, New York, San Francisco, 1975
2. P.H. Emmett, S. Brunauer, J. Am. Chem. Soc. 59, 310, 1937
3. J.J.F. Scholten, Thesis Delft University of Technology, 1959
4. R.B. Anderson et al., J. Am. Chem. Soc. 70, 2465, 1948
5. S. Brunauer and P.H. Emmett, J. Am. Chem. Soc. 59, 310, 1937
6. P. Zwietering and R. Westrik, Proc. Kon. Akad. Wet. 56, 492, 1953
7. P.H. Emmett and Nis Skau, J. Am. Chem. Soc. 65, 1029, 1943
8. J.J.F. Scholten and J.A. Konvalinka, Trans. Faraday Soc. 65, 2456, 1969
9. R.M. Dell, F.S. Stone and P.F. Tiley, Trans. Faraday Soc. 49, 201, 1953
10. G. Ertl, Surface Science 6, 208, 1967
11. G. Ertl, Z. Phys. Chem. 50, 46, 1966
12. Th.J. Osinga, B.C. Linsen and W.P. van Beek, J. Catalysis 7, 277, 1967
13. T.N. Rhodin, Adv. Catalysis, 5, 39, 1953
14. B. Dvorák and J. Pašek, J. Catal. 18, 108, 1970
15. J.J.F. Scholten, J.A. Konvalinka and F.W. Beekman, J. Catal. 28, 209, 1973
16. K.M. Kholyavenko and M. Ya. Rubanik, Kinet. Katal. 5, 505, 1964
17. W.M.H. Sachtler, Catal. Rev. 4, 27, 1970
18. W. Herzog, Ber. Bunsenges. Phys. Chem. 74, 216, 1970
19. Malcolm E. Schrader, J. of Colloïd and Interface Sc. 59, 456, 1977
20. J.H. Sinfelt, W.F. Taylor and D.J.C. Yates, J. Phys. Chem. 69, 95, 1965
21. K. Christmann, O. Schober, G. Ertl and M. Neumann, J. Chem. Phys. 60, 4528, 1974
22. F. Sweet and Sir Eric Rideal, Actes du Deuxième Congrès International de Catalyse, Paris 1960, page 175, Edition Technip, Paris 1960
23. J.F. Lynch and T.B. Flanagan, J. Phys. Chem. 77 2628, 1973
24. C. Benndorf and F. Thieme, Z. für Phys. Chem. N.F. 87, 40, 1973
25. N. Hayashi and K. Kawasaki, J. Catal. 48, 243, 1977
26. P.C. Aben, J. Catal. 10, 224, 1968
27. J.A. Konvalinka and J.J.F. Scholten, J. Catal. 48, 374, 1977
28. B.E. Sundquist, Acta. Met. 12, 67, 1964 and 12, 585, 1964

29. J.J.F. Scholten and A. van Montfoort, J. Catal. **1**, 85, 1962
30. P.A. Sermon, J. Catal. **24**, 460, 1972
31. R.J. Cvetanovic and Y. Amenomiya, Advan. in Catal. **17**, 103, 1967
32. P.C. Aben, H. van der Eijk and J.M. Oelderik, Proceedings of the fifth international congres on catalysis, West-Palm Beach, Florida, Aug. 1972 North-Holland Publishing Company, Amsterdam 1973, page 717
33. J. Freel, J. Catal. **25**, 139, 1972, and **25**, 149, 1972
34. J. Prasad, K.R. Murthy and P.G. Menon, J. Catal. 1978, to be published
35. F.M. Dautzenberg and H.B.M. Wolters, J. Catal. 1978, to be published
36. P.A. Sermon and G.C. Bond, Catal. Rev. **8**, 211-39, 1973
37. P. Mars, J.J.F. Scholten and P. Zwietering, Actes du deuxième congres international de catalyse, Paris 1960, Edition Technip, Paris, 1961, page 1245
38. W.M.H. Sachtler and R.A. van Santen, Adv. in Catalysis **26**, 69, 1977
39. J.J.F. Scholten, J.C. Rasser and Ming Lee, unpublished results

NOTE ADDED IN PROOF

Platinum

The study by Dautzenberg and coworkers (our reference 35) is now published. See :
Dautzenberg, F.M. and Wolters, H.B.M., J. Catal. **51**, 26 (1978).
den Otter, G.J. and Dautzenberg, F.M., J. Catal. **53**, 116 (1978).

According to these authors our equation (3) should read :

$$O_{(s)} + 2H_{2(g)} \rightarrow 2H_{(s)} + H_2O.$$

After reduction of Pt-on-γ-alumina catalysts at relatively high temperatures (500 and 650°C), the hydrogen chemisorption capacity decreases. This is rationalized by Dautzenberg and coworkers in terms of an alloy model, by assuming that hydrogen chemisorption, which does not take place on Al, is dissociative and needs two adjacent Pt sites. On alloy particles, surface enriched with alumina, the number of Pt dual sites is very low and hence the hydrogen chemisorption is suppressed.

Hydrogen spillover is not considered by these authors as a possible reason for deviating stoichiometries.

DISCUSSION

M. BAERNS : You mentioned the difficulty of reaching agreement on the definition of monolayer coverage to determine the active metal surface, that is to say whether one should use a definite pressure (for instance 50 Torr) or whether one should extrapolate the chemisorption data to zero pressure, (the latter even implying that there is in fact no equilibrium). Would it not be useful to compare maximum coverage in order to have a measure independent of pressure ?

J.J.F. SCHOLTEN : The difficulty lies in the fact that the nature of weak chemisorption (also called "hydrogen in excess" in the literature) is unknown. This type of weak adsorption is reported on Pd (ref. 23 and 27) and on Pt (see Fig. 9, the peak at 273 K). We also found it with Ni (not yet published).

According to ref. 23, we are dealing with "interstitial hydrogen" in between the strongly chemisorbed hydrogen atoms. In ref. 27, we found evidence in the case of Pd that this is subsurface hydrogen in the octahedral interstices just below the surface.

The difficulty lies in the precise separation of these two forms of sorbed hydrogen. den Otter and Dautzenberg (J. Catal. $\underline{53}$, 116 (1970) take $\theta = 1$ at the point where the strong chemisorption comes to saturation. Then $H/Pt_{surface} = 1$. The pressure at which this occurs is not exactly known.
On the basis of this definition of full coverage, the particle size calculated from it is in reasonable accordance with electron microscopy results.

P.G. MENON : One reason for the inconsistent results for H_2-O_2 titrations of supported platinum catalysts, reported in the past, arises from taking the very first H_2 chemisorption measurement on a freshly reduced catalyst as the basis for the stoichiometry calculation. Recently we have found (1) that if the reduced catalyst is first given a few H_2-O_2 cycles at room temperature to anneal or "homogenize" it, and if the calculations are based on H_2 titration (and not on the first H_2 chemisorption) then the stoichiometry HC:OC:HT is consistently 1:1:3, independent of the pretreatment of the catalyst and independent of Pt crystallite size.

On exposure of a Pt-alumina catalyst to temperatures above 500°C in H_2, the stronger chemisorption of H_2 occurring seems to be responsible for the apparent "loss" or "inaccessibility" of a part of the surface Pt and the sharp decrease in the hydrogenolysis activity of the catalyst. Temperature programmed desorption (TPD) of H_2 and the desorption-readsorption of H_2 in a 5% H_2-Argon stream during programmed heating (TPR) of Pt-Al_2O_3, as also the very similar results obtained for Pt-Al_2O_3 and platinum black, support this hypothesis of a stronger chemisorption of H_2 (and spill-over on to Al_2O_3) at higher temperature

(see ref. 2).
1. J. Prasad; K.R. Murthy and P.G. Menon, J. Catalysis, $\underline{52}$, 515 (1978).
2. P.G. Menon and G.F. Froment, J. Catalysis (submitted).

J.J.F. SCHOLTEN : Obviously your results are not in accordance with those of Dautzenberg and coworkers. I have no personal experience with this method and hence cannot give a final judgement.

H. CHARCOSSET : I have two questions and two remarks on your communication. (1) I agree that (Pt, Al) formation takes place easier than it was thought before. We have shown that reduction of Al_2O_3 to Al is inhibited, due to alloying of Pt with Re or Ir.
2) I recall that hydrogen-oxygen titrations of bimetallic catalysts give no true data on the dispersion of the two metals nor on their interaction (see our paper at this symposium).
3) When measuring the free metallic area in monometallic catalysts, a phenomenon to be considered is a possible partial reoxidation of the metal by the support during the evacuation of H_2 at high temperature. Could you comment on this ?
4) H_2-TPD is a very suitable method to characterize metal catalysts. I wonder if there is no experimental artefact during desorption under inert gas flow. For instance in Fig. 9 of your paper, I would think that part of the hydrogen desorbed at low temperature is readsorbed, giving rise to H_2 desorption at higher temperatures. What do you think about this effect ?

J.J.F. SCHOLTEN : Thank you for your comments.
As to your second question, I know that surface hydroxyl groups from the silica support may migrate to the nickel surface and oxidize it. This was shown long ago by Schuit and van Reijen (see their chapter on this subject in "Advances in Catalysis, 10). However , this is a very slow process, and the lower the hydroxyl coverage of the silica the lower its rate.

Our Ni catalyst was reduced in a 30 l/h hydrogen stream, purified via a deoxo-purifier and molecular sieves. Finally it was treated at 773 K in ultra pure argon for 12 hrs. The argon first passes through a cartridge filled with a large amount of a highly reduced Ni catalyst. Under these circumstances the OH-coverage of the silica becomes very low. Moreover, the TPD run was started immediately after pulsing, and pulsing is done immediately after pretreatment.

As to your last question, I do not agree with your point of view. Our TPD measurements are performed under such conditions that the rate of hydrogen desorption is only a little bit higher than the rate of adsorption (see the explanation in ref. 27, and the article preceeding it). Hence, we measured the rate of the adsorption-equilibrium shift. Once the population of a weakly adsorbed type of hydrogen is totally desorbed, it is impossible that it should readsorb in a higher temperature range

Z. PAAL : I highly appreciate your ideas of combining pulse titration and TPD. There are however other factors making the situation more complicated. This is hydrogen retention studied thoroughly by Thomson et al. (1-4). Metals after reduction contain some hydrogen even after a long exposure to argon. Supported catalysts may exhibit spill-over. A very strongly held hydrogen was observed with platinum black which was identified as absorbed hydrogen. This may be slowly exchanged by gas phase hydrogen even at elevated temperature (several hours are required for its total exchange by tritium) and very probably does not desorb even at high temperatures.

The deficit in your hydrogen balance may well be explained by such an absorbed (interstitial) hydrogen. Subsurface hydrogen may only be its "upper layer". So its presence and eventual slow equilibrium with adsorbed surface hydrogen should be considered.

(1) Taylor, G.F., Thomson, J.J., Webb, G., J. Catal. $\underline{8}$, 388 (1967).
(2) Taylor, G.F., Thomson, J.J., Webb, G., J. Catal. $\underline{12}$, 191 (1968).
(3) Altham, J., Webb, G., J. Catal. $\underline{18}$, 133 (1968).
(4) Paal, Z., Thomson, S.J., J. Catal. $\underline{30}$, 96 (1973).

J.J.F. SCHOLTEN : I always have had difficulties in accepting the existence of such very tightly bound hydrogen in metals, but I will study the papers cited by you on this subject again. I don't think that the deficit in our hydrogen balance can be made understandable from the occurrence of very strong hydrogen bonding. The reason is that in a series of measurements (pulse-TPD-pulse-TPD-pulse etc.), the deficit is found after each pulse. This can probably be understood only in terms of H_2 spill-over.

L. GUCZI : I highly appreciate Dautzenberg's idea about the double bonded H atom to two Pt atoms. This agrees with Eley's data on double bonded H atom on Ni with a high heat of formation. However, it would mean that with increasing dispersion the probability to find a proper double Pt site will decrease. Consequently, the H/O ratio in the titration will decrease and this is in contradiction with Hall's data who found it in the reverse way. Would you comment this ?

J.J.F. SCHOLTEN : Highly dispersed Pt particles still have a high number of dual sites ! For instance a 10 Å Pd particle contains about 50 Pd atoms at its surface, on a total number of 63 Pd atoms. Then, the ratio $Pd_{surface}/Pd_{total}$ is nearly one (very high dispersion), but the number of dual sites has hardly decreased.

A. OZAKI : Why H_2 chemisorption on Fe and CO chemisorption on Ni at -196°C is not adopted since even Fe can form carbonyl ?

J.J.F. SCHOLTEN : CO chemisorption is traditionally used. It is a very valuable method, which was designed by Emmett and co-workers. I have no experience with hydrogen on iron, but in principle it might be useful as well.

N. PERNICONE : With respect to the ratio R of the volume of CO chemisorbed to that of N_2 monolayer in pure Fe, I would remark that it depends on the evacuation temperature after the first CO adsorption (ref. 1) and also on the reduction temperature of Fe_3O_4 (ref. 2). In ref. 2, R is about 1.2 when Fe_3O_4 is reduced at 300°C while the usual reduction temperature of Al_2O_3 promoted Fe catalysts are in the range 400-500°C. We have found that, if Fe_3O_4 is reduced at 450°C, the R value is in the range 0.5-0.6, in good agreement with the value extrapolated form the data of ref. 2. While there may be some doubts about the purity of Fe_3O_4 used in ref. 2, this is not the case for a very pure sample.

This problem is very important for the correct determination of the Fe surface area in oxide-promoted Fe catalyst. Therefore a reexamination of the whole problem would be timely.

(1) F.V. Kozenevskaya and A.Y. Rozovski, Kinetics Catalysis

(English Transl.), 7 (1966), 951.

(2) R. Westrik and P. Zwietering, Proc. Koninkl. Ned. Akad. Wet. B 56 (1953), 492.

J.J.F. SCHOLTEN : Thank you for your interesting remark.

H. SHINGU : As to the determination of the free-metal surface area of copper and silver either by the nitrous oxide or the oxygen adsorption method, I would like to point out the dubiousness of the definition "free-metal surface area". Indeed the adsorptive behaviour (to oxygen) and therefore, the surface structure and/or microscopic texture of the metallic layer, are so much variable in wide ranges, depending upon the preparation parameters of the metal, especially supported metal samples, that even the "activated nature" of the adsorption of N_2O cannot be used to distinguish between these surface structures which are so much related to their catalytic behaviours.

J.J.F. SCHOLTEN : On page 2 of our paper we stated already that the determination of the free-metal surface area is only one aspect of a more full characterization of a catalyst. Of course, the surface crystallography may influence the catalytic behaviour, but no methods are available to study this aspect for metals on supports.

Above diameters of 5 nm, the variation of the surface-plane distribution of supported metal crystallites as a function of the diameter is very limited. Perhaps the method of preparation may have an influence on this distribution. It is, however, known that in the epoxidation of ethylene over the (111), (100) and (110) planes of silver, the final activity and selectivity is exactly the same for all three planes.

It is likely that during the reaction recrystallization occurs, resulting in an equilibrium surface plane distribution (see W.M.M. Sachtler in "Catalysis Reviews", $\underline{4}$(1), 27-52 (1970)).

A. BOSSI : Concerning the determination of copper surface area I think it could be interesting to mention the work of Vasilevich et al. (Kinetics and Catalysis, $\underline{16}$(6), 1571, Russian ed.). These authors determined the oxygen adsorption from room temperature to the liquid nitrogen temperature by using a static volumetric apparatus. They found a minimum in the adsorption capacity of the copper at about -136°C. At higher temperatures a bulk oxidation of copper takes place while at lower temperatures chemisorption and physical adsorption were observed. By comparing oxygen adsorption results and nitrous oxide decomposition data they suggest a method for the determination of the copper surface area in $Cu/ZnO/Al_2O_3$ catalysts based on the oxygen chemisorption at $-136°C$.

J.J.F. SCHOLTEN : In the paper by A.A. Vasilevich and co-workers, I did not find a real experimental proof that the oxygen uptake at $-136°C$ is only chemisorption and nothing else. It might be right, but I am not yet convinced. The equality of S_{metal}, from oxygen chemisorption, and S_{BET}, from krypton physisorption, as claimed by Vasilevich, has to be checked with a really pure Cu surface, and not with Cu prepared from a "analytical grade" copper compound, whatever this may be. Moreover, the comparison between both results is not very clearly and quantitatively demonstrated in Vasilevich's paper. Conclusion; reinvestigation is highly desirable.

P.A. SERMON : You have described weak chemisorption (in excess of monolayer capacity) for CO and H_2 upon various metals, which complicates chemisorption measurements of metal areas but enhances catalytic activity. (i) What physical or chemical model do you have for chemisorbed CO and H_2 (particularly in view of the paper of Guczi, this symposium, describing subsurface hydrogen as strongly chemisorbed). (ii) What thermal or basic conditions can we use to minimise weak chemisorption in adsorption experiments, and what conditions will allow us to maximise it in catalytic reactions ?

J.J.F. SCHOLTEN : I have no experience with CO adsorption "in excess". With hydrogen, however, I found this phenomenon for the case of palladium and platinum, whereas for nickel strong indications for its existence are demonstrated during the presentation of my paper.

It is well known that the position of hydrogen atoms in the bulk of metals is quite difficult to study (neutron diffraction). Hydrogen on a metal surface is even much more difficult and we have mainly to rely on indirect evidence.

Arguments for the subsurface position of "hydrogen in excess" may be found in my paper with J.A. Konvalinka, J. Catal. $\underline{48}$, 374 (1977), but no definite proof is available. A hydrogen atom in an octahedral subsurface interstice can have a chemical interaction with a chemisorbed hydrogen atom on the surface. Therefore, I think the subsurface position of the hydrogen atom is distinct from the position in a bulk octahedral hole, and is energetically more favourable.

According to Lynch and Flanagan (J. Phys. Chem. 77, 2628 (1973)) hydrogen taken up in excess of a monolayer is adsorbed in interstitial surface sites, and this hydrogen should be a precursor to interstitially absorbed hydrogen.

I have no experience with extremely strongly bound subsurface hydrogen. For the case of palladium we know for sure (neutron diffraction) that H atoms are absorbed in the octahedral interstices only. This dissolved hydrogen is more weakly bound than the chemisorbed hydrogen ($\Delta H \simeq 24$ KJ/mol versus 90 kJ/mol for the strongest form of H_2 chemisorption). For me it is difficult to accept that there should exist a special type of very strong bonding in the octahedral holes.

Indeed it is to be expected that the weakly chemisorbed forms of hydrogen play an important role in catalysis. A good example of this may be found in a study of benzene hydrogenation over platinum by Oelderik and co-workers (Proc. of the Vth Int. Congr. on Catalysis, Florida, Aug. 1972, paper n° 49). The extent of weakly adsorbed hydrogen is increased by increasing the hydrogen pressure. Weakly bound hydrogen is not perceptible in LEED studies, presumably due to the low pressures applied in this technique.

V.D. YAGODOVSKI : I would like to make some general remarks on the BET equation. It is well known that BET equation is not quite suitable to determine the surface area of the catalysts. There are two reasons to this. Firstly, in the course of the most correct statistical derivation of the BET equation (for example, by T. Hill) we cannot substitute the ratio P/P_s into the X value of this equation. This does not allow us to obtain rigorously the ordinary form of this equation, which is used for the calculation of the surface area and the energetic parameter C from experimental isotherms. Secondly, even if we neglect this circumstance, we will have to take into account that in the BET model the homogeneity of the surface as well as the absence of the interaction between adsorbed molecules is assumed. But the porous catalysts have heterogeneous surfaces.
For this reason the BET equation is not correct for such catalysts. Since the experimental adsorption isotherms often obey the BET equation, we should consider them only as empirical and therefore its

parameters have not the physical meaning which is usually attributed to them. However, the situation is improved in the case of noble gases adsorption at low temperatures, because it is slightly sensitive to the inhomogeneity of the surface. Therefore, it would be better to measure the isotherms at three or four temperatures to be sure that the dependence of isosteric heat of adsorption on adsorbed amount is negligeable.

In my opinion, the BET equation is suitable only for semiquantitative estimation of the specific surface area of a series of catalysts of the same nature, but is not correct for accurate calculations of the absolute value of the surface area.

PROGRESS REPORT ON THE WORK OF THE SCI/IUPAC/NPL WORKING PARTY ON
CATALYST REFERENCE MATERIALS

C.C. BOND [a], R.L. MOSS [b], R.C. PITKETHLEY [c], K.S.W. SING [d] and R. WILSON [e]

a Department of Industrial Chemistry, Brunel University, Uxbridge, Middlesex (UK)
b Warren Spring Laboratory, Stevenage, Hertfordshire (UK)
c ex BP Research Centre, Sunbury, Middlesex (UK)
d School of Chemistry, Brunel University, Uxbridge, Middlesex (UK)
e Division of Chemical Standards, National Physical Laboratory, Teddington, Middlesex (UK)

INTRODUCTION

After the successful conclusion of the Society of Chemical Industry (SCI)/IUPAC/ National Physical Laboratory (NPL) project on surface area reference materials (ref. 1), it was suggested by IUPAC Commission I 6 on Colloid and Surface Chemistry that SCI should initiate a similar project on catalyst reference materials. This was agreed by the SCI Colloid and Surface Chemistry Group and a working party consisting of the authors with Professor Sing as chairman, Dr. Wilson as secretary and Professor Kemball (University of Edinburgh) representing IUPAC Commission I 6 was set up and the first meeting took place in January 1976, when it was decided that the project should be based at the NPL which already had a general programme on reference materials.

The decision to establish a set of heterogeneous catalyst reference materials was justified for the following reasons:-

(A) The variety of methods used in different laboratories for the preparation of catalysts makes it difficult to compare published results. The availability of catalyst reference materials would greatly improve the situation and should thereby aid progress in relating the activity of catalysts to their composition and structure.

(B) Readily available catalyst reference materials would facilitate the validation of procedures for catalyst characterisation and performance testing and permit the comparison of the activity and efficiency of other catalysts.

(C) For the training of research students and experimental workers.

Eight catalysts were chosen as being potentially suitable as reference materials. Three of these, (i) 13% Al_2O_3/SiO_2 acid-type cracking catalyst, (ii) $Co/Mo/Al_2O_3$

desulphurisation catalyst and (iii) 10% Ni/SiO$_2$ hydrogenation catalyst where selected for immediate attention. Others under consideration included (iv) bismuth molybdate, (v) chromium oxide gel, (vi) 0.5% or 0.75% Pt on Al$_2$O$_3$, (vii) 2% Pd on γ-Al$_2$O$_3$, (viii) 0.5 - 0.75% Pt/Re/Al$_2$O$_3$. Selection of the catalyst was based on industrial relevance and commercial availability and in drawing up detailed specifications for catalysts (i), (ii) and (iii), the requirements of chemical engineers in addition to those of chemists were considered.

PROGRESS

Approximately 100 kg quantities of catalyst (i), (ii) and (iii) have been donated by Joseph Crosfield and Sons Ltd, Laporte Industries Ltd and Akzo Chemie Nederland bv respectively. Representative, approximately 100 g, samples have been abstracted using a specially constructed 20 litre rotating riffle. Such a sampling procedure was considered essential as it cannot be assumed that the bulk materials are completely homogeneous with respect to their physical and chemical properties; the catalysts are in the form of powders or granules with mean particle sizes of approximately 90 μm for (i), 1 mm for (ii) and 80 μm for (iii) and there is a tendency for the finer and coarser material to separate.

Questionnaires asking for advice on characterisation and test procedures on catalysts (i), (ii) and (iii) have been distributed to over sixty experts, mainly in the UK. About one third of those contacted have offered advice and practical assistance, and agreed to participate in the physical characterisation and testing of the materials.

After careful consideration of this advice, the Working Party is writing characterisation and test programmes. Details of the procedures to be used, including pretreatment conditions, will be specified and representative samples of the catalysts are being distributed to the participating laboratories. In the case of Ni/SiO$_2$, it was necessary to ask a few of the participating laboratories to make a preliminary assessment of the "ease of reduction" before the pretreatment conditions could be specified.

Routine measurements of chemical analysis, surface area and pore size distribution, have been specified for all three catalysts and will be made by a number of the participating laboratories. It was more difficult to specify methods for the measurement of catalyst activity. However, after careful consideration, it was agreed that cumene cracking should be specified for (i), thiophene desulphurisation for (ii) and benzene hydrogenation for (iii).

Other methods to be used by some of the participants will include selective chemisorption, electron microscopy, X-ray line broadening, Auger and photo-electron spectroscopy and secondary ion mass spectroscopy.

When the Working Party is satisfied that the selected catalysts are suitable for use as reference materials, detailed results will be made available as soon as possible and the certified samples will be made available via the NPL.

Further offers of active participation would be welcome. The Working Party would also be grateful for assistance in obtaining bulk quantities of potential catalyst reference materials, particularly supported noble metal catalysts.

ACKNOWLEDGMENTS

We are very grateful to the three companies, Joseph Crosfield and Sons Ltd, Laporte Industries Ltd and Akzo Chemie Nederland bv for their kind donations of bulk quantities of catalyst materials. We are also indebted to the laboratories that are participating in the measurement programme.

REFERENCE

1 D.H. Everett, G.D. Parfitt, K.S.W. Sing and R. Wilson, J.appl.Chem. Biotechnol. 24 (1974) 199-219.

ORGANIZATION AND FUNCTIONS OF ASTM COMMITTEE D-32 ON CATALYSTS

ARTHUR H. NEAL

Exxon Research and Development Laboratories, Baton Rouge, LA   (U.S.A.)

ABSTRACT

ASTM Committee D-32 on Catalysts was organized in 1975 with the objectives of developing standard test procedures for characterizing catalysts and related materials and of stimulating research in these areas. Members of the committee represent a broad cross-section of the catalyst industry. The technical activities of the committee are carried on in four technical subcommittees covering the major areas of catalyst characterization: Physical Chemical; Physical-Mechanical; Chemical Analysis and Catalytic Properties. Task groups within each of these subcommittees are working on a broad variety of individual test methods which are beginning to evolve as new ASTM standard procedures.

OBJECTIVES AND ORGANIZATION

ASTM Committee D-32 on catalysts was officially organized on January 14, 1975. At that time officers were elected, objectives were decided upon, and a working organization was developed. Some of the steps leading to the formation of this committee were reviewed at the first symposium in this series, in 1976, by Dr. J. R. Kiovsky of the Norton Company. Although only three years old, this committee has nearly 100 members from the United States and several other countries. Membership has been drawn from catalyst manufacturing companies, petroleum refining companies, the chemical process industries and government and academic laboratories. The objectives of the committee are to develop standard test methods for the characterization of catalysts and related materials and to stimulate research in these areas. For all practical purposes, "catalysts" refers to the usual heterogeneous catalysts broadly used in industry. Consideration has already been given, however, to other forms of catalysts such as supported enzymes and the organometallics. It is likely that, in time, committee activities will be broadened to consider standards for these and other new materials as well.

In order to meet its objectives, it was decided to organize the committee into technical subcommittees, each dealing with major areas of catalyst characterization. The four technical subcommittees are the following.

- Physical Chemical
- Physical-Mechanical
- Chemical Analysis
- Catalytic Properties

Other ways of organizing the technical work of the committee were considered, for instance, dividing it into areas of application such as petroleum refining and petrochemical processing. It was felt, however, that the organization chosen was the most effective, since it allows each member to concentrate his efforts in those areas where his expertise is used to best advantage.

In addition to these technical subcommittees, there are other supporting subcommittees, which assist the technical subcommittees in the collection, handling and presentation of data, in the details of writing standard procedures and in interfacing and cooperating with other committees. Four of these supporting subcommittees are presently active as follows.

- Editorial
- Nomenclature and Definitions
- Statistics and Data Handling
- Liaison

It should be pointed out that it is the intention of Committee D-32 to see that its evolving standard procedures are written in metric terms using SI units.

TECHNICAL ACTIVITIES

Within each of the technical subcommittees, the development of individual test methods is largely carried on through "task groups". Typically such a group will contain 10 or 15 individuals, particularly interested in and skilled in a particular area of characterization. The current activities within Committee D-32 are best illustrated by recounting the task groups which are active in each of the subcommittees. First, however, it might be of interest to consider the criteria by which the current projects have been selected from the many areas of testing which were considered. In deciding among the many types of tests which might be worked on, two questions are usually asked. These are:

1. Is a standard test really needed at the present time?
2. Can a standard procedure be developed with reasonable effort?

This tends to eliminate procedures which are of interest only in catalyst research, and not really necessary for the buying and selling of commercial catalysts. This also gives low priority to very sophisticated analytical and characterization procedures which would be difficult to standardize, and where the required capabilities might exist in a limited number of laboratories or installations. Essentially all of the areas of current activity, enumerated below, clearly meet these criteria. Of course, some of the tests currently being developed will require a great deal of work and it could be several years before they lead to standard procedures.

The present projects under study by the Physical Chemical Subcommittee are the following. One group, working on surface area determination by nitrogen adsorption has completed development of a multi-point method, and only the full society ballot is required for final approval of this as an ASTM standard. It is likely that this will be the second published standard from D-32. This group has now turned its attention to a single-point method and will soon begin to consider dynamic techniques. A second group is working on methods for pore size distribution and is very close to writing a draft of the proposed method. Another group is working on the measurement of zeolite content of cracking catalysts by x-ray techniques. Finally, chemisorption techniques for the determination of metals dispersion have been studied both for nickel and for noble metal catalysts.

The Subcommittee on Physical-Mechanical Properties has active task groups working on methods for bulk density determination, particle size measurement, attrition and abrasion. Another group is working on both single-particle and bulk crush strength methods. In this area it is quite obvious that standardized techniques are highly desirable, since specifications on these properties are commonly used in the purchase of most industrial catalysts.

The Chemical Analysis Subcommittee has produced the first actual published standard procedure arising from D-32 activities. This is a method for the determination of cobalt in cobalt-alumina catalysts by a potentiometric technique. The same task group is now working on a draft for a procedure for the measurement of molybdenum which may be applied to the widely used CoMo and NiMo catalysts. Task groups have also been active in the analysis of noble metals as found in reforming and selective hydrogenation catalysts and in the oxidation catalysts on many late model automobiles. Under development at the present time are wet chemical procedures for assay type analyses, as well as rapid x-ray fluorescent procedures which might be used for plant control. It is especially important in the area of chemical analysis that liaison be maintained with other ASTM Committees, and with other bodies interested in chemical analysis, since many of the analytical procedures already developed by other groups can be easily adapted to the special problems of catalyst analysis.

Finally, the Catalytic Properties Subcommittee is interested in developing standardized procedures for measuring the actual performance of catalysts. Describing the activity, selectivity or other performance characteristics of a catalyst obviously involves testing it for a particular process application. This is a very difficult area technically, because of the many variables which must be standardized in any given test. Also, the details of such tests often are considered proprietary by the companies using such tests. In this area, it has been necessary to take a much longer range viewpoint, in expecting to come up with actual standardized procedures. This group, however, has already selected one catalyst activity test which obviously meets the criteria mentioned before, of need and standardizability. This is a microactivity test for fluid cracking catalysts. A test of this type is already broadly used in the petroleum industry throughout the world. This has been a particularly appropriate place to start trying to develop broadly useful standard catalyst activity tests. The successful development of such a microactivity test for cracking catalysts should lead to a succession of other such standard procedures. Certainly there are many types of catalysts in common industrial use where standardized performance tests are necessary and achievable. In addition to the efforts toward developing standard activity tests, other groups within this subcommittee are concentrating on developing standard reactor design criteria and also standard approaches to analyzing and reporting performance data.

To date, ASTM Committee D-32 has been fortunate to have the participation of several individuals and companies from Europe. In addition, we have had contact with, and offers of cooperation from, several national and international groups with similar interests in developing standard testing procedures. ASTM D-32 welcomes the participation of any individuals who are willing to lend their efforts toward the objectives of ASTM D-32, and will strive to cooperate with any groups working along similar lines.

DISCUSSION

J.W. GEUS : Can you give any details about the method(s) ASTM is using to calculate the size-distribution of mesopores from the experimental nitrogen sorption data ?

A.H. NEAL : Work toward standardizing pore size distribution by nitrogen sorption is in an early stage of development. Circulation of a single 82-point isotherm among several laboratories with each calculating pore size distribution by their own method showed considerable disagreement. This indicated the need to reach consensus on data treatment, before proceeding with the standardization of the experimental procedures. This will be the next step, but this is not presently being worked on, since the same task group is concentrating on mercury penetration methods.

MEASUREMENT OF THE ACTIVITY OF SOLID STATE CATALYSTS

G.K. BORESKOV
Institute of Catalysis, Novosibirsk, U S S R

This communication is devoted to a consideration of catalyst test methods for quality control of industrial catalysts, the improvement of test methods and the development of new catalysts. It is not expedient and, in most cases, quite impossible to test the catalysts under conditions duplicating their industrial application. Customarily the procedure is to measure activity under isothermal conditions. This is then used as a basis for estimation of results to be obtained in large scale use. The great variety of catalytic processes precludes any possibility of creating all-purpose installations. However, for certain groups of catalytic reactions common ideas and similar designs of individual units may be employed.

Most catalytic processes are carried out under steady state conditions with essentially constant catalytic properties. Therefore, during tests the steady state reaction conditions conforming to a given temperature and reaction mixture composition must be achieved. For these processes we recommend a gradientless method for measuring the activity with rapid circulation of the reaction mixture. This method permits direct measurement of the reaction rate, excluding the effects of external diffusion and non-uniform flow of the reaction mixture through the catalyst bed. Furthermore, variations in the temperature of the catalyst bed as a result of the reaction (even for processes with sizeable heat effects) may be reduced to a minimum. Electromagnetic pistons or membrane pumps may be used for circulation. It is thus possible to test random-sized catalyst grains and in the case of individual large grains to study the internal diffusion of reaction mixture components. These results may be used directly to calculate the operating parameters of the catalyst in industrial reactors. When developing catalyst preparation methods it is useful to measure catalytic activity in the kinetic region as well. This is done with finely divided particles of the catalyst under study. Such permits estimation of the effect of porous structure, thus determining optimal pore size distribution. For catalyst evaluation, measurements should be carried out at average rates of conversion, sufficiently far from equilibrium. Regulation of component feed, temperature, and the choice of sample for analysis should be automated. Computers should be used for the calculation of results. In the case of complex multi-path reactions the most advantageous scheme involves feed-back between the computer and the operating unit to control variations in parameters.

For a limited number of steady state catalytic processes, the gradientless method is difficult. Problems arise because of side-reactions, e.g. polymerization in the reactor, oxidation of ammonia to nitrogen oxides, or methanol oxidation on silver catalysts, proceeding via external diffusion. Any attempt to carry out these reactions in the kinetic region results in a decrease in selectivity. Results cannot then be characteristic of the catalyst working under industrial conditions. For these reactions the catalyst should be tested in flow-circulation installations operating via external diffusion.

For catalysts working under non-steady state conditions, the testing procedure is much more complicated. In this case it is necessary to determine the catalytic activity for different catalyst states reached in the course of operation as well as the rate of catalyst-state variation when changing the reaction mixture composition and temperature. Here the best approach is a pulse method involving determination of the catalytic activity and calculation of the change in catalyst composition from the material balance. If analysis of the reaction mixture proceeds very rapidly, the change in activity may be investigated using the gradientless and flow methods. It is extremely important to avoid a lag between variations in the reaction mixture composition and variations in the catalyst properties. For this reason the volume of the reaction system including the circulation loop must be minimal.

Tests of catalyst thermostability and duration of operation reduce to a determination of the change in catalytic activity after exposure to the conditions in question.

W.J. THOMAS comment: I would like to concur with and emphasize one remark made in Prof. Boreskov's contribution on the measurement of catalyst activities. Some important industrial processes proceed via concurrent or consecutive paths and are also affected by intra- and inter-particle transport resistances. Any attempt to standardise catalysts for such reactions should take cognizance of the effect of transport resistances on reaction selectivity. At the University of Bath we have recently completed some work which demonstrates that reaction selectivity may be either suppressed or enhanced for certain classes of reactions by changes in inter- and intra-particle resistances to heat and mass transfer. Thus activities should be compared for closely prescribed conditions in the diffusion limited region rather than the kinetic region when normalising catalysts for certain reactions.

THE COUNCIL OF EUROPE RESEARCH GROUP ON CATALYSIS

E.G. DEROUANE

Facultés Universitaires de Namur, rue de Bruxelles, 61, B-5000 Namur. Belgium.

At the beginning of 1975, under the auspices of the Committee on Science and Technology of the Parliamentary Assembly of the Council of Europe, the Research Group on Catalysis (of the Study Group on Surface Chemistry and Colloids) was founded. Its aim was to generate an efficient link between individuals and laboratories willing to share information and cooperate in the fields related to the "surface reactivity and catalytic properties of highly dispersed metals and alloys".

Contacts with the Council of Europe (CE) Research Group on Catalysis can be made either through:

Professor Eric G. Derouane,   Doctor J.P. Massué,
Facultés Universitaires de Namur,   Conseiller Scientifique,
Laboratoire de Catalyse,   or Secrétariat de la Commission de
Rue de Bruxelles, 61,   la Science et de la Technologie,
B-5000-Namur. Belgium   Conseil de l'Europe,
                                       F-67006-Strasbourg Cedex. France.

Members of the CE Research Group on Catalysis are:
Prof. H. Gruber (Austria), Prof. B. Delmon, Prof. E. Derouane, Dr. A. Frennet, Prof. G. L'Homme (Belgium), Prof. J.J. Fripiat, Prof. G. Gault, Prof. B. Imelik, Prof. R. Maurel, Dr. C. Naccache, Dr. J.C. Védrine (France), Dr. J.K.A. Clarke (Ireland), Prof. J.W. Coenen, Prof. J.W. Geus, Prof. V. Ponec, Prof. J. van Hooff (Netherlands), Prof. D.L. Trimm (Norway), Prof. R. Larsson, Dr. H. Lervik (Sweden), Prof. G.C. Bond, Prof. C. Kemball, Dr. R. Joyner, Prof. M.W. Roberts, and Dr. P.B. Wells (United Kingdom).

Among the common programmes of the CE Research Group on Catalysis is the European Reference Catalysts project (Eurocat project).
The aims of this joint effort are as follows:
i. to have a number of common reference catalysts that will be distributed among the laboratories affiliated to the Group and eventually made available to other laboratories upon request and according to their availability. All the reference catalysts are supported metals. A 6% Pt on silica and a 20% Ni on silica are presently available and their characterization is in progress. At a later date, the Group intends to study a 2% Pt on carbon and a 2% Pt on alumina.

ii. to characterize in depth these reference catalysts by all the methods which are currently available in the laboratories participating in the project. Among others, the following characterization tests and methods are used:
   - Analysis by various chemical and physical methods : metal and impurities contents.
   - Adsorption : physisorption, chemisorption, titration, and porosity measurements.
   - Physical techniques: electron microscopy (TEM, SEM), electron spectroscopies (UPS, XPS, AES), magnetic measurements (static and FMR absorption), EXAFS.
   - Catalytic activity and selectivity tests for hydrogenation (mainly benzene) and exchange (olefins and paraffins) reactions.
iii. to compare and discuss critically the results obtained in the various laboratories in order to evaluate and compare the accuracy and reliability of the techniques and testing conditions that are used and optimize these.
iv. to use these catalysts as reference materials, i.e., as standardizing catalysts (common reference points) for a better and quantitative evaluation of catalytic data pertaining to a limited and selected number of catalytic reactions; the attention will be most particularly focussed on reactions and mechanisms pertaining to the synthesis and conversion of hydrocarbons.

Results and their discussion will be made available to the catalytic community by publication in the relevant journals and through a detailed report to be issued by the Committee on Science and Technology of the Council of Europe.

CONCLUDING REMARKS

A scientific symposium has the primary function to convey to the participants, as directly and efficiently as possible, the scientific information detained by the authors of the lectures and communications and by the other participants, and to promote, during the discussions and conversations, the development of new concepts, new interpretations and new lines of research. It also has other less obvious and visible functions, which reach, well beyond purely scientific problems, more general questions related to the social and economical role of fundamental and applied research. Mgr E. Massaux, Rector of the Université Catholique de Louvain, Professor G. Smets, President of the Union of Pure and Applied Chemistry (IUPAC), and Dr. A. Lecloux, President, Société Chimique de Belgique mentioned various aspects of these functions. The very lively Round Table Discussion which took place at the end of the Symposium put into light many other interesting ideas.

The present concluding remarks will be an attempt to speculate on what has been, or what could ideally have been, the "other" functions of the Second International Symposium on the Scientific Bases for the Preparation of Heterogeneous Catalysts.

In many countries (and, most probably, in **all** countries), a major problem is to foster the contact between fundamental science and applied or industrial science. Innumerable methods were imagined, on the one hand, to help that industry benefits completely of the progress of science and, on the other hand, for encouraging scientists to study problems of interest for industry. When he took the IUPAC presidential charge, Professor G. Smets said that "there should not be a separation of the applied field from the fundamental". "Today, there is a much greater overlap and interplay among the different branches of chemistry". If the distinction, or the border, between the scientific and technological domains becomes less clear, the real problem is to foster contacts between scientists working in universities or academic institutes, and scientists and engineers working in industry. In principle, a symposium can provide such contacts.

The Rector of the Université Catholique de Louvain, in his welcoming address, cited the sentence describing the general philosophy of the Symposium, emphasizing three words : the Symposium concerned "selected <u>scientific</u> domains involved in the preparation of <u>real</u>, i.e. <u>industrial</u> heterogeneous catalysts". With the subject described this way, the Symposium attracted investigators working in pure science as well as engineers. The half-day session on "Catalyst Normalization" had the same effect. And, indeed, a good balance in the number of scientists and engineers among the registrants, and in their respective contributions (in the form of lectures, communications or discussions) was achieved, with slightly more than half the participants coming from industry and only slightly less than half of the contributions being presented by them. This approximately equal representation tends to characterize many meetings on catalysis nowadays.

A meeting like our Symposium may serve as a forum for contact between fundamental science and applied or industrial science. But there is no guarantee that making the people meet together will really initiate communications between them.

On a purely scientific and rational basis, there are some reasons why communication might be difficult. The main reason is the pluridisciplinarity of applied science, namely the fact that each field of technology makes use of scientific facts coming from many domains. Enumerating the scientific domains involved in the preparation of heterogeneous catalyst might be long, with inorganic and possibly organo-metallic chemistry, surface chemistry, colloid chemistry and solid-state chemistry being some of the major headings. There is some paradox in the fact that University, which is naturally inclined to investigate all the directions of human thinking and to protect and develop all the disciplines of science, which, in summary, strives to <u>universality</u>, actually houses and produces primarily monodisciplinary minds. It is one task of meetings like our Symposium to initiate mutual communication between different disciplines. The various discussions during the Symposium, and especially during the Round-Table, indicated how imperfect the result still is, except possibly for some aspects of solid state chemistry and colloid chemistry, where fruitful contributions begin to appear.

The role of University is to teach. One participant said that "Industry had come [to the Symposium] more to listen than to tell". Although scientists should be modest vis à vis their colleagues in industry, there is some feeling that a meeting like our Symposium "may contribute to the diffusion of knowledge among people absorbed by the demanding tasks of industrial activity". Teaching supposes that the teacher assembles disparate science in self-consistent, logically organized bodies of knowledge (theories or interpretations); indeed, real teaching is not only teaching how to reproduce results, but mainly teaching the facts and concepts which will serve as new starting points for obtaining new results, for the attainment of more distant goals. In catalyst preparation, the need of such a teaching is immense, because, in this field, as a participant said "there is very little in terms of rational quantitative science and [this] reminds one of the Middle Ages", adding humoristically that, "even then, there was more theory in the Middle Ages!"

There are other reasons, of human or sociological nature, for which communication might be difficult between participant to interdisciplinary meetings on an applied field of science.

There are first the prejudices that scientists and engineers nourish against each other. They are common-place : research in the University is "academic", the making of catalysts is "no science", etc. What might be surprising for an outsider is that the prejudices are not the privilege of the "other" category. University scientists sometimes feel that engineers are not multidisciplinary enough, and during the Symposium Round-Table, an equal number of scientists from University and from Industry expressed the view that (let us keep the past tense) "there was many recipes found over many years, and many facts and pseudo-facts which may or which may not be true" in catalyst formulation and manufacture, or : "catalysis is, on the whole, empiricism still". Some discussion concerning these prejudices was made in the General Remarks of the First Symposium of this young series and during the Round-Table, but we believe the mutual imperfect appreciation of each other contribution is not a major factor against contact between basic research and industrial practice. The participants to the present Symposium, as those to all meetings on catalysis, know each other quite well, and generally recognize the value of the various kinds of contributions.

Much more of a nuisance might be the problem of secrecy, which was already mentioned in the General Remarks of the First Symposium, and which gave way to lively intervention during the Round-Table of the second one. Some quite interesting ideas were expressed, and, conspicuously, they all came from Industry. First, there is the general feeling that "the real knowledge of catalyst preparation is presumably located in industry, with some research also going on in universities, and may be expanding in recent years... A lot of [that knowledge] cannot be protected by patents, and giving it away free would help the competition possibly more than the Company which was holding it out". Even keeping in mind that "industrial people have not told everything about their best catalyst" that "they have not told how to do and what to do", one participant remarked that "perhaps they have told how not to do and what not to do [to their University Colleagues] and that is enough to have a meeting". Another correction to these view is that knowledge detained by industry is of another nature than the one which is normally transmitted in scientific meetings : "when we are talking about industrial know how, in the field of catalysis, ... what we are often talking about is formulations which work, as opposed to an organized body of knowledge... There is a lot of work which can be shared with the universities on obtaining an organized body of knowledge. I do not say we are hiding it in industry. I do not think it is hidden in industry in the field of catalysis. It may be hidden in some other subjects, but not catalysis".

Even supposing that a large amount of knowledge is hidden by industry, the participants rightly recalled that the extent to which knowledge is hidden is necessarily limited. On one hand, a catalyst manufacturing company must "publish as much knowledge as possible, for two reasons. One reason is that the customers, in order to use [our] catalysts, have to know a lot of what we could call proprietary knowledge. Another [reason], which should not be forgotten, is the value of public relation by distributing knowledge, by publishing papers, preferably of a scientific nature". On the other hand, a senior scientist of a major oil company reminded his "academic colleagues that there is in fact a tremendous amount [of information] available on ... the latest thing that this or that company is doing. Unfortunately, this is in the patent literature, and is not the accepted kind of literature, that one gets in the mail, as a scientific journal ... It is almost a necessity for having a valid patent to have things in there in such a way that anyone that is knowledgeable in the field can repeat it".

Most gratifying for the organizers of the Symposium was the following remark from Industry, which related directly to the functions of a Symposium grouping University scientists and Industry engineers : "If you did not have a conference of this nature, there is no way to flush out the information that we have".

The idea emerged from the Round-Table discussions that when considering specific problems concerning a specific catalyst, and more general ones, common to many systems, the secrecy constrainsts apply only to the former. It is our personal feeling that scientific research in University is profitably <u>inspired</u> by the practical problems of Industry, but that only problems having a reasonable degree of generality are of interest. A scientist from Industry confirmed : "all our problems [in Industry] are, basically, not specific problems. I feel that I do not learn relatively much out of another paper on a specific catalyst ...". If this is true, there is much hope that communication could function satisfactorily. But we have the impression that University scientists and Industry engineers need cooperation for extracting the general problems from the specific scientific studies and specific practical difficulties.

We might conclude this part of our considerations by saying that the discussions confirmed, more clearly than we had thought, the central role of a meeting like our Symposium in helping communication between Industry and University.

But this kind of Symposium might well have, in addition, a quite different function. There are several reasons, pertaining to the structure of our societies (here, eastward or westward) which accentuate the natural predilection of University scientists to stick to his or her scientific domain, to keep away of pluridisciplinarity to avoid new budding fields and to ignore the industrial aspects of science. Working in well established fields is comfortable, in many respects - too comfortable! Intellectually, this obviously demands less effort than grasping new areas, new relationships between branches of science. For the financial welfare of a laboratory, long-established frames and circuits of money allocation are often more secure than hypothetical and timid new programs, unless the new field is obviously part of a fashionable subject of national or international interest. On psychological grounds, a scientist will feel more secure in his field or micro-field than when discussing with specialists of other disciplines with a different jargon, and his convivial spirit will be more gratified by meeting again and again his old fellow-specialists, who usually do not forget to cite them in communications and discussions. It is easy to describe ironically this situation. But, is it sure that the system the scientists are part of is not responsible for that situation ? The career of a scientist largely depends on the recognition of his work by his peers. Let us assume that his value on the job market is measured by this $\underline{R}$ecognition, e.g. by his Science Citation Index Number, that we shall simply note R; R is approximately given by :

$$R = N \times A \times V$$

where N is the $\underline{N}$umber of people working in the field (the fellow-specialists), A the time the field has been established, e.g., its $\underline{A}$ge, and V the $\underline{V}$alue of the cited publication. Two factors out of three are maximalized if the scientist keeps working in his relatively old, monodisciplinary field.

We feel that one of the major function of a meeting like our Symposium is to help create recognition for work done outside the well established fields. It is to attract more interest from the University scientists in multidisciplinary fields and in fields highly related to industrial activities. It is to raise these fields from the level of practical knowledge to hopefully fashionable domains, ultimately recognized as "fully scientific". Scientific meetings on multidisciplinary topics which, sometimes, look little scientific, because they are inspired by practical or industrial problems, certainly contribute to the recognition of the corresponding investigations as truely scientific. These meetings can incite University laboratories to make the high quality research which the respective countries have the right to expect, in view of the amount of money they spend.

This is the point of view of an University scientist. We presume that a meeting like our Symposium also has additional functions, specific for fulfilling the needs of Industry investigators and engineers.

In view of the above considerations and of the clearly expressed hope of the participants, the organizers of the Symposium feel that a third one should be added to the series, probably in 1982. They will try to retain what was estimated to be good in 1978. But clearly, a continued effort should be made for ensuring maximum pluridisciplinarity to the future Symposium.

B. DELMON

PARTICIPANTS

ADAMIS V.	Petrobras Ilha do Fundao Quadra 7 (CENPES)	BRASIL
ALBERT R.	Degussa Wolfgang Technical Service Catalysts Division P.O. Box 602 6450 Hanau 1	W.GERMANY
AL-CHALABI H., Dr.	Kuwait Institute for Scientific Research P.O. Box 12009 Safat	KUWAIT
ALLEY S.K., Jr., Dr.	Union Oil Company of California Research Center Box 76 Brea, CA 92621	U.S.A.
ALZAMORA L.E.	University of Bradford Bradford W. Yorks. BD7 1DP	ENGLAND
ANDERSSON A.	Dept. of Chemical Technology Kemicentrum Box 740 S-22007 Lund 7	SWEDEN
ANDREW S.P.S., Dr.	Imperial Chemical Industries Ltd. Agricultural Division P.O. Box 6 Billingham Cleveland TS23 1LD	ENGLAND
ANTONELLI G.	Euteco S.p.A. Via Montesanto 23 Sesto S. Giovanni Milano	ITALY
ANTOS G.J., Dr.	UOP, Inc. Corp. Research Center Algonquin & Mt. Prospect Rds. Des Plaines, Ill. 60016	U.S.A.
ANUNDSKAS A., Dr.	Norskhydro A.S. Research Centre 3901 Porsgrunn	NORWAY
AOMURA K., Prof.	Faculty of Engineering Hokkaido University North 13, West 8 Sapporo 060	JAPAN
ARNTZ D., Dr.	Degussa Wolfgang Abt.FC-0 Postfach 602 D-6450 Hanau 1	W.GERMANY

BACHELIER J.	Université de Caen
Ensica
Avenue d'Edembourg
F-14000 Caen	FRANCE

BADIE P.	Société Norsolor
B.P. 108
F-57503 Saint Avold	FRANCE

BADILLA-OHLBAUM, R.	Imperial College
Dept. of Chemical Engineering
London SW7 2BY	ENGLAND

BAERNS M., Prof.	Ruhr-Universität Bochum
Postfach 102148
D-4630 Bochum	W.GERMANY

BANKS, R.G.S.	British Gas Corporation
Midlands Research Station
Wharf Lane, Solihull
W. Midlands	ENGLAND

BARANSKI A., Prof.	Jagellonian University
Institute of Chemistry
Krupnicza 41
30-060 Cracow	POLAND

BARBY D., Dr.	Unilever Research Lab.
Port Sunlight, Wirral
Merseyside	ENGLAND

BARRACLOUGH R.N., Dr.	National Res. and Dev. Corporation
P.O. Box 236, Kingsgate House,
66/74 Victoria Street
London SW1	ENGLAND

BARRAL R.	Produits Chimiques Ugine Kuhlmann
Centre de Recherches de Lyon
F-69310 Pierre-Bénite	FRANCE

BATTY C.R.	British Petroleum Co. Ltd.
Research Center
Chertsey Road, Sunbury-on-Thames
Middx.	ENGLAND

BECKER E.R.	Air Products & Chem. Inc.
P.O. Box 538
Allentown, Pa. 18104	U.S.A.

BEECROFT T.	Laporte Industries Ltd.
Moorfield Road
Widnes,
Cheshire, WA8 OJU	ENGLAND

BERETS D.J.	American Cyanamid Co.
1937 Westmain Street
Stamford, Conn. 06903	U.S.A.

BERREBI G.	Pro-Catalyse
21, rue Jean Goujon
75008 Paris	FRANCE

BETTAHAR M., Dr.	O.N.R.S. Abri-Hydra Alger	ALGERIA
BIRKENSTOCK U., Dr.	Bayer AG OC-P A-Fabrik, Geb. 919 509 Leverkusen	W.GERMANY
BLANCHARD G.	Institut de Recherches sur la Catalyse 79 Bd. du 11 novembre 1918 69626 Villeurbanne	FRANCE
BLASER H.U.	Ciba-Geigy AG R-1060 312 4002 Basel	SWITZERLAND
BLINDHEIM U., Dr.	Central Institute for Industrial Research Forskningsv. 1 P.O. Box 350 Blindern, Oslo 3	NORWAY
BOERMA H.	Unilever Research Olivier van Noortlaan 120 Vlaardingen	THE NETHERLANDS
BOERSMA M.A.M., Dr.	Koninklijke Shell Laboratorium Badhuisweg 3 Amsterdam-N	THE NETHERLANDS
BOHLBRO H.	Haldor Topsøe A/S Nymøllevej 55 2800 Lyngby	DENMARK
BORDES	Université de Compiègne B.P. 233 60206 Compiègne Cedex	FRANCE
BOSSI A., Dr.	Montedison S.p.A. Istituto Donegani Via G. Fauser, 4 Novara	ITALY
BOUDART M., Prof.	Stanford University Dept. of Chem. Engineering Stanford University Stanford, CA 94305	U.S.A.
BRADSHAW D.I., Dr.	British Gas Corporation London Research Station Michael RD, Fulham London SW6 2AD	ENGLAND
BRANDENBURG J.A., Dr.	Unilever Research P.O. Box 114 3130 AC Vlaardingen	THE NETHERLANDS
BRAUN G.	Bessunger Strasse 194 D-6100 Darmstadt	W.GERMANY

BREMER H., Prof.　　　　　T.H. Carl Schorlemmer
　　　　　　　　　　　　　42 Merseburg
　　　　　　　　　　　　　Leuna-Merseburg　　　　　　　　D.D.R.

BRITSCH R.　　　　　　　　Inst. für Angew.Physikalische
　　　　　　　　　　　　　Chemie
　　　　　　　　　　　　　Im Neuerheimer Feld 253
　　　　　　　　　　　　　D-6900 Heidelberg　　　　　　W.GERMANY

BROEKHOFF J.C.P., Dr.　　Unilever Research
　　　　　　　　　　　　　P.O. Box 114
　　　　　　　　　　　　　3130 AC Vlaardingen　　THE NETHERLANDS

BROOKS C.S.　　　　　　　United Technologies
　　　　　　　　　　　　　Research Center
　　　　　　　　　　　　　East Hartford, Conn.06108　　　U.S.A.

BRUNELLE J.P., Dr.　　　Rhône-Poulenc Industries
　　　　　　　　　　　　　Centre de Recherches d'Aubervilliers
　　　　　　　　　　　　　12-14, rue des Gardinoux
　　　　　　　　　　　　　F-93308 Aubervilliers　　　　FRANCE

BUKOWIECKI S., Dr.　　　Chemische Fabrik
　　　　　　　　　　　　　Uetikon,
　　　　　　　　　　　　　CH-8707 Uetikon a.See　　SWITZERLAND

BULENS M., Dr.　　　　　　Boulevard du Souverain 259
　　　　　　　　　　　　　1160 Bruxelles　　　　　　　　BELGIUM

BUSCH-PETERSEN B.　　　Katalysatorenwerke Houdry-Hüls GmbH
　　　　　　　　　　　　　Postfach 1320
　　　　　　　　　　　　　D-4370 Marl　　　　　　　　W.GERMANY

CAHEN R., Dr.　　　　　　Labofina S.A.
　　　　　　　　　　　　　Chaussée de Vilvorde 98-100
　　　　　　　　　　　　　1120 Bruxelles　　　　　　　BELGIUM

CANDIA R.　　　　　　　　Haldor Topsøe A/S
　　　　　　　　　　　　　Nymøllevej 55
　　　　　　　　　　　　　DK-2800 Lyngby　　　　　　　DENMARK

CARBUCICCHIO O., Prof.　Istituto Fisica Universita
　　　　　　　　　　　　　Via H. d'Azeglio 85
　　　　　　　　　　　　　Parma　　　　　　　　　　　　　ITALY

CHABOT J., Mrs.　　　　　CEA-CEN SACLAY
　　　　　　　　　　　　　DRA-SAECNI, B.P. 2
　　　　　　　　　　　　　Gif-sur-Yvette　　　　　　　　FRANCE

CHADWICK D., Dr.　　　　Imperial College
　　　　　　　　　　　　　Dept. of Chemical Engineering
　　　　　　　　　　　　　London SW7 2BY　　　　　　ENGLAND

CHARCOSSET H., Dr.　　　Institut de Recherches sur la
　　　　　　　　　　　　　Catalyse
　　　　　　　　　　　　　79, bd du 11 novembre 1918
　　　　　　　　　　　　　F-69626 Villeurbanne　　　　FRANCE

CHENEBAUX M.T., Mrs.	Institut Français du Pétrole Av. de Bois-Préau 4 B.P. 311 F-92506 Reuil-Malmaison	FRANCE
CHENOWETH J.G.	Imperical Chemical Industries Ltd. Petrochemicals Division HQ P.O. Box 90, Wilton, Middlesbrough, Cleveland TS6 8JE	ENGLAND
COENEN J., Prof.	Unilever Research Olivier van Noortlaan 120 Postbus 114 Vlaardingen	THE NETHERLANDS
COGNION A.	Produits Chimiques Ugine Kuhlmann Centre de Recherches de Lyon F-69310 Pierre-Bénite	FRANCE
COOPER B.H.	Haldor Topsøe A/S Nymøllevej 55 DK-2800 Lyngby	DENMARK
COSGROVE L.A., Dr.	Air Products & Chemicals Inc. P.O. Box 427 Marcus Hook, PA 19061	U.S.A.
COSYNS J.	Institut Français du Pétrole Av. de Bois-Préau 4, B.P. 311 F-92506 Reuil-Malmaison	FRANCE
COURTINE P.	Université de Compiègne B.P. 233 60200 Compiègne	FRANCE
COURTY Ph.	Institut Français du Pétrole Av. de Bois-Préau 4, B.P. 311 F-92506 Reuil-Malmaison	FRANCE
DAUMAS J.C., Dr.	Rhône-Poulenc Industries 12, rue des Gardinoux 93308 Aubervilliers	FRANCE
DE BEER V.H.J., Dr.	Eindhoven University of Technology Dept. Inorg. Chem. and Catalysis P.O. Box 513 Eindhoven	THE NETHERLANDS
DECLERCK Cl.	Solvay et Cie. 310 rue de Ransbeek B-1120 Bruxelles	BELGIUM
DECLIPPELEIR G., Dr.	Labofina S.A. Chaussée de Vilvoorde 98-100 B-1120 Bruxelles	BELGIUM
DELLER, Dr.	Degussa Wolfgang Postf. 602 Abt. FC-PH 6450 Hanau 1	W.GERMANY

DELMON B., Prof.	Groupe de Physico-Chimie Minérale et de Catalyse Place Croix du Sud 1 B-1348 Louvain-la-Neuve	BELGIUM
DELOBEL R.	Laboratoire de Catalyse E.N.S.C.L. B.P. 40 59650 Villeneuve d'Ascq	FRANCE
DEMMERING G., Dr.	Henkel KGaA von Galenstr. 40 565 Solingen-Gräfratte	W.GERMANY
DEROUANE E.G., Prof.	Facultés Universitaires Notre Dame de la Paix Laboratoire de Catalyse rue de Bruxelles 61 B-5000 Namur	BELGIUM
DE VOS R.	Kemisk Reaktionstekn. Chalmers Tekn. Högsk. Fack. S-40220 Göteborg	SWEDEN
DEWOLFS R., Dr.	Universitaire Instellingen Antwerpen Department Scheikunde Universiteitsplein 1 B-2610 Wilrijk	BELGIUM
DIJKHUIS C.G.M., Dr.	DSM-Central Laboratory Department of Catalysis P.O. Box 18 Geleen	THE NETHERLANDS
DISTELDORF J., Dr.	Veba Chimie AG Werksgruppe Herne Postfach 2840 4690 Herne-2	W.GERMANY
DJALALI M., Dr.	Iran/Teheran-Rey NIOC Research Center Catalyst Unit	IRAN
DODWELL C.H.	British Petroleum Co. Ltd. BP Research Centre Chertsey Road, Sunbury-on-Thames Middx. TW16 7LN	ENGLAND
DOESBURG E.B.M.	Technische Hogeschool Delft Lab. voor Anorg. en Fysische Chem. Julianalaan 136 Delft	THE NETHERLANDS
DUROCHER	BASF AG Ammonlabor 67 Ludwigshafen	W.GERMANY
DVORAK B., Dr.	Technical University Praha Department of Organic Technology 166.28 Praha 6	CZECHOSLOVAKIA

EDERER H.J., Dr.	Inst. für Angew. Physikalische Chemie IM Neuenheimer Feld 253 D-6900 Heidelberg	W.GERMANY
EDMONDS T., Dr.	British Petroleum Co. Ltd. BP Research Centre Chertsey Road, Sunbury-on-Thames Middx. TW16 7LN	ENGLAND
ENGELBACH H., Dr.	BASF AG D-6700 Ludwigshafen	W.GERMANY
ENGLER M.	Institut für Anorganische Chemie und Analytische Chemie Johannes Gutenberg Universität D-6500 Mainz	W.GERMANY
FATTORE V., Dr.	Snamprogetti S.p.A. Via Fabiani 1 S. Donato Milanese	ITALY
FENELONOV B., Dr.	Institute of Catalysis Academy of Sciences of USSR Siberian Branch Novosibirsk	U.S.S.R.
FERINO I., Dr.	Universita di Cagliari Istituto Chimico Policattedra Via Ospedale 72 09100 Cagliari	ITALY
FERRARIO M.	Grace Italiana Viale Certosa 222 Milano	ITALY
FIGUEIREDO C.	Petrobras-Cenpes Ilha do Fundao Quadra 7 - Cidade Universitaria Rio de Janeiro	BRASIL
FISCHER L., Dr.	Chemische Werke Hüls GB 1/Anorg. Abt. D-437 Marl	W.GERMANY
FISHEL N.A., Dr.	Monsanto 800 North Lindbergh St Louis, Mo. 63166	U.S.A.
FONTANA P., Dr.	Alusuisse Forschung & Entwicklung CH-8212 Neuhausen	SWITZERLAND
FOUQUET G., Dr.	BASF AG 6700 Ludwigshafen	W.GERMANY
FRENNET A., Dr.	Labo. Catalyse FNRS Chimie Générale-ERM 30 Av. de la Renaissance Bruxelles	BELGIUM

GAAF J.	Shell Laboratory Badhuisweg 3 1031 CM Amsterdam	THE NETHERLANDS
GABELICA Z., Dr.	Facultés Universitaires Notre Dame de la Paix Laboratoire de Catalyse rue de Bruxelles B-5000 Namur	BELGIUM
GALIASSO R., Dr.	Intevep Sede Central Intevep Apartado Postal 149 Los Teques	VENEZUELA
GAMAURY J.F., Dr.	Universal Matthey Rue Epinal Calais 62100	FRANCE
GATI G., Dr.	High Pressur Res. Inst. H-2443 Szazhalombatta Pf 32	HUNGARY
GEMBICKI S.A.	UOP Process Division 20 UOP Plaza Algonguin & Mt Prospect Road Des Plaines, Ill. 60016	U.S.A.
GEORGE Z.M.	Alberta Research Council 11315-87 Avenue Edmonton Alberta T6G 2C2	CANADA
GEUS J.W.	State University Utrecht Anal. Chem. Lab. Croesestr. 77A Utrecht	THE NETHERLANDS
GRANGE P., Dr.	Groupe de Physico-Chimie Minérale et de Catalyse Place Croix du Sud 1 B-1348 Louvain-la-Neuve	BELGIUM
GREMMELMAIER C., Dr.	Ciba-Geigy AG Ref. PE 3.32/CH/mc Postfach CH-4002 Basel	SWITZERLAND
GROENENDAAL W.	SICC CIMS/4 Shell Centre London SE1 7PG	ENGLAND
GRONSTRAND M.	Neste Oy Research Center SF-06850 Kulloo	FINLAND
GRYAZNOVA Z.V., Dr.	Moscow University, Chemical Department Lenin's Hill Moscow	U.S.S.R.

GRZECHOWIAK J.	Polish Chem. Soc. Gdanska 7/9 str. 50-344 Wroclaw	POLAND
GUBITOSA G., Dr.	Montedison S.p.A. Istituto Donegani Via G. Fauser, 4 Novara	ITALY
GUCZI L., Prof.	Institute of Isotopes P.O. Box 77 H-1525 Budapest 114	HUNGARY
GUIL J.M., Dr.	Instituto Rocasolano Serrano 119 Madrid-6	SPAIN
HABERSBERGER K., Dr.	Heyrovsky Institute of Physical Chemistry and Electrochemistry Czechoslovak Academy of Sciences Machova 7 CS-121 38 Praha 2	CZECHOSLOVAKIA
HAEGH G.S.	Central Institute for Industrial Research Forskningsv. 1 P.O. Box 350 Blindern, Oslo 3	NORWAY
HALCOUR K., Dr.	Bayer AG ZB ZF Petrochemie und Verfahrensentwicklung Q 18 D-5090 Leverkusen	W.GERMANY
HEGEDUS L.L., Dr.	General Motors Co. Research Laboratories Warren, Mich. 48090	U.S.A.
HEINRICH, Dr.	Chemische Werke Hüls AG 2BFE/33 PB 15 Postfach 1180 D-4370 Marl	W.GERMANY
HENRION P.N., Dr.	CEN-SCK B-2400 Mol	BELGIUM
HENSEL J., Dr.	Degussa Wolfgang Gruenaustr. 3 D-645 Hanau 9	W.GERMANY
HERMANN W.	Elektrokemiska AB Fack 445 01 Surte	SWEDEN
HIGHTOWER J.W., Prof.	George R. Brown School of Engineering Rice University Houston, Texas 77001	U.S.A.

HOLT A.	Universal Matthey Products Ltd. Underbridge Way, Brimsdown Enfield, Middx.	ENGLAND
HUTCHINGS G., Dr.	Imperial Chemical Industries Ltd. Petrochemicals Division HQ Wilton, P.O. Box 90 Middlesbrough Cleveland	ENGLAND
IANNIBELLO A., Dr.	Stazione Sperimentale Combustibili Via A. De Gasperi 3 S. Donato M.	ITALY
INACKER O., Dr.	Dornier System, NTF Postfach 1360 D-7990 Friedrichshafen	W.GERMANY
IRGANG M., Dr.	BASF AG D-6700 Ludwigshafen	W.GERMANY
IRVINE E.A., Dr.	Imperial Chemical Industries Ltd. Agricultural Division Q. Buildings (Catalysis and Chem. Group) Billingham, Cleveland	ENGLAND
ISAEV O., Dr.	Institute of Chemical Physics Academy of Sciences of USSR 117334 Vorob'evskoe Shosse 2-6 Moscow	U.S.S.R.
JACOBS P.A., Dr.	Katholieke Universiteit Leuven Laboratorium voor Oppervlakte en Colloidalescheikunde de Croylaan 42 B-3030 Heverlee	BELGIUM
JACQUES	Rhone Poulenc Ind. 30100 Salindres	FRANCE
JANNES G., Dr.	CERIA/IIF-IMC Avenue Gryson 1 B-1070 Bruxelles	BELGIUM
JANSSEN M.M., Dr.	Shell Research Postbus 3003 Amsterdam	THE NETHERLANDS
JENNINGS J.R., Dr.	Imperial Chemical Industries Ltd. Agricultural Division Billingham, Cleveland	ENGLAND
JENSEN E.J.	Haldor Topsøe A/S Nymøllevej 55 DK-2800 Lyngby	DENMARK
JOHANSEN K.	Haldor Topsøe A/S Nymøllevej 55 DK-2800 Lyngby	DENMARK

JOHANSSON L.E.	Dept. of Chem. Techn. Kemicentrum Box 740 S-22007 Lund	SWEDEN
JOHNSON M.M., Dr.	Phillips Petroleum Co. 347 RB # 1 Bartlesville, Okl. 74003	U.S.A.
JOUSTRA A., Drs.	Shell Research P.O. Box 3003 Amsterdam	THE NETHERLANDS
JOVER B., Dr.	Elte TTK Slervetlen es Anal. Kem. Tan 52 P.O. Box 123 1143-Budapest	HUNGARY
JUHASZ J.	Elte TTK Slervetlen es Anal. Kem. Tan 52 P.O. Box 123 1143-Budapest	HUNGARY
KAWAGUCHI T.	Department of Chemistry Tokyo Gakugei University Koganei-Shi, Tokyo 184	JAPAN
KERKHOF F.P.J.M.	Laboratorium voor Chemische Technologie Pl. Muidergracht 30 1010 TV Amsterdam	THE NETHERLANDS
KIEFFER, Dr.	Institut de Chimie rue Blaise Pascal, 1 F-67008 Strasbourg	FRANCE
KIOVSKY J.R., Dr.	Norton Company P.O. Box 350 Akron, Ohio 44309	U.S.A.
KLOCKENKEMPER	Universität Münster Inst. für Physikal. Chemie Schlossplatz 4 44 Münster	W.GERMANY
KOCH H.	Institut für Technische Chemie Technische Universität Berlin Strasse des 17 Juni 135 1000 Berlin 12	W.GERMANY
KOCHLOEFL K., Dr.	Süd-Chemie AG Katalysatorwerk D-8052 Moosburg	W.GERMANY
KOHLER M.	Solvay & Cie. rue de Ransbeek 310 B-1120 Bruxelles	BELGIUM

KOOPMAN P.G.   Technische Hogeschool Delft
                            Lab. voor Organische Chemie
                            Julianalaan 136
                            Delft   THE NETHERLANDS

KOSTKA H., Dr.   Siemens AG
                            Bau 32
                            Günter Scharowski-str.
                            D-8520 Erlangen   W.GERMANY

KOSTROV V.V., Dr.   Chem. & Eng. Institute of Ivanovo
                            Engels St., 1
                            153460 Ivanovo   U.S.S.R.

KOTTER M.   Institut für Chem.Verfarenstechnik
                            Universität Karlsruhe
                            P.O. Box 6380
                            D-75 Karlsruhe   W.GERMANY

KRAUS M., Dr.   Institute of Chemical Process
                            Fundamentals
                            Czechoslovak Academy of Sciences
                            Suchdol
                            16502 Praha 6   CZECHOSLOVAKIA

KREMENIC G., Dr.   Inst. Catalisis y Petroleoquim.
                            Serrano 119
                            Madrid 6   SPAIN

KRENZKE L.D., Dr.   CNRS
                            Centre de Recherche sur les
                            Solides à Organisation
                            Cristalline Imparfaite
                            rue de la Ferollerie 1B
                            F-45045 Orléans Cedex   FRANCE

KRICSFALUSSY Z., Dr.   Bayer AG
                            D-5090 Leverkusen   W.GERMANY

KRISHNAMURTHY K.R., Dr.   Indian Petrochemicals Corp. Ltd.
                            Research Center
                            Baroda 391346   INDIA

KRISTIANSEN L.A., Dr.   Norsk Hydro A/S
                            Research Center
                            3901 Porsgrunn   NORWAY

KRIVORUCHKO O.P., Dr.   Institute of Catalysis
                            pr. Nayki, 5
                            630090 Novosibirsk 90   U.S.S.R.

KRUISSINK E.C., Drs.   Techn. Univ. Delft
                            Lab. Inorgan. and Phys. Chem.
                            Julianalaan 136
                            Delft   THE NETHERLANDS

KRYLOVA A.V., Dr.   Chem. Eng. Mendeleev's
                            Institute of Moscow
                            Moscow A-47   U.S.S.R.

KUBASOV A.A., Dr.	Moscow State University Moscow	U.S.S.R.
KUCK M.A.	Stauffer Chemical Co. Eastern Research Center Dobbs Ferry, N.Y. 10522	U.S.A.
KUNO K.	Central Research Laboratory Showa Den Ko 2-24-60 2-Chome, Tamagawa Ohtaku Tokyo	JAPAN
LAMBERT P., Dr.	UCB, S.A. Secteur Chimique Chaussée de Charleroi 4 B-1060 Bruxelles	BELGIUM
LAURER P., Dr.	BASF AG D-6700 Ludwigshafen	W.GERMANY
LECLERE C., Dr.	Compagnie des Métaux Précieux bd P. Vaillant Couturier 74 F-94200 Ivry	FRANCE
LECLERCQ L., Dr.	Université de Poitiers rue du Haut des Sables 174 F-86000 Poitiers	FRANCE
LECLOUX A., Dr.	Solvay & Cie rue de Ransbeek 310 B-1120 Bruxelles	BELGIUM
LEIGH D., Dr.	Engelhard Industried Ltd. Valley RD, Cinderford Glous. - GL14 2PB	ENGLAND
LE PAGE J.F., Dr.	Institut Français du Pétrole Av. Bois-Préau, 4 B.P. 311 F-92506 Rueil-Malmaison	FRANCE
LE PAGE,M.	Rhone Poulenc Industries 12, Rue des Gardinoux 93308 Aubervilliers	FRANCE
L'HOMME G., Prof.	Université de Liège Chaire de Génie Chimique rue A. Stévart 2 B-4000 Liège	BELGIUM
LIDEFELT J.O.	Kemisk Reaktionsteknik Chalmers Tekn. Hogskola S-40220 Göteborg-9	SWEDEN
LOHRENGEL G.	Ruhr-Universität Bochum D-4630 Bochum	W.GERMANY

LOPEZ F.J., Dr.	Escuela de Quimica Facultad de Ingenieria Universidad de Carabobo Valencia	VENEZUELA
LUCIEN J.	Société Shell Française Centre de Recherches de Grand Couronne 76530 Grand Couronne	FRANCE
LUDENA E., Dr.	IVIC Apartado 1827 Caracas	VENEZUELA
LUDWIG G.	Katalysatorenwerke Houdry-Hüls Postfach 1320 D-437 Marl	W.GERMANY
MAAT H.J.	AKZO Chemie Box 15 Amsterdam-N	THE NETHERLANDS
McCULLOCH A.	Imperial Chemical Industries Ltd. Mond Div. R & D Dept. P.O. Box 8, The Heath Runcorn Cheshire WA7 4QD	ENGLAND
MALINOWSKI S.,Prof.	Institute of Organic Chemistry and Technology Technical University Politechnike 00-662 Warszawa Kozykowa 75	POLAND
MANDY T.	High Pressure Res. Inst. H-2443 Szazhalombatta P. Box 32	HUNGARY
MARION J.	Produits Chimiques Ugine Kuhlmann C.R.L. rue Henri Moissan F-69310 Pierre-Bénite	FRANCE
MARSDEN C.E., Dr.	Imperial Chemical Industries Ltd. Mond Division The Heath, Runcorn Cheshire WA7 4QD	ENCLAND
MARTAN M., Dr.	Chemical Processes Weizmann Institute Plastic Department Rehovot	ISRAEL
MARTINEZ N.P.	I.U.T. Cumana Aptdo. 255 Cumana, Edo. Sucre	VENEZUELA
MARTOS J.	Enpetrol Centro de Investigacion de Enpetrol Escombreras 183 Cartagena (Murcia)	SPAIN

MATERNOVA J., Dr.   Institute of Chemical Process
                           Fundamentals
                           Czechoslovak Academy of Sciences
                           16502 Praha G-Suchdol   CZECHOSLOVAKIA

MATHEWS J.F., Prof.   University of Saskatchewan
                           Dept. of Chemical Engineering
                           Saskatchewan
                           Saskatoon S7N OWO   CANADA

MATIJEVIC E., Prof.   Institute of Colloid and
                           Surface Science
                           Clarkson College
                           Potsdam, N.Y. 13676   U.S.A.

MEIDER H., Dr.   Institute Rudjer Boskovic
                           Bijenicka 54
                           41000 Zagreb   YUGOSLAVIA

MEIKLE J.   U.S. Army
                           DARCOM Stitsur.
                           Abrams Bldg. Rm 740
                           D-6000 Frankfurt/M   W.GERMANY

MENON P.G., Dr.   Universiteit Gent
                           Labor. Petrochemische Techniek
                           Krijgslaan 271
                           B-9000 Gent   BELGIUM

MENZER R.   KFA Jülich GMBH
                           Postfach 1913
                           D-5170 Jülich IRB   W.GERMANY

MILBERGER E., Dr.   The Standard Oil Co.
                           4440 Warrensville Ctr.
                           Cleveland, Ohio 44128   U.S.A.

MILES R.   Shell Research Ltd.
                           Thornton Research Centre
                           P.O. Box 1
                           Chester CH1 3SH   ENGLAND

MITCHELL P.C.H., Dr.   Dept. of Chemistry
                           University of Reading
                           Whiteknights
                           Reading RG6 2AD   ENGLAND

MOL J.C., Dr.   Lab. Chemical Technology
                           Plantage Muidergracht 30
                           1010 TV Amsterdam   THE NETHERLANDS

MONE R.   AKZO Chemie
                           Postbus 15
                           Amsterdam   THE NETHERLANDS

MONTARNAL R.   Institut Français du Pétrole
                           Avenue Bois-Préau 4
                           B.P. 311
                           F-92500 Rueil-Malmaison   FRANCE

MOORE C.L.   UOP Process Division
Catalyst and Adsorbent Division
20 UOP Plaza
Des Plaines, Ill.   U.S.A.

MORAIS-ANES J.M., Dr.   CNP
Rua Artilharia, UM 79-7°
Lisboa   PORTUGAL

MORAVSKY A.P., Dr.   Moskowskaja Oblast
Chernogolovka, Otdelenie ICP
Moscow   U.S.S.R.

MORIMOTO T.   Chiyoda Chemical Engineering
and Construction Company
3-13 Moriya-cho, Kanagawa-ku
Yokohama 221   JAPAN

MORMINO V., Dr.   Snamprogetti S.p.A.
Via Fabiani 1
20097 S. Donato Milanese   ITALY

MOSCOU L.   AKZO Chemie Nederland
Research Centre Amsterdam
P.O. Box 15
1000 AA Amsterdam   THE NETHERLANDS

MOUREAUX   Société Shell Française
Centre de Recherches de Grand Couronne
F-76530 Grand Couronne   FRANCE

MROSS W., Dr.   BASF AG
Wak, M310
D-6700 Ludwigshafen   W.GERMANY

MULLER W.H., Dr.   Hoechst AG
Forschg Aliphat. Zwipro D569
D-6230 Frankfurt/Mainz 80   W.GERMANY

MURRELL D.L.   Exxon
P.O. Box 45
Linden, N.J. 07036   U.S.A.

NEAL A.H., Dr.   Exxon R & D Labs
P.O. Box 2226
Baton Rouge, LA 70821   U.S.A.

NEUKERMANS H., Dr.   Catalysts & Chemicals Europe
Place du Champ des Mars 2, bte 3
1050 Bruxelles   BELGIUM

NEWSON E.J., Dr.   Haldor Topsøe A/S
Nymøllevej 55
DK-2800 Lyngby   DENMARK

NICLAES H.J.   UOP Processes Internat.
250 Avenue Louise, Bte 101
B-1050 Bruxelles   BELGIUM

NIELSEN P.E.H.	Haldor Topsøe A/S Nymøllevej 55 DK-2800 Lyngby	DENMARK
NIERLICH F., Dr.	Chemische Werk Hüls AG ZBFE/33, PB 15 Postfach 1180 D-4370 Marl	W.GERMANY
NILSEN B.P.	Central Institute for Industrial Research Forskningsv. 1 P.B. 350 Blindern, Oslo 3	NORWAY
NOWECK K., Dr.	Condea Chemie GMBH Postfach 132191 D-2000 Hamburg 13	W.GERMANY
OCHOA DE O.	Inst. Venezolano de Investigaciones Cientificas Carretera Panamericana Km. 11 Aptdo. 1827 Caracas	VENEZUELA
ODENBRAND, L.	Dept. of Chemical Technology Kemicentrum, Box 740 S-22007 Lund	SWEDEN
OZAKI A., Dr.	Tokyo Institute of Technology 4259 Nagatsuta, Midori-ku Yokohama 227	JAPAN
PAAL Z., Dr.	Inst. Isotopes P.O. Box 77 H-1525 Budapest	HUNGARY
PAPEE D.	Pro-Catalyse rue Jean Goujon 21 F-75008 Paris	FRANCE
PASSERIEUX R.	H.P.C. Produits Chimiques B.P. 540 87011 Limoges Cedex	FRANCE
PAZOS J.M.	Intevep BP Research Center Process Branch Sunbury-on-Thames Middx.	ENGLAND
PEPE F., Dr.	Universita di Roma Istituto di Chimica Generale Roma	ITALY
PERNICONE N.	Montedison S.p.A. DIPI/Attività Catalizzatori Via G. Fauser 4 28100 Novara	ITALY

PETRO J., Dr.	Organic Chem. Technology Dept. Müegyetem H-1521 Budapest	HUNGARY
PIRARD J.P., Dr.	Université de Liège Inst. de Chimie Industrielle rue A. Stévart 2 B-4000 Liège	BELGIUM
PIVOT J.C.	Société A.D.G. B.P. N° 1 69230 Saint Genis Laval	FRANCE
POISSON R., Dr.	Rhone-Poulenc Industries 12, Rue des Gardinoux 93308 Aubervilliers	FRANCE
PONCELET G., Dr.	Groupe de Physico-Chimie Minérale et de Catalyse Place Croix du Sud 1 B-1348 Louvain-la-Neuve	BELGIUM
POSPISIL M., Dr.	Czech. Technical University Dept. of Technical and Nuclear Engineering Brehova 7 11519 Praha 1	CZECHOSLOVAKIA
PRESCHER G., Dr.	Degussa Wolfgang Postfach 602 Abt. FC-0 D-6450 Hanau 1	W.GERMANY
PRICE R.	N.C.B. C.R.E. Stoke Orchard Cheltenham Glos.	ENGLAND
QUET C., Dr.	SNEA-(P)-France Centre de Recherches de Lacq B.P. 34 Lacq F-64170 Artix	FRANCE
RABO J.A., Dr.	Union Carbide Corp. Tarrytown Technical Center Old Saw Mill River Road Tarrytown, N.Y. 10591	U.S.A.
RAO V.N.P., Dr.	Dupont Company Chemicals, Dyes & Pigments Dept. Experimental Station Wilmington, Delaware 19898	U.S.A.
REES L.V.C., Dr.	Imperial College Chemistry Dept. London SW7 2AY	ENGLAND

REID R.	Imperial Chemical Industries Ltd. Corporate Lab. P.O. Box 11 The Heath, Runcorn Cheshire	ENGLAND
RICHARDSON J.T., Prof.	University of Houston Dept. Chem. Eng. Houston, Texas 77004	U.S.A.
RICHARDSON P.J., Dr.	Imperial Chemical Industries Ltd. Organics Division Hexagon House, Manchester M9 3DA	ENGLAND
RIEKERT L., Prof.	Universität Karlruhe Inst. Chem. Verfarenstechnik P.O. Box 6380 D-7500 Karlsruhe	W.GERMANY
RIESSER G.H., Dr.	Shell Development Co. Box 1380 Houston, Texas 77001	U.S.A.
RIJNTEN H.Th.	AKZO Chemie BV Stationstraat 48 3818 LW Amersfoort	THE NETHERLANDS
ROBINSON E., Dr.	Lambeg Industrial Research Assoc. Lambeg, Lisburn Co Antrim	NORTHERN IRELAND
ROBINSON W.D.	Monsanto Co. 800 N Lindbergh Blvd St. Louis, Mo. 63166	U.S.A.
ROONEY J.J., Dr.	Queens University Chemistry Department David Keir Building Delfast	NORTHERN IRELAND
ROOZEBOOM F., Drs.	Techn. University Twente P.O. Box 217 Enschede	THE NETHERLANDS
ROSS J.R.H., Dr.	School of Chemistry University of Bradford Bradford BD7 LDP W. Yorks.	ENGLAND
ROTERUD P.T.	Sintef Applied Chemistry Division N-7034 Trondheim NTH	NORWAY
RUIZ PANIEGO A., Dr.	Instituto Rocasolano Serrano 119 Madrid 6	SPAIN
RUPPERT W.	Institut für Techn. Chemie I Egerlandstr. 3 852 Erlangen	W.GERMANY

RUSEK M.	Ciba-Geigy AG R 1055.608 4002 Basel	SWITZERLAND
RYNDIN J., Dr.	Institute of Catalysis Academy of Sciences of USSR Siberian Branch Novosibirsk 90	U.S.S.R.
SANTACESARIA E.	Istituto di Chimica Fisica ed Elettrochimica Plazza Leonardo da Vinci 32 Milano	ITALY
SAUM, Dr.	BASF AG 6700 Ludwigshafen	W.GERMANY
SAWYER W.H., Dr.	Exxon Res. and Dev. Labs. P.O. Box 2226 Baton Rouge, LA 70821	U.S.A.
SCHACHNER H., Dr.	Institut Battelle 7, Route de Drize CH-1227 Carouge/Genève	SWITZERLAND
SCHARF B., Dr.	Chemische Werke Hüls AG ZBFE-33, Postfach 80 D-4370 Marl	W.GERMANY
SCHEVE J., Dr.	Zentralinstitut für physik. Chemie der AdW der DDR Rudower Chaussee 5 DDR-1199 Berlin-Adlershof	D.D.R.
SCHIEFLER J., Dr.	Condea Chemie GMBH Postfach 132191 D-2000 Hamburg 13	W.GERMANY
SCHMIDT F., Dr.	Degussa Postfach 602 D-645 Hanau	W.GERMANY
SCHOLTEN J.J.F., Prof.	DSM Central Laboratory Geleen	THE NETHERLANDS
SCHREIBER, Dr.	Henkel KGAA Postfach 1100 D-4000 Dusseldorf 1	W.GERMANY
SEIPENBUSCH R.	VEBA-Chimie AG Werksgruppe Herne Postfach 2840 4690 Herne 2	W.GERMANY
SERMON P.A., Dr.	Brunel University School of Chemistry Uxbridge Middx. UB8 3PH	ENGLAND

SETINEK K., Dr.	Institute of Chemical Process Fundamentals Czechoslovak Academy of Sciences 16502 Praha-6-Suchdol CZECHOSLOVAKIA
SHINGU H., Dr.	Kyoto University 38 Higashikishimoto, Shimogamo, Sakyo, Kyoto 606 JAPAN
SHUTT J.R., Dr.	Essochem Europe Inc. Nijverheidslaan 2 B-1920 Diegem BELGIUM
SIEG R.P., Dr.	Chevron Research Co. 576 Standard Ave. Richmond, CA 94807 U.S.A.
SING K.S.W., Prof.	Brunel University Dept. Of Applied Chemistry Uxbridge Mddx UB8 3PH ENGLAND
SOKOLOF V., Dr.	Moscow Leninskiy Prospect 47 Institute of Organic Chemistry Moscow U.S.S.R.
SPEAKMAN J.G., Dr.	British Petroleum Chemicals Ltd. Bo'ness Road, Grangemouth Stirlingshire ENGLAND
SPEK T.G., Dr.	Shell Research Postfach 3003 Amsterdal THE NETHERLANDS
STACEY M.H., Dr.	Imperial Chemical Industries Ltd. R & D Department, Mond Division P.O. Box 8 The Heath, Runcorn Cheshire ENGLAND
STEFANI G., Dr.	Ftalital PCS SpA Via E. Fermi 51 24020 Scanzorosciate Bergamo ITALY
STEINER K., Dr.	Hoffmann-La Roche & Co Bldg. 62 Grenzacherstr. 124 4002 Basel SWITZERLAND
STERLING E., Dr.	Imperial Chemical Industries Europe Everslaan B-3078 Everberg BELGIUM
STONE F.S., Prof.	University of Bath School of Chemistry Bath BA2 7AY ENGLAND

STRATZ A., Dr.   Degussa Wolfgang
Geschaftbereich Katalysatoren
Postfach 602
D-6450 Hanau 1   W.GERMANY

STRELETS V., Dr.   Institute of Chemical Physics
Academy of Sci. of U.S.S.R.
Chernogolovka, Moscowskaja oblast,
142432   U.S.S.R.

SZABO G.   Compagnie Française de Raffinage
Centre de Recherches
B.P. 27
F-76700 Harfleur   FRANCE

SZABO Z.G., Prof.   Eötvos University
Inst. for Inorg. and Anal. Chem.
Muzeum krt 4/b
1443 Budapest   HUNGARY

SZASZ G., Dr.   General Electric Co.
Pelikanstr. 37
CH-8001 Zurich   SWITZERLAND

TANG S.C., Dr.   Shell Development Co.
Westhollow Research Center
P.O. Box 1380
Houston, Texas 77001   U.S.A.

TENNISON S.R.   British Petroleum Co. Ltd.
Research Centre
Chertsey Rd.
Sunbury-on-Thames
Middx.   ENGLAND

THIBAULT C., Dr.   SNEAP
Centre de Recherche
BP 34 Lacq
F-64170 Artix   FRANCE

THOMAS W.J., Prof.   University of Bath
School of Chemical Engineering
Clavertondown, Bath   ENGLAND

TOPCHIEVA K.V., Prof.   Moscow State University
Department of Chemistry
Lenin's Hills
Moscow   U.S.S.R.

TOPSOE H., Dr.   Haldor Topsøe
Nymøllevej 55
DK-2800 Lyngby   DENMARK

TOURNAYAN L.   Institut de Recherches sur la
Catalyse
79, Bd. du 11 novembre 1918
69626 Villeurbanne Cedex   FRANCE

TRAINA F.	Montedison S.p.A. DIPI/Attività Catalizzatori Via G. Fauser 4 28100 Novara	ITALY
TRIFIRO F., Prof.	Universita Bologna Facolta di Chimica Industriale Viale Risorgimento, 4 Bologna	ITALY
TRIKI A.	Compagnie Française de Raffinage B.P. 27 CFR F-76700 Harfleur	FRANCE
TRIMM D.L., Prof.	NTH Industriel Kjemi 7034 Trondheim	NORWAY
TWIGG M.V., Dr.	Imperial Chemical Industries Ltd. CAT/CHEM GRP Res. & Dev., Agric Div. Billingham Cleveland	ENGLAND
UNGER K., Prof.	Institut für Anorganische Chemie und Analytische Chemie Johannes Gutenberg Universität D-6500 Mainz	W.GERMANY
URWIN D., Dr.	Tioxide International Ltd. Portrack Lane Stockton-on-Tees Cleveland	ENGLAND
VACCARIELLO A.U.	Norton Chemical Process Products Ltd. King Street, Fenton Stoke-on-Trent Staffs	ENGLAND
VALIGI M.	Universita di Roma Istituto di Chimica Gen. ed Inorg. Città Universitaria Roma	ITALY
van de IJL H.	Dow Chemical P.O. Box 48 Terneuzen	THE NETHERLANDS
van de MOESDIJK I.	DSM Holland Central Laboratory P.O. Box 18 Geleen	THE NETHERLANDS
van de MOND, Th.	DSM Holland Central Laboratory P.O. Box 18 Geleen	THE NETHERLANDS

van den BERG G.H.	AKZO Chemie Nederland bv P.O. Box 15 Amsterdam	THE NETHERLANDS
van OMMEN J., Drs.	Technische Universiteit Twente P.O. Box 217 Enschede	THE NETHERLANDS
van REISEN C.A.M., Drs.	AKZO Zout Chemie Nederland bv P.O. Box 25 Hengelo	THE NETHERLANDS
van REYEN L.L., Prof.	Technische Hogeschool Delft Lab. voor Anorg. en Fys. Chemie Julianalaan 136 Delft	THE NETHERLANDS
van ROMPAY P., Prof.	Instituut voor Chemie- Ingenieurstechniek K.U.L. de Croylaan 2 B-3030 Heverlee	BELGIUM
van SINT FIET T.H.M., Dr.	Dow Chemical P.O. Box 48 Terneuzen	THE NETHERLANDS
VERMEULEN A.C., Drs.	State Univ. Utrecht Anal. Chem. Lab. Coesestr. 97A Utrecht	THE NETHERLANDS
VERSLUIS F.	Harshaw Chemie bv P.O. Box 19 De Meern	THE NETHERLANDS
VOGELS C., Dr.	Solvay & Cie. rue de Ransbeek 310 1120 Bruxelles	BELGIUM
VOGT W., Dr.	Hoechst AG Werk Knapsack D-5030 Hürth	W.GERMANY
VOIGT C., Dr.	Degussa Wolfgang Postfach 602 D-6450 Hanau	W.GERMANY
WARD J.W.,Dr.	Union Oil Co. of California P.O. Box 76 Brea, CA 92621	U.S.A.
WASSERMANN M., Dr.	Condea Chemie GMBH Postfach 132191 D-2000 Hamburg 13	W.GERMANY
WEBB M., Dr.	Unilever Research Lab. Port Sunlight, Wirral Merseyside	ENGLAND

WEBSTER D.E., Dr.	Johnson Matthey Chemicals Orchard Road Royston Herts.SG8 5HE	ENGLAND
WEIL C.	Coppee-Rust S.A. Avenue Louise 251 B-1050 Bruxelles	BELGIUM
WEISZ P.B., Dr.	Mobil Res. & Dev. Corp. Central Research Division P.O. Box 1025 Princeton, N.J. 80540	U.S.A.
WHITE J.L., Prof.	Purdue University Dept. of Agronomy West Lafayette, Ind. 47907	U.S.A.
WRIGHT C.J., Dr.	Aere Harwell Bldg. 521, Didcot Oxon	ENGLAND
WUNDER F.H., Dr.	Hoechst AG Fo Aliph. Zwipro D-569 Frankfürt	W.GERMANY
YAGADOVSKI, Prof.	Lumumba's People's Friendship University Ordzowikidze 3 Moscow	U.S.S.R.
YATES D.J.C., Dr.	Exxon Research and Engineering Co. P.O. Box 45 Linden, N.J. 07036	U.S.A.
YOUNG D.A.	Union Oil Co. of California P.O. Box 76 Brea, CA 92621	U.S.A.
ZIRKER G., Dr.	BASF AG D-6700 Ludwigshafen	W.GERMANY

AUTHOR INDEX

Alzamora, L.E. 143
André, J.M. 585
Andreu, P. 493
Angelov, S. 605

Baerns, M. 41
Baranski, A. 353
Barbee, T. 627
Becker, E.R. 159
Blanchard, G. 179
Bond, C.C. 715
Boreskov, G.K. 723
Bossi, A. 405
Boudart, M. 627
Broekhoff, J.C.F. 663
Brunelle, J.P. 211
Burriesci, N. 479

Cahen, R.M. 585
Cairati, L. 279
Cale, T.S. 131
Candia, R. 479
Carbucicchio, M. 279
Charcosset, H. 197
Chenebaux, M.T. 197
Chou, T.S. 171
Christiansen, L.J. 353
Clausen, B.S. 365, 479
Coenen, J.W.E. 89
Courty, Ph. 293
Crump, J.G. 131

Damyanov, D. 605
Dass, M.R. 41
Debus, H.R. 585
Delmon, B. 439
Derouane, E.G. 365, 727
Desai, P. 131
Doesburg, E.B.M. 143
Dubus, R.J. 131
Dumesic, J.A. 365

Edmonds, T. 507
Engler, M. 29
Eszterle, M. 391

Fenelonov, V.B. 233

Galiasso, R. 493
Garbassi, F. 405
Geus, J.W. 113
Guczi, L. 391

Haase, R. 615
Hagan, A.P. 417
Hegedus, L.L. 171
Hermans, L.A.M. 113
Hombek, R. 595
Houalla, M. 439

Iannibello, A. 65, 469
Imura, K. 627
Inacker, O. 29
Inui, T. 245

Jerschkewitz, H.-G. 615

Kerkhof, F.P.J.M. 77
Kheifets, L.I. 233
Kijenski, J. 595
Kiraly, J. 391
Kotter, M. 51
Kruissink, E.C. 143

Lagan, M. 353
Leclercq, L. 627
Le Page, J.F. 293
Lofthouse, M.G. 417
Lohrengel, G. 41

Malinowski, S. 595
Manninger, I. 391
Marengo, S. 65
Matijevic, E. 555
Matusek, K. 391
Mehandjiev, D. 605
Mitchell, P.C.H. 469
Montarnal, R.E. 519
Mørup, S. 365, 479
Moscou, L. 659
Moss, R.L. 715
Moulijn, J.A. 77
Murrell, L.L. 307

Neal, A.H. 719
Neimark, A.V. 233
Nuttall, T.A. 159

Ochoa, O. 493
Öhlmann, G. 615
Orlandi, A. 405
Orr, S. 143
Osterwalder, U. 131
Oudejans, J.C. 77
Ozaki, A. 381

Pattek, A. 353
Pernicone, N. 321
Petrini, G. 405
Petro, J. 641
Pitkethley, R.C. 715
Plog, C. 29
Potter, N.M. 171
Primet, M. 197

Ramaswamy, A.V. 185
Ratnasamy, P. 185
Reizer, A. 353
Richardson, J.T. 131
Richter-Mendau, J. 615
Riekert, L. 51
Rijnten, H.T. 265
Ross, J.R.H. 143
Ruggeri, O. 279

Samakhov, A.A. 233
Scheve, J. 615
Scholten, J.J.F. 685
Seidl, M. 29
Shimazaki, K. 381
Shingu, H. 245
Sing, K.S.W. 715
Sivasanker, S. 185
Stevens, G.C. 507
Stone, F.S. 417
Sumiya, S. 381
Summers, J.C. 171

Thomas, R. 77
Topsøe, H. 353, 365, 479
Topsøe, N. 365
Traina, F. 321
Trevethan, M.A. 417
Trifiro, F. 65, 279
Trimm, D.L. 1

Unger, K. 29
Urabe, K. 381

Van den Berg, G.H. 265
Van Reijen, L.L. 143
Van Veen, G. 143
Villa, P.L. 65
Villadsen, J. 365

Wilson, R. 715

Yates, D.J.C. 307
Yoshida, S. 627

Zanderighi, L. 405

LIBRARY